CHARACTERIZATION OF MATERIALS IN RESEARCH

Ceramics and Polymers

SAGAMORE ARMY MATERIALS RESEARCH CONFERENCE PROCEEDINGS
Published by Syracuse University Press

Fundamentals of Deformation Processing
Walter A. Backofen et al., eds.
(9th Proceeding)

Fatigue—An Interdisciplinary Approach
John J. Burke, Norman L. Reed, and Volker Weiss, eds.
(10th Proceeding)

Strengthening Mechanisms—Metals and Ceramics
John J. Burke, Norman L. Reed, and Volker Weiss, eds.
(12th Proceeding)

Surfaces and Interfaces I
Chemical and Physical Characteristics
John J. Burke, Norman L. Reed, and Volker Weiss, eds.
(13th Proceeding)

Surfaces and Interfaces II
Physical and Mechanical Properties
John J. Burke, Norman L. Reed, and Volker Weiss, eds.
(14th Proceeding)

Ultrafine-Grain Ceramics
John J. Burke, Norman L. Reed, and Volker Weiss, eds.
(15th Proceeding)

Ultrafine-Grain Metals
John J. Burke and Volker Weiss, eds.
(16th Proceeding)

Shock Waves and the Mechanical Properties of Solids
John J. Burke and Volker Weiss, eds.
(17th Proceeding)

Powder Metallurgy for High-Performance Applications
John J. Burke and Volker Weiss, eds.
(18th Proceeding)

Block and Graft Copolymers
John J. Burke and Volker Weiss, eds.
(19th Proceeding)

Sagamore Army Materials Research Conference
(20th; 1973; Raquette Lake, N.Y.)

Characterization of Materials in Research

CERAMICS and POLYMERS

EDITORS

JOHN J. BURKE

Associate Director, Army Materials and
Mechanics Research Center

VOLKER WEISS

Professor, Syracuse University

Proceedings of the 20th Sagamore Army Materials Research Conference. Held at Sagamore Conference Center, Raquette Lake, New York, September 11, 12, 13, and 14, 1973. Sponsored by the Army Materials and Mechanics Research Center. Organized and directed by the Army Materials and Mechanics Research Center in cooperation with Syracuse University.

SYRACUSE UNIVERSITY PRESS 1975

Library of Congress Cataloging in Publication Data

Sagamore Army Materials Research Conference, 20th,
  Raquette Lake, N.Y., 1973.
  Characterization of materials in research.

  (Proceedings of the 20th Sagamore Army Materials
Research Conference)
  Includes bibliographical references.
  1. Ceramic materials—Congresses. 2. Ceramics—
Congresses. 3. Polymers and polymerization—Con-
gresses. I. Burke, John J. II. Weiss, Volker,
1930– III. United States. Army Materials and
Mechanics Research Center. IV. Syracuse University.
V. Title. VI. Series: Sagamore Army Materials
Research Conference. Proceedings; 20.

UF526.3.S3 no. 20 [TA430] 623′.028s [620.1′92]
ISBN 0-8156-5040-X   75-5272

Manufactured in the United States of America
Composed and printed by Science Press, Ephrata, Pa.
Bound by Vail-Ballou Press, Inc., Binghamton, N.Y.

SAGAMORE CONFERENCE COMMITTEE

Co-Chairmen
JOHN J. BURKE, Army Materials and
Mechanics Research Center
and
VOLKER WEISS, Syracuse University

Program Director
VOLKER WEISS, Syracuse University

Secretary
SAMUEL VALENCIA, Army Materials and
Mechanics Research Center

Conference Coordinator
ARAM TARPINIAN, Army Materials and
Mechanics Research Center

Program Committee
CHARLES F. BERSCH, Naval Air Systems
Command
FRED W. BILLMEYER, JR., Rensselear
Polytechnic Institute
JOHN J. BURKE, Army Materials and
Mechanics Research Center
WENZEL DAVIDSOHN, Army Materials and
Mechanics Research Center
DONALD GROVES, National Materials
Advisory Board
R. NATHAN KATZ, Army Materials and
Mechanics Research Center
KENNETH J. SMITH, New York State
University College of Environmental Science
and Forestry
VOLKER WEISS, Syracuse University

Arrangements at
SAGAMORE CONFERENCE CENTER
JAMES REID, Syracuse University

Contents

Foreword

The Army Materials and Mechanics Research Center has conducted the Sagamore Army Materials Research Conference in cooperation with the Materials Science Group of the Department of Chemical Engineering and Materials Science of Syracuse University since 1954. The purpose of the conference has been to gather together scientists and engineers from academic institutions, industry, and government who are uniquely qualified to explore in depth a subject of importance to the Army, the Department of Defense, and the scientific community.

The principles of characterization have already led to the development of well-defined materials. This volume, *Characterization of Materials in Research—Ceramics and Polymers,* addresses the areas of the concept of characterization, physics of nonmetallic solids, ceramic characterization, molecular characterization of polymers, characterization of polymers in bulk, and, finally, case histories.

The dedicated assistance of Mr. Edward J. Lemay of the Army Materials and Mechanics Research Center throughout all stages of the conference planning and, finally, the publication of the Sagamore Conference proceedings is deeply appreciated. The support of the Technical Reports Office under the supervision of Mrs. A. V. Gallagher, and the Technical Information Office under the supervision of Miss M. M. Murphy of the Army Materials and Mechanics Research Center in preparing the final manuscript is acknowledged.

The continued active interest and support of these conferences by Dr. Alvin E. Gorum, Director, LTC Robert B. Henry, Commander/ Deputy Director, of the Army Materials and Mechanics Research Center, is appreciated.

Syracuse University   The Editors
Syracuse, New York

SESSION I

Keynote Address

THE CONCEPT OF CHARACTERIZATION

NATHAN E. PROMISEL
National Materials Advisory Board
Washington, D.C.

Chapter 1

The Concept of Characterization

ABSTRACT

The concept of "characterization" is expressed in terms of a pragmatic definition and yet basic through inclusion of atomic character, microcharacter and macrocharacter, emerging as features of composition and structure. The importance of understanding and agreeing on a concept is discussed, with resultant significance in research, exchange of information, technology transfer, standardization, and other facets of the materials field. The hazards and waste of effort resulting from negligence in materials characterization are illustrated. A program of activity and research and development to enhance the status and adequate practice of characterization is suggested.

Introduction

This conference, sponsored by the Army Materials and Mechanics Research Center and Syracuse University, will concern itself for the next three days with current programs, problems, and gaps in the characterization of ceramic and polymeric materials. Your attendance here hopefully is evidence of a growing national interest in the field of characterization, a field which encompasses with basic significance all materials and materials science and engineering. The implications of failure to characterize properly are dramatic.

Let me start this discussion with some numbers. The Federal Government alone—*i.e.,* exclusive of private organizations, industry, academe, etc.—in FY 1971 spent $190 million for documentation, reference, and information services dealing with scientific and technical information. One hundred and fifty-four million dollars were spent just to acquire, store, index, and loan the publications containing this information, a large portion of which dealt with materials properties; 4.9 million dollars

3

were devoted to collecting and evaluating undigested data. One is tempted to ask how much of the money spent on data collection, storage, and disposal is really unproductive and not warranted by the quality and reliability of the data. Certainly there is no question but that a substantial amount of data collected on materials, because of lack of characterization, can reasonably be described as unworthy of permanent record and reference. This is particularly unfortunate considering the pervasive and fundamental importance of materials.

Man owes his earthly survival to materials; they are literally vital to him—his food, shelter, clothing, transportation, protection, luxuries, etc. His early recognition and use of stone materials as tools separated and distinguished him from the other animal life around him. Historically, the production of ceramics and some metal articles, such as copper beads, date from the beginning of the 9th millennium B.C.; smelting of minerals to around 5,000 B.C.; many alloys were made by the 3rd millennium B.C.; steel was made by 1200 B.C.; and cast iron by 500 B.C. The great engineering achievements of the Romans; the wonders of Byzantine architecture; the Industrial Revolution; and even most of the nineteenth-century engineering were for the most part all based on materials of types that had been found and put to use by the ancients some 4,000 years earlier. I emphasize the word "types," because the copper of the 9th millennium B.C., the steel and cast iron of the past millennium, and other ancient materials were not specific, individual, pedigreed, reproducible products as we know them now—they really were types, for the most part.

A seventeenth-century philosopher, by the name of John Granvill, probably was one of the first to really put his finger prophetically on the current situation in the context of the need for understanding materials better in order to enhance their usefulness. Granvill wrote: "Iron seemeth a simple metal . . . but in it's nature are many mysteries . . . and men who bend to them their minds shall, in arriving days, gather therefrom greater profit, not to themselves alone but to all mankind."

Some men did indeed "bend their minds" a bit, and the very framework upon which our modern materials science is built received a major forward thrust at the end of the eighteenth century with the discovery of the presence of carbon in varying amounts in wrought iron, steel, and cast iron. It was the "gleam in the eye" for the conception of the iron–carbon diagram, on which our vast steel industry is based.

By 1900, metal behavior was beginning to be more explainable, not on the basis of chemical composition alone as formerly but by relating the shape, size, relative distribution, and interrelationships of distinguishable microstructural features of the material.

In the past three-quarters of a century, hundreds of thousands of new alloys and nonmetallic materials have been developed, and new and sophisticated analytical techniques for probing ever deeper into their basic structure, on the finest scale, have come into existence. We now appreciate that some of the minute variations in composition, structure, and defects—features of the atomic world inside these materials—may greatly affect the behavior and properties of materials; variations that are often too small to be detected with present techniques. We can observe the effects, although we cannot always understand them or identify the causes. Because of this, we cannot really be predictive of behavior based on understanding, or tailor-make many needed materials with assurance of reliability or reproducibility. Empiricism, and empirical predictive testing, are still the dominant way of technical life. Thus it can be safely said that a true science of materials has yet to evolve.

The seriousness of this state of ignorance—even full appreciation of the problem of how uniquely to describe or characterize a material—was recognized in the early 1960s. And so it was that the Materials Advisory Board of the National Academy of Sciences, in 1964, formed a Committee on Characterization of Materials to undertake a relevant detailed study. Their findings were reported in March 1967 in an impressive document entitled, not surprisingly, "Characterization of Materials" (MAB Report 229-M).

First, and since characterization has meant many things to many people and indeed is still rather loosely used, it was necessary for the Committee to develop a working definition of the term, which it did as follows:

> *Characterization describes those features of the composition and structure (including defects) of a material that are significant for a particular preparation, study of properties, or use, and suffice for reproduction of the material.*

This definition distinguishes between an intrinsic or internal set of characteristics and the external manifestations of them. The intrinsic set are of a compositional, structural, and defect nature. Included are crystal structure, stoichiometry, valence state, location and distribution of impurities, dislocations, etc. Examples of these characteristics are plentiful. The transistor depends on a very sophisticated characterization of purity of germanium or silicon with controlled, characterized impurities. A given piece of steel may show a whole range of properties because of variations in structure obtained by heat treatment. Perfect single crystals of sapphire whiskers should show tensile strength up to 30 million psi, dropping to a fraction of that depending on the crystal defect structure. The techniques for determining these characteristics include the use of X-rays, electron microprobe, scanning electron microscope, neutron and electron diffraction spectroscopy, wet chemistry techniques, polarography, electron spin resonance, and many others. As you all know, there are limitations to the use, applicability, and even credibility of some of these techniques; there are many needs for improvements, and there is a need for completely novel techniques in this area of science.

The behavioristic manifestations resulting from the above internal characteristics are reflected in the engineering properties of materials— melting point, mechanical strength, ductility, solubility, conductivity,

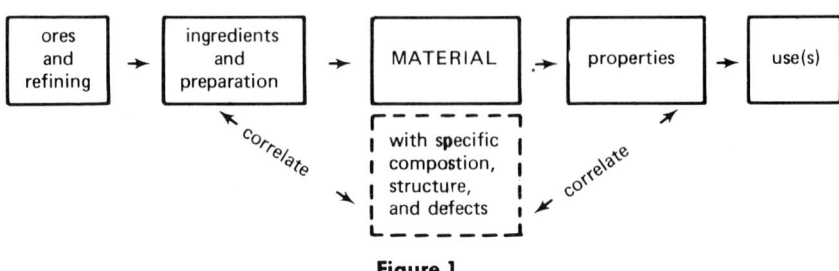

Figure 1.

magnetic and optical properties, etc. The test methods used for determining such properties of materials are many and varied. While many of these are also in need of improvement and standardization, material property measurements are not in bad shape—with a few exceptions.

Diagrammatically, we might describe the above as in Figure 1. One progresses from the natural materials (*e.g.*, the ores and minerals) to a set of ingredients or intermediate products that are processed into the usable forms of materials. These are characterized by their specific composition, structure, and defects, which in turn are determined by the starting ingredients and the processing procedures. These specific characteristics determine the materials properties; the converse is not true, *i.e.*, the properties do not uniquely characterize the material. Finally, of course, the properties determine the end uses. Today there are such detailed specialists for each step—and, indeed, intermediate steps—in this sequence that correlation among them is too often difficult and fragmented. There should be better communication among the synthesizers, characterizers, property measures, users, etc. The weakest link in the chain is so often characterization. It is interesting to note that, given adequate characterization, information on the previous stages for producing the materials need not be known; they are essentially irrelevant. Furthermore, given adequate characterization and a true understanding of the correlation with properties, properties become predictable, and susceptible to being tailor-made for specific applications.

It must be recognized that often methods of preparation and properties are used to describe a material. Such description is inadequate and hazardous and serves only as a pragmatic expedient and substitute in the absence of true characterization, which must be constantly sought.

As an example of a compromise, however, and recognizing the state of affairs in our real world, material "description" has been treated in a manner consistent with the above in a study of ceramic processing, where lack of characterization was identified as one cause of ceramic irreproducibility. In that study, the material description is given by D in Equation (1). The total character C_T is given in Equation (2).

$$D = xC_T + (1 - x)(P + H) \qquad (1)$$

where

D represents material description
x is a factor between 0 and 1
C_T represents total character of the material
P represents properties not relatable to character features
H represents the processing history of the material

$$C_T = C_A + C_m + C_M \qquad (2)$$

where

C_T is total character
C_A is atomic character
C_m is microcharacter
C_M is macrocharacter, including size and shape

When all properties and behavior characteristics $(P + H)$ are adequately understood and, ideally, capable of being expressed in terms of these features of character, the x in Equation (1) becomes unity and the need for the supplementary term disappears. At the other extreme, when true characteristics are relatively unknown, then x is small and the material description must be expressed, for practical purposes, in terms of measurable properties and processing history.

Referring to the character equation, Equation (2), traditional ceramic processing has generally been concerned with problems of achieving a desired size and shape C_M. These objectives involve good engineering design of operational techniques and equipment for starting-material treatment, batch preparation, forming, drying, and firing. The scientific input is the fundamental understanding of the response of the material to these processes. Lack of this understanding has resulted in unrealistic limitations on size, shape, and character of the material, as well as lack of uniformity in the product.

However, the processor could place additional emphasis on the achievement and maintenance of a certain specified character of material $(C_A + C_m)$. This objective has been receiving attention in recent years because of the desire for new compositions and structures. Studies have indicated the resulting dependence on knowledge of solid-state chemistry and physics, both thermodynamic and kinetic in nature. These situations have led to the realization of the need for a science of ceramic processing, covering the whole spectrum from the basic characterization of the starting material through each phase of modification and processing, with complete characterization and understanding of what is happening at each processing step, such as variation in microstructure. Such thoroughness in characterization will surely lead to, and is the only way to achieve, optimized properties in the final product, with that degree of uniformity and reproducibility that are absolutely necessary for expanded design use of high-quality, high-performance ceramics.

Turning now more generally to the significance of characterization, we find clear significance in practically every phase and facet of our science and technology. In research, it would appear to be self-evident that the researcher must have a precise knowledge of the specimen material used in his investigation if he is to deduce definitive and reproducible conclusions about a particular property or phenomenon. Unfortunately, our technical literature is glutted with reports of investigations using inadequately characterized specimens of materials, yielding inconclusive results, reproducible only by accident, or at best by rigid empirical duplication. By way of illustration, even in research concerning health—where one would assume the investigators would be extremely careful—inadequate characterization of polymers used as blood containers have given some erratic research results, hampering progress. The transfer of uncharacterized information through publication is itself unsatisfactory. How can different investigators compare results or duplicate experiments if there is no assurance that they are using the same basic material, because the material really isn't known, *i.e.,* hasn't been characterized? It is important that adequate characterization data be included in technical papers, even as test methods are included; otherwise, the paper should be considered technically unacceptable. Stating this in a more pragmatic sense, technology transfer cannot be achieved successfully—for research, fabrication, or production—without effective characterization of the materials involved. Particularly is the above true for international information exchange, because of increased difficulty in communication.

One corollary to the above relates to standardization and specifications. Specifications cannot be intelligently written to insure procurement of a given material with assurance that it will behave as expected if its basic characteristics are unknown or inadequately understood. Metal specifications have made good progress in this respect, but specifications involving polymers are particularly in need of sophistication. Rarely are we in a position to describe polymeric materials in terms of adequate intrinsic characterization and therefore in terms of *basic,* direct performance requirements. The simpler but less satisfactory method of using indirect, *empirical* behavioristic features is used; often, in fact, simply describing a particular proprietary product meeting certain empirical tests. The result, for example in government procurement, is expenditure of millions of dollars that could be saved if we could but write more basic polymer specifications. What is needed, first of all, is basic research to enhance our understanding of the correlation between intrinsic characteristics and behavior; and then a more intelligent, scientific, and sophisticated approach to writing specifications for polymer-containing materials and products.

Another aspect of characterization significance has to do with the conservation and more efficient use of materials, particularly those that are critical or strategic, where there is now a real threat of crucial shortages. The connection is obviously through efficient design, which, in turn, requires accurate and reliable properties data, and therefore well-characterized materials, if, for example, overdesign and wastage of material are to be avoided. Here, too, progress in materials science studies, which could lead to improved materials and efficient, tailor-made materials, requires precise materials characterization.

One could recite other fields of characterization significance, but the point has been made adequately. Before concluding with a few recommendations, let us illustrate, from past events, only two of many characterization-oriented effects. The following are taken from the Materials Advisory Board Report on this subject: (1) A few years ago, the Army experienced a catastrophic failure of a large-bore artillery gun tube. Analysis indicated the direct cause as being surface damage inside the tube. It appeared that absorption of carbon, oxygen, and nitrogen interstitially, coupled with mechanical stresses, had produced early surface damage. Lack of nondestructive characterization of the surface, correlating changes in composition and structure with failure modes, was basically responsible for the catastrophe. (2) Highly purified ZnS was required to prepare cathodoluminescent phosphors for radar and television kinescopes, requiring copper, for example, to be below 1 part per billion. Existing analytical techniques were inadequate for compositional characterization, so the limit was empirically established by addition of 1 part per billion to highly purified ZnS and by subsequent demonstration of successful operation, using blue and green luminescense emissions as criteria. Later, using refined X-ray diffraction techniques, these emissions were correlated with silver-activated and copper-activated hexagonal ZnS, respectively, and superiority shown over cubic ZnS. This correlation between behavior and structure and compositional characterization has been extremely useful in radar and television.

In conclusion, it may be said that, in spite of studies and accumulated evidence of the importance of materials characterization, there is discouragingly slow progress in the establishment and implementation of an adequate and comprehensive national program. There is still a strong need for research and development to produce new tools and techniques for both more intensive and more cost-effective characterization; a need for standardization of materials and analytical and test techniques; a need for better understanding of character–property relationships; a need for more extensive publicity on the subject and assignment of responsibility on a broader scale; a need for recognition of the subject in university curricula; and a need for emphasizing the importance of international cooperation on the subject, although some of this does exist and is well done. The Materials Advisory Board report referenced above contains many general as well as specific recommendations. Although they were written about six years ago, they are still valid, in spite of the partial progress that has been made.

The technical papers and panel discussions that will be heard in the next few days at this meeting are all pointed in the right direction. It is a commendable effort. But I fear that after this conference the subject will relax to another low level of activity, as has happened historically.

Therefore, I would recommend that, before you finally adjourn, you discuss and appoint a small task force to continue deliberations on how to foster an active, cooperative, national, and even international, program in this field, involving government and nongovernment, universities, and the professional societies. The National Bureau of Standards, with its National Standards Reference Data program, could work with such a task force. Also, cooperation could be obtained from the Numerical Data Advisory Board and the National Materials Advisory Board of the National Academy of Sciences, organizations that are already involved. Within the Interagency Council for Materials, a Planning Group is already concerned with this theme. Certainly, accelerated progress will not be easy, in this period of austere government budgets, critical industrial economies, and confrontation with demanding societal problems. But the incentives are great, especially over a long term; indeed, characterization can contribute importantly to the solution of some of these very problems.

I congratulate the Army Materials and Mechanics Research Center and Syracuse University on this twentieth anniversary in a series of outstanding conferences, which, collectively, cover such an impressive gamut of science and technology. I also congratulate the organizers of this program on both the selection of the topic and the agenda and menu for the meeting. I am sure that you will all make important contributions to the subject, and I wish you a constructive, successful, as well as enjoyable, meeting. Finally, I want to acknowledge with appreciation the important assistance rendered me by Mr. Donald Groves, of our NMAB staff, in the preparation of this paper.

SESSION II

Physics of
Nonmetallic Solids

Moderator: Charles F. Bersch
Naval Air Systems Command

PHILIP F. KANE
Texas Instruments, Inc.
Dallas, Texas

Chapter 2

The Determination of Defects in Solid-State Materials

ABSTRACT

Some recent developments in the determination of defects in solid-state materials, particularly single-crystal semiconductors, are reviewed. Both structural and compositional defects are discussed and emphasis is placed on the specific identification of point defects. The growing importance of area defects places increased demands on surface studies.

Introduction

The highly significant report from the Materials Advisory Board[1] "Characterization of Materials," was issued in 1967, over six years ago, and at that time the MAB made a number of suggestions on work which it felt was needed in the various areas of this subject. It would be of considerable interest to review the progress made since that report, particularly the report from the panel on defects, but to attempt this in a comprehensive manner is a far more ambitious undertaking than is possible within the confines of this presentation. Consequently, the viewpoint has been restricted to semiconductor materials and to a few selected topics which hopefully will give some feel for the progress made during the last few years.

Classification of Defects

Table I shows the classification of defects used in the report. As far as electronic devices are concerned, Type B defines the fundamental attributes which govern the electrical properties and their relative numbers and characteristics and are what semiconductor manufacturing is all

13

TABLE I
Types of Defects

A.	Point Defects	C.	Line Defects
	Vacancies		Dislocations
	Interstitial Atoms		
	Substitutional Atoms	D.	Area Defects
	Antistructure Defects		Solid–Solid Interfaces
			Gas–Solid Interfaces
B.	Electronic Defects		Liquid–Solid Interfaces
	Electrons		
	Holes		
	Excitons		

about. The point defects tailor these properties; as the report points out, they can be defined as dopants or impurities only if their effect is wanted or unwanted. The line defects are usually dislocations in otherwise perfect crystals. The area defects represent interfaces, and it is undoubtedly this type that has grown most in importance since 1967.

Another, and perhaps more convenient, way of classifying these defects is by the two major classes, structural and compositional, in line with the MAB definition of materials characterization. In point of fact, this does not quite cover all of the defect properties and a third class, electrical, is a useful addition. Using this classification, the defects appear as shown in Table II. There is a good deal of interaction, of course, between these three classes; for example, a vacancy is often associated with an interstitial atom in irradiated materials, but perhaps this just reinforces the argument that structure and composition are just different aspects of the same discipline.

Electrical Defects

In determining the electrical properties of bulk material, resistivity and Hall effect are still the basic tools and provide the fundamental char-

TABLE II
Types of Defects

A.	Electrical	C.	Compositional
	Electrons		Interstitial Atoms
	Holes		Substitutional Atoms
			Antistructure Defects
B.	Structural		
	Vacancies		
	Dislocations		
	Solid–Solid Interfaces		
	Gas–Solid Interfaces		

acteristics of carrier concentration, mobility, and type. A considerable amount of information can be obtained from Hall measurements as a function of temperature. An example is given[2] in Figure 1. The log of carrier concentration calculated from the resistivity and Hall coefficient is plotted against the reciprocal of temperature for a sample of silicon before and after heat treatment. The original material was lightly doped with boron to form a p-type material; the open circles are experimental, the line is a computer-generated curve based on the parameters at top right. The dots are experimental points determined after the heat treatment. The material was now n-type, and the curve was generated from a model based on the parameters at lower left. The changes are attributed to the presence of oxygen which, when subjected to this heat treatment, becomes electrically active providing two donor levels, one of which essentially compensates the boron at lower temperatures, the other generating electrons at the higher temperatures.

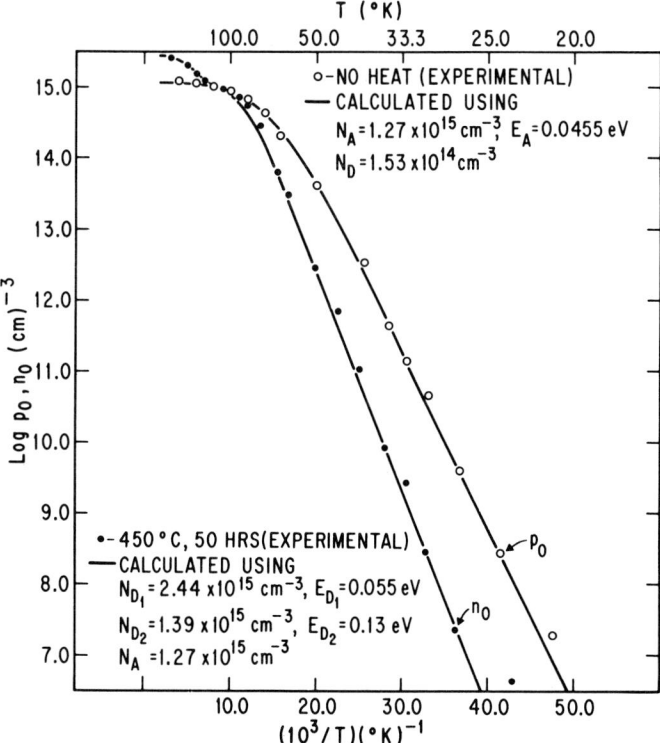

Figure 1. The variation of carrier concentration with temperature before and after heat treatment of a silicon sample. p_0 and n_0 are the number of p- and n-type carriers, respectively; N_A, the number of acceptors; N_D, the number of donors with subscripts D_1 and D_2 indicating two levels; and E_A, E_{D1}, and E_{D2}, the corresponding energy levels.[2]

Lifetime can be determined by the classical method of injecting carriers by a light pulse and determining the decay curve of conductivity with time[3]. For epitaxial films, the methods must be modified. For example, resistance is measured by the spreading resistance probe[4], shown in Figure 2. The usual four-point probe induces a field which extends into the substrate. This field can be reduced by the specially fabricated two-point system shown, although even here problems can arise with very thin films. Also shown in Figure 2 is a plot of resistivity against distance across an angle-lapped epitaxial slice[2]. The change at the epi-substrate interface is easily detected, as is the junction between the diffused layer and the main bulk of the substrate. (DUF is a TI term for diffusion under film, a diffused region formed in the substrate prior to growth of the epitaxial film.) The spatial resolution of this method is quite good, about 2 microns (reduced to about 0.1 micron vertically by the angle lap), but it is not an absolute method. It must be standardized against a material of known resistivity.

The problem in determining electrical properties of the epitaxial film is related, of course, to the difficulty in separating the contribution of the substrate. Lifetime has been one of the more difficult things to determine since the carriers generated by the pulse, either optical or electrical, can rapidly diffuse to a substrate which is often much more heavily doped, obscuring the effect. One of the more informative electrical techniques

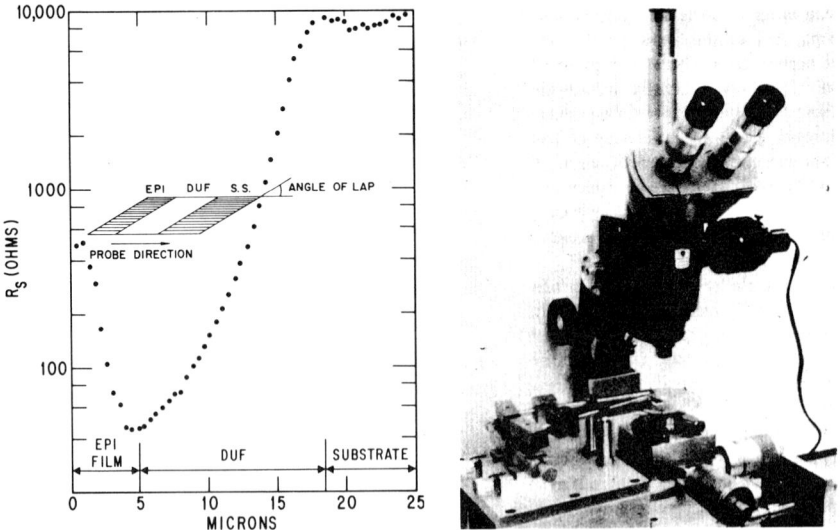

Figure 2. The spreading resistance probe. The graph on the left is a resistivity profile through an epitaxial layer obtained by taking measurements with the instrument (right) down an angle-lapped section. A DUF (diffusion under film) layer is a diffused region in the surface of the substrate under the epitaxial film[2].

recently introduced is the C-V technique, the generation of capacitance—voltage curves using an MIS (metal–insulator–semiconductor) structure. An excellent review of this technique has been given by Zaininger and Heiman[5]. The structure is shown in Figure 3. The MOS field effect transistor has a similar structure to this but has a contact on each side, a collector and an emitter. By adjusting the voltage on the gate, the majority carrier concentration can be adjusted at the interface to allow more or less current to flow. In this C-V technique, the effect is the same but capacitance is measured instead of current. A dc voltage is applied to the gate and a small ac signal imposed; this latter is used to measure the capacitance. There is, of course, no collector or emitter, and the other "plate" of the capacitor is the bulk semiconductor "sea." If the semiconductor is n-type and a negative voltage is applied to the gate, then the field set up drives electrons away from the interface, forming a depletion layer. If enough are repelled, then the minority carriers become a majority and inversion occurs. A change in carrier concentration causes a change in the dielectric constant of the material which is reflected in the capacitance measured. Figure 4 shows a typical C-V curve for n-type silicon. In the accumulation region of the curve, the capacitance is merely that of the insulator layer, an oxide in this case (*i.e.,* C/C_{ox} is 1). If the dielectric constant is known, as it often is at least approximately, the thickness can be determined. As the gate voltage becomes more negative, the electrons are repelled and the capacitance decreases as the effective depth of the dielectric layer increases. Finally, an inversion layer is formed and the capacitance becomes essentially a minimum since the ef-

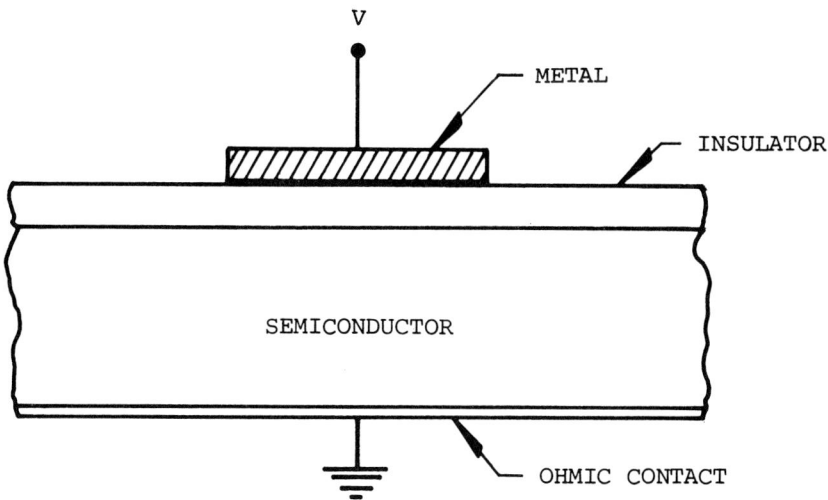

Figure 3.   Metal-oxide-semiconductor (MOS) capacitor structure.

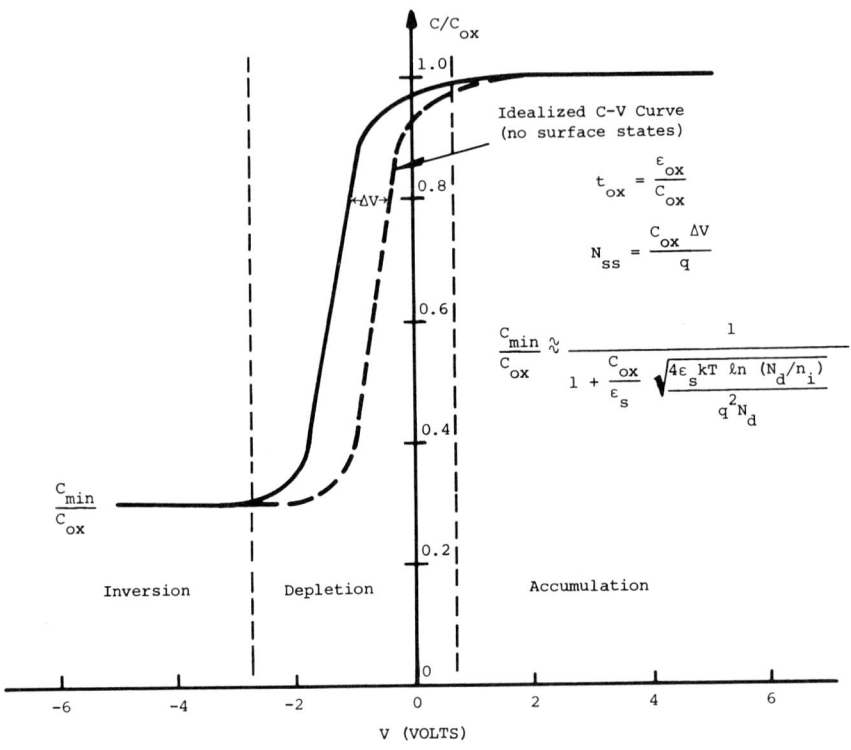

Figure 4. Normalized high-frequency C-V curve for n-type silicon.

fective depth of the depletion layer is not changed; any further change in gate voltage merely adds holes to the inversion layer. This minimum capacitance, or the ratio of C_{min}/C_{ox}, is a function of the donor concentration N_d, so that the carrier concentration of the material can be determined. The curve for p-type material would be reversed, with the inversion section on the positive side of the gate voltage, so that this also indicates type. In addition, comparison with an idealized C-V curve based on the carrier concentration will reveal surface states at the oxide–semiconductor interface. ΔV represents the additional voltage necessary to overcome this potential barrier.

This method will also determine lifetime by determining the time it takes an inversion layer to form. The curve shown in Figure 4 was generated with a slow sweep, about 5–10 seconds; with a rapid sweep, 5–10 microseconds, the curve would have continued down with a much less pronounced inflection. Figure 5 shows the transient response curve for the silicon sample of the last slide. The gate is biased with a step voltage, negative in the case of this n-type material, and the curve of capacitance against time is recorded. The slope is a function of the time required to generate sufficient minority carriers in the depletion region and for these

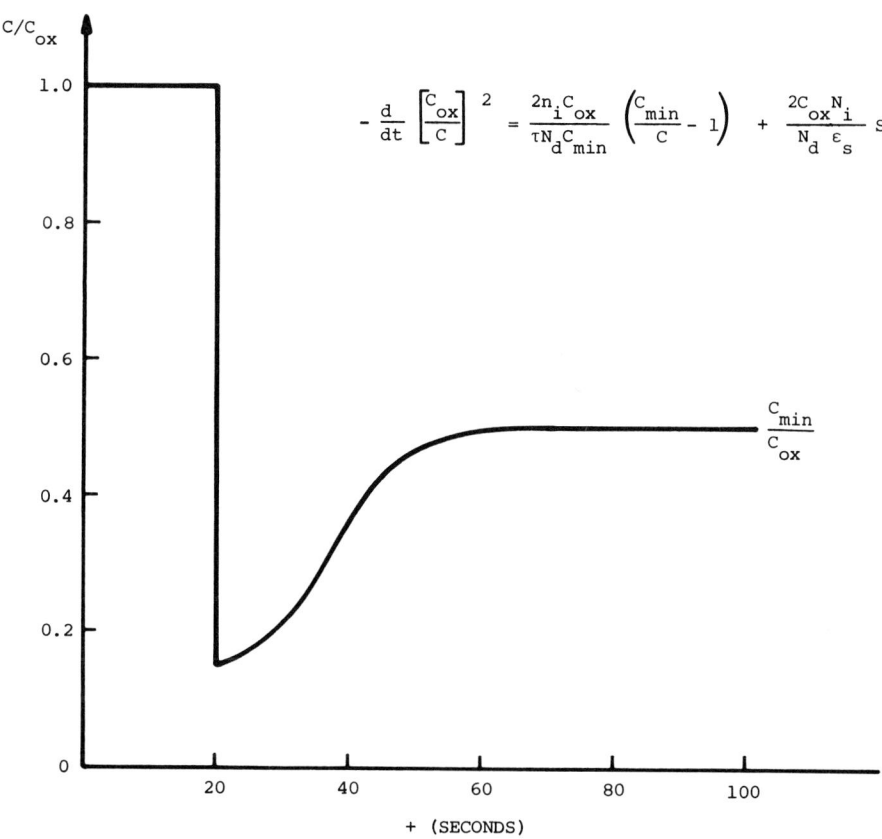

$$-\frac{d}{dt}\left[\frac{C_{ox}}{C}\right]^2 = \frac{2n_i C_{ox}}{\tau N_d C_{min}}\left(\frac{C_{min}}{C} - 1\right) + \frac{2C_{ox} N_i}{N_d \varepsilon_s} S$$

Figure 5. Transient response of MOS capacitor. ($V_{acc} \rightarrow V_{inv}$)

carriers to reach the interface under the influence of the applied field. If we plot the square of the normalized capacitance against time, a straight-line relationship results, as shown in the equation, and the lifetime can be calculated from the slope of this line. The second term is a function of the surface states at the interface and S, the surface recombination velocity, can be determined from the intercept.

Actually, this is a very simplified account of this most versatile technique. We have assumed that the oxide layer is a simple insulator, for example; often this is not the case and some contribution from, for example, ions can be detected. All these measurements we have discussed used high-frequency capacitance measurements, a MHz or more; other information, *e.g.,* on the inversion layer, is available at lower frequencies. The fast-response curve we mentioned earlier can give information on concentration profiles; for example, on the thickness of an epitaxial film. As can be seen, this can be a very valuable tool in determining the basic electrical characteristics of both bulk and epitaxial material.

Physical Defects

One of the most important parameters of the epitaxial film is its thickness, that is, the location of the solid–solid interface. Figure 2, spreading resistance, showed one method but this required angle lapping, a destructive technique. Actually, the exact location of an interface can be the basis of considerable philosophical discussion since, like almost any other boundary in nature, this is not absolute. Some diffusion of the dopants in the two regions must occur leading to a transition region. Hopefully, this will not be too diffuse, or control of the subsequent device characteristics will not be possible. As is usually the case, the thickness has been defined by a method, the infrared interferometric method[6]. Using a conventional IR spectrophotometer, an interference pattern is generated by reflection from the surface and the interface, as shown by the sketch in Figure 6. The determination is complicated somewhat by the phase shifts at A and

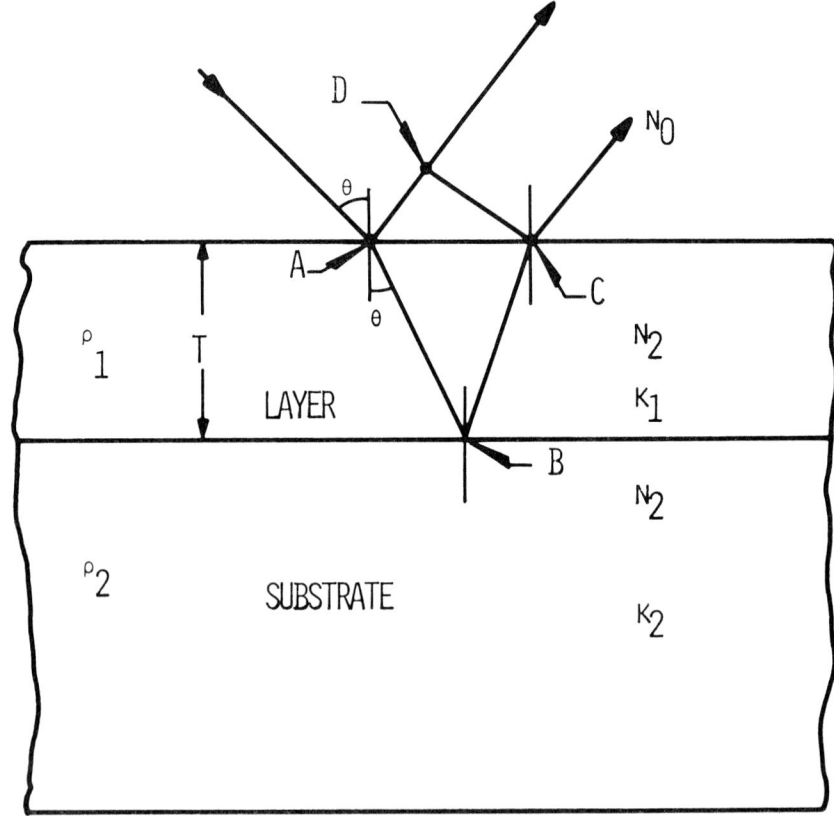

Figure 6. Geometry of IR method for epitaxial film thickness[6].

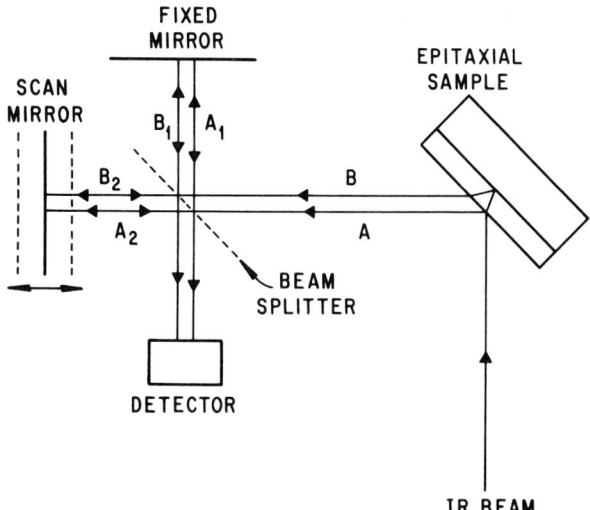

Figure 7. Schematic drawing of the radiation path involving an epitaxial sample and a Michelson interferometer[7].

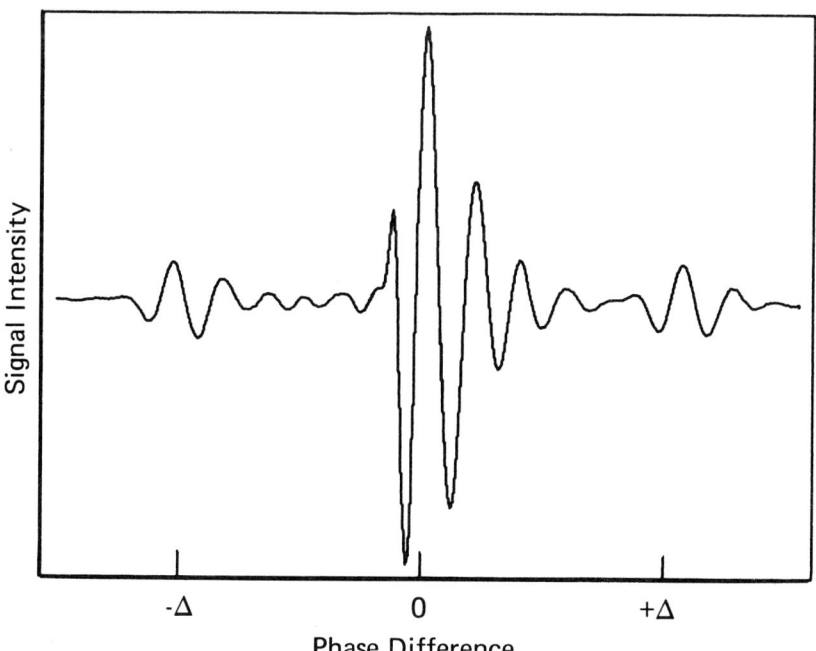

Figure 8. Unprocessed interferogram of reflection from a 12.5 μm epitaxial film[7].

B, which are dependent on the optical characteristics of the epitaxial layer and substrate, respectively. They are related to the resistivities which must be known. This method is subject to several errors, particularly in large-volume measurements and a new method, the Epilog system, has recently been introduced[7]. It is shown schematically in Figure 7. The optics consists of a Michelson interferometer with a Globar source. For a reflection from one surface, the minimum interference will obviously occur at the center of the mirror travel when both beams are traveling the same path; for all other positions, some interference must occur. A curve is generated which is characterized by the major peaks shown in Figure 8. They are somewhat off-symmetry due to the time lag in the detector. With an additional reflection from the interface, satellite peaks appear on either side, as can be seen, and the difference between these two satellites is a function of the film thickness. In addition, a relationship has been established between the amplitude of these satellite peaks and the phase shift at the interface, as shown in Figure 9. All the data can be corrected

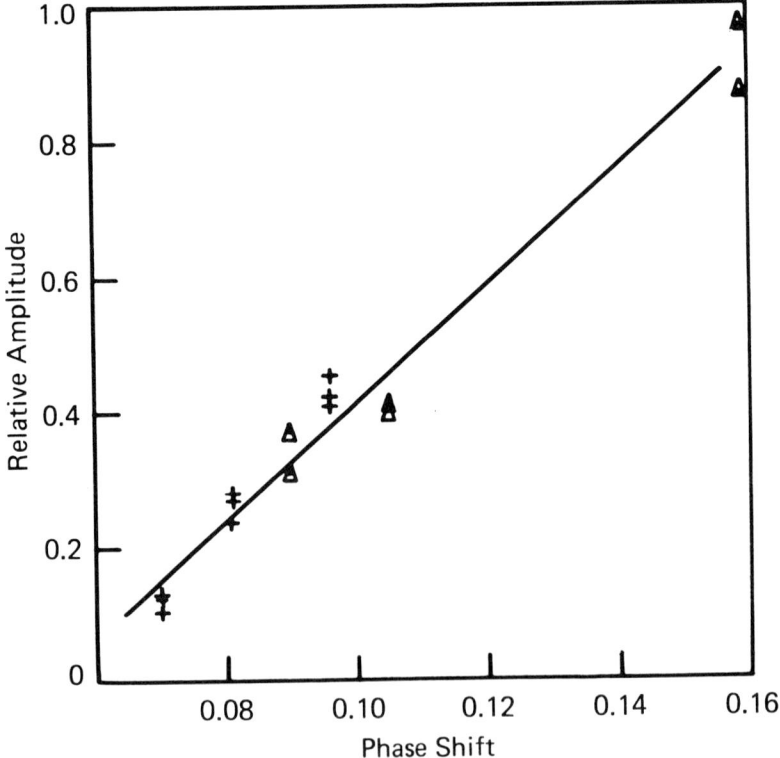

Figure 9. Correction of layer-substrate phase shift with satellite amplitude[7].

TABLE III
Comparison of Epitaxial Film-Thickness Measurements [7]

	IR	Epilog I	Epilog II
Thickness range	1.0–100 μm	30–70.0 μm	0.75–70.0 μm
Sampling Time	30 sec–15 min	5 sec	6 sec
Precision (3)	$\pm(0.25\,\mu m + 0.25t)$	$\pm(0.15\,\mu m + 0.015t)$	$\pm(0.05\,\mu m + 0.005t)$
Profiling capability	Yes	No	Yes

and presented by a minicomputer within 3–6 seconds. A comparison of this method with the ASTM standard method is shown in Table III.

The physical evaluation of bulk material has, in general, not progressed particularly over the last few years, due perhaps to the fact that in the case of silicon and, to a lesser extent, of the III-Vs, this bulk material is merely a substrate; the active devices are in the epitaxial film. However, one very simple technique[8] might be worth mentioning which is quite useful for checking the grain structure of boules, for example, of HgCdTe. It may be termed Laue topography, and the principle is shown in Figure 10. The X-ray beam is not collimated and is large enough in cross section to cover the whole slice or cross section of the boule. If the material is single-crystal, a Laue pattern of discs results. If it is polycrystalline, several Laue patterns are formed, each corresponding to a particular grain

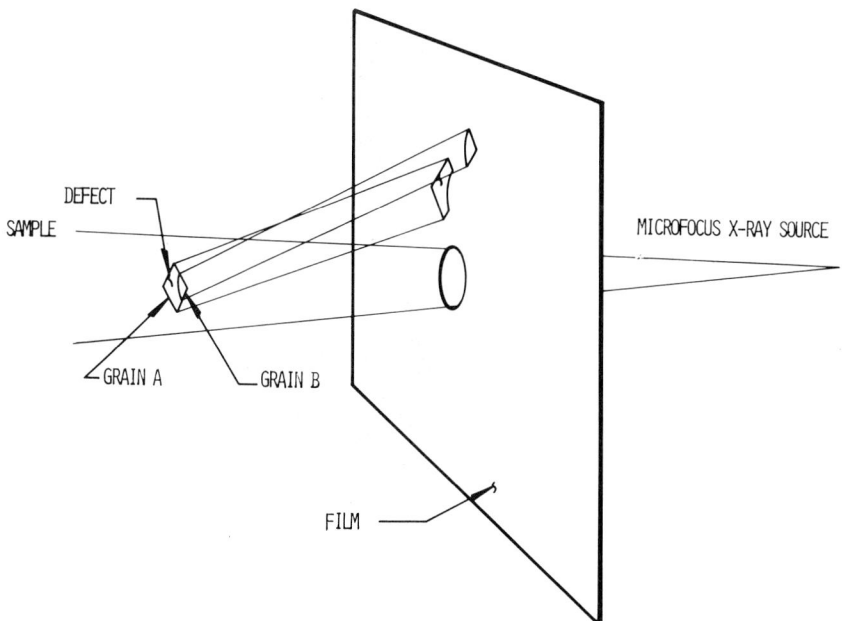

Figure 10.   Laue topography.

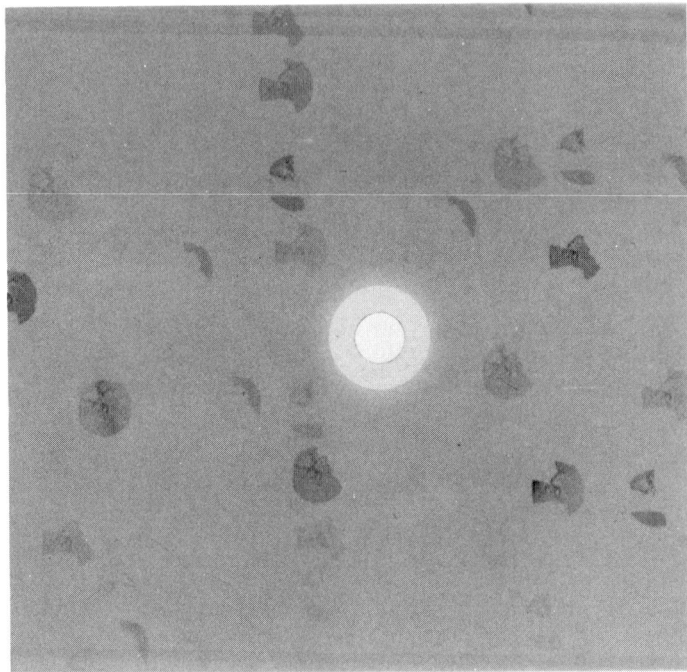

Figure 11.   A Laue topograph of a poor-quality HgCdTe crystal.

since their orientations will be different. An example is given in Figure 11. This is a comparatively rapid and inexpensive method of monitoring the quality of research materials, particularly the compound materials used as detectors.

In the general area of dislocations, there has really been very little done over the past few years. The Lang X-ray topographs and the variations on this method[3] are still the most useful for locating, counting, and identifying line defects, but there seems to have been very little work done in correlating these types of defects with the electrical properties. Some effort has been aimed at radiation-hardened devices and some interesting work has been done by the people at NBS[9] on correlating IR response of lithium-drifted germanium diodes with the various vacancy complexes. Otherwise, most of the structural-defect studies have been aimed at correlating defects in the epitaxial layer with device yield[10], outside the scope of this paper.

Chemical Defects

Undoubtedly, there is a great deal to learn about the relationship between the physical defects (dislocations, vacancies, and similar but grosser types of damage) and the electrical properties. A good deal more is known about the chemical defects since the band-structure theory has been largely devoted to the identification of various dopant levels and recombination centers arising from the presence of foreign elements in the lattice. Accordingly, there has been a considerable effort directed towards the determination of very low-level impurities in these high-purity materials. One of the earliest new techniques arising from this effort was the solids mass spectrograph[11], hailed at the time as a breakthrough. In point of fact, it did give an increase of two or three orders of magnitude over emission spectrography and proved to be a useful survey tool. However, it left a lot to be desired as a quantitative tool and the progress over the last 10 years or so has been very disappointing.

In fact, activation analysis is a much more useful technique even though it does not have the same universal application. With the introduction of semiconductor detectors and the application of computers to the data handling, the analysis of silicon has become almost automatic[12]. A typical scan for a silicon sample is shown in Figure 12. This has received no chemical treatment other than a surface etch. The sample was irradiated at $10^{13}n$ cm^2/sec and allowed to decay for 24 hours prior to carrying out this gamma-ray spectroscopy. The system consists of a pair of lithium-drifted germanium detectors connected to a 4096-channel analyzer. The data is handled by a TI960 minicomputer. The sensitivity is adequate for this particular application but could be much improved by gamma-gamma coincidence counting to eliminate the background due to Compton scattering. The resolution, about 2 or 3 keV, allows us to determine most elements without any preliminary separations. An application to films[13] is shown in Figure 13 which shows a distribution profile of several elements through an epitaxial layer. A spatial resolution of about a micron can be obtained by careful lapping techniques.

A very useful adjunct to this technique is provided by autoradiography. Figure 14 shows autoradiograms[2] taken at various levels of an arsenic-diffused 12.5 micron-thick epitaxial silicon film. "Piping" is evident in these autoradiograms, indicative of faults in the lattice.

An associated technique which has recently been improved to the sensitivity necessary for semiconductor quality material is X-ray fluorescence[14]. Figure 15 shows a spectrum excited by an X-ray tube source. The material is a tin-doped gallium arsenide and the read-out system is very similar to that used for gamma-ray spectroscopy. The tin is easily detected at about 0.4 ppma using a lithium-drifted silicon detector.

In a study of the energy levels in semiconductors, one of the most direct methods of determining these quantitatively is the use of infrared spectroscopy. Obviously, any absorption of energy must relate to some energy change within the lattice or its component atoms. Consequently, considerable work has been done[15] relating the peaks seen in IR spectra to the transitions taking place. To reduce the bandwidth of the transitional peaks, measurements are made at liquid-nitrogen or, preferably,

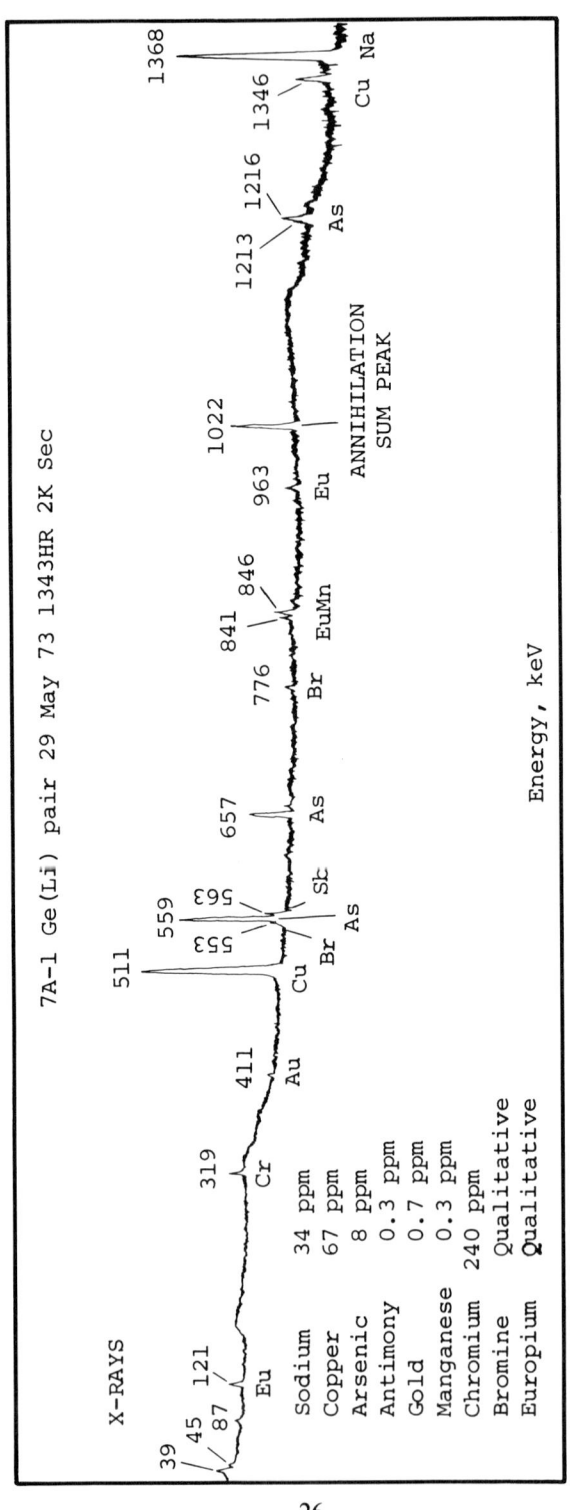

Figure 12. A gamma-ray spectrum from an activated silicon sample.

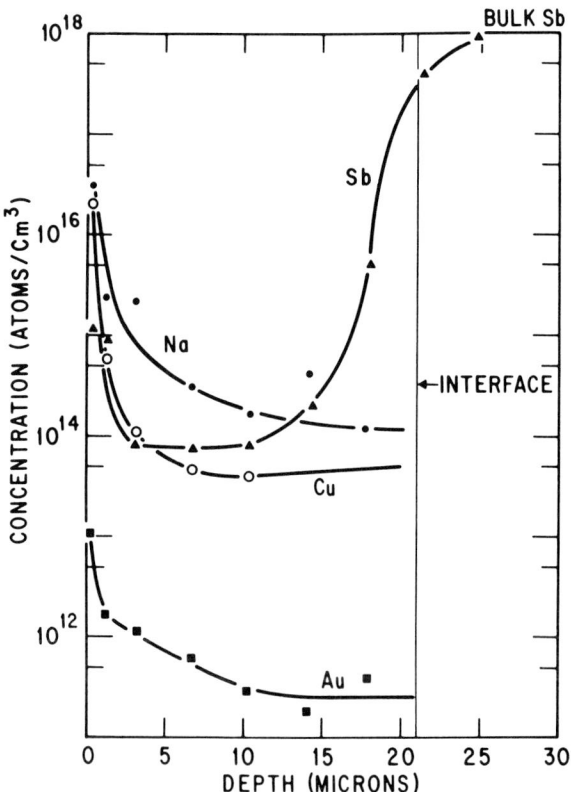

Figure 13. The distribution of impurities in an epitaxial layer, determined by activation analysis[13].

Figure 14. Arsenic distribution through a 12.5 μm-thick silicon epitaxial film.[2]

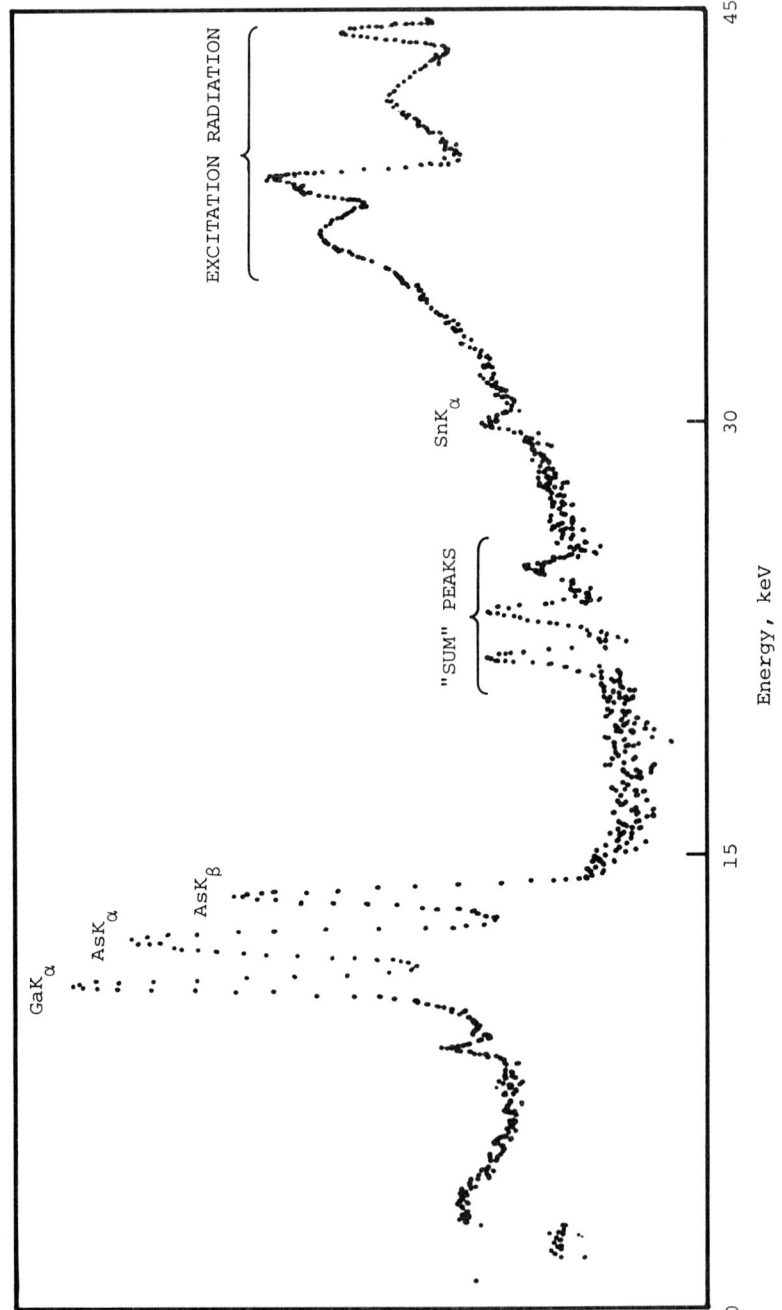

Figure 15. Tube excited x-ray fluorescence spectrum of tin-doped GaAs using a Si (Li) detector. Tin concentration 0.4 ppmw.

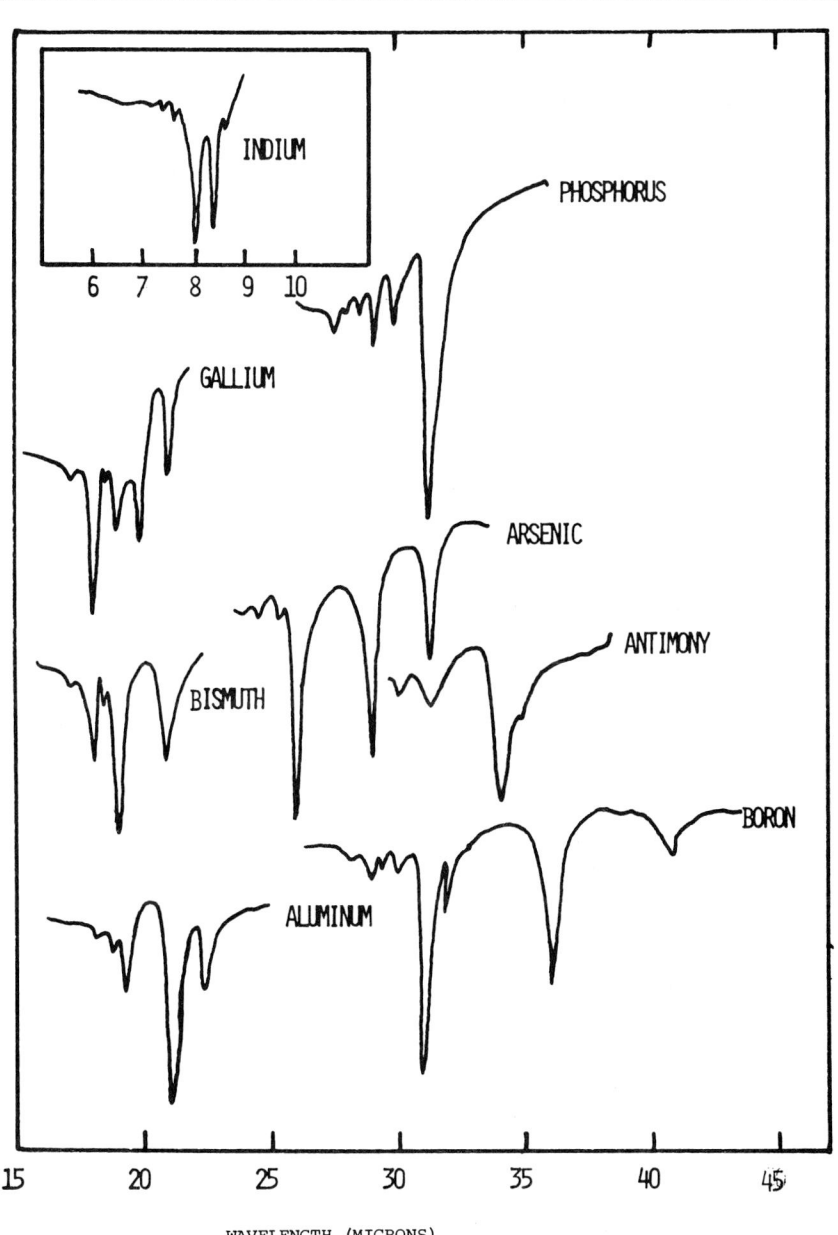

Figure 16. Some absorption peaks in silicon using a conventional IR spectrophotometer.

liquid-helium temperatures. Even then, these spectra can be quite complex and difficult to interpret. Figure 16 shows a series of peaks obtained using a conventional spectrophotometer, and it will be seen that a number of elements can be identified through their characteristic absorption peaks. Alternatively, the presence of peaks will indicate unsuspected energy-band transitions arising in the crystal. Although the resolution at higher temperatures is really a function of the crystal, there is a point at which the resolution is instrument-limited. Consequently, fast Fourier transform spectroscopy[16] is being investigated for application to this problem. This is another application of the technique used in the Epilog instrument. Figure 17 is a comparison of a conventional and an FFT spectrum. Note in particular the oxygen peak, not identifiable on the conventional spectrometer even at the slowest speed but easily recognizable on the FFT spectrometer.

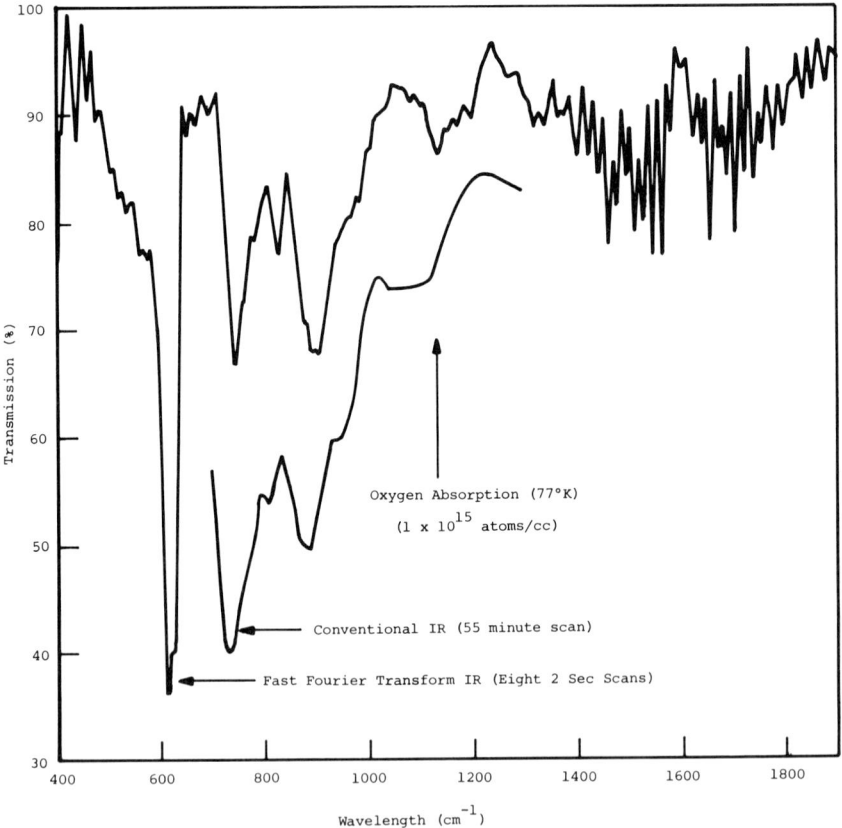

Figure 17. Oxygen absorption in silicon using FFT and conventional IR spectrometers.

As was pointed out earlier, perhaps the greatest emphasis over the last few years has been, as was forecast by the MAB report, on the characterization of surfaces and the topographic distribution of defects in these surfaces. Consequently, a good deal of attention has been devoted towards these surface characterization techniques. Of the more generally applicable methods, scanning electron microscopy and electron probe microanalysis are really applicable only to devices and consequently are outside the scope of this chapter. Transmission electron microscopy has been used extensively[17], of course, to study dislocations although really this is a method for bulk material. Field ion microscopy[18] is an intriguing technique but it is difficult to get enough current to the tip to apply it to semiconductors. Of the "true surface" methods, two have considerable promise as applied to solid-state materials.

Low-energy electron diffraction has had a somewhat undistinguished history in semiconductor characterization and this has led, since the instrumentation has much in common, to an interest in Auger spectroscopy[19]. Figure 18 shows the principle. Relatively low-energy (about 3 keV) electrons are used to bombard the surface, removing electrons from inner shells. Electrons falling from higher energy orbitals evolve energy which is either given off as radiation (X-ray fluorescence) or is transferred to electrons on even higher energy orbitals, which may then be ejected as so-called Auger electrons with energies characteristic of the transitions giving rise to them. This is a useful qualitative tool for detecting foreign atoms in the surface. Figure 19 shows a spectrum from a

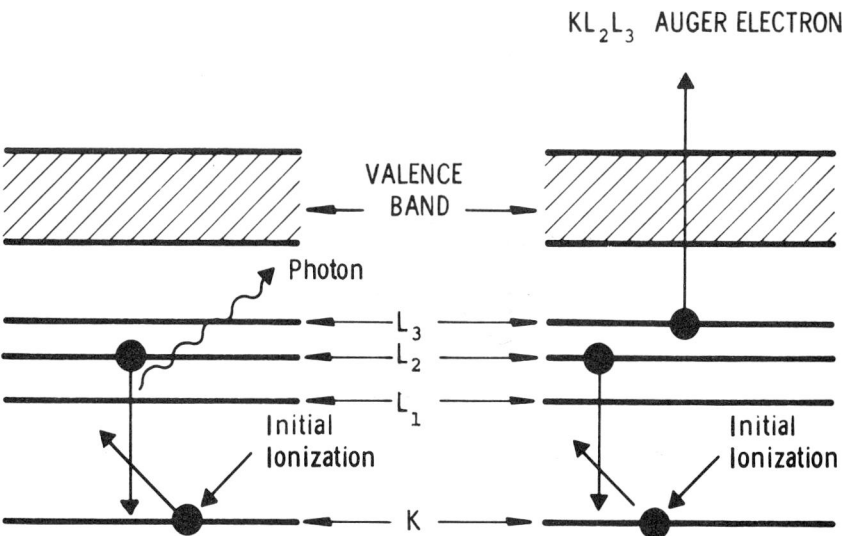

Figure 18. Energy-band transitions for (left) x-ray fluorescent and (right) Auger electron emission.

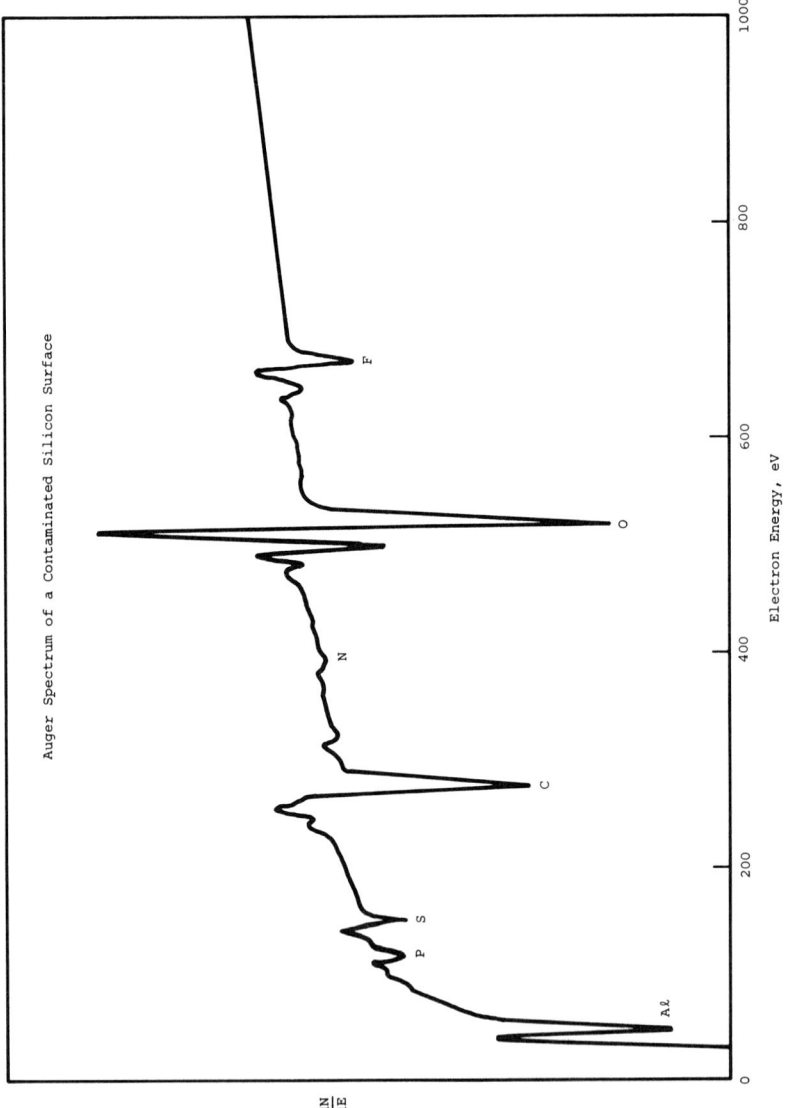

Figure 19. Auger spectrum of contaminated silicon surface.

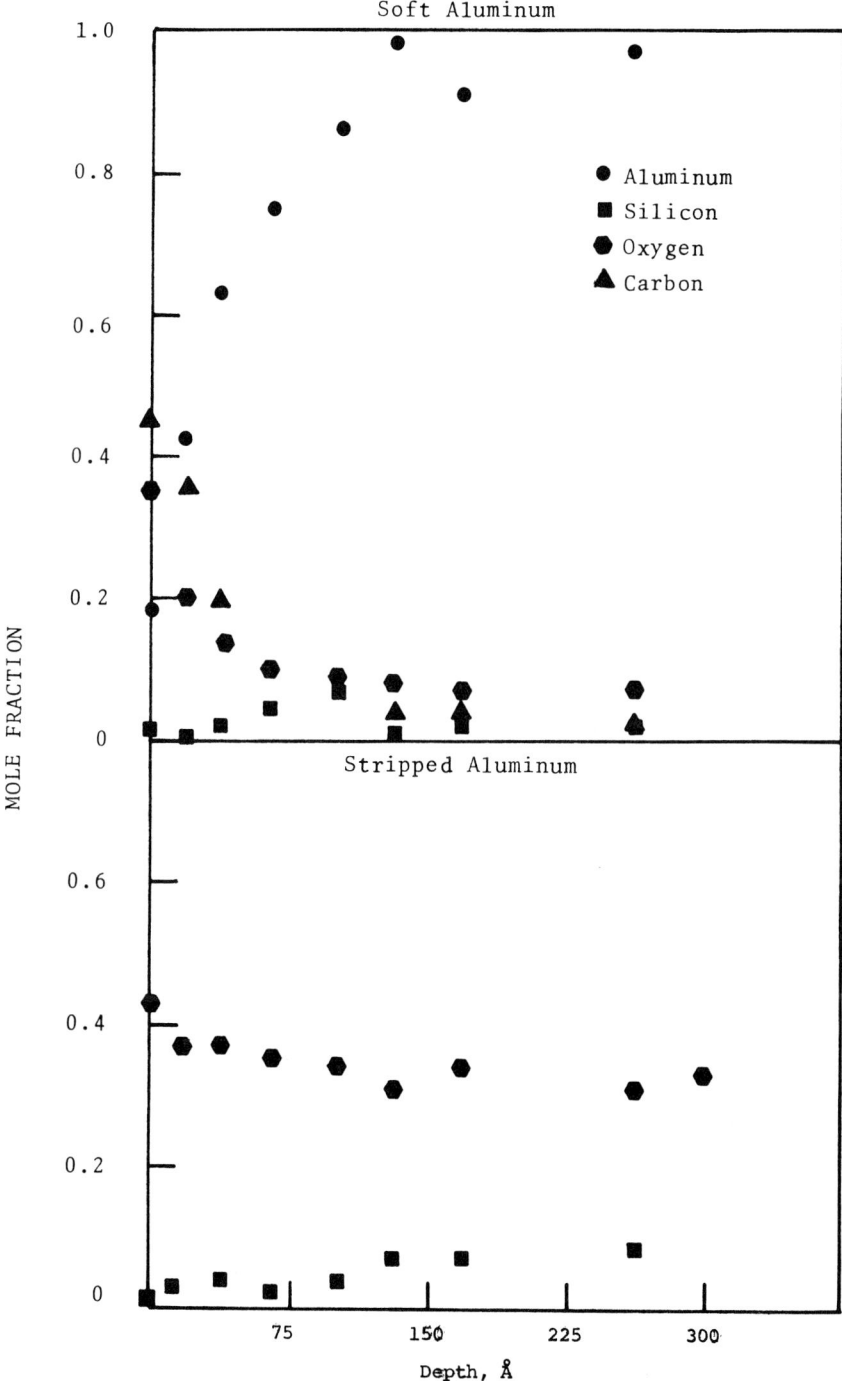

Figure 20. Auger analysis of aluminum films.

33

silicon surface; note particularly the peaks for carbon, nitrogen, oxygen, sulfur, and fluorine, all elements very difficult to detect by any other method.

As can be seen, the energy associated with these Auger electrons is quite low (a few hundred eV) so that they must originate quite close to the surface—within a few angstroms. This implies a sample depth of the order of three or four monolayers, so in absolute terms the sensitivity is extremely high. However, putting even an approximate value on it is difficult. Nevertheless, the technique can be most valuable in following surface treatments in a relative fashion.

A useful extension of this technique is obtained by adding a sputtering source of, say, argon ions to the spectrometer so that Auger spectra can be taken in a series as the material is sputtered away. Figure 20 shows profiles through two samples of aluminum, both having undesirable properties as metallization materials. The first was too soft to allow satisfactory ball bonding and, as can be seen, was contaminated to a depth of about 75 Å with carbon and oxygen. The other gave very poor adhesion to the underlying silicon dioxide and, in fact, was essentially a suboxide right through. This technique has considerable promise, and another dimension was recently added to this instrument by the announcement[20] of a scanning Auger spectrometer with a spatial resolution of about 5 microns.

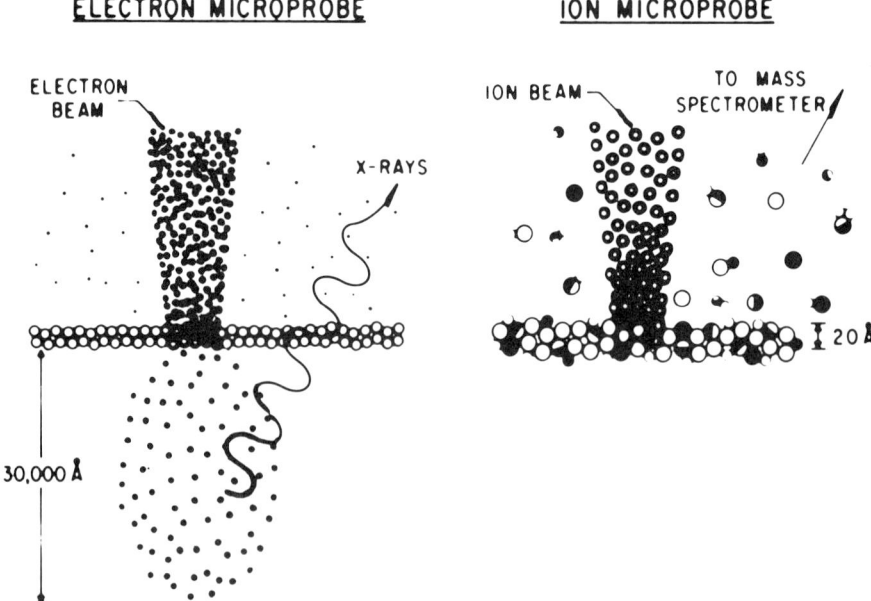

Figure 21.   The interaction of (left) electrons and (right) ions with materials.

Perhaps the most significant technique yet developed for the detection and location of chemical defects is the ion probe mass analyzer. This has developed along two lines represented by the ARL probe, which is analogous to the electron probe microanalyzer, and the Cameca, which is

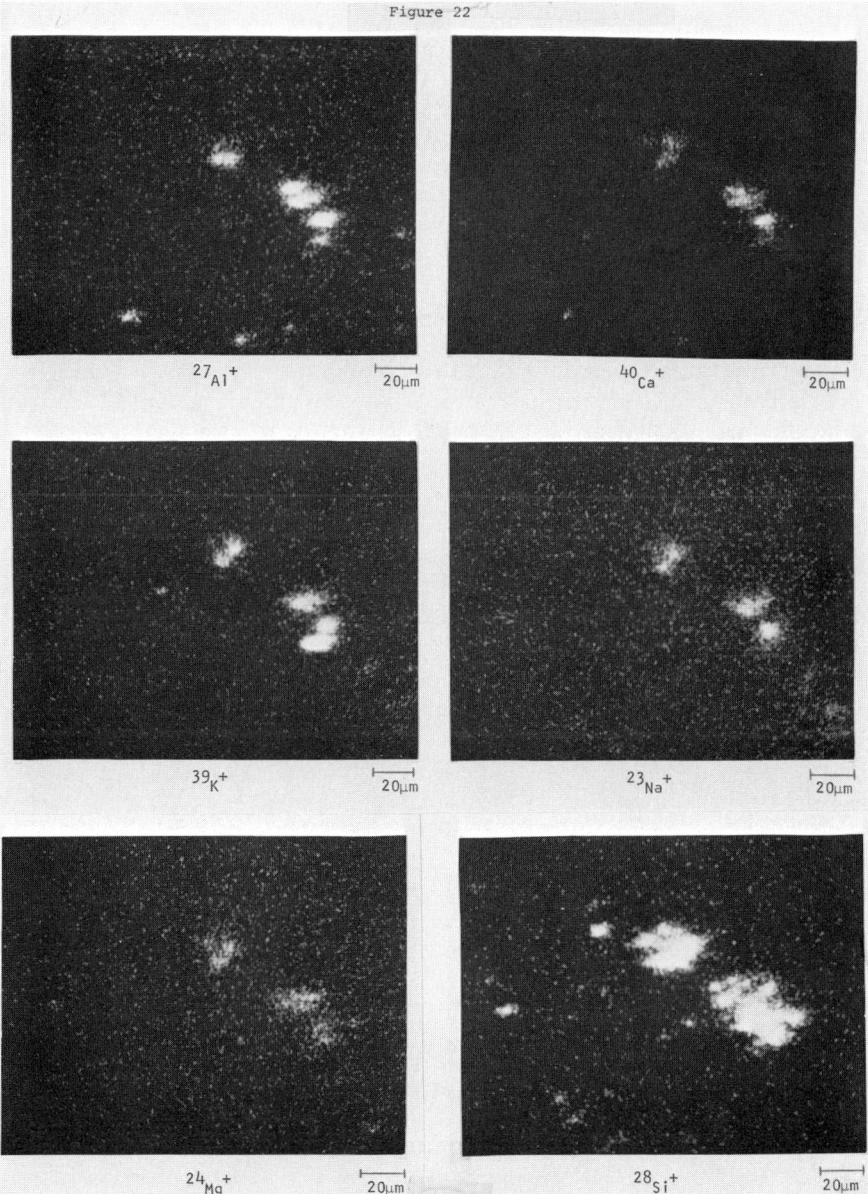

Figure 22. Ion probe topographs of a contaminated silicon surface.

analogous to the electron microscope. Both use a relatively high energy
ion beam, 10–20 kV, as the source, and the principle is shown in Figure 21
compared with the electron-beam source. The depth of penetration is
considerably less—20 Å as compared with 30,000 Å—which makes it
much more of a surface method, potentially at least. The secondary ions
are analyzed and the sensitivity is such that doping levels of impurity can
be detected although not, of course, with a 20 Å-deep sample. The beam
diameters are comparable—about a micron or less. In the scanning
mode, the instrument will give topographical information on the surface
which can be presented as a series of distribution patterns, just like an
electron probe. Figure 22 shows a series of such pictures taken from the
surface of a silicon slice using an oxygen ion beam. They represent the
analysis of a defect on the surface. Due to various work-function effects,
the silicon yield from SiO_2 is higher than from silicon itself so the top
right shows oxide film on the surface. The various elements present suggest this may be an alumino-silicate of some kind, possibly talc.

Since the ion beam is removing atoms from the surface, it will in fact
drill into the material so that, with time, we can obtain a profile through
the material. Since it also has the scanning capability, it is possible to
build up a three-dimensional distribution for an impurity. Figure 23
shows[21] two depth profiles for boron-doped material using a stationary
oxygen ion probe. On the left is a boron diffusion with the high surface

BORON IN SILICON

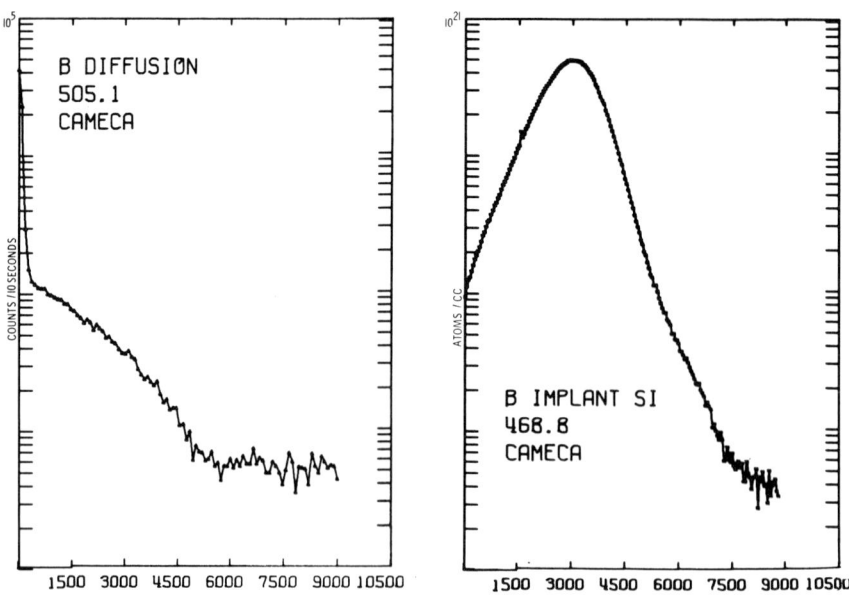

Figure 23.   Ion microprobe determinations of boron profiles in silicon[21].

concentration and the exponential curve typical of this technique. On the right, an ion-implanted sample with, this time, the Gaussian curve typical of these materials. Perhaps a more interesting case is given in the examination of a bipolar structure, shown in Figure 24. This is difficult to form by conventional diffusion technology since it must be done in two stages at about 1100°C. However, after boron has been diffused, the subsequent arsenic diffusion leads to deformation of the front. Figure 25(a) shows an ion probe profile of a diffused structure; Figure 25(b) is a profile from an ion-implanted structure. First, boron is implanted at 40 kV to give the correct sheet resistance near the surface, followed by an arsenic

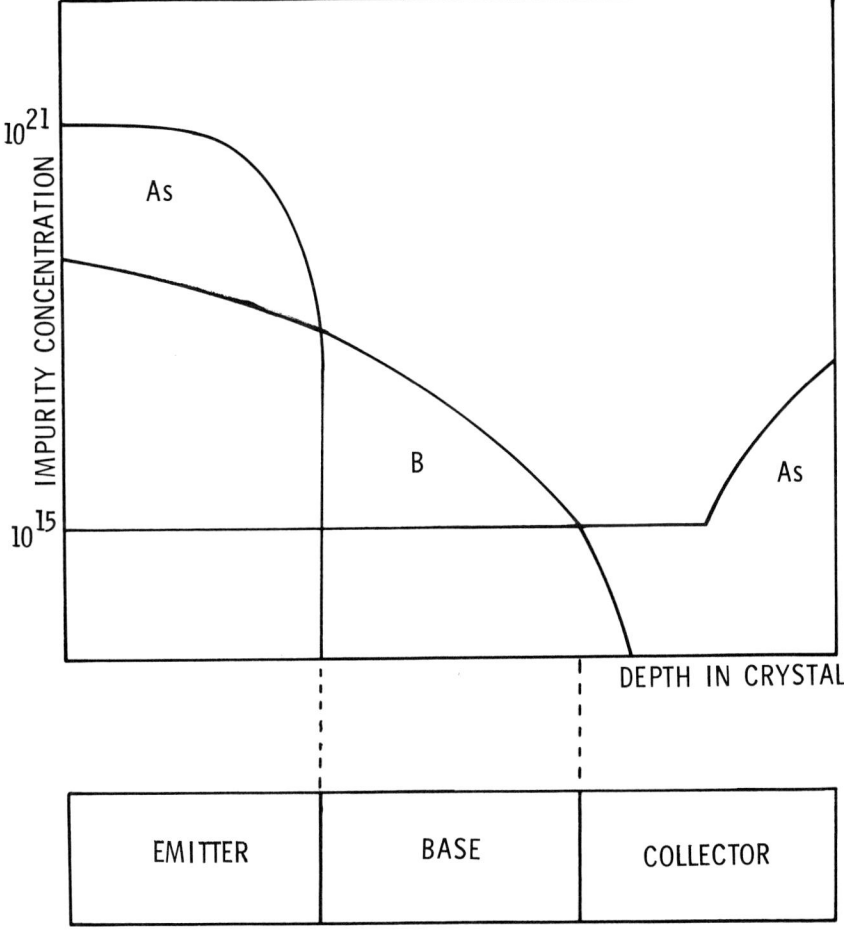

BIPOLAR TRANSISTOR

Figure 24. Idealized concentration profiles for a bipolar transistor[21].

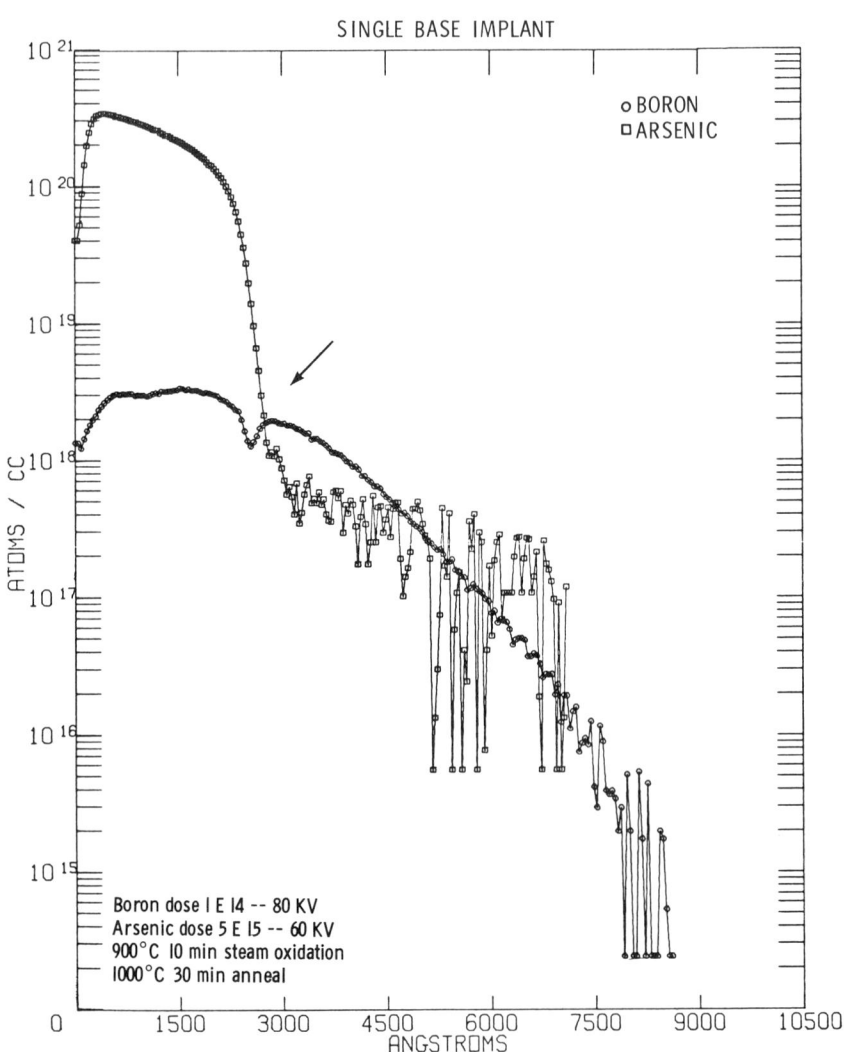

Figure 25(a). Ion microprobe determinations of concentration profiles in a conventional bipolar structure[21].

implantation at 60 kV to form the emitter. After annealing at 1000°C to drive the emitter to the correct depth, another boron implantation is carried out at 80 kV, followed by a 900°C treatment to form the base region. At 900°C, arsenic does not move significantly so that the structure is more readily controlled. The little inflection shown by the arrow is interesting; apparently the boron is depleted at the arsenic interface, possibly due to some field effect.

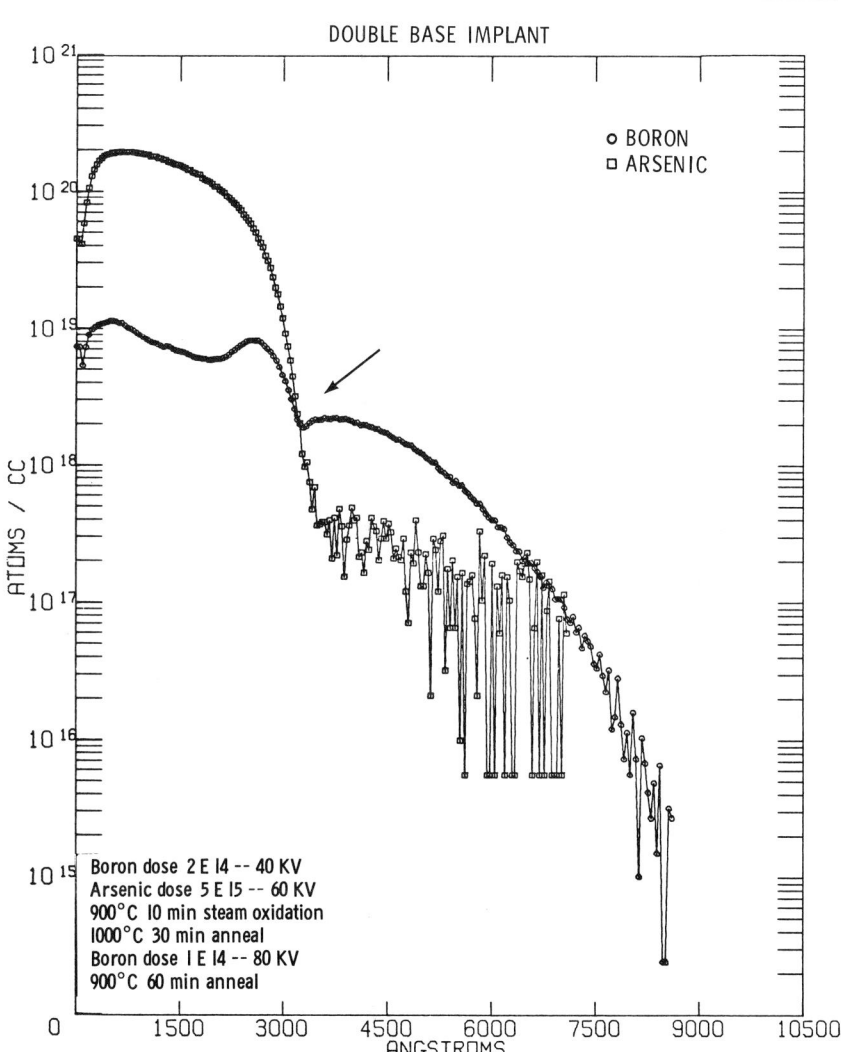

Figure 25(b). Ion microprobe determination of concentration profiles in an ion-implanted bipolar structure[21].

Conclusion

In reviewing the progress made in the characterization of solid-state materials over the last six years, it is instructive to look again at the recommendations made in the MAB report[1]. These were summarized for each of the five panels, and those relevant to semiconductor materials are from the panels on composition, structure, and defects.

TABLE IV
Composition[1]

1.	Development of . . . techniques for . . . stoichiometry beyond the ±0.1% level.
2.	Improved techniques for . . . O, N, C, B, S, halogens.
3.	Better techniques for . . . distribution . . . of impurities with spatial resolution at the micron level . . . ; the assessment of sample homogeneity. . . .
4.	Improved techniques for . . . location and distribution of impurities. . . .
5.	New methods . . . for . . . the valence state. . . .
6.	Improved survey techniques for ≤ 1 ppm.

The recommendations on composition are shown summarized in Table IV. The first is still a pressing need; there is still no method better to more than 1 part in 10^4 for determining the ratio of III-V compounds. For items 2, 3, and 4, I believe we have made significant progress; IR methods for oxygen and carbon have achieved sensitivities of the ppm level and Auger spectroscopy has proved very sensitive to the light elements. The probe methods have resulted in substantial gains in determining distribution of impurities and certainly the ion probe has tremendous potential. ESCA may contribute further information on the valence state. Item 6 is still a problem. Solids mass spectrography appears to have the potential of a true survey method but progress in improving the reliability has been very slow.

Table V summarizes the recommendations of the Panel on Structure. There do not appear to have been any great breakthroughs arising from the first suggestion except that there has been a renewed interest in Raman spectroscopy using lasers for the exciting radiation. In the X-ray field,

TABLE V
Structure[1]

7.	Research in optical methods . . . by the new coherent light sources. . . .
8.	Among the X-ray methods, . . .
	(a) Increase . . . ASTM powder data file.
	(b) Optimize . . . information . . . from powder data.
	(c) Decreased limitations of single-crystal accuracy.
	(d) Develop precise . . . X-ray apparatus.
9.	In *electron microscopy and diffraction* . . .
	(a) Scanning electron microscopy exploitation.
	(b) Improvement of . . . techniques . . . for thinning of samples. . . .
	(c) Purchase of . . . (>1 meV) . . . microscopes.
10.	Pulsed *neutron* time-of-flight *spectrometry*. . . .
11.	. . . new approaches to determining . . . *noncrystalline* solids. . . .

TABLE VI
Defects[1]

12.	Improved techniques for determination. . . .
13.	. . . the equilibrium and kinetic behavior . . . formation and annihilation.
14.	Better understanding of the interrelationships between measured physical phenomena and defects.
15.	Improved techniques for characterizing surfaces.

since the ASTM file was put on computer tape, many laboratories have devised programs for searching these files more effectively and certainly additions have been made to it. In electron microscopy, progress has been made on all these items. Certainly SEMs have mushroomed in both academic and industrial fields, and the more recent addition of nondispersive detectors considerably expands their utility. Thinning by ion bombardment has made this technique much more controllable and the big microscopes—at least a few of them—are now installed and beginning to show results.

Finally, Table VI summarizes the recommendations of the Panel on Defects. This is probably the area where the least progress has been made. Certainly, there has been little or no improvement in the techniques for the study of dislocations and little in the semiconductor field in kinetic studies. There is an urgent need to follow up Item 14; there is really very little information on the effect of physical defects on the electrical properties, for example. Techniques for determining chemical defects in surfaces have made substantial progress but the X-ray reflection topograph is still the best technique for physical defects.

In summary, progress of a substantial nature has been made in determining the nature and location of chemical impurities in semiconductors and this, of course, implies their location and identification in surfaces. We have been much less successful in the area of physical defects, and it is in this field that much of the effort must be expended in the immediate future.

References

1. Materials Advisory Board, Publication MAB-229-M, NAS-NAE, Washington, D.C., (1967).

2. P.F. Kane, Chem. Tech., 1, 532 (1971).

3. P.F. Kane and G.B. Larrabee, "Characterization of Semiconductor Materials," McGraw-Hill, N.Y., 1970.

4. P.A. Schumann, Jr., and J.F. Hallenback, Jr., J. Electrochem. Soc., 110, 538 (1963).

5. K.H. Zaininger and F.P. Heiman, Solid State Tech., (May) 49, (June) 46 (1970).

6. ASTM Standard F95-68T, ASTM, Philadelphia, 1968.

7. P.F. Cox, and A.F. Stalder, J. Electrochem. Soc., 120, 287 (1973).

8. L.N. Swink and M.J. Brau, Metall. Trans., 1, 629 (1970).

9. W.M. Bullis, ed., NBS Technical Note 733, Washington, D.C., 1972.

10. G.H. Schwuttke, Microelectronics and Reliability, 9, 397 (1970).

11. N.B. Hannay and A.J. Ahearne, Anal. Chem., 26, 1056 (1954).

12. J.A. Keenan and G.B. Larrabee, Chem. Instrum., 3, 125 (1971).

13. G.B. Larrabee and J.A. Keenan, J. Electrochem. Soc., 118, 1351 (1971).

14. R.D. Giaugue, F.S. Goulding, J.M. Jaklevic and R.M. Petal, Anal. Chem., 45, 671 (1973).

15. D.L. Greenaway and G. Harbeke, "Optical Properties and Band Structure of Semiconductors," Pergamon Press, Oxford, 1968.

16. L. Mertz, Appl. Optics, 10, 386 (1971).

17. S. Amelinckx, "The Direct Observation of Dislocations," Academic Press, N.Y., 1964.

18. E.W. Müller and T.T. Tsong, "Field Ion Microscopy; Principles and Applications," Elsevier, N.Y., 1969.

19. G.L. Connell and Y.P. Gupta, Mater. Res. Stand., 11 (1), 8 (1971).

20. Physical Electronic Industries Inc., PHI Date Sheet 1034, 7-73.

21. R.D. Dobrott, F.N. Schwettmann and J.L. Prince, Proc. Eighth Nat. Conf. on Electron Probe Analysis, New Orleans, La., August 13–17, 1973.

D. R. UHLMANN
Massachusetts Institute of Technology
Cambridge, Massachusetts

Chapter 3

Amorphous Solids

ABSTRACT

The structure of amorphous solids on various scales, including atomic structure, submicrostructure, microstructure, and macrostructure, will be discussed. Particular attention will be directed to recent advances in our knowledge in each of these areas, and to the developments in both experimental techniques and means of analyzing data which have made these advances possible. Both inorganic and polymeric glasses will be considered, although greater emphasis will be directed to oxide glasses because of the greater availability of reliable structural information on such glasses.

In the area of the atomic structure of oxide glasses, attention will be directed to the relation between experimental data (and distribution functions derived therefrom) and the predictions of various structural models. In the area of submicrostructure, attention will be focused on the widespread occurrence of phase separation, as well as on the relation between the multiphase structures observed and the mechanism of phase separation. In the case of polymeric glasses, attention will be directed to the available information on atomic distributions, to the occurrence of nodular structures, and to the characteristics of orientation and local order.

Introduction

Information provided by a wide variety of experimental techniques has been used to characterize the structures of amorphous solids. These techniques and their applications to both inorganic oxide and polymeric materials will be discussed in considerable detail in the following chapters. This chapter will therefore be directed to summarizing our present state of knowledge about the conditions required to form glasses by cooling from the liquid state and about the structural features of amorphous

43

solids, with emphasis on information obtained from recent advances in experimental techniques and means of analyzing data. Particular attention will be directed to inorganic glasses formed by cooling from the liquid state, since these have been the most extensively investigated.

It should be recognized at the outset that one can speak of the structure of a particular glassy material only within the range imposed by its thermal history. That is, a particular material may persist as a glass for extended periods of time in a large number of states characterized by different physical and thermodynamic properties—and correspondingly different structures. For example, the range of specific volumes obtainable by cooling glass-forming materials at different rates is in the range of a few percent; and beyond such a range one cannot describe the structure of a glass without specifying the conditions under which it was formed.

In addition to cooling from the liquid state, amorphous solids can be formed by a variety of other techniques. These include: (1) condensation from the vapor onto a substrate, with the vapor generated by methods such as thermal evaporation, sputtering, or electron beam evaporation; (2) chemical reactions, which may occur in the vapor phase, in solution or adjacent to a substrate; (3) electrodeposition; (4) oxidation; (5) casting a film from solution, etc. The variations in structure on various scales obtainable with such widely different preparation techniques should exceed those obtainable by simple variations in cooling rate; but in no case does a systematic study of these variations appear to have been carried out.

The dimensional scales on which structural information has been obtained for amorphous solids may be grouped into three categories: (1) atomic structure, *i.e.*, 1 Å to 5 or 10 Å; (2) submicrostructure, *i.e.*, 30–50 Å to a few hundred or a few thousand Å; and (3) micro-macrostructure, *i.e.*, microns to mm or more. The scale of structure between 5–10 Å and 30–50 Å remains almost completely unexplored, and it seems that experimental techniques to provide structural information on this scale are not presently available.

Glass Formation

Early discussions of glass formation focused attention on types of bonding and molecular configurations, and achieved some success within limited classes of materials. However, it is by now well established that there are materials which can be prepared as glasses within every category of bonding (including covalent, ionic, metallic, van der Waals, and hydrogen) and every type of molecular configuration.

These observations are usefully coupled with the fact that nearly any glass-forming material—with the exception of atactic polymers—will form a crystalline rather than an amorphous solid if held for extended periods of time in the temperature range where crystallization can occur. Together they lead to the question not of whether a material will form an amorphous solid when cooled in bulk from the liquid state, but rather how fast a given liquid must be cooled in order that detectable crystallization be avoided. In turn, the estimation of a necessary cooling rate reduces to two questions: (1) how small a volume fraction of crystals embedded in a glassy matrix can be detected and identified; and (2) how can the volume fraction of crystals be related to the kinetic constants describing the nucleation and growth processes, and how can these kinetic constants in turn be related to readily measurable parameters?

This approach has been adopted in a recent kinetic treatment of glass formation[1] in which a volume fraction crystallized of 10^{-6}, distributed randomly through the bulk of the liquid, was identified as a just-detectable concentration of crystals. The fraction crystallized was related to the nucleation frequency and growth rate using the formal theory of transformation kinetics[2]. Concern was directed to the minimum cooling rates for glass formation, and hence homogeneous nucleation alone was treated. The cooling rates required to avoid a given fraction crystallized were estimated by constructing time–temperature–transformation (T-T-T) curves.

Examples of such curves are shown in Figure 1 for salol and salol-like materials having various melting points. The noses in the T-T-T curves, corresponding to the least time for the given volume fraction to crystallize, result from a competition between the driving force for crystallization, which increases with decreasing temperature, and the atomic mobility, which decreases with decreasing temperature. The cooling rate required for glass formation is rather insensitive to the assumed volume fraction crystallized, x, since the time at any temperature on the T-T-T curve varies only as the one-fourth power of x.

This approach is then based on the view that nearly any liquid is a potential glass-former, but that large variations exist in the cooling rates which must be employed to obtain specimens as amorphous rather than crystalline solids.

The analysis has been applied to a number of materials, including SiO_2, GeO_2, $Na_2O \cdot 2SiO_2$, salol, H_2O, Ag, Ni, and several lunar compositions. In all cases, the results are in favorable agreement with experimental experience.

The results of the calculations carried out to date indicate that the most favorable conditions for glass formation involve a large viscosity at the melting point of the crystalline phase and/or a rapidly rising viscosity with falling temperature below the melting point. The important oxide glass-formers are generally characterized by the former, while organic glass-formers are usually marked by a large $|d\eta/dT|$. The importance to glass formation of a low melting point relative to a given viscosity–temperature relation is illustrated by estimates for the salol-like materials shown in Figure 1, for which the estimated critical cooling rates are re-

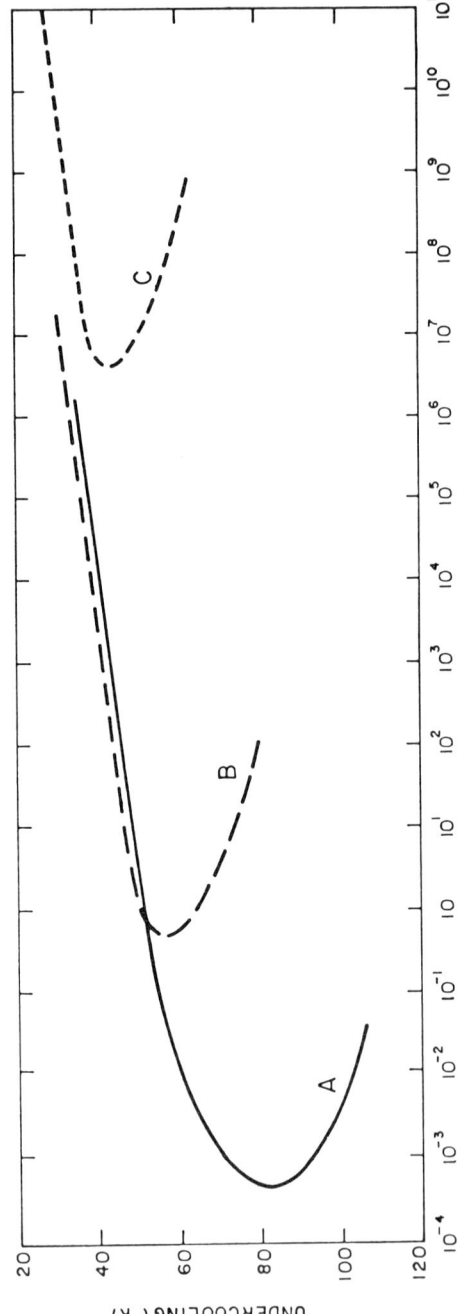

Figure 1. Time-temperature-transformation curves for salol-like materials having various melting points. Volume fractions crystallized of 10^{-6}. (A) $T_E = 356.6°$ K; (B) $T_E = 316.6°$K; (C) $T_E = 276.6°$K.

spectively about 10^5, 50, and 5×10^{-6}°K sec^{-1}. A low melting point, T_E, relative to the glass transition implies for a given class of materials a relatively large η at T_E and a relatively large $|d\eta/dT|$. The importance of near-eutectic compositions in favoring glass formation in alloy systems is related to these same considerations—and also to the redistribution of material required for crystallization to proceed in cases where a single crystalline phase is not stable relative to the amorphous phase.

In a very recent extension of this analysis[3], use has been made of a crystal distribution function $\psi(r,t,R)$ defined such that the number of crystallites in a volume dv at r having radii between R & $R + dR$ at time t is $\psi dv dR$. This function contains essentially complete statistical information about the state of crystallinity of a material. It can be used to predict the state of crystallinity when the thermal history is known or to evaluate the thermal history of a sample from post-mortem measurements of crystallinity.

With either analytical approach, the essential features of the problem of glass formation are identified as kinetic in character; and nearly all materials are expected to form glasses if cooled sufficiently rapidly to a sufficiently low temperature that detectable crystallinity is avoided. The identification of the viscosity and its variation with temperature, together with the location of the melting point relative to the glass transition temperature, as critical features of glass formation is instructive. When combined with information about the distributions of nucleating heterogeneities in particular systems, it should provide detailed insight into the general problem of forming glasses.

Atomic Structure

Information about the atomic structure of amorphous solids has been provided by a number of experimental techniques, which will be discussed at length in the following chapters. Of these, the most important and widely used has been X-ray diffraction. In contrast to the precision with which crystal structures can be determined, structural studies of amorphous solids primarily provide information with which proposed structural models must be consistent, and never (to date) determine the detailed atomic configurations in the structure.

Structural Models

A number of models have been suggested to represent the atomic structure of glasses. Prominent among these are the following.

Crystallite Model

The diffraction patterns from glasses often exhibit broad peaks centered in the range where sharp peaks are seen in the diffraction patterns of the corresponding crystals. Such observations led[4,5] to the suggestions that glasses are composed of very small crystals, termed crystallites, and that the observed breadth of the glass diffraction pattern results from particle size broadening. In the case of multicomponent glasses, the crystallites were suggested to have the compositions of compounds in the system. This model was originally advanced for crystallites, several unit cells in extent, in oxide glasses; but it is now widely discussed, for nearly all types of glasses, only in the sense of local atomic arrangements on a scale of less than 5–10 Å.

Random Network Model

The structures of glasses have also been pictured[6,7] as three-dimensional networks lacking periodicity, in which no unit of the structure is repeated at regular intervals. In the case of multicomponent glasses, the modifying cations are pictured as randomly distributed through the structure, near anion configurations such as singly bonded oxygens and BO_4 tetrahedra. This model was originally advanced for oxide glasses in which three-dimensional networks of triangles or tetrahedra can be formed. In the more general form of a "random array" model, in which the structural units are randomly arranged, the model can be used to describe a wide variety of liquid and glass structures, including those in which three-dimensional networks are not possible nor are appropriate representations.

Pentagonal Symmetry Models

Structural units which cannot fill space, particularly those having fivefold symmetry, have also been suggested as important elements in the structures of glasses and liquids[8–12]. The symmetry of the coordination cells observed in hard sphere models is predominantly pentagonal, and models based on pentagonal dodecahedral arrays of tetrahedra have been suggested for silicate glasses. Models of this type, which view the glass as composed only of pentagonal units, seem less reasonable than those which merely emphasize the significance of such units in overall glass structures composed of units of various symmetries (as 5-member, 6-member, and 7-member rings).

Micelle, Paracrystal, and Nodule Models

These models view glasses as composed of structural units having a degree of order intermediate between that of a perfect crystal and that of a random array. These units have been variously denoted as micelles or paracrystals or nodules[13–15], each having a different formal definition. The degree of order in the units is viewed as large enough for their mutual misorientation to be discerned in electron microscopy but small enough for sharp Bragg reflections to be avoided in X-ray diffraction studies.

Radial Distribution Functions and Their Determination

The distinguishing structural features of amorphous solids are the characteristic short-range order on a scale of a few Å and the absence of long-range order. The short-range order is usefully represented in terms of an atom-centered coordinate system, and is generally described in terms of radial distributions and correlation functions. Because of their greater familiarity and wider usage, the discussion in this chapter will be largely concerned with radial distribution functions.

The radial distribution function $\rho(R)$ is defined as the atom density in a spherical shell of radius R from the center of a selected atom in the liquid or glass. The radial distribution function for fused silica, determined in an early X-ray diffraction study, is shown in Figure 2. Modulations are seen in the radial density of atoms for interatomic separations of the order of a few Å. At distances in the range of 5–10 Å, the observed atom density approaches the average value, ρ_o. Since such modulations in $\rho(R)$ represent the structure of the glass, the scale where they are seen (up to 5–10 Å) indicates the scale of short-range order in the glass.

Radial distribution functions are determined from X-ray diffraction data by carrying out Fourier inversions of the data. This can have the effect of introducing spurious satellite peaks in the radial distribution function, resulting from the fact that the experimental data are measured only out to some value of $4\pi\sin\theta/\lambda$, designated k_{max}, not out to infinity as required for the Fourier integral. Here θ is the diffraction angle and λ is the wavelength of the X-radiation employed. Such effects of termination can be reduced in several ways[17,18]; but in all cases it is desirable to obtain intensity data out to k_{max} values of 15–20 Å^{-1} or more. This is usually accomplished by using CuKα radiation together with MoKα, RhKα or AgKα.

Of greater concern is the fact that at large values of $\sin\theta/\lambda$, for glasses composed of elements of small atomic number (such as most polymers and many oxide glasses), the Compton modified intensity which contains no structural information may comprise the bulk (perhaps as much as 80 percent or more) of the measured intensity. This large modified scat-

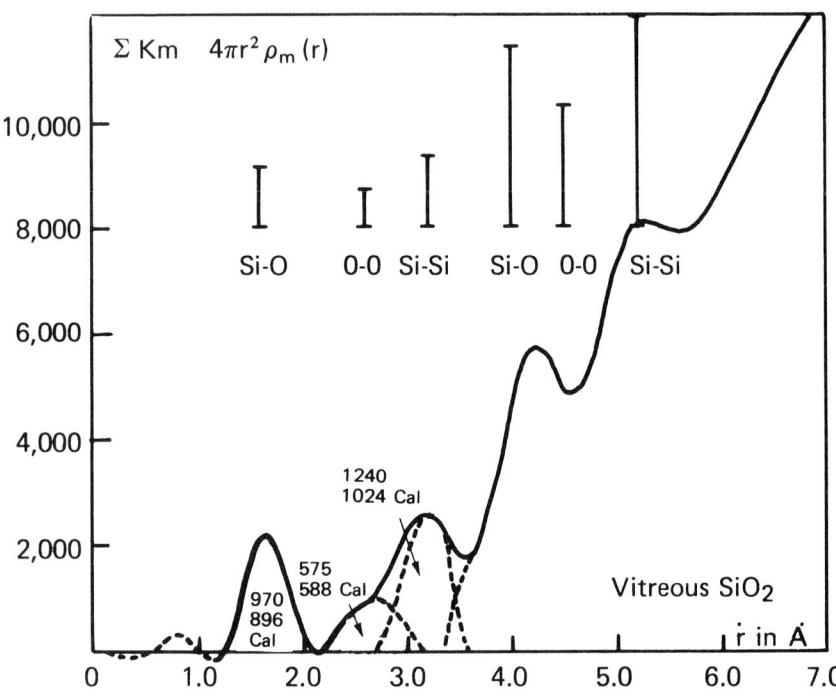

Figure 2. Radial Distribution Function of Fused Silica. After [16].

tering must be separated from the coherently diffracted intensity, which contains the structural information, by a theoretical correction which introduces a degree of uncertainty into the precise form of the data at large $\sin\theta/\lambda$. Experimental developments, involving the use of double-crystal monochromators or fluorescence detectors, permit the Compton scattering to be reduced or eliminated from the measured intensity[19]; and these offer particular promise for future investigations.

A critical limitation in most previous diffraction studies of multi-component glasses has been use of the so-called "proportionality of scattering factors" approximation. This approximation, which is necessary for separating variables to effect the Fourier inversion, assumes that the $\sin\theta/\lambda$ dependence of the scattering factors is the same for all atoms in the glass. Unfortunately, this is a poor assumption for many glasses, *e.g.*, SiO_2. With the advent of modern computation techniques, however, this approximation can be eliminated by avoiding the Fourier inversion, carrying out the integral only to k_{max}, and describing the structure in terms of pair correlation functions (see discussion in [17]). The combination of new experimental techniques for eliminating Compton scattering and analytical methods for avoiding the proportionality of scattering factors approximation appear to have opened a new era of glass diffraction studies.

Oxide Glasses

To date, the full power of the new techniques has been applied only to two simple oxide glasses, SiO_2 and B_2O_3. In the case of SiO_2, the improvements of the new experimental techniques are shown by Figure 3, where the second and third peaks—corresponding to the O–O and Si–Si distances, respectively—are clearly resolved. In contrast, the earlier study of Figure 2 shows appreciable overlap of these peaks. The new data indicate that the essential randomness of the glass structure results from a significant variation in Si–O–Si angles, from about 120° to 180°, centered about 145°. The results are well represented by a random network of SiO_4 tetrahedra, and are inconsistent with structures based on pentagonal dodecahedra or cristobalite crystallites. Present discussion focuses attention on the range over which significant structure exists in fused silica (should significance be attached to modulations in the pair function beyond 6–7 Å?), and on whether the structure should properly be viewed as random beyond the linking between tetrahedra.

Even for a random network structure, local variations in density and structure are expected. The extent of such fluctuations in the case of SiO_2 (and GeO_2 as well) is close to that expected from thermodynamic fluctuation theory for liquids applied at the glass transition temperature[21,22].

The quality of this agreement is demonstated by the results shown in Table I.

This agreement between the SAXS data and liquid fluctuation theory seems to provide support for the view that the structures of glasses should be described in the same terms as liquids and that a representation in terms of specific arrays (such as crystallites), which differ significantly and systematically from the interconnecting material, is unlikely. The latter suggestion is also supported by the essentially identical mean square density fluctuations found in SiO_2 and GeO_2. (The fractional difference in density between the most likely crystallites and the glass in the case of GeO_2 is larger by about a factor of 3 than the corresponding difference in the case of SiO_2.)

In the case of B_2O_3, both X-ray diffraction and nuclear magnetic resonance results indicate a structure composed of BO_3 triangles. The recent diffraction data[23,24] seem inconsistent with a random network of such triangles; and the question of random arrays of boroxyl units versus distorted ribbon structures (suggestions advanced in the two recent studies) remains to be resolved. This disagreement illustrates, in fact, the caution which must be employed in making structural assignments for amorphous materials.

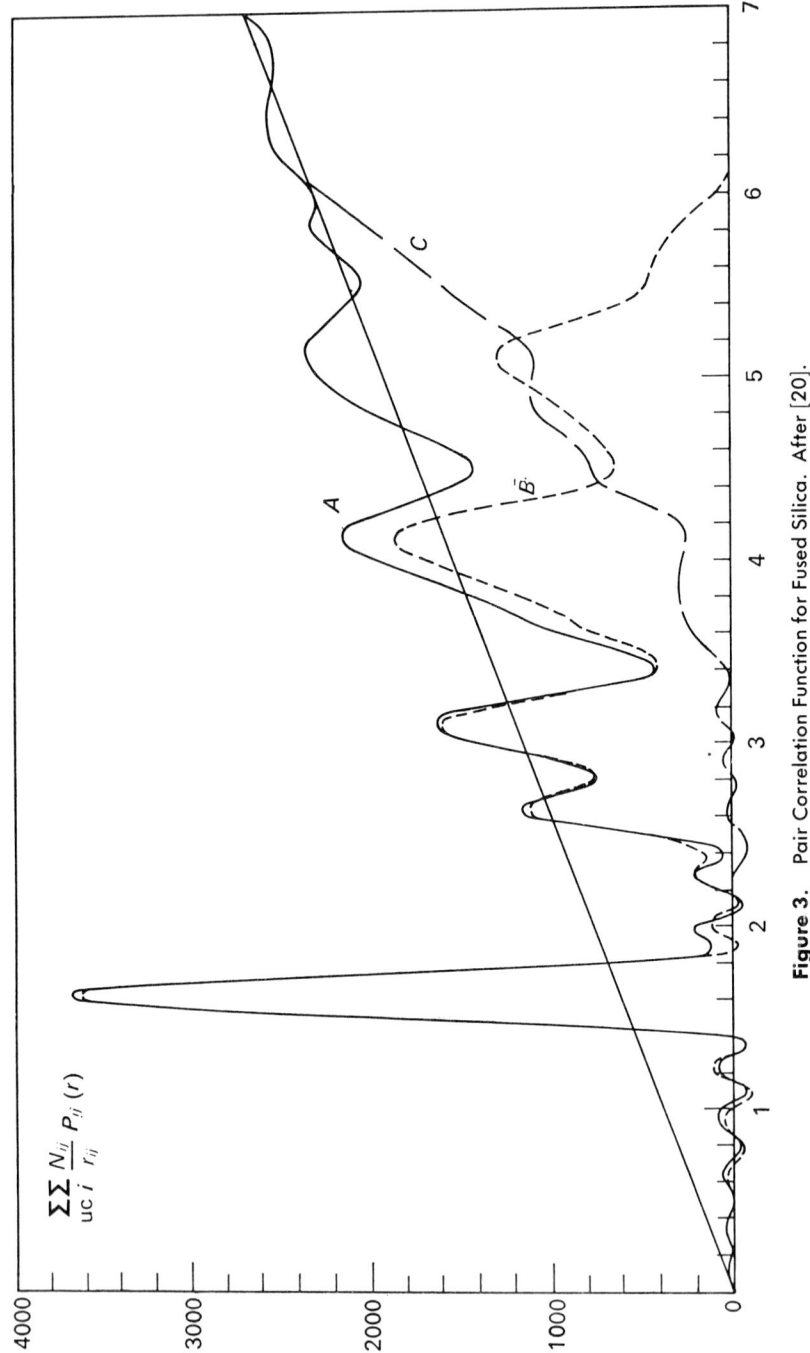

Figure 3. Pair Correlation Function for Fused Silica. After [20].

TABLE I
Density Fluctuations from SAXS Data and Thermodynamic
Fluctuation Theory. After [22].

Material	$\dfrac{<(\Delta\rho)^2>^{1/2}}{\rho}$ exp $(cm^{-3/2})$	$\dfrac{<(\Delta\rho)^2>^{1/2}}{\rho}$ fluctuation theory $(cm^{-3/2})$
GeO_2	$10^{-12} V^{-1/2}$	$7.5 \times 10^{-13} V^{-1/2}$
SiO_2	$1.07 \times 10^{12} V^{-1/2}$	$6.9 \times 10^{-13} V^{-1/2}$

On adding alkali or alkaline earth oxides to SiO_2, the three-dimensional network is broken up with the formation of singly bonded oxygens. Little is known, however, about the cation environments in such glasses; and even the extent of clustering remains a subject of considerable discussion. These questions should be resolved when the results of ongoing diffraction studies are completed[25].

It has been established[26] by nuclear magnetic resonance studies that the addition of alkali or alkaline earth oxides to B_2O_3 results in the formation of BO_4 tetrahedra. The fraction of 4-coordinated borons varies with the concentration of alkali oxide as shown in Figure 4. The smooth curve shown in the figure represents the condition that each of the added oxygens converts two triangles to tetrahedra. For alkali oxide additions beyond about 30 percent, singly bonded oxygens—presumably associated

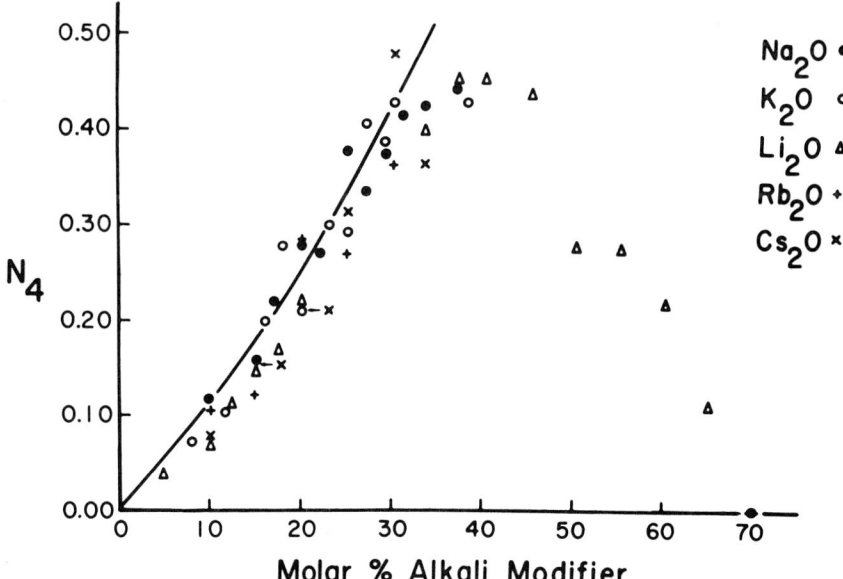

Figure 4. Fraction N_4 of tetrahedrally coordinated boron atoms in alkali borate glasses. After [26].

with BO_3 triangles—are apparently produced in appreciable numbers. Almost nothing is known, however, about the structural state of the modifying cations in these glasses; again, X-ray diffraction studies should provide elucidation.

Most familiar phosphate glasses consist of rings or chains of PO_4 tetrahedra. Chromatographic studies[27] indicate systematic changes in average chain length with composition; but little information is available on many compositions of technological interest. Glassy GeO_2 is apparently composed of GeO_4 tetrahedra, but the details of its structure remain to be elucidated satisfactorily. The local environments of transition-metal and rare-earth ions in glasses have also been investigated. Many of the results are summarized in a useful review[28] and will not be repeated here. It may be emphasized, however, that the most useful results in this area are often obtained by employing a combination of experimental techniques rather than relying upon a single tool.

Nonoxide Inorganic Glasses

The structures of a number of nonoxide glasses—but few polymers—have been determined (see bibliography of these and other structural investigations in [29]). Nearly all, however, were carried out before the advent of the new experimental and analytical techniques in diffraction studies, and here we shall consider only Ni–P alloys, Si and Ge, and Se.

The structures of Ni–P metal alloy glasses appear to be rather well represented by hard sphere models, particularly the model of dense random packing[30]. The glasses appear to have a continuous structure rather than one composed of highly ordered regions separated by less ordered boundaries. These conclusions appear to be on rather firm ground, since the scattering factors of Ni and P vary in a similar manner with $\sin\theta/\lambda$ and a Fourier inversion of the diffraction data could readily be effected.

For Si and Ge, tetrahedral arrays are found in both the crystalline glassy forms. Of various models which have been advanced for the structures of the glasses, Polk's description of a random network[31] seems the most satisfactory. It provides a close representation of the observed density deficit between crystal and glass as well as the measured radial distribution functions[32]. In this model, all bonds are included in at least one 5- or 6-member ring, but the rings are distorted from their symmetrical shapes. The only significant objection to this model is provided by high-resolution lattice-imaging electron microscope observations on glassy Ge[33]. These observations indicate regular fringes corresponding closely to the close-packed interplanar spacings of the crystalline form in regions about 14 Å in extent. While they may be taken as evidence for crystalline arrays, the fringes could also be associated with astigmatism in the lens system or other instrumental difficulties, and further work in the area seems indicated.

Glassy Se has been the subject of several structural investigations. Both long polymer chains and 8-member rings are apparently present in the glass, with a ring component of 30–50 percent being inferred from data on dissolution behavior and Raman spectroscopy[34,35], rather than from X-ray diffraction data, which is less conclusive. Devitrification of amorphous Se generally leads to the hexagonal form, which consists entirely of chains; and appreciable reconstruction of the structural units at the interface seems clearly indicated.

Polymers

Only a handful of detailed X-ray diffraction studies appear to have been carried out on glassy polymers. This is in large part associated with a a number of difficulties inherent to most of these materials. First, because of the small atomic numbers of the constituent atoms, the Compton modified component is quite large at large values of $\sin\theta/\lambda$; second, the "proportionality of scattering factors" approximation must often be employed in obtaining a radial distribution function; and third, the intensity data are often obtained only out to modest values of $\sin\theta/\lambda$ (in the range of 0.6), and the effects of termination are then correspondingly large. In some cases, modulations in the intensity data are not evident at large values of $\sin\theta/\lambda$; this could reflect a broad distribution of interatomic distances in the glasses, but may simply represent a small coherent intensity superimposed on a much larger Compton modified intensity in the range of large $\sin\theta/\lambda$.

The classic radial distribution function analysis of a polymeric material was that of Simard and Warren[36] on unstretched natural rubber. These workers found four peaks in the $\rho(R)$ function at distances smaller than about 6 Å. These peaks could all be interpreted in terms of distances in the individual chains; and no consideration of chain conformations and interchain distances was required to interpret the experimental $\rho(R)$.

A number of more recent diffraction studies have been carried out without materially improving the experimental techniques and means of analyzing data. Quite recently, a study of polystyrene appeared[37] in which improved analytical procedures were employed. Both isotactic and atactic materials were investigated in both quenched and annealed states (the isotactic material crystallized to a significant extent upon annealing). The radial distribution functions for the glassy atactic and isotactic material are closely similar, but some significant differences are noted between the glassy and partly crystalline isotactic material. The published crystal structure of isotactic polystyrene was used to identify the peaks in

the radial distribution functions, and it was suggested that both intermolecular and intramolecular distances contribute to the peaks in the
polystyrene $\rho(R)$ for R greater than 3.7 Å.

This study represents a significant advance in the area. It remains,
however, subject to a number of limitations—particularly the use of only
CuKα radiation, which results in a k_{max} value of only about 8, and the
correction for the Compton modified scattering by analytical methods,
rather than its elimination experimentally. Studies are presently under-
way[38] on a number of glassy polymers in which the technique of
fluorescence detection is used to eliminate the modified scattering, and in
which data are obtained out to a k_{max} value exceeding 20. It is hoped that
these studies will provide new and useful insights into the structure of
glassy polymers. It should be particularly emphasized in discussing
polymers, however, that the principal result of diffraction studies is to
provide data with which proposed models must be consistent, rather than
to establish with any uniqueness the structure of the materials. At the
very least, the availability of reliable correlation functions should provide
impetus for new and insightful modeling of the structure of these
materials.

Submicrostructure

The wide acceptance of the random-network model for oxide glasses,
which followed the classic work of Warren in the 1930s, received a critical
challenge when electron microscopy was introduced in the 1950s as a tool
for studying the submicrostructure of glasses. Using both direct transmission and replication techniques, features on a scale of a few hundred Å
were observed in many glasses. These have now been observed in silicate,
borate, chalcogenide, and fused salt glasses, and have been shown to result from a process of phase separation[41], in which a liquid which is
homogeneous at high temperatures separates into two or more liquid
phases on cooling into a miscibility gap. In certain polymeric systems, the
separation can occur on heating into a miscibility gap having a lower
consolute temperature.

Oxide Glasses

The characteristics of submicrostructure in glasses are well illustrated by
the system BaO–SiO$_2$. The miscibility gap in this system is metastable
(Figure 5a), and the submicrostructural features have been investigated
using direct transmission electron microscopy[39,40]. For compositions
on the SiO$_2$-rich side of the miscibility gap [Figure 5(b)], the submicrostructure consists of discrete spherical particles of a BaO-rich phase em-

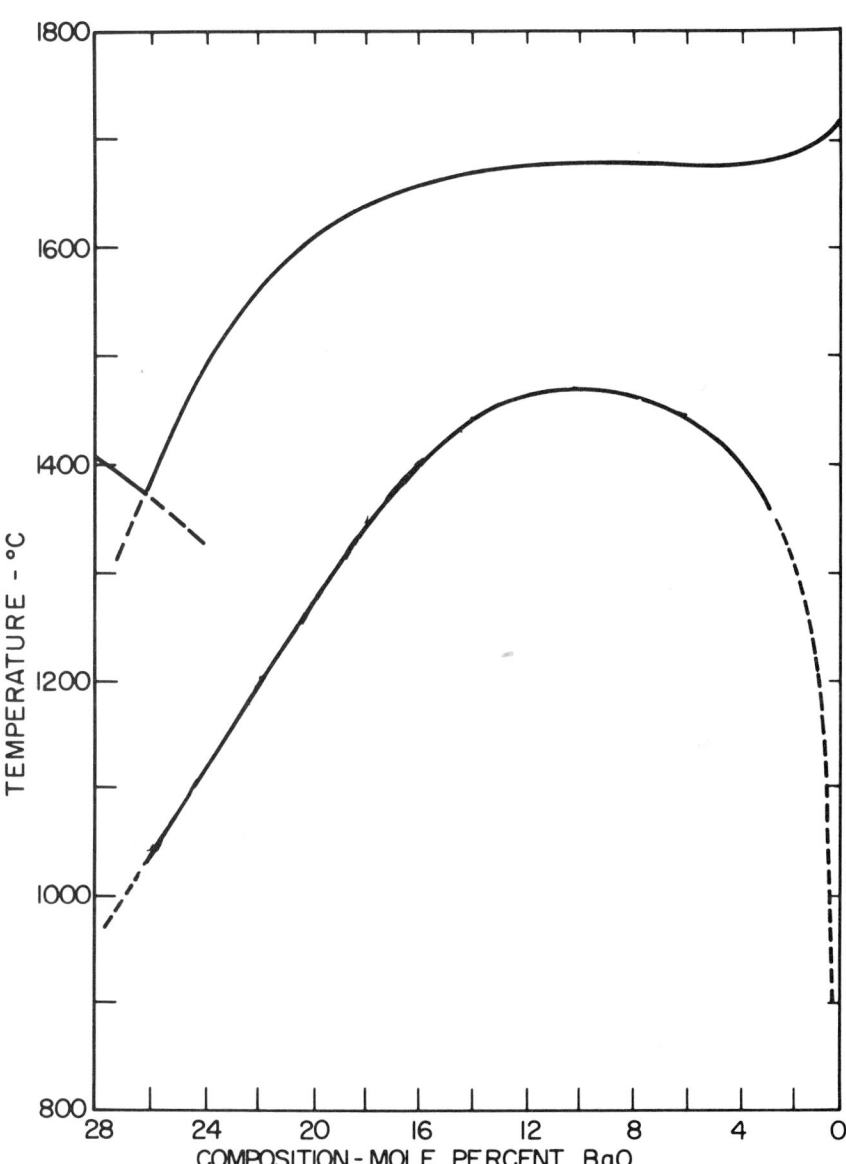

Figure 5a. Miscibility gap in the System BaO-SiO₂. After [39].

bedded in a continuous matrix of a SiO_2-rich phase. For compositions on the BaO-rich side of the miscibility gap [Figure 5(c)], the submicrostructure consists of spherical SiO_2-rich particles embedded in a continuous BaO-rich matrix. For compositions near the center of the misci-

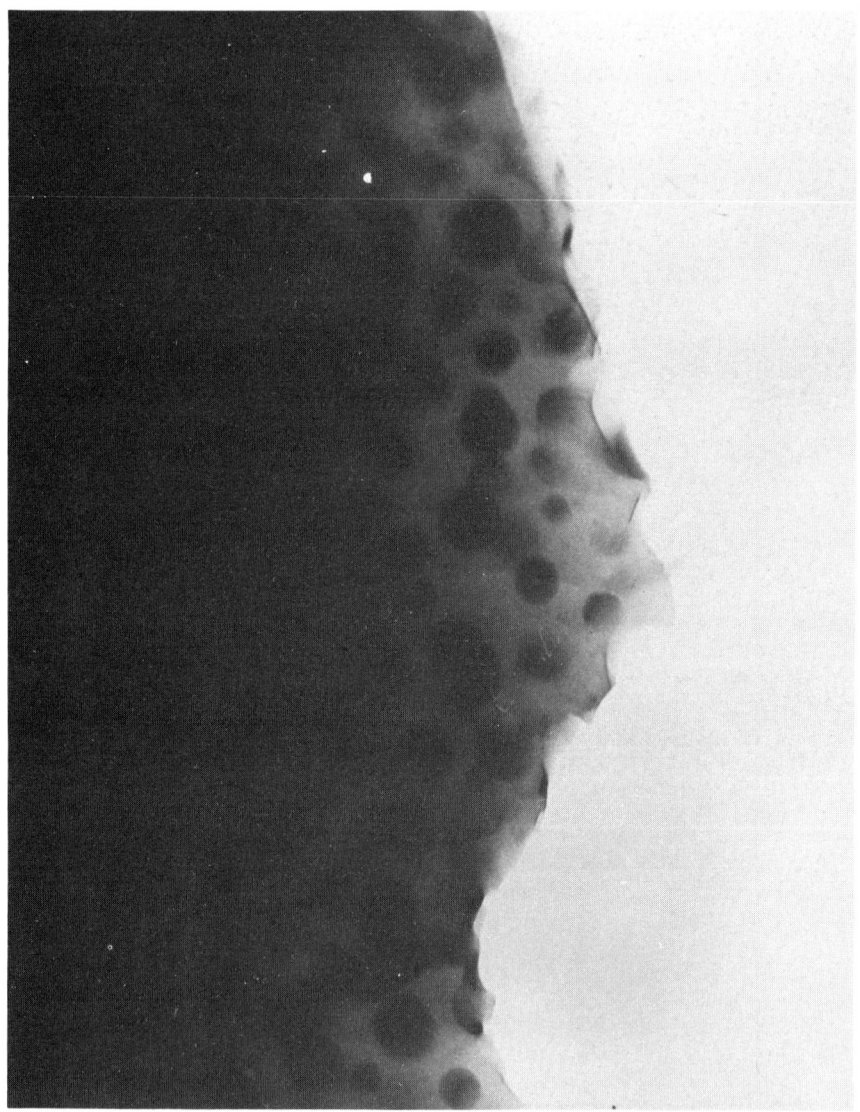

Figure 5b.   Submicrostructure of 0.04 BaO–0.96 SiO$_2$ glass. After [39].

bility gap (Figure 5d), the submicrostructure frequently consists of two phases, each of which is three-dimensionally interconnected. The phases in Figures 5(b)–5(d) are all amorphous as determined from electron diffraction.

The system BaO–SiO$_2$ was selected for the large difference in electron and atom density between the phases, which led to the contrast shown in

Figure 5c.   Submicrostructure of 0.24 BaO–0.76 SiO₂ glass. After [39].

the electron micrographs. Similar submicrostructural variations are anticipated in other systems as miscibility gaps are traversed. Among the features seen in Figures 5(b)–5(d), the interconnected submicrostructures seem particularly interesting. Such submicrostructures are often, but not always, seen where both phases are present in large volume fractions. For comparison, Figure 6 shows an electron micrograph of a discrete-particle

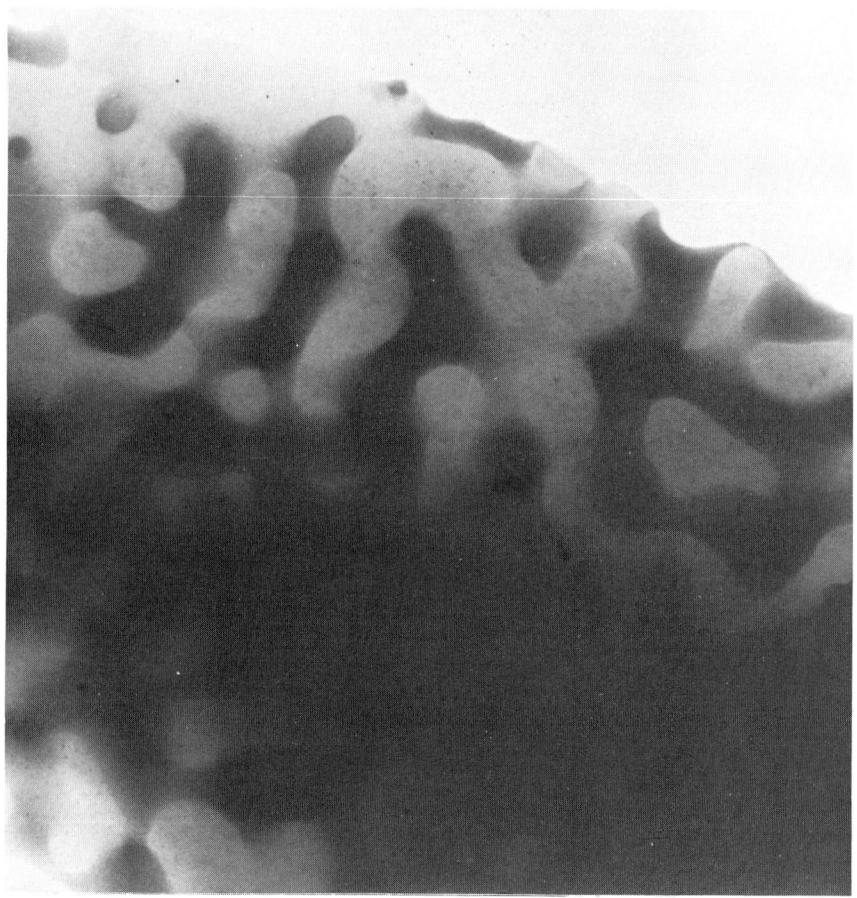

Figure 5d.   Submicrostructure of 0.10 BaO–0.90 SiO₂ glass. After [39].

structure noted in the central region of the miscibility gap in the system PbO–B₂O₃.

The origin of interconnected submicrostructures has been the subject
of considerable discussion in recent years.   Submicrostructures calculated on the basis of the linearized theory of spinodal decomposition[42]
indicate morphologies in which both phases are three-dimensionally interconnected (Figure 7).   These results led several workers to identify such
structural features with the process of spinodal decomposition.   It has
been noted[43], however, that the most important higher order terms in
the relevant diffusion equation for spinodal decomposition result in a
sharpening of interfaces and a breaking off of connectivity.   In recent
calculations[44] it has been demonstrated that a concentration-depend-

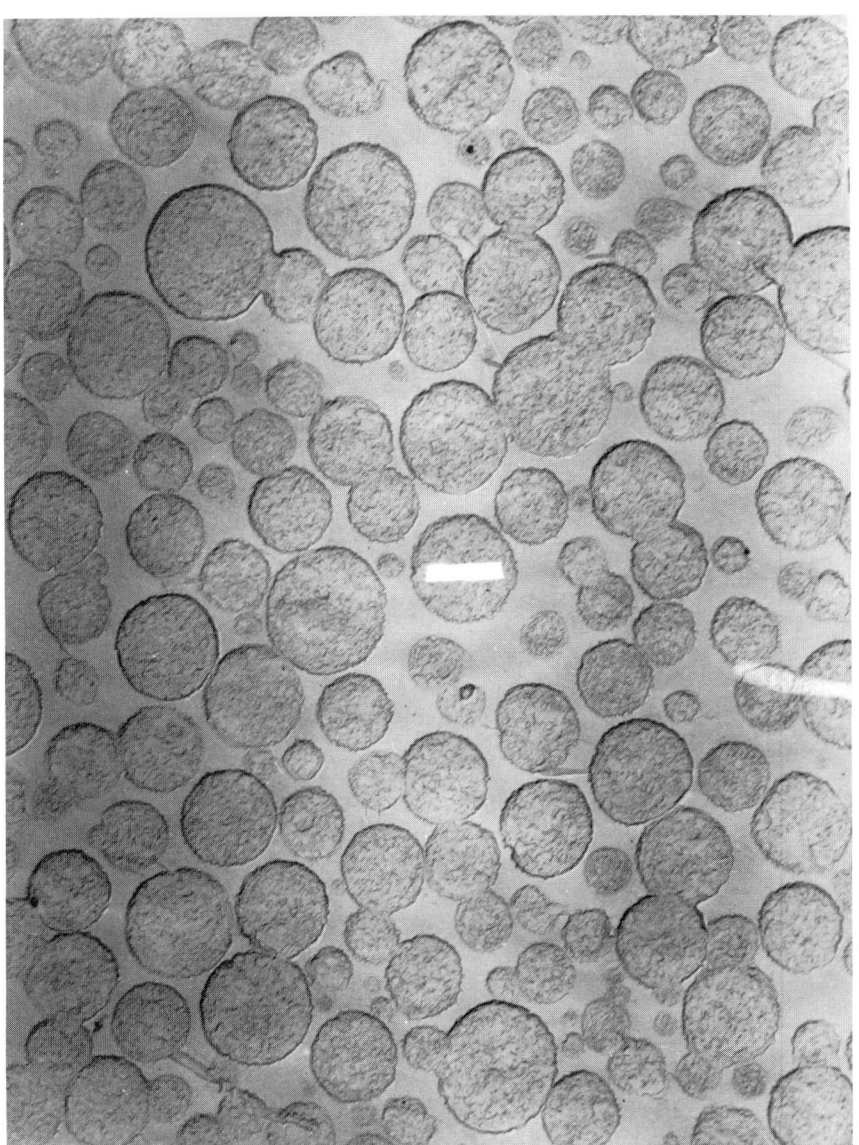

Figure 6. Submicrostructure of 0.11 PbO–0.89 B_2O_3 glass showing B_2O_3-rich discrete particles in a PbO-rich matrix. After [41]. BAR = 1 Micron.

Figure 7.   Computed sections through a two-phase linear spinodal structure in which both phases occupy a volume fraction of 50%. After [42].

ent mobility can lead to the formation of discrete particle structures by a spinodal decomposition process.

Interconnected submicrostructures can also form by the coalescence of discrete particles. This view was advanced by Haller[45], who carried out calculations in which a random array of point nuclei simultaneously began to grow at a constant rate until they occupied a specified volume fraction. For large volume fractions, a high degree of connectivity is seen in calculated cross sections through the structure (Figure 8).

It has been noted[46] that overlapping diffusion fields in the region of approach between growing particles would reduce the diffusive flux to the region and lead to flattened particles rather than interconnected structures. It has recently been demonstrated[47], however, that coalescence can take place in reasonable times despite interparticle interference when the effects of surface tension are included in the calculations. Coalescence was predicted to be most likely when many small (30–100 Å) second-phase particles occupy a large volume fraction. The development of interconnected submicrostructures by the formation, growth, and coalescence of discrete second-phase particles has been directly observed in one study[40] and inferred from morphological observations in others[48].

Interconnected submicrostructures can therefore result either from the coalescence of second-phase particles or from a spinodal decomposi-

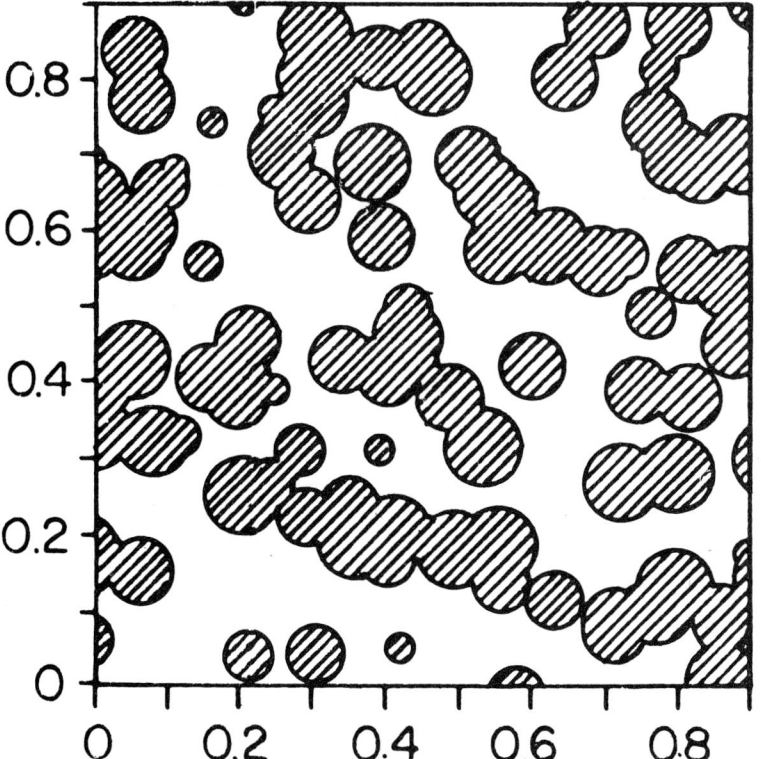

Figure 8. Computed section through a three-dimensional array of equal spheres ($r = 0.05$). After [45].

tion process in which higher order terms are unimportant. To complicate the situation further, interconnected submicrostructures have in some cases been observed to coarsen while retaining a high degree of connectivity, and in other cases to coarsen with a break-up of the connectivity[39,49]. These observations, together with those cited in the preceding paragraph, indicate the caution which must be exercised in deducing mechanisms of phase separation from morphological observations alone.

A further complicating factor in representing phase morphologies is the occurrence of secondary phase separation. This refers to the development of finer secondary phases within each of the primary separated phases as the system is cooled from the temperatures of primary separation and each of the separated phases becomes supersaturated. Such secondary structures depend strongly upon factors such as the shape of the miscibility gap and the thermal history of the specimens. They are frequently observed in phase-separating oxide glass systems[39,50,51].

The occurrence of phase separation itself is widespread in oxide systems. Metastable miscibility gaps are found in the Li_2O-SiO_2, Na_2O-SiO_2, and $BaO-SiO_2$ systems; and stable gaps occur in the other alkaline earth silicate systems. Metastable gaps are found in all the alkali borates and stable gaps in all the alkaline earth borates. Immiscibility is also known to occur in the TiO_2-SiO_2 and $PbO-B_2O_3$ systems, as well as in the commercially important $Na_2O-B_2O_3-SiO_2$ and $Na_2O-CaO-SiO_2$ systems. With such widespread occurrence of immiscibility in glass-forming oxide systems, it can no longer be assumed that optical transparency is indicative of homogeneity (separation often occurs on a scale of 30 to a few hundred Å and would then not result in translucency or opacity).

Single-Component Glasses

Submicrostructural features similar to those cited above, but appreciably weaker in contrast than those shown in Figures 5b–5d, have also been reported for the single-component glasses, SiO_2 and GeO_2; and such heterogeneities have been suggested as characteristic of glasses in general[52]. A recent study of SiO_2 has indicated, however, the absence of such features as structural characteristics of the glass[53]. Heterogeneities on a scale of 30–100 Å were sometimes seen; but when seen, they were found to depend upon the techniques used in preparing the sample for electron microscope examination and were unchanged by extended heat treatment in the region of the glass transition.

By considering the sources of contrast in transmission electron microscopy, it has also been noted[53] that the variations in density and internal potential required to account for the observed heterogeneities far exceed those expected from fluctuation theory or those observed in small-angle X-ray scattering studies of SiO_2. The heterogeneities were then associated with surface effects rather than bulk structural features of the glass. The origin of the surface features in particular cases remained to be elucidated satisfactorily. In any event, the heterogeneities sometimes seen in electron micrographs of fused silica—and presumably other single-component glasses as well—cannot be taken as structural characteristics of the bulk glass.

Studies of small-angle X-ray scattering from fused silica have yielded the results discussed in the section on Atomic Structure above (see Table I). The agreement between measured density fluctuations and those predicted from fluctuation theory seems inconsistent with any significant microstructure composed of pronounced high-density and low-density regions—at least in the specimens studied. It is anticipated, however, that samples prepared in other ways, as by condensation from the vapor onto a cold substrate, would exhibit significant scattering from void structures (see discussion of glassy Ge in the following section).

Non-oxide Inorganic Glasses

Liquid–liquid immiscibility is also a frequently occurring phenomenon in chalcogenide systems[54]. Discussions of the area are limited, however, by the fact that in almost no case has the miscibility gap been mapped out in detail. This in turn reflects the difficulty of carrying out controlled studies of phase separation under conditions where crystallization is a closely competing process. Phase separation has also been observed in a number of fused salt and salt-water systems[55].

The submicrostructures in glasses prepared by condensation from the vapor onto a cold substrate can differ substantially from those in glasses formed by cooling from the liquid state. In addition to differences in density and compositional fluctuations, and possibly in phase separation itself, in the solid phase, the presence of voids in the vapor-deposited material can form an important submicrostructural feature. Particularly interesting in this regard are vapor-deposited films of Si and Ge, which generally differ in density by 10–30 percent from the crystal density. The density of the glassy phase is, however, within a few percent of that of the crystal; and the discrepancy seems to result from the presence of pores or other low-density regions in the evaporated samples. A small-angle X-ray scattering investigation[56] of relatively thick evaporated Ge films has indicated very long rod-shaped low-density regions, with approximate dimensions of 22 Å and 46 Å in the film plane and 2,200 Å normal to the plane. Other pore morphologies have, however, been suggested by other measurements on other samples, and the differences may well be indicative of differences in film thickness and conditions of deposition. Further work in the area is clearly indicated, and should pay useful dividends in clarifying the possible importance of extrinsic variations on transport and optical properties.

Polymer Glasses

The submicrostructural features of glassy polymers have been the subject of renewed attention in recent years. Much of this attention has been occasioned by the observation, using electron microscopy, of heterogeneities on a scale of about 50–100 Å in a number of polymers. These have been termed nodular or micellar structures, and seem at variance with the random coil picture of the structure of amorphous polymers. Since this picture has been used with considerable success in representing the properties of amorphous polymers, the evidence for such nodular structures seems deserving of close scrutiny.

Observations of nodular structures, 50–100 Å in size, have been reported for a number of nominally glassy polymers, including polyethylene terephthalate, polycarbonate, natural rubber, isotactic and atactic polystyrene, and polymethyl methacrylate (see[15,57–59] as well as discussion and other references in[60]). The principal types of observations associated with these structures are the following: (1) they are observed in both direct transmission and replication electron microscopy; (2) ordered regions of approximately the same size as the nodular regions are noted in dark-field electron microscopy; (3) the nodular structures are observed to increase in size and/or rearrange upon annealing and to align upon stretching; (4) in the case of polycarbonate, the surfaces are not etched uniformly by ion bombardment but develop a granularity on the scale of the nodular structures; (5) the nodules do not vary much in size from one polymer to another, or from one form of a given polymer to another, or from a given polymer prepared under different conditions; (6) in the case of polystyrene, electron irradiation has a pronounced effect on a diffraction halo corresponding to a Bragg law d spacing which is then associated with intermolecular scattering; and (7) during cyrstallization, the nodules are suggested to merge into patches which aggregate to form lamellar planar structures.

Many of the morphological observations of nodular structures in glassy polymers are similar to those of submicrostructures in oxide glasses which have been discussed above. Even the size range of the heterogeneities in the polymers (50–100 Å) is similar to those sometimes seen in fused silica. Since replication electron microscopy is often questionable for critical study of structural features smaller than 100 Å, much of the support for the existence of these structures must be based upon the direct transmission electron microscope observations. With quantitative small-angle X-ray scattering data (on an absolute basis) on the respective polymers, it would be possible to estimate if the contrast required for the nodules to be seen in the electron microscope is consistent with the measured variations in density in the materials. While such data are presently being obtained[61] on most of the glassy polymers in which nodular structures have been reported, information is presently available only on polycarbonate[62]. The results of the latter investigation indicate an intensity which increases continuously with decreasing scattering angle. This stands in contrast to the decrease toward an asymptotic value at low angles observed for fused silica. The range of inhomogeneity in the polycarbonate samples was estimated as about 130 Å; and on the assumption of a nodular structure, the measured mean square density fluctuation corresponds to a density difference between nodules and matrix in the range of 1.6–1.8 percent. This is appreciably smaller than that (21 percent) anticipated on the basis of crystalline nodules, and seems to present a challenge to the dark-field electron microscope observations (coherently diffracting arrays with a density closely similar to that of the amorphous phase would be required for consistency between the observations).

The association of diffraction maxima in amorphous materials with Bragg-law distances seems fraught with difficulties. The diffraction pattern represents a Fourier transform of the structure; and for amorphous materials there is no general relation between diffraction peaks and frequently occurring distances in the glass. This may be illustrated by results on the best-characterized glass, fused silica, where the first three peaks in the diffraction pattern occur at $4\pi\sin\theta/\lambda$ values of about 1.5, 4.4, and 5.2 Å^{-1}. The corresponding Bragg law distances are respectively 4.2, 1.4, and 1.2 Å. As seen from Figure 3 above, only the 4.2 Å distance occurs near a peak in the pair function, and this peak itself represents primarily the contribution from silicon-second oxygen distances.

Other observations of the nodular structures seem also subject to possible reservations. The close similarity among different polymers and the spherical shapes sometimes observed seem difficult to rationalize; it is far from clear how the coalescence of nodular structures is consistent with the observed kinetics of crystallization; the nonuniform etching of surfaces could reflect simply a heterogeneous surface structure, and the observation of similarity in nodular structures for different preparation conditions seems curiously at variance with the systematic changes on annealing seen in other studies. Finally, it has been well established that polymers can experience radiation damage to a significant extent during the time required for observation and photographic recording.

The strongest evidence in support of the existence of nodules as structural features of bulk polymer glasses remains the dark-field electron microscopy observations. While it is not clear how they are consistent with the small-angle scattering results, it is also unclear how they can result from radiation damage to the materials or other artifacts (save perhaps resolution of the electron microscope).

More generally, the question of nodular structures in glassy polymers is one of considerable potential importance to understanding many aspects of polymer behavior. In the case of stereoregular polymers, it seems reasonable that regions approaching crystalline order would exist in the material; but it seems unreasonable that they would occupy as large a volume fraction as the nodular structures or that they would occur on so uniform a scale as the nodular structures in materials having widely different kinetic constants and thermal histories. In the case of nominally atactic polymers, it seems reasonable that regions will exist in the materials where local tacticity permits the formation of regular arrays of some small extent. It also seems reasonable that such formation would be most likely in the case of highly oriented polymers. It seems less reasonable, however, that pronounced differences would not be seen between atactic and stereoregular polymers, or that the local order would extend over a scale of 50–100 Å or more.

Immiscibility in polymer systems has been known for many years, but has received less detailed attention than in the case of oxide glasses. This seems to be another fruitful area for future research activities, utilizing the combination of experimental techniques (electron microscopy, scanning electron microscopy, small-angle x-ray scattering, and light scattering) which have proved valuable in the study of other systems. Of particular interest is the difference between upper-consolute-temperature miscibility gaps invariably seen in oxide systems (associated with enthalpies of solution) and the lower-consolute-temperature gaps possible in polymer systems (associated with solution entropies).

Macrostructure and Concluding Discussion

The discussion of the preceding sections should make abundantly clear that amorphous solids are far from simple homogeneous materials. Many optically transparent glasses are phase separated on a scale of 30 Å to a few hundred Å; all glasses have fluctuations in density; and all multicomponent glasses have fluctuations in composition.

On a scale of microstructure and macrostructure, the heterogeneity of glasses depends on a variety of factors. In the case of inorganic glasses, these include the batch raw materials used, the rate at which glass is removed from the melter relative to its size, the nature of the lining of the melter, the highest-melting phases retained in the melt, the degree of stirring employed during melting, and the forming technique used in fabrication. This area, like that of submicrostructure, has received increased attention in recent years because of the interest in low-loss optical waveguides. By using high-purity raw materials, selected compositions (e.g., TiO_2-doped SiO_2 and GeO_2-doped SiO_2), vapor-phase transport, and care in forming the fibers and their subsequent treatment, clad fibers with total losses in the range of 2–4 db/km have already been obtained. In future developments, attention will be directed to other characteristics of the waveguides such as dispersion, strength, susceptibility to radiation damage, and the like; and in achieving the desired combination of properties, other compositions may well be prepared. In any case, developments in the area are already taxing the technology of characterizing amorphous solids.

In the case of polymer glasses, the heterogeneity on a scale of micro-structure and macrostructure depends on factors such as stabilizers, tints and process aids, forming techniques and conditions, as well as the thermal history. For many years it has been possible to form ophthalmic quality plastic lenses by appropriate control of the material and process parameters. In applications where homogeneity is critical, however, oxide glasses are generally preferred.

The recent trends in characterizing amorphous solids, both inorganic and polymeric, have emphasized the importance of utilizing combinations of experimental techniques. In many cases, the information potentially obtainable on such materials seems considerably less than that which can be provided for crystalline materials. The glass or polymer technologist will then have to work with a lower level of characterization, which in turn—perhaps perversely—contributes to the challenge and excitement of the activity.

Acknowledgments

Financial support for the present work was provided by the National Science Foundation. This support is gratefully acknowledged.

References

1. Uhlmann, D.R., "A Kinetic Treatment of Glass Formation," *J. Non-Cryst. Solids,* 7 (1972), 337–48.

2. Christian, J.W., *The Theory of Transformations in Metals and Alloys,* New York: Pergamon Press, 1965.

3. Hopper, R.W., Scherer, G. and Uhlmann, D.R., "Crystallization Statistics, Thermal History and Glass Formation," *J. Non-Cryst. Solids,* 15 (1974), 45–62.

4. Randall, J.T., Rooksby, H.P. and Cooper H.S., "The Structure of Glasses: the Evidence of X-Ray Diffraction," *J. Soc. Glass Technol.,* 14 (1930), 219–29.

5. Valenkov, N. and Porai-Koshits, E.A., "X-Ray Investigation of the Glassy State," *Nature (London),* 137 (1936), 273–76.

6. Warren, B.E., "X-Ray Determination of the Structure of Liquids of Glass," *J. Appl. Phys.,* 8 (1937), 645–52.

7. Zachariasen, W.H., "Atomic Arrangement in Glass," *J. Amer. Chem. Soc.,* 54 (1932), 3841–51.

8. Tilton, L.W., "Noncrystal Ionic Model for Silica Glass," *J. Res., Nat. Bur. Stand.,* 59 (1957), 139–48.

9. Bernal, J.D., "The Structure of Liquids," *Proc. Roy. Soc., London, Sect. A,* 280 (1964), 299–322.

10. Frank, F.C., "Supercooling of Liquids," *Proc. Roy. Soc., London, Sect. A.,* 215 (1952), 43–46.

11. Bagley, B.G., "A Dense Packing of Hard Spheres with Five-Fold Symmetry," *Nature (London)*, 208 (1965), 674-75.

12. Finney, J.L., "Random Packings and the Structure of Simple Liquids," *Proc. Roy. Soc., London, Sect. A*, 319 (1970), 495-507.

13. Zarzycki, J. and Mezard, R., "A Direct Electron Microscope Study of the Structure of Glass," *Phys. Chem. Glasses*, 3 (1962), 163-66.

14. Hosemann, R. and Bagchi, S.N., *Direct Analysis of Diffraction by Matter*, Amsterdam: North Holland Publishing Company, 1962.

15. Yeh, G.S.Y., "Order in Amorphous Polystyrenes as Revealed by Electron Diffraction and Diffraction Microscopy," *J. Macromol. Sci., Phys.*, B6 (1972), 451-64.

16. Warren, B.E., Krutter, H. and Morningstar, O., "Fourier Analysis of X-Ray Patterns of Vitreous SiO_2 and B_2O_3," *J. Amer. Ceram. Soc.*, 19 (1936), 202-206.

17. Warren, B.E., *X-Ray Diffraction*, Reading, Mass.: Addison-Wesley Publishing Company, 1969.

18. Konnert, J.H. and Karle, J., "The Computation of Radial Distribution Functions for Glassy Materials," to be published.

19. Warren, B.E. and Mozzi, R.L., "Corrections for Intensity Measurements from Glass Samples," *J. Appl. Crystallogr.*, 3 (1970), 59-65.

20. Mozzi, R.L. and Warren, B.E., "The Structure of Vitreous Silica," *J. Appl. Crystallogr.*, 2 (1969), 164-72.

21. Levelut, A.M. and Guinier, A., "Diffusion des rayons X au petits angles par les substances homogenes," *Bull. Soc. Franc. Mineral. Crystallogr.*, 90 (1967), 445-51.

22. Pierre, A., Uhlmann, D.R. and Molea, F.N., "Small Angle X-Ray Scattering Study of Glassy GeO_2," *J. Appl. Crystallogr.*, 5 (1972), 216-21.

23. Mozzi, R.L. and Warren, B.E., "The Structure of Vitreous Boron Oxide," *J. Appl. Crystallogr.*, 3 (1970), 251-57.

24. Dunlevy, F.M. and Cooper, A.R., "X-Ray Scattering from Hydrostatically Compacted B_2O_3 Glass," paper presented at the 74th Annual Meeting of the American Ceramic Society, Washington, D.C., 8-10 May 1972. (Abstract: *Bull. Amer. Ceram. Soc.*, 51 (1972), 374)

25. Wicks, G.S. and Uhlmann, D.R., "The Structure of Sodium Silicate and Potassium Silicate Glasses," to be published.

26. Bray, P.J., "Magnetic Resonance Studies of Bonding, Structure, and Diffusion in Crystalline and Vitreous Solids," in *Interaction of Radiation with Solids* A. Bishay, ed., New York: Plenum Press (1967), 25-54.

27. Westman, A.E.R., "Constitution of Phosphate Glasses," in *Modern Aspects of the Vitreous State, Vol. 1*, J.D. Mackenzie, ed., London: Butterworths (1960), 63-91.

28. Bates, T., "Liquid Field Theory and Absorption Spectra of Transition-Metal Ions in Glasses," in *Modern Aspects of the Vitreous State, Vol. 2*, J.D. Mackenzie, ed., Washington: Butterworths (1962), 195-254.

29. Hendren, J.K., "Bibliography on X-Ray Diffraction Studies of Liquids," *Amer. J. Phys.*, 40 (1972), 1343.

30. Cargill, G.S., "Dense Random Packing of Hard Spheres as a Structural Model for Non-Crystalline Metallic Solids," *J. Appl. Phys.*, 41 (1970), 2248-50.

31. Polk, D.E., "Structural Model for Amorphous Silicon and Germanium," *J. Non-Cryst. Solids*, 5 (1971), 365-73.

32. Moss, S.C. and Graczyk, J.F., "Evidence of Voids Within the As-Deposited Structure of Glassy Silicon," *Phys. Rev. Lett.*, 23 (1969), 1167-71.

33. Rudee, M.L. and Howie, A., "The Structure of Amorphous Si and Ge," *Phil. Mag.*, 25 (1972), 1001–07.

34. Schottmiller, J., Tabak, M., Lucousky, G. and Ward, A., "The Effects of Valency on Transparent Properties in Vitreous Binary Alloys of Selenium," *J. Non-Cryst. Solids*, 4 (1970), 80–96.

35. Ward, A.T., "Molecular Structure of Dilute Vitreous Selenium-Tellurium Alloys," *J. Phys. Chem.*, 74 (1970), 4110–15.

36. Simard, G.L. and Warren, B.E., "X-Ray Study of Amorphous Rubber," *J. Amer. Chem. Soc.*, 58 (1936), 507–509.

37. Wecker, S.M., Davidson, T. and Cohen, J.B., "A Structural Study of Glassy Polystyrene," *J. Mater. Sci.*, 7 (1972), 1249–59.

38. Wicks, G. and Uhlmann, D.R., "X-Ray Diffraction Studies of Glassy Polymers," to be published.

39. Seward, T.P., Uhlmann, D.R. and Turnbull, D., "Phase Separation in the System BaO-SiO$_2$," *J. Amer. Ceram. Soc.*, 51 (1968), 278–85.

40. Seward, T.P., Uhlmann, D.R. and Turnbull, D., "Development of Two-Phase Structure in Glasses, with Special Reference to the System BaO-SiO$_2$," *J. Amer. Ceram. Soc.*, 51 (1968), 634–43.

41. Shaw, R.R. and Uhlmann, D.R., "Effects of Phase Separation on the Properties of Simple Glasses. I. Density and Molar Volume," *J. Non-Cryst. Solids*, 1 (1969), 474–98.

42. Cahn, J.W., "Phase Separation by Spinodal Decomposition in Isotropic Systems," *J. Chem. Phys.*, 42 (1965), 93–99.

43. Cahn, J.W., "Later Stages of Spinodal Decomposition and the Beginnings of Particle Coarsening," *Acta Met.*, 14 (1966), 1685–92.

44. Hopper, R.W. and Uhlmann, D.R., "Morphology in the Later Stages of Spinodal Decomposition," paper presented at the 1973 Annual Meeting of the International Commission on Glass, Bedford, Pa., 9–12 October 1973. (Abstract: *Bull. Amer. Ceram. Soc.*, 52 (1973), 700.

45. Haller, W., "Rearrangement Kinetics of the Liquid–Liquid Immiscible Microphases in Alkali Borosilicate Metals," *J. Chem. Phys.*, 42 (1965), 686–93.

46. Goldstein, M., "Interparticle Interference Effects in Diffusion-Controlled Growth," *J. Cryst. Growth*, 3-4 (1968), 594–99.

47. Hopper, R.W. and Uhlmann, D.R., "Coalescence of Second Phase Particles in Phase Separations," *Discuss. Faraday Soc.*, 50 (1970), 166–74.

48. MacDowell, J.F. and Beall, G.H., "Immiscibility and Crystallization in Al$_2$O$_3$-SiO$_2$ Glasses," *J. Amer. Ceram. Soc.*, 52 (1969), 17–25.

49. Elmer, T.H., Nordberg, M.E., Carrier, G.B. and Korda, E.J., "Phase Separation in Borosilicate Glasses as Seen by Electron Microscopy and Scanning Electron Microscopy," *J. Amer. Ceram. Soc.*, 53 (1970), 171–75.

50. Porai-Koshits, E.A. and Averjanov, V.I., "Primary and Secondary Phase Transformation of Sodium Silicate Glasses," *J. Non-Cryst. Solids*, 1 (1968), 29–38.

51. Shaw, R.R. and Breddis, J.F., "Secondary Phase Separation in Lead Borate Glasses," *J. Amer. Ceram. Soc.*, 55 (1972), 422–25.

52. Roy, R., "Alternative to the Random Network Structure for Glass: Nonuniformity as a General Condition," in *Advances in Nucleation and Crystallization of Glasses*, L.L. Hench and S.W. Freiman, eds., Columbus, Ohio: American Ceramic Society (1972).

53. Seward, T.P. and Uhlmann, D.R., "On the Existence of Submicrostructure in Fused

Silica," in *Amorphous Materials,* R.W. Douglas and B. Ellis, eds., New York: John Wiley and Sons (1972), 327–35.

54. Phillips, S.V., Booth, R.E. and McMillan, P.W., "Structural Changes Related to Electrical Properties of Bulk Chalcogenide Glasses," *J. Non-Cryst. Solids,* 4 (1970), 510–17.

55. Easteal, A.J. and Angell, C.A., "Viscosity of Molten $ZnCl_2$ and Supercritical Behavior in Its Binary Solutions," *J. Chem. Phys.,* 56 (1972), 4231–33.

56. Cargill, G.S., "Anisotropic Microstructures in Evaporated Amorphous Germanium Films," *Phys. Rev. Lett.,* 28 (1972), 1372–75.

57. Yeh, G.S.Y. and Geil, P.H., "Crystallization of Polyethylene Terephthalate from the Glassy Amorphous State," *J. Macromol. Sci. Phys.,* B1 (1967), 235–49.

58. Frank, W., Goddar, H. and Stuart, H.A., "Electron Microscope Investigations on Amorphous Polycarbonate," *J. Polym. Sci., Part B,* 5 (1967), 711–13.

59. Siegmann, A. and Geil, P.H., "Crystallization of Polycarbonate from the Glassy State. Part I. Thin Films Cast from Solution," *J. Macromol. Sci., Phys.,* B4 (1970), 239–72.

60. Yeh, G.S.Y., "A Structural Model for the Amorphous State of Polymers: Folded-Chain Fringed Micellar Grain Model," *J. Macromol. Sci., Phys.,* B6 (1972), 465–78.

61. Renninger, A. and Uhlmann, D.R., "Small-Angle X-Ray Scattering from Glassy Polymers," to be published.

62. Lin, W. and Kramer, E.J., "Small-Angle X-Ray Scattering from Amorphous Polycarbonate," *J. Appl. Phys.,* 44 (1973), 4288–92.

K. F. J. HEINRICH, D. E. NEWBURY
and H. YAKOWITZ
Institute for Materials Research
National Bureau of Standards, Washington, D.C.

Chapter 4

New Techniques for the Surface Analysis of Nonmetallic Solids

ABSTRACT

Several modern techniques for the characterization and analysis of surfaces and shallow layers of nonmetallic materials are discussed, with particular emphasis on their relevance to technological problems. These include electron and ion probe microanalysis, scanning electron microscopy, ion scattering analysis, and secondary ion mass spectroscopy of surfaces.

Introduction

It is a long-standing practice to subject materials of technological or scientific interest to composition, structure and performance tests. Techniques such as chemical analysis, X-ray diffraction, and measurements of diverse physical properties have been used extensively for this purpose. Progress in science and technology has increased the requirements, as well as the arsenal of tools, for such characterization tests. In particular, there is an increasing awareness of the importance of the state of the solid matter on a microscopic scale. A good, but not unique, illustration of this trend is the wide utilization of electron probe microanalysis, applied to a wide range of metallic and nonmetallic materials. In mineralogy, for instance, the acceptance of a newly discovered mineral requires its characterization by the electron probe microanalyzer. Thus, the microscopic characteristics first observed by means of optical and electron microscopes can be complemented by compositional information. Structural information on a comparative scale of dimensions can be provided by electron diffraction.

Many processes of technological importance take place, or are initiated, at the very skin of matter. The same is true for biological processes of great scientific interest. Among such technological problems are

73

catalysis and the poisoning of catalysts, oxidation and related corrosive processes, the formation and propagation of cracks, thermionic emission, and adhesion. In addition, the manufacture of delicate solid-state electronic devices includes processes such as ion implantation and etching, deposition from vapor and liquid phases, and diffusion through the surfaces of materials and diverse additives. It is therefore not surprising that there is great interest in the characterization of surfaces and of the distribution in depth of elements in materials treated by surface processes. A closely related field of great practical importance is the study of thin films [1].

The first difficulty in a general discussion of characterization of surfaces is the definition of the depth of a material surface and of the thickness of a "thin" film. The physical properties of a homogeneous solid differ from the bulk properties only in the first few atomic layers. The depth of a surface defined by these changes is thus on the order of 10^{-8}–10^{-9} m (100–10 Å). On the other hand, the distribution of dopants in semiconductors after surface diffusion may be in the order of 10^{-7} m, and in other technological areas—such as in the study of corrosion—layers 10^{-6} m thick are considered thin.

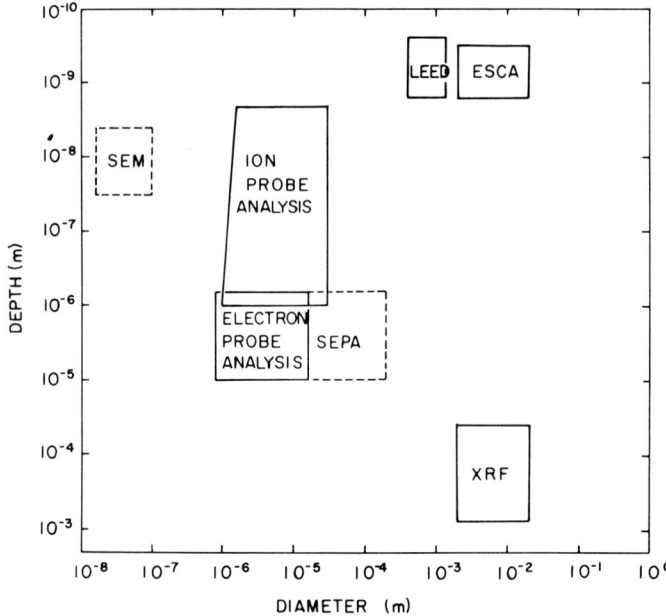

Figure 1. Spatial and depth resolution of various analytical techniques for surfaces. Key: SEM = Scanning electron microscopy, LEED = Low energy electron diffraction, ESCA = Electron spectroscopy for chemical analysis, XRF = X-ray fluorescence spectrometry, SEPA = Scanning electron probe analysis.

It is important to recognize that the tools of investigation which can be applied to the study of surfaces also have a large range of effective depths, and that this parameter is critical in the selection of a method of investigation (Figure 1). In general, the methods which sample a shallow depth in the specimen give less precise or accurate results; there is frequently a compromise to be made between resolution and analytical accuracy. On the other hand, any method characterized by spatial resolution is related to microscopy as well as to chemical analysis. It is therefore advisable to combine—or even to obtain simultaneously—compositional, topographic, and structural information. The techniques which we will discuss here will illustrate this complex nature of the investigation of surfaces.

Scanning Electron Microscopy

We will start with a relatively new topographic characterization technique which is applied in practically all domains of solid-state materials technology. The scanning electron microscope (SEM) uses a very simple technique to form an image of a microscopic region of the specimen surface. An electron beam or probe—which may range in size from 5 nm to several μm—is scanned in a well-defined raster pattern (usually square) across the specimen. The interaction of the electron beam with the specimen produces a series of phenomena, such as backscattering of electrons (of high energy), secondary electrons (of low energy), absorption of electrons (measurable as specimen current), X-rays, and, in some cases, visible light (cathodoluminescence). Any one of these signals can be continuously monitored by detectors. The detector signal is amplified and used to modulate the brightness of a cathode ray tube (CRT), the beam of which is scanned in synchronism with the electron beam impinging upon the specimen. A correspondence between each scanned point at the specimen surface and a corresponding point on the CRT screen is thus established. Usually the area scanned upon the specimen is very small compared with the corresponding area on the CRT screen. The magnification of the image on the screen—or a photograph thereof—is the ratio of a distance on the screen and the corresponding distance on the specimen.

The components of the SEM system are illustrated schematically in Figure 2. The electron gun uses an electrically biased thermal emitter, usually a tungsten filament, to produce a flow of electrons. The electron gun produces a focused image of the emitting area, called the crossover. The function of the electromagnetic lenses is to demagnify the beam diameter so that the beam width of approximately 50 μm at the gun crossover is reduced to 10 nm or less at the specimen, if the smallest possible beam is required. By suitable adjustment of the lenses, a range of beam diameters

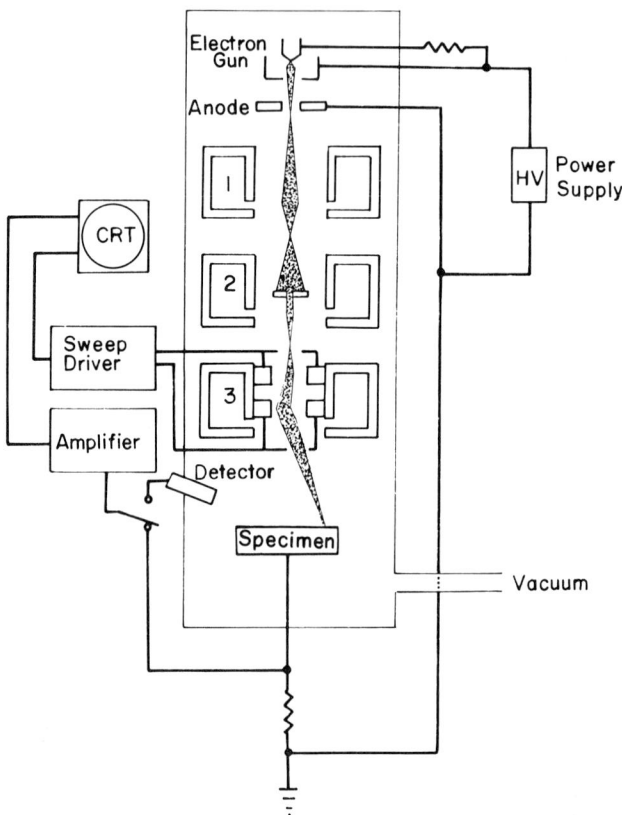

Figure 2. Schematic view of the scanning electron microscope. The lenses are numbered 1, 2, 3, respectively. Double deflection scan coils are located in the bore of lens 3.

and beam currents can be produced. A set of electromagnetic coils, located within the bore of the final lens, scans the beam in a raster on the specimen. An array of detectors sensitive to electrons, X-rays, or light can be placed in the specimen chamber. If the specimen is electrically insulated from ground, the current absorbed from the beam—which is affected by the geometry of the specimen, its mean atomic number, and other factors—can be measured and used to form an image. The specimen state is usually quite elaborate to allow manipulation of the specimen. The entire path of the electrons from gun to specimen must be within a vacuum enclosure. The largest allowable pressure is typically 10^{-5} torr. Particulars on SEM design can be obtained from references [2,3].

Three main reasons account for the widespread use of the SEM in materials science. First, the available sources of signal which relate to different modes of interaction of the electron beam with the specimen can provide the investigator with a large and varied range of information. Second, because these interactions are usually confined to a region of the specimen near the impact of the beam, and because the beam itself can be made very small, the spatial resolution of the technique is high, as much as 20 times better than that of the light microscope. Third, the large depth of focus of the SEM—about a factor of 50 over the optical microscope, and about equal to the diameter of the scanned area—provides an impressive three-dimensional realism to the SEM images (see Figure 3) and permits a very effective use of stereo techniques.

A survey of the literature[4] shows that the range of potential applications of SEM is very wide. However, the SEM is not a panacea for high-resolution surface studies, and its practical use has its limitations, particularly in the study of polymeric and ceramic materials, which are somewhat difficult to study by the SEM because of their low electric and thermal conductivity.

The limit of resolution in an SEM investigation of a material is not always well understood by the user of the instrument. Commercially

Figure 3. SEM micrograph of fracture in pure iron. Specimen fractured at 77°K. Micrographs show regions of cleavage and intergranular failure.

available instruments carry a guaranteed resolution, typically of 10 nm (100 Å). To relate this specification to the performance in practice, it should be noted that this optimum resolution is achieved only on certain specimens of optimum characteristics. Furthermore, the resolution is usually defined as the minimum distance between two objects which can be distinguished on a photograph. This does not necessarily mean that one object of the same size could be detected against a featureless background.

The resolution achieved in an observation does not depend on the instrument alone, but also on the achievable contrast, and on the duration of the scan. For a feature to be observed, the signal produced by it must be distinguished from the random signal fluctuations (noise level) inherent to the signal formed by individual electrons (or photons). The effect of these random fluctuations diminishes with increasing electron-beam current and with the length of exposure. However, the minimum achievable beam diameter is strongly dependent on the beam current. For a typical mode of operation, such as a photographic recording of a frame during 50 seconds, an approximate expression can be used to determine the lowest current, I_{th}, required to observe a contrast level of contrast ratio, C, $(C = I_{max}/I_{min})$[5]:

$$I_{th} = \frac{1.6 \times 10^{-12}}{(C - 1)^2} A \qquad (1)$$

This equation is valid for any electron source. At the usual rates of scan for visual observation with CRT phosphors of long persistence, the current required may increase by a factor of 10 above this estimate.

Once the threshold current has been determined, the smallest beam diameter achievable with this current can be calculated from the relation between the brightness of the electron source β, the divergence of the electron optical system α (both of which are constant for a given experiment and may be known or determined), the beam diameter d, and the beam current I_b[6]:

$$\beta = \frac{0.4 I_b}{d^2 \alpha^2} \qquad (2)$$

The application of these equations to practical situations shows the crucial role played by the contrast in the definition of resolution. The factor C is, with secondary electron signals, a complicated function of the specimen and detector characteristics, and of the geometric arrangement, and it frequently varies with the orientation of the specimen. Under unfavorable conditions, the limits of detection for a given observation may exceed the specifications of the instrument by a factor as large as 10.

Another important limitation of spatial resolution stems from the diffusion of the impinging electrons within the specimen, which may produce signals at distances from the point of beam impact much larger than the beam diameter. Of all signals mentioned before, the secondary electrons, because of their low energy, are confined most closely to the area of beam impact. Primary backscattered electrons can escape from far larger distances, as can X-ray or light photons. Therefore, the inherent limit of resolution with these signals is larger than with secondary electrons, by at least an order of magnitude. Due to the two limitations outlined above, the minimum discernable size for many types of objects is 20–500 nm (200–5,000 Å), or more, depending on the specimen and the character of the contrast mechanism.

The observation of electric insulators, such as polymers and ceramics, is complicated by the build-up of electrostatic charges when such specimens are irradiated with an electron beam. If these charges are sufficiently high, the image may become distorted, and periodic discharges due to dielectric breakdown will cause streaks and discontinuities in the photograph. Several techniques can be used to overcome this problem. The specimen can be coated by vacuum deposition of a thin (ca. 20 nm) layer of carbon or gold. This layer provides a conducting path which eliminates the charge build-up. However, this procedure may obscure the observation of details in high-resolution work. It is not objectionable for magnifications below 10,000. Another solution is to work with low electron accelerating potential ("operating voltage"). At 2 k V or below, the emission of secondary electrons tends to balance the charge input due to the electron beam, and many specimens can thus be observed uncoated (Figure 4). Unfortunately, the brightness of electron guns falls drastically at low accelerating voltages, and useful magnifications are limited to about 5,000. Low voltages may, however, be advantageous in the investigation of organic materials which are less susceptible under such conditions to radiation damage. A third technique for insulating specimens is to heat the specimen in a heating stage. The surface conductivity of many materials rises rapidly above room temperature, and heating the specimen to about 100°C is often sufficient to allow its examination at high beam potentials and thus greater gun brightness. Finally, it is now possible to limit the irradiation of the specimen to a single frame scan, with storage and later processing of the signal. Frequently, the first scan can be terminated before the onset of the charge effects.

Special Techniques

In the remainder of this discussion of SEM we will detail contrast mechanisms of special interest to ceramics and polymer studies.

Electron Channeling Contrast

Electrons which penetrate a crystal travel farther along certain channeling directions (Figure 5)[7]. In these particular directions, a small fraction of electrons—5 percent or less—advance up to 100 nm before their

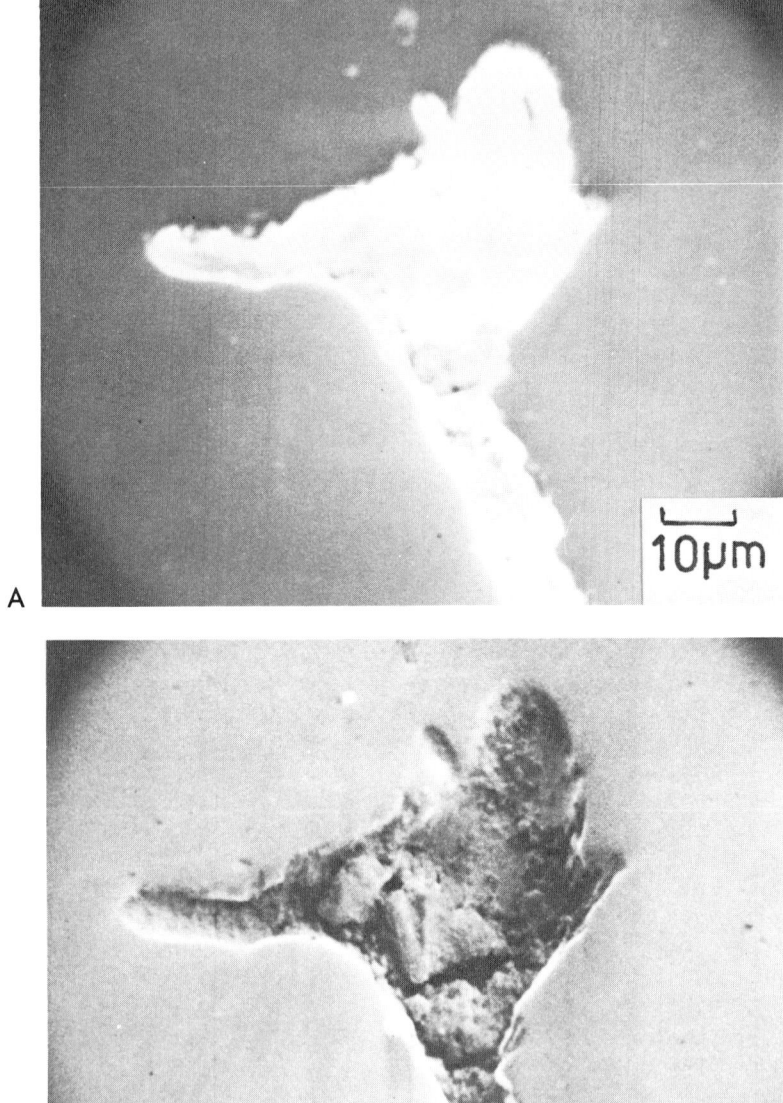

Figure 4.   SEM micrographs of uncoated lithium niobate. Figure 4a shows the effects caused by accumulation of surface charge at a beam accelerating potential of 10 kV. Figure 4b shows the same region with a beam accelerating potential of 2 kV.

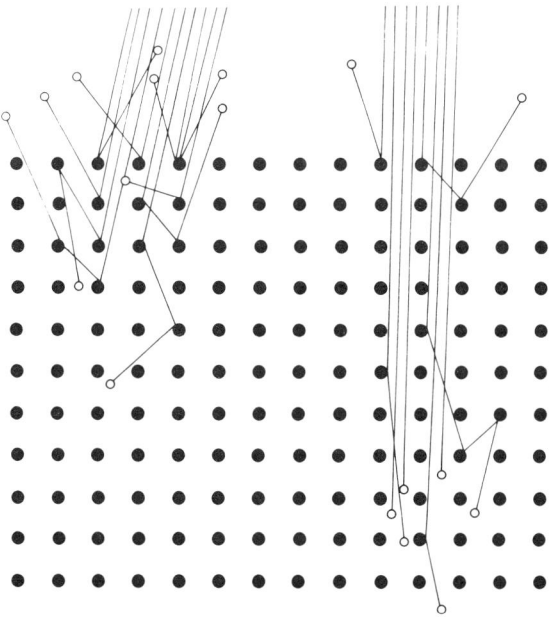

Figure 5.   Schematic representation of the electron channeling process in a crystal.

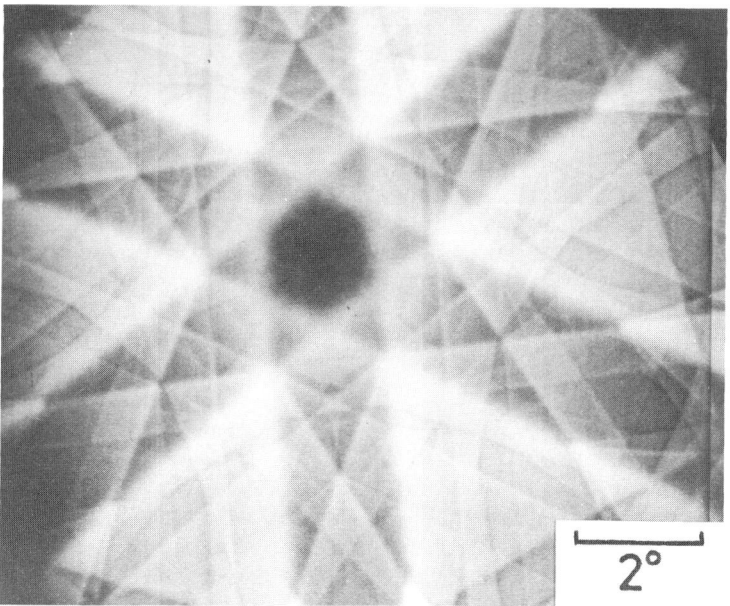

Figure 6.   Electron channeling pattern of a silicon crystal oriented with (111) plane
nearly normal to the electron beam axis.

Figure 7.   SEM micrograph showing contrast caused by electron channeling effects in polycrystalline nickel. Optically, the specimen is a mirror.

first scattering event; the remainder of the electrons scatter at normal distances. This phenomenon can be used to gain information on the crystallographic characteristics of the specimens. If the electron beam is made to rock through all directions within a conical solid angle, the reemerging backscattered and secondary electrons produce a pattern, with contrast up to 5 percent, known as an electron channeling pattern. Thus we obtain information about the orientation and perfection of the crystalline lattice of a region as small as 5 μm in diameter and less than 100 nm in depth (Figures 6 and 7).

Magnetic Contrast

Two different contrast mechanisms can be used in the SEM to observe magnetic domains[8,9]. The first procedure makes use of the deflection by leakage magnetic fields above the domains of secondary electrons, after they have left the specimen. Contrast arises from the fact that some domains deflect the secondary electrons toward the detector, and those of opposite polarity deflect them away from it. An example of this mech-

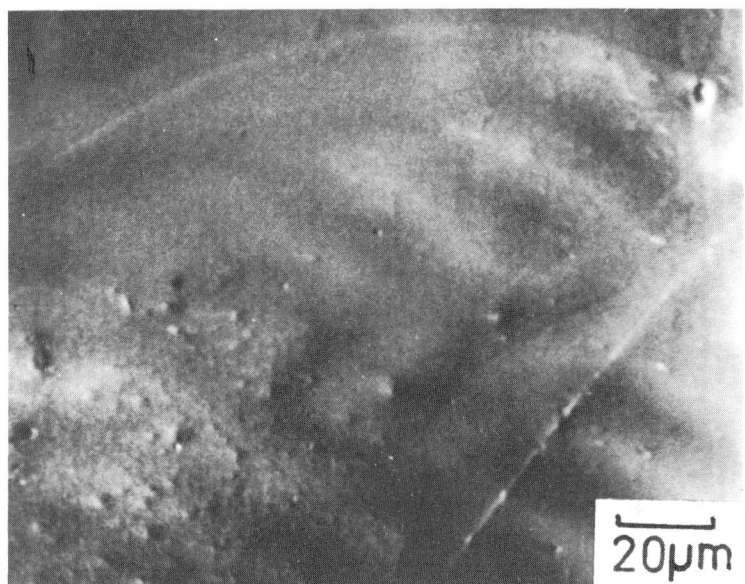

Figure 8. SEM micrograph showing magnetic domains in yttrium orthoferrite (YFeO$_3$) single crystal. Contrast made visible through alteration of secondary electron paths by the magnetic leakage field above the specimen surface.

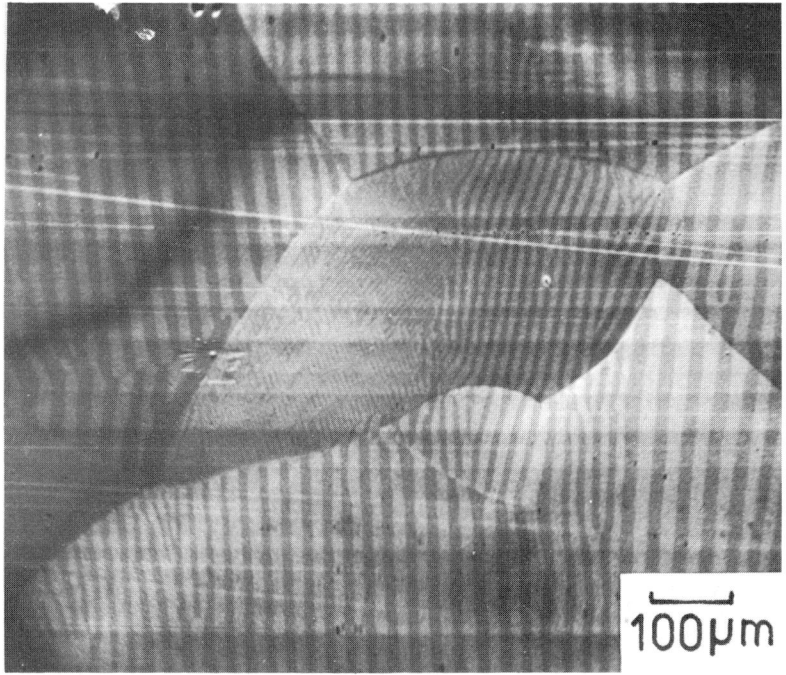

Figure 9. SEM micrograph showing magnetic domains in commercial transformer sheet alloy (Fe-3 1/4% Si). Contrast made visible by interaction of the beam electrons within the sample with the internal magnetic field. Optically, the specimen is a mirror.

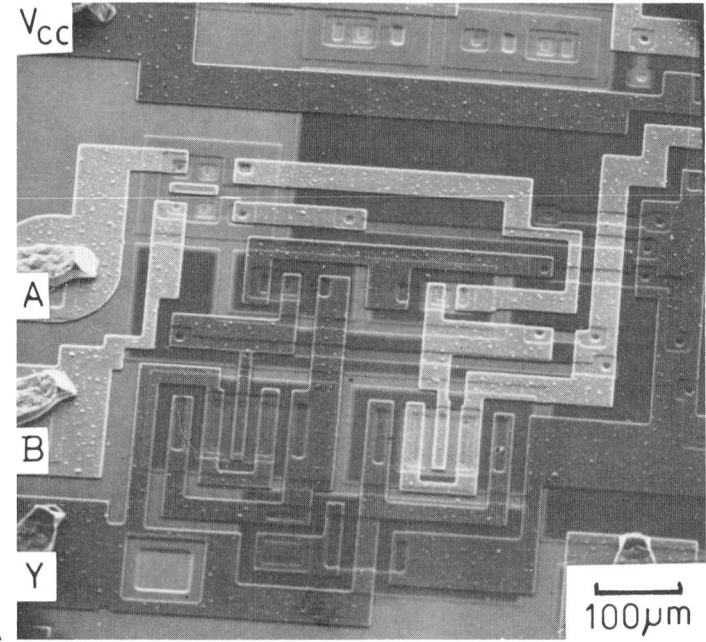

A

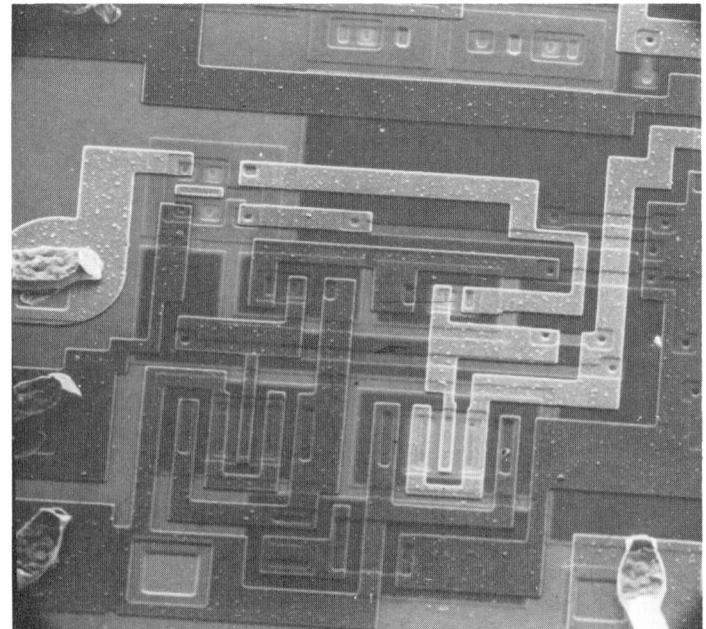

B

Figure 10

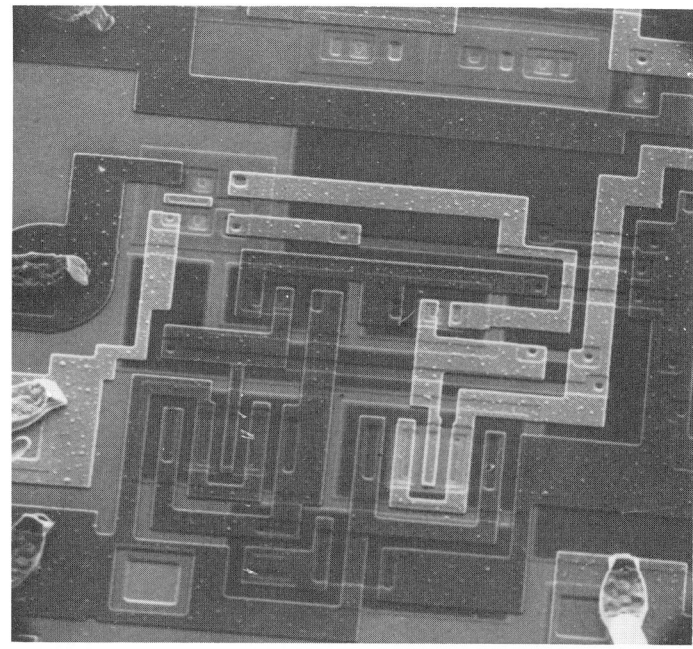

C

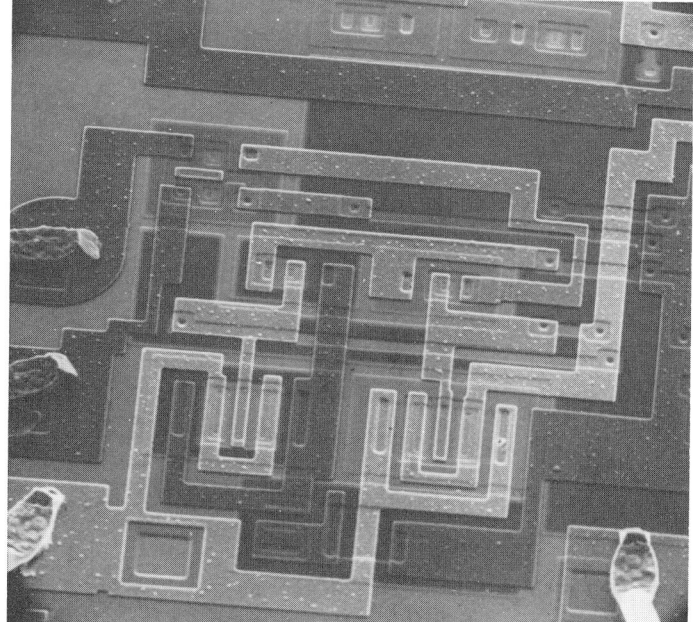

D

Figure 10

anism is shown in Figure 8, showing domains in yttrium ortho-ferrite ($YFeO_3$). The other type of contrast (Figure 9) is due to the magnetic deflection of primary electrons within the specimen, which produces differences in the backscatter coefficients from domains of opposite magnetization. This mechanism has been observed in metallic alloys, but it should also be observable from appropriate ceramic materials.

Voltage Contrast

When bias is applied to part of a specimen, such as a functional element in a semiconductor device, the secondary electron emission of this region can be observed to vary. If the bias is negative, the emission is enhanced, and a positive bias diminishes or completely suppresses the electron emission. This technique is widely used in the study of the performance of semiconductor devices (Figure 10). It has been observed that a variation in bias of one volt can produce a change of 33 percent in the emission of secondary electrons[10]. The method is thus quite sensitive to small local changes of potential. Voltage contrast can also arise from naturally occurring electric fields within specimens, such as ferroelectric domains[11]. Domains in lithium niobate are shown in this manner in Figure 11.

Cathodoluminescence

When an electron beam strikes a suitable specimen, light in the visible, infrared, or ultraviolet range may be emitted in addition to X-rays. This emission can be detected in a photomultiplier and analyzed in a wavelength spectrometer[12]. The light is usually emitted due to recombina-

Figure 10. Operation of integrated circuit observed using voltage contrast (circuit is a quadruple 2-input positive nand gate, SN 74H00). The dark areas are at a positive potential relative to light areas. The operation of the nand gate for all possible conditions is given in a truth table:

Condition	Inputs		Output
	A	B	Y
a	0	0	1
b	0	1	1
c	1	0	1
d	1	1	0

An input "zero" means that the designated input is at ground potential, while a "one" indicates that the input is at the supply potential V_{cc}, of +5.5 V. The predicted variations in potential are clearly observed. (Figure 10 courtesy of W. J. Keery and K. O. Leedy, Institute of Applied Technology, NBS.)

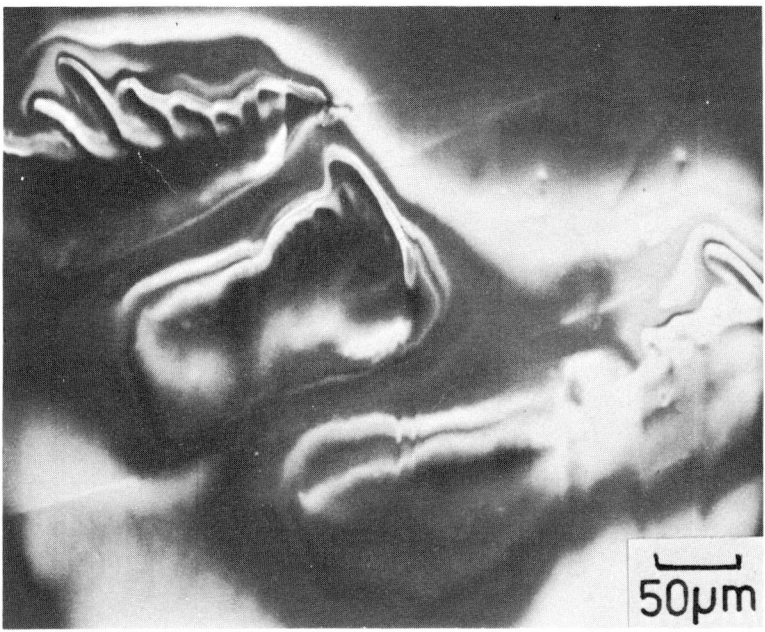

Figure 11. SEM micrograph showing ferroelectric domains in lithium niobate. Sample was uncoated and subjected to an accelerating potential of 30 kV. Contrast made visible by build-up of charge carriers (due to impinging electrons) in various regions.

tion of an electron-hole pair close to the surface. A suitable specimen for the production of cathodoluminescence must therefore be capable of producing electron-hole pairs and be somewhat transparent to the generated light. Many ceramic materials, as well as some polymeric substances, fall into this category (Figure 12). Cathodoluminescence has been shown to be very sensitive to the presence of certain elements at very low concentrations, and it is used fairly frequently in the study of minerals.

Electron Probe Microanalyzer

It remains now to discuss the use of X-ray emission from the specimen bombarded by electrons. Of all signals produced in the scanning electron probe, this is most specifically related to the elemental composition of the excited area. X-rays are generated subsequent to the ionization of an electron located in the inner shells of the atom. When, after ionization, an electron from another shell farther from the nucleus descends into the vacancy created by the ionization, the energy freed in the process can be emitted in the form of an X-ray photon. Since the corresponding energy

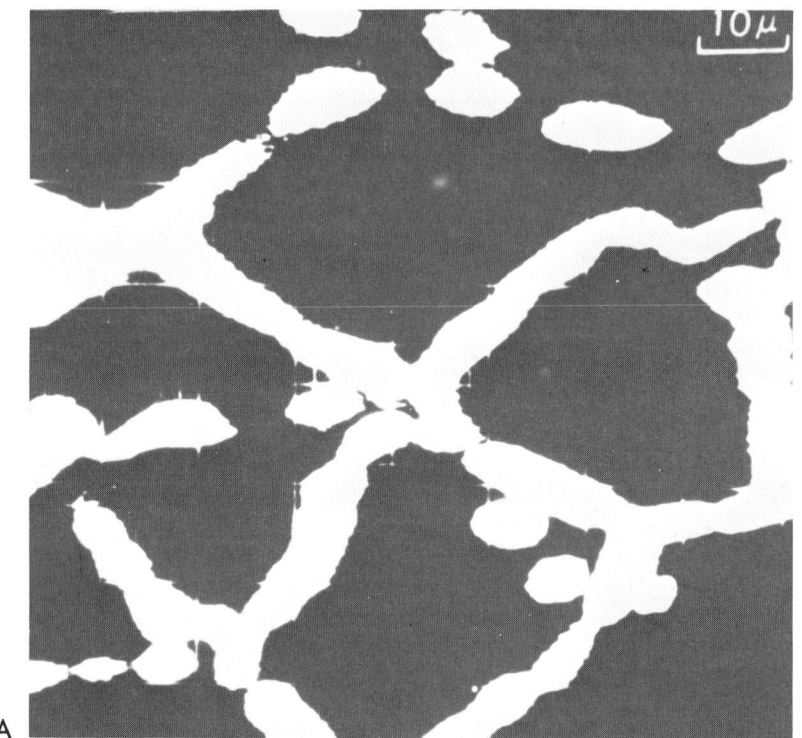

A

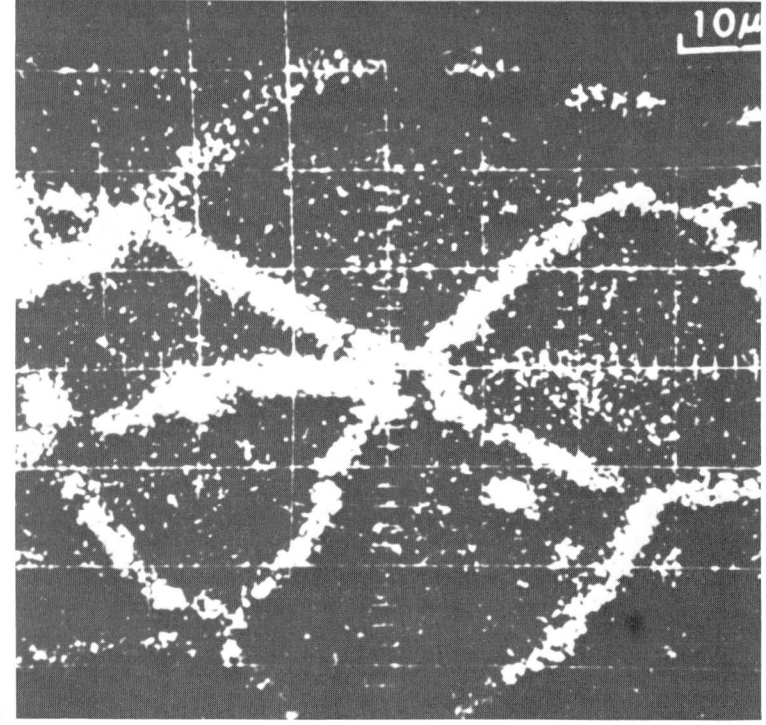

B

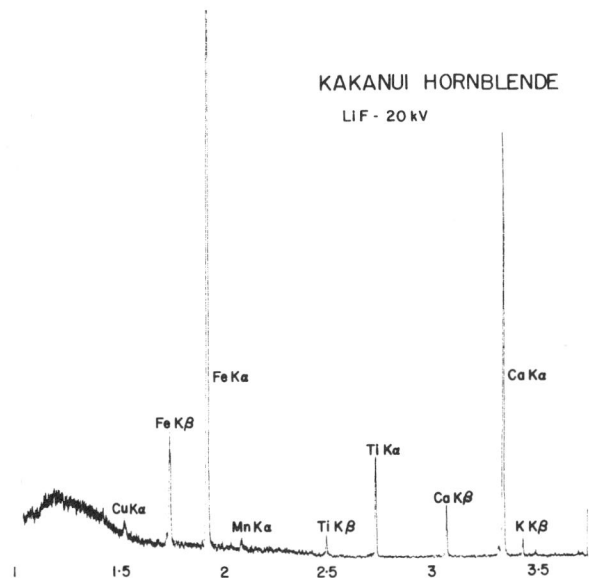

Figure 13. Wavelength scan of Kakanui Hornblende prepared using 20 kV accelerating potential and a LiF diffracting crystal. Elements present are indicated.

levels depend on the atomic number of the ionized atom, the energies and wavelengths of the emitted X-rays are specific to the emitting element (qualitative analysis). The intensities of the characteristic X-ray lines can, after applying appropriate corrections[13,14], be used to determine the mass fractions of the emitted elements, commonly called concentrations (quantitative analysis). The characteristic X-ray spectrum is thus eminently an analytical tool (Figure 13). On the other hand, X-ray photons are particularly inadequate for image formation in scanning microscopy for two reasons. X-rays can penetrate materials over long distance, and therefore emerge from deeper regions than most other signals generated by the electron beam. Therefore, X-ray images have poor resolution. Furthermore, the number of X-ray photons produced by an electron beam (10^{-3}–10^{-5} photons per incident electron) is much lower than that of secondary or backscattered electrons. Hence, the images obtained from X-rays are affected by statistical fluctuations according to Poisson statistics[15]. X-rays are therefore rarely used for purely topographic purposes.

Figure 12. X-ray area scan and cathodoluminescence images of a high temperature alloy in which ZrO_2 has accumulated in the grain boundaries. 12a. X-ray area scan using Zr-Lα at 20 kV. 12b. Cathodoluminescence image of same region as 12a.

X-ray spectra can be obtained by means of spectrometers based on the diffraction of X-rays by crystals, according to Bragg's law:

$$n\lambda = 2d \cdot \sin \theta \qquad (3)$$

In the above equation, λ is the wavelength of the X-ray line, d is the distance between diffracting planes of atoms within the analyzer crystal, θ is the angle of X-ray incidence upon the crystal—which must be equal to the angle of reflectance—and n is an integer called the order of diffraction. For further discussion of crystal spectrometers, see [16].

Alternatively, X-rays can be analyzed by means of proportional detectors, which produce electric pulses of height proportional to the energies of the detected photons. The most useful detector of this kind is the lithium-drifted silicon detector [17,18]. The electrical signal must be amplified and sorted out in a multichannel analyzer. The resulting spectra have a much poorer resolution than those obtained by means of crystals (Figure 14). However, in energy-dispersive analysis the proportional detector analyzes simultaneously the entire wavelength spectrum of interest while a crystal spectrometer based on Bragg's law observes only one wavelength at a given moment. Therefore, spectra showing the major components of the specimen can be frequently obtained in one or two minutes, while comparable spectra—although of much better resolution—from crystal spectrometers require a time interval 10–30 times longer. Both types of spectrometers are usually limited in wavelength range to about 1–10 Å. Diffracting devices for the observation of X-ray spectra from 10–100 Å are available, but they are far less efficient, and the X-ray emission in the long-wavelength region is less intense. As a consequence, X-ray spectrometers can efficiently detect and measure the X-rays emitted by elements of atomic number from 11 to the transuranic elements, but the first 10 elements are inefficiently observed, and hydrogen and helium do not emit X-rays at all.

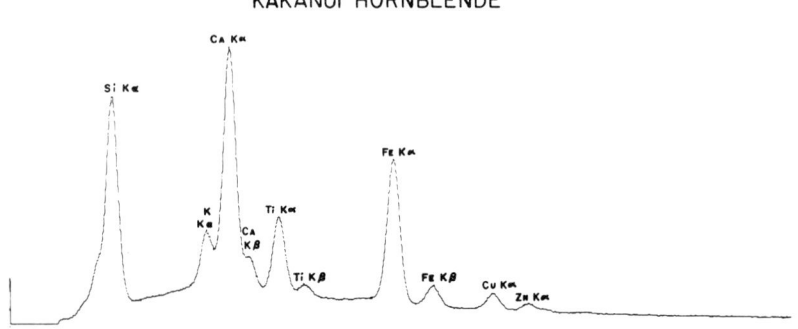

Figure 14. X-ray spectrum prepared with a Li drifted silicon detector and multichannel readout. All elements present of atomic number greater than 11 are indicated.

Crystal spectrometers for X-rays emitted from a point source are complex, expensive, and bulky. For proper focusing, they require the incorporation, in the instrument, of a light microscope for the adjustment of the specimen height. Furthermore, their efficiency is relatively low, so that high beam currents—at least 10^{-8} A—are required for quantitative measurements. As pointed out before, such currents preclude high spatial resolution in the SEM. For these reasons, SEMs do not usually carry crystal spectrometers and are frequently equipped with silicon X-ray detectors. However, for best quantitative analyses, instruments incorporating several crystal spectrometers are necessary. In such instruments, small beam diameters (below 0.2 μm) are not important, while good beam stability, efficient crystal X-ray spectrometers, and exact positioning of the specimen surface are required. Instruments fulfilling these conditions, and built primarily for the purpose of analysis, are called *electron probe microanalyzers* (Figure 15).

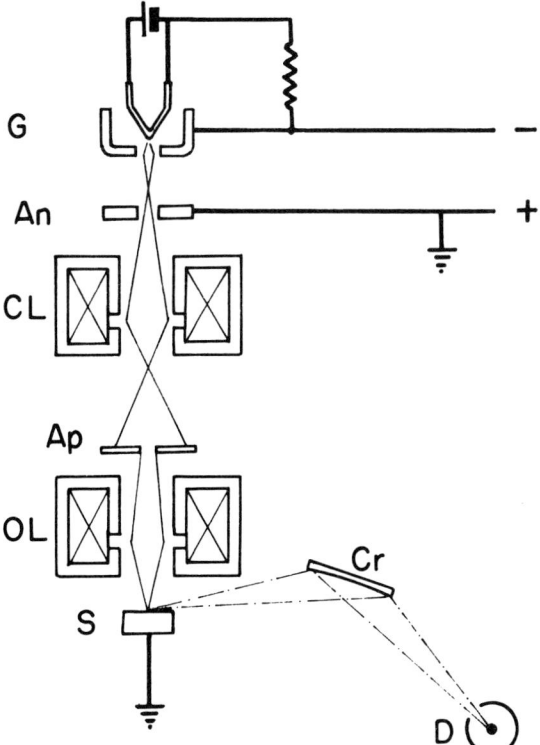

Figure 15. Schematic representation of an electron probe microanalyzer. G = electron gun; An = anode; CL = condenser lens; Ap = aperture; OL = objective lens; S = specimen; Cr = crystal; D = detector.

The region of X-ray emission in the electron probe microanalyzer is typically 1–3 μm wide and deep. Such a volume corresponds to a range from $10^{-10}-10^{-11}$ g, depending on the density of the specimen. The minimum detectable masses for most elements are on the order of $10^{-14}-10^{-15}$ g. It follows that the limits of detection are on the order of 0.001–0.1 percent. Electron probe analysis is thus a microanalytical, but not a trace analytical, method.

Quantitative electron probe analysis is based on the comparison of X-ray intensities obtained from the specimen with those emitted from a suitable standard. If standards of composition similar to the specimen region being analyzed are available, the relative X-ray intensity is very close to the ratio of concentrations between specimen and standard. Empirical corrections of simple nature can be used to improve the accuracy of the measurement[19,20]. If, however, such standards are not available, the analysis can be performed with the aid of pure elements or simple compounds used as standards. In such a case, a more elaborate correction scheme is required. To yield the mass fractions of the elements present, the relative X-ray intensities must be adjusted to account for differences in the mean atomic number, for absorption of X-rays within the specimen, and for possible secondary (fluorescent) excitation of X-rays. The arithmetic procedures are cumbersome, and digital computers

TABLE I

Analysis of the silicates: Johnstown Hypersthene
and Diopside 65— Jadeite 35, using FRAME4

	Chemical analysis of minerals used						
Mineral	Na	Mg	Al	Si	Ca	Ti	Fe
NaCl	0.393						
SiO$_2$				0.467			
Fayalite	—	—	—	0.137	—	—	0.525
Diopside · 2% TiO$_2$	—	0.110	—	0.254	0.181	0.012	—
Garnet 110752	—	0.043	0.120	0.188	0.129$_5$	0.002	0.088
Diopside 65—Jadeite 35	0.0398	0.073	0.0467	0.266	0.120	—	—
Johnstown Hypersthene	—	0.164	0.002	0.251	0.0099	0.0013	0.121

Standards used:
NaCl for Na in Diopside 65— Jadeite 35
SiO$_2$ for Si in both
Fayalite for Fe in Johnstown Hypersthene
Diopside · 2% TiO$_2$ for Mg, Ca, Ti, in Johnstown Hypersthene
Garnet 110752 for Mg, Al, Ca, in Diopside 65— Jadeite 35

Al$_2$O$_3$ used as BKG specimen for Fe, Si, Mg, Ca, Ti } Scheme I
MgO used as BKG specimen for Na, Al, Ca

Off-peak (above and below used for all elements)   Scheme II

Table I.   Continued

Analysis of Johnstown Hypersthene**

E_o = 15 kV

	Mg	Si	Ca	Ti	Fe
k_m^*— S1	0.114	0.202	0.011	0.0008	0.104
k_m^*— S2	0.117	0.202	0.010	0.0007	0.104
$C_{CALC}^\dagger$— S1	0.165	0.260	0.0110	0.0009	0.121
$C_{CALC}^\dagger$— S2	0.169	0.260	0.0110	0.0008	0.121
C— Chemical Analysis	0.164	0.251	0.0099	0.0013	0.121

Analysis of Diopside 65— Jadeite 35

E_o = 10 kV

	Na	Mg	Al	Si	Ca
k_m^*— S1	0.0281	0.0593	0.0363	0.236	0.111
k_m^*— S2	0.0275	0.0595	0.0363	0.236	0.111
$C_{CALC}^\dagger$— S1	0.0384	0.0716	0.0433	0.268	0.121
$C_{CALC}^\dagger$— S2	0.0375	0.0718	0.0433	0.268	0.121
C— Chemical Analysis	0.0398	0.073	0.0467	0.266	0.120

**Analysis carried out calculation oxygen by stoichiometry was exactly the same as that carried out calculating oxygen by difference.

S1— Background taken on Al_2O_3 or MgO
S2— Background taken off-peak
k_m^*— Measured intensity ratio corrected for background and dead-time effects.
C$\dagger$— Mass fraction

must be used to reduce the experimental data. Procedures are now available which permit such data reduction on-line, with the aid of a small computer attached to the electron probe microanalyzer[21]. For further discussion of quantitative analysis and computer programs, see[22,23]. A typical result of the analysis of a silicate of known composition is shown in Table I.

Concerning the qualitative observation of element distributions on the specimen surface, the same comments made for the SEM are also valid for X-ray analysis. On insulators, conductive coatings can be used freely, since, at the scale of dimensions applied to electron probe analysis, the presence of the coating is unimportant. When quantitative analysis is performed, further precautions are necessary, however. To assure that the effects of X-ray absorption and electron backscattering can be predicted, the specimen surface must be oriented and, if needed, prepared so that the angles of electron beam incidence and X-ray emergence are well known. If, for this purpose, it is necessary to grind, polish, or cut the specimen surface, such operation must be performed in such a way as not to alter the surface composition by smearing, deposition of impurities, or preferential etching[24]. If conducting coatings are applied to the specimen surface, the standard should be coated simultaneously, so that the electron and X-ray attenuations in the coatings are identical. Finally, it should be noted that the procedures for quantitative analysis usually fail if the region excited by the electron beam exceeds that of the intended sampling region in volume or depth; for instance, in the analysis of thin coatings and of submicron particles.

All electron probe microanalyzers of recent manufacture have provisions for scanning and for the use of backscattered, secondary, or absorbed electrons (target current) for image formation. X-ray line and area scans can also be performed, and the area scans are particularly useful for disclosing the element distribution in specimens of complex structure (Figure 16). The X-ray scan procedures are discussed in detail in[25]. It is sometimes useful to combine the scanning images for several elements by means of a color addition procedure[26].

The analysis of oxidic and organic materials offers special challenges to the analyst, as they are not only electrical insulators, but also have low thermal conductivity. Both heat and electrical charges are injected into the specimen through the electron-beam interaction. Although surface coatings dissipate these effects and may prevent the occurrence of gross charging effects and damages—such as local fusion of glasses and ceramic

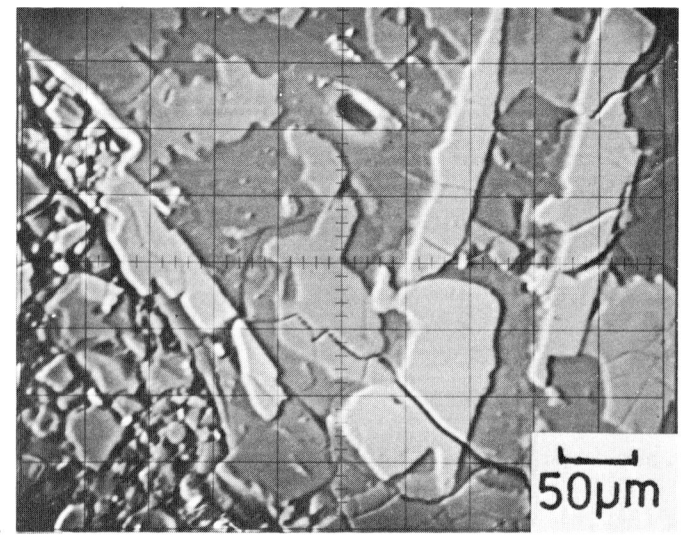

A

Figure 16. X-ray area scans at 20 kV of lunar specimen containing silicates and ilmenite. (16a) specimen current image; (16b) silicon distribution; (16c) titanium distribution. 1 cm = 50 μm.

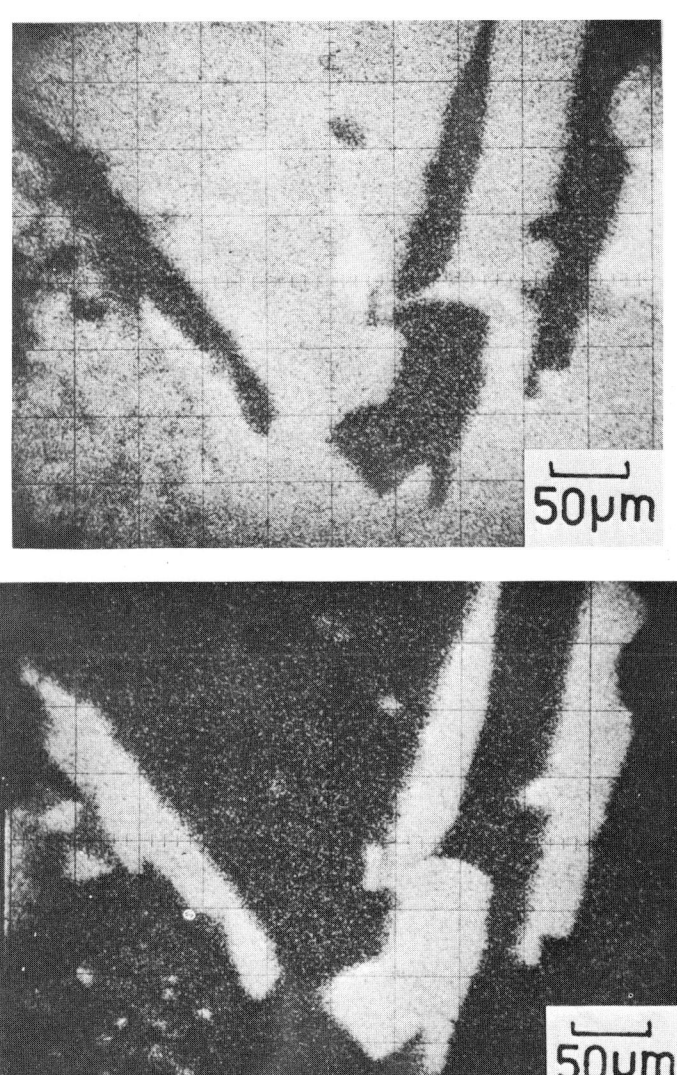

materials, and decomposition of organic specimens—more subtle changes may be caused in the material beneath the conductive coating. In particular, artifacts due to movement of ions of sodium and potassium in glasses have been observed.

Notwithstanding these difficulties, electron probe analysis has proven to be of great usefulness in the study of ceramics and glasses, as recently described by Kane[27]. In the field of organic polymers, progress has been less spectacular, possibly because of the low concentrations in which most inorganic additives are present.

Other Techniques Related to Inner-Shell Ionization

Although the energies of the inner shells depend primarily on the atomic number of the atom, we can observe in the long-wavelength region effects of chemical composition on the shapes and wavelengths of X-ray lines and bands[28]. These manifestations of binding energies have been studied extensively; however, the resolution of available X-ray spectrometers sets a limit to what can be achieved. Some limitations also arise from the high absorption of soft X-rays in matter, and from the low fluorescence X-ray yields for the radiation of elements of low atomic number.

It should be recalled (Figure 17) that both the ionization (*e.g.,* ionization due to primary X-rays) and the Auger process cause the emission of electrons of low energy (tens to hundreds of electron-volts)[29]. The kinetic energies of these electrons are related to the energies of the inner shell levels in an analogous manner as the X-rays emitted from an ionized atom. But, electron spectroscopy has developed to a point where the achievable resolutions, on a comparable scale, are considerably superior to those obtainable in X-ray analysis. Therefore, the study of chemical bonding energy by means of the analysis of the fine structure and position

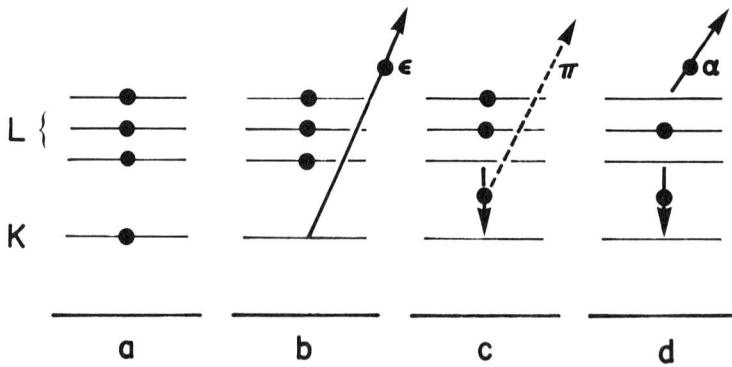

Figure 17. Schematic representation of phenomena related to inner-shell ionization. (a) Neutral state. All levels are occupied by electrons. (b) Ionization. An electron from an inner shell is ejected. If the ionization was caused by a photon, the electron ϵ is called a photoelectron. A vacancy is created in the K-shell. (c) X-ray photon emission. An electron from a high level descends into the vacant level; the energy difference is invested in the photon π. (d) Auger electron emission. Same as (c), but the energy excess is used to remove the Auger electron α from its orbit, and to impart to it kinetic energy.

of X-ray lines has been largely displaced by electron spectroscopy for chemical analysis (ESCA) and by Auger electron spectroscopy.

ESCA thus provides information on bonding in molecules, derived from the analysis of the energy spectra of electrons emitted under irradiation of the specimen by X-rays. In general, soft X-rays from magnesium or aluminum targets are used, and photoelectrons from all elements of atomic number 3 or above can be obtained. Such important elements as carbon, oxygen, and nitrogen can be easily observed, and, therefore, the method can be applied to the recognition of organic molecules.

Owing to the high absorption of low-energy electrons in matter, the signals obtained by ESCA come from a very shallow depth within the specimen, not exceeding 20–30 Å, if appropriate experimental conditions are chosen. This shallow sampling is a characteristic which is very desirable in the analysis of surfaces; at the same time, the strong absorption of the emitted signal makes it very difficult to develop quantitative procedures of analysis by ESCA.

Ion Microprobe Mass Analysis (IMMA)

Elemental mass spectrometry of high sensitivity can be performed by means of the mass spectrometer. In order to apply this technique to surface phenomena, it is necessary to use a method of excitation which is limited to a shallow layer. Such a method is obtained if the specimen to be analyzed is bombarded with primary ions, which produce the emission of secondary ions from the first few atomic layers at the specimen surface. This technique is not, in principle, nondestructive, since the specimen suffers a material loss due to the sputtering action of the primary ions, and only a very small fraction of the atoms removed from the specimen can be analyzed. We will briefly discuss two variants of secondary ion mass spectrometry: ion probe microanalysis[30], and static secondary ion mass spectroscopy (SIMS)[31].

The instrumental arrangement in the ion microprobe mass analyzer is, in principle, similar to that of the electron probe microanalyzer (Figure 18). The instrument has a source of ions—which is equivalent of the electron gun in the electron probe—focusing optics for the primary ions, and an analyzer which receives the secondary ions. As in the electron probe, provisions for raster scanning of the primary beam can be incorporated. The main differences with respect to the electron probe are the following. The ion gun can emit ions of different elements, and also ions of different mass numbers formed from the same element. Therefore, it is useful to add to the primary focusing optics an analyzer which selects a single ion species for the primary beam. The optics are electrostatic rather than electromagnetic, in view of the large masses of the ions. The secondary spectrometer, rather than an X-ray analyzer, is a

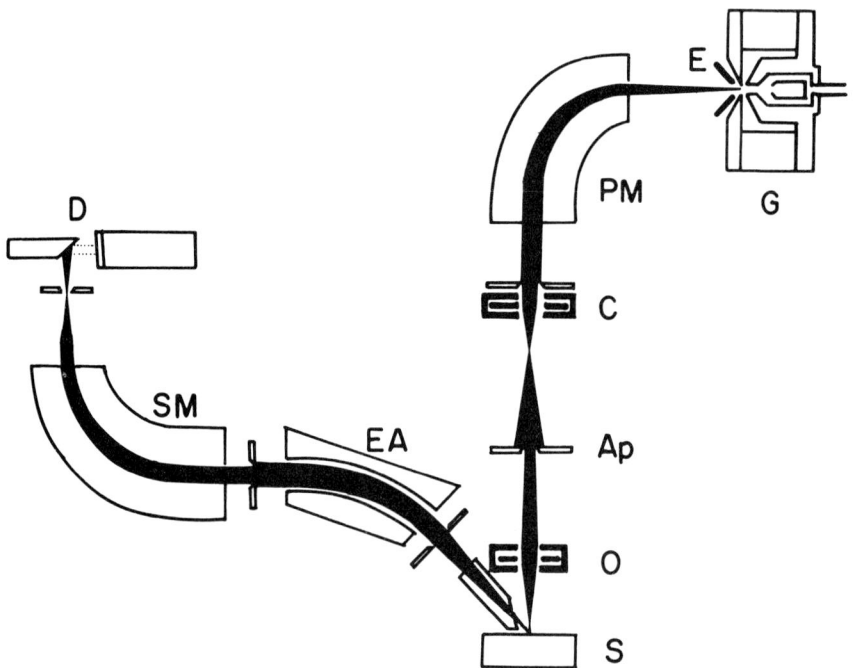

Figure 18.   Schematic representation of ion microprobe mass analyzer. G = primary ion gun; E = ion extractor; PM = primary magnet; C = condenser lens; Ap = limiting aperture; O = objective lens; S = specimen; EA = electrostatic analyzer; SM = secondary magnet; D = detector.

mass spectrometer, typically containing both an electrostatic and an electromagnetic sector.   Provisions for rapidly switching mass numbers for measuring ratios of isotopes can be added.   The detector may be either a multiplier tube or the Daley detector[32], based upon ion-electron conversion outside the photomultiplier, and fluorescence produced by the accelerated electrons.

The width of the analyzed area can be as small as 2 µm.   This dimension is quite comparable with that of the spot analyzed by the electron probe.   However, the depth of the analyzed region, which depends essentially on the amount of sputtering necessary to collect a significant signal, is typically 100 Å, a hundred times shallower than that in the case of electron probe microanalysis.

As in electron probe microanalysis, the electrostatic charges accumulating in the specimen present some problems.   The advantages of ion probe microanalysis include the possibility of analyzing the lowest atomic numbers, of establishing isotope ratios, and of the very shallow range in depth.   The sensitivity for many elements is also much higher.   Furthermore, it is possible to make good use of the erosion due to the

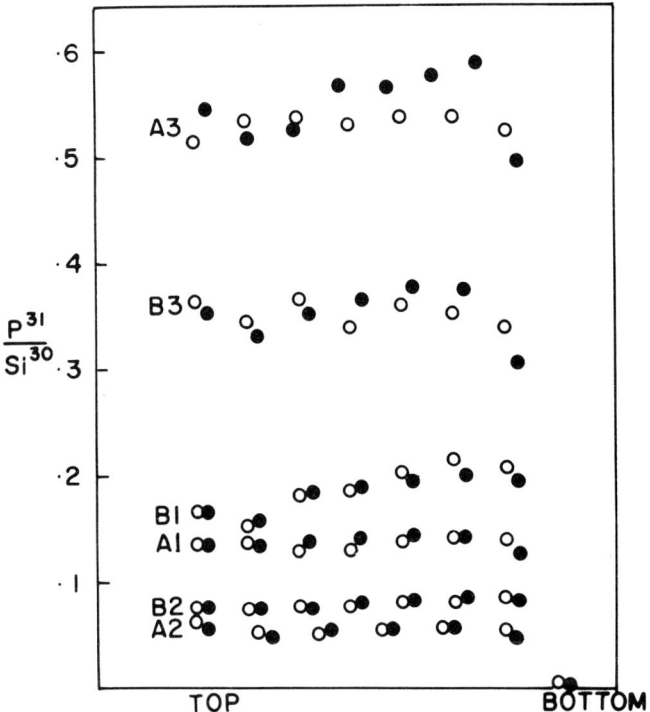

Figure 19. Ratio of phosphorus to silica as a function of depth in silica layers approximately 1.5 μm thick. The points were taken on six specimens at various depths, after etching a ramp with $^{16}O_2^+$.

sputtering to serially analyze layers of increasing depth beneath the surface (Figure 19). The quantitative treatment of ion-probe data is, however, much more uncertain at present, and may never reach the accuracy which can be achieved with the electron probe microanalyzer.

The application of the ion probe to analytical problems is still in its initial stages, with much of the attention focused on the investigation of semiconductor materials. Although the high price of the instrument is a drawback to its rapid diffusion in industrial and scientific laboratories, the interest in this tool for surface analysis is rapidly increasing.

Static Secondary Ion Mass Spectroscopy (SIMS)

In practice, the resolution in depth of ion microprobe mass analysis is limited by the speed of etching of the specimen. This speed increases, in the depth dimension, as the beam diameter decreases. Conversely, a very shallow sampling can be obtained if the primary ions irradiate a large surface. Benninghoven[31] has performed experiments with an ion current of 1 nA/cm^2, over 0.1 cm^2 of specimen surface. The depth of the ion interaction is reduced by using relatively low ion energies (100–3,000 eV). Under such conditions, information concerning the first monolayer of the specimen can be obtained. The method can detect about 10^{-14} g of many elements, and molecular fragments are observed in abundance; this fact allows the investigator to draw conclusions concerning the compounds present in the specimen.

The sensitivity of this method varies greatly depending on the nature of the surface studies, so that careful calibration is required for quantitative data. Huber, Selhofer, and Benninghoven also proposed combining SIMS in the same instrument with provisions for Auger spectroscopy[33], and Benninghoven, Loebach, and Treitz[34] described an instrument in which SIMS, electron impact desorption, and flash-filament desorption are combined. These are good examples of the synergistic action among various techniques. The individual techniques of surface characterization require such an interplay of disciplines if the structures and event located at the specimen surface are to be described efficiently and precisely.

Summary

Several techniques have been described with emphasis on their usefulness for surfaces analysis in nonmetallic solids. Scanning electron microscopy provides a high resolution, three-dimensional image of the surface. Electron probe microanalysis, ESCA, IMMA, and SIMS provide direct analytical information from the sampled region of the surface; these techniques are not high-resolution imaging methods. Future work will almost certainly combine one or more of the analytical techniques with scanning electron microscopy. An example is the use of energy dispersive X-ray analytical equipment with the scanning electron microscope.

Crucial considerations in applying any of these analytical methods are the resolution in depth from which analytical data are obtained and the spatial (lateral) resolution. All of the analytical techniques can be made to yield quantitative information. However, the choice of which method to use depends on the nature of the specimen. For example, some techniques can be used in analyzing for hydrogen while others cannot. In addition, not all of the techniques are capable of trace analysis. Thus, all of these factors must be carefully weighed before investing in a particular technique.

References

1. Murt, E.M. and Gulover, W.G., eds., *Physical Measurement and Analysis of Thin Films,* New York: Plenum Press, 1969.

2. Oatley, C.W., Nixon, W.C. and Pease, R.F.W., "Scanning Electron Microscopy," in *Advances in Electronics and Electron Physics,* L. Marton, ed., New York and London: Academic Press (1965), 181-247.

3. Oatley, C.W., *The Scanning Electron Microscope,* New York: Cambridge University Press, 1972.

4. Johari, O., ed., *Scanning Electron Microscopy/1968-1973,* Proceedings of the 1st-6th Annual Scanning Electron Microscopy Symposia, Chicago: IIT Research Institute, 1968-1973.

5. Thornton, P.R., *Scanning Electron Microscopy,* London: Chapman and Hall, Ltd. (1968), 200.

6. Joy, D., "The Scanning Electron Microscope—Principles and Applications," in *Scanning Electron Microscopy/1973.* Proceedings of the 6th Annual Scanning Electron Microscopy Symposium. O. Johari and I. Corvin, eds., Chicago: IIT Research Institute (1973), 743-50.

7. Booker, G.R., "Scanning Electron Microscopy, The Instrument," and "Scanning Electron Microscopy, Applications," in *Modern Diffraction and Imaging Techniques in Material Science,* S. Amelinckx, R. Gevers, G. Remaut, and J. Van Landuyt, eds., New York: American Elsevier Publishing Company (1970), 553-63.

8. Fathers, D.J., Joy, D.C. and Jakubovics, J.P., "Magnetic Contrast in the SEM," in *Scanning Electron Microscopy/1973.* Proceedings of the 6th Annual Scanning Electron Microscopy Symposium. O. Johari and I. Corvin, eds., Chicago: IIT Research Institute (1973), 259-66.

9. Newbury, D.E., Yakowitz, H. and Myklebust, R.L., "Monte Carlo Calculations of Magnetic Contrast from Cubic Materials in the Scanning Electron Microscope," *Appl. Phys. Lett.,* 23 (1973), 488-90.

10. Yakowitz, H., Ballantyne, J.P., Munro, E. and Nixon, W.C., "The Cylindrical Secondary Electron Detector as a Voltage Measuring Device in the Scanning Electron Microscope," in *Scanning Electron Microscopy/1972,* O. Johari and I. Corvin, eds., Chicago: IIT Research Institute (1972), 33-40.

11. Kittel, C., *Introduction to Solid State Physics,* 2d ed. New York: John Wiley & Sons (1956), 200.

12. Thornton, P.R., *Scanning Electron Microscopy: Applications to Materials and Device Science,* London: Chapman and Hall, Ltd., (1968), 244-77.

13. Heinrich, K.F.J., *Quantitative Electron Probe Microanalysis,* National Bureau of Standards Special Publication No. NBS-SP-298, Washington, D.C.: U.S. Government Printing Office, 1968.

14. Heinrich, K.F.J., "Errors in Theoretical Correction Systems in Quantitative Electron Probe Microanalysis—A Synopsis," *Anal. Chem.,* 44 (1972), 350-54.

15. Liebhafsky, H.A., Pfeiffer, H.G., and Zemany, P.D., "Precision in X-Ray Emission Spectrography," *Anal. Chem.* 27 (1955), 1257-58.

16. Birks, L.S., *X-Ray Spectrochemical Analysis,* New York: Interscience Publishers, Inc., 1959.

17. Russ, J.C., ed., *Energy Dispersion X-Ray Analysis: X-Ray and Electron Probe Analysis,* ASTM Special Technical Publication No. 485, Philadelphia: American Society for Testing and Materials, 1971.

18. Woldseth, R., *All You Ever Wanted to Know about X-Ray Energy Spectrometry,* Burlingame, Calif.: Kevex Corporation, 1973.

19. Ziebold, T.O. and Ogilvie, R.E., "Quantitative Analysis With the Electron Microanalyzer," *Anal. Chem.,* 35 (1963), 621–27.

20. Bence, A.E. and Albee, A.L., "Empirical Correction Factors for the Electron Microanalysis of Silicates and Oxides," *J. Geol.,* 76 (1968), 382–403.

21. Yakowitz, H., Myklebust, R.L. and Heinrich, K.F.J., *"FRAME: An On-Line Correction Procedure for Quantitative Electron Probe Microanalysis,"* National Bureau of Standards Technical Note No. NBS-TN-796, Washington, D.C.: U.S. Government Printing Office, 1973.

22. Beaman, D.R. and Isasi, J.A., *Electron Beam Microanalysis,* ASTM Special Technical Publication No. 506, Philadelphia: American Society for Testing and Materials, 1972.

23. Henoc, J., Heinrich, K.F.J. and Myklebust, R.L., *"A Rigorous Correction Procedure for Quantitative Electron Probe Microanalysis (COR 2),"* National Bureau of Standards Technical Note No. NBS-TN-769, Washington, D.C.: U.S. Government Printing Office, 1973.

24. Birks, L.S., Gilfrich, J.V. and Yakowitz, H., "Report of the Washington Electron Probe Users' Group," in *Fifty Years of Progress in Metallographic Techniques,* ASTM Special Technical Publication No. 430, Philadelphia: American Society for Testing and Materials (1968), 343–53.

25. Heinrich, K.F.J., "Scanning Electron Probe Microanalysis," National Bureau of Standards Technical Note No. NBS-TN-278, Washington, D.C.: U.S. Government Printing Office, 1967.

26. Yakowitz, H. and Heinrich, K.F.J., "Color Representation of Electron Microprobe Area—Scan Images by a Color Separation Process," *J. Res., Nat. Bur. Stand., Sect. A,* 73A (1969), 113–20.

27. Kane, W.T., "Applications of the Electron Microprobe in Ceramics and Glass Technology," in *Microprobe Analysis,* C.A. Andersen, ed., New York: John Wiley & Sons (1973), 241–70.

28. Henke, B.L., Newkirk, J.B. and Mallett, G.R., eds., *Advances in X-Ray Analysis, Vol. 13,* New York: Plenum Press, 1970.

29. Siegbahn, K., *et al., ESCA: Atomic, Molecular and Solid State Structure by Means of Electron Spectroscopy.* Uppsala: Almquist & Wiksells Boktryckeri Ab, 1967.

30. Andersen, C.A. and Hinthorne, J.R., "Ion Microprobe Mass Analyzer," *Science,* 175 (1972), 853–60.

31. Benninghoven, A., "Surface Investigation of Solids by the Statical Method of Secondary Ion Mass Spectroscopy (SIMS)," *Surface Sci.,* 35 (1973), 427–57.

32. Dietz, R.A., "Detection of Single Ions By Pulse Counting: Application to Ion Microscope Mass Analyzer," in *Advances in X-Ray Analysis, Vol, 15,* K.F.J. Heinrich, C.S. Barrett, J.B. Newkirk, and C.O. Ruud, eds., New York: Plenum Press (1972), 36–55.

33. Huber, W.K., Selhofer, H. and Benninghoven, A., "Analytical System for Secondary Ion Mass Spectrometry in Ultra High Vacuum," *J. Vac. Sci. Technol.,* 9 (1972), 482–86.

34. Benninghoven, A., Loebach, E. and Treitz, N., "Simultaneous SIMS, EID, and Flash-Filament Investigations of the Interaction of Gases with a Tungsten Surface," *J. Vac Sci. Technol.,* 9 (1972), 600–602.

H. K. HERGLOTZ
E. I. du Pont de Nemours & Co.
Wilmington, Delaware

Chapter 5

Characterization of Polymers by Unconventional X-Ray Techniques

ABSTRACT

To recognize the basic features of polymer molecules from the rheological behavior of solutions was an accomplishment worthy of the Nobel Prize (H. Staudinger, 1953). However, mature understanding of the properties of solid polymers is unthinkable without the contribution of X-ray diffraction. While polymers often disappoint the classical X-ray structure analyst who wants well-defined single crystals and high orders of diffraction, the X-ray physicist is delighted to have the novel phenomenon of small-angle diffraction. X-ray methods using rigorous collimation of the beam and/or long-wavelength features have furnished new insights into polymeric solids on a 100–1000 Å scale, going beyond those obtainable by electron microscopy. These features, and particularly the merits of long-wavelength X-ray techniques, will be demonstrated on some examples.

Furthermore, the novel information obtainable by ESCA, an offspring of X-ray techniques will be discussed. ESCA permits analysis of the surface apart from the bulk, and yields bonding information.

The complexity of the polymeric arrangement in the solid precludes any single method from providing complete characterization. Incremental understanding furnished by novel methods such as those discussed in this chapter are therefore appreciated.

Introduction

The word "polymer" usually carries the connotation of a modern space-age material. It is too often forgotten, however, that Mother Nature has provided an ample supply of polymers—from cellulose, a lightweight structural material for plants, to the message-carrying biopolymers common to all living cells. Recognizing the molecular structure of natural polymers was man's first step in this field, then imitating nature, and

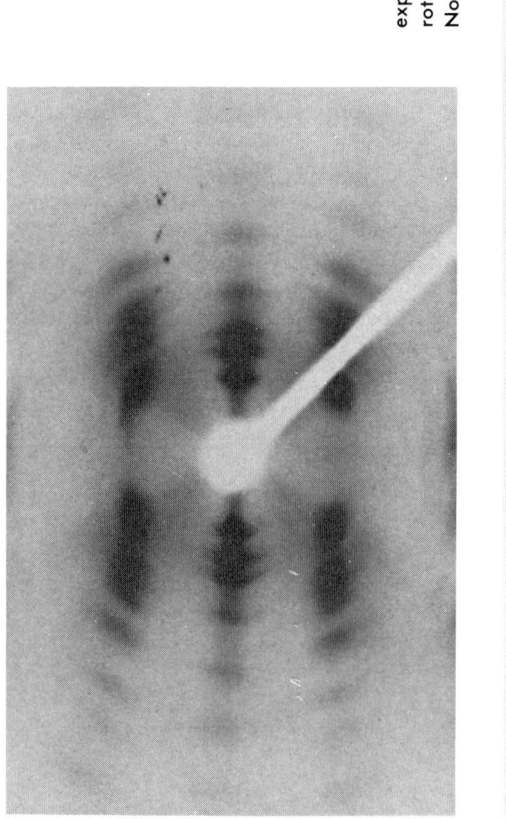

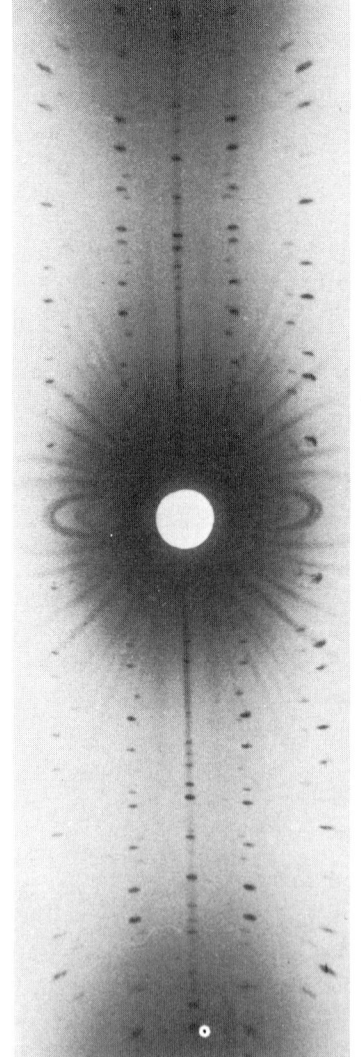

Figure 1. Comparison of X-ray diagrams of experimental polymeric fiber (A) and quartz crystal rotated by the C-axis (B). Both copper radiation. Note the layer lines in both diagrams.

A

B

finally controlling the synthesis of "tailor-made" polymers to fill modern man's specific needs. In all these efforts, characterization, *i.e.,* analysis in its broadest sense, was one of the essential steps. The long-chain character was deduced from viscous behavior of solutions, and the man with the acumen to recognize molecular shape and structure from macroscopic properties earned the Nobel Prize (Hermann Staudinger, 1953). Much more refined information was provided by a submicroscopic probe. The scattering of X-rays on polymers in solution was used by many researchers, but it seems fair to refer to O. Kratky and his school who devoted a lifetime to unraveling macromolecular structure. The wavelength of X-rays makes them an equally potent tool for exploring the structure of solid polymers. At first glance, they are a disappointment to the classical X-ray crystallographer. There are no single crystals that he can orient on a goniometer. If he gets a defined X-ray pattern at all, the reflections are of low indices. Figure 1 demonstrates this by comparing a stationary polymeric fiber diagram with a rotating quartz crystal diagram. The equivalence of the two types of pattern is obvious; so is the paucity of reflections with higher h, k, in the layer lines hk0, hk1 in the polymer, compared with quartz. However, this dearth in the classical X-ray pattern is compensated for by large periodicities, unusual arrangements of crystallites, and other new phenomena, to the delight of the X-ray physicist. It is beyond the scope of this article to elaborate on these, particularly since they have been summarized in the literature[1,2].

This chapter describes some contributions of the Engineering Physics Laboratory, E. I. du Pont de Nemours & Co., to polymer characterization by X-rays. None of these developments replaces older methods, but rather they provide useful additional information.

Soft X-Ray Diffraction and Scattering

The term "soft X-rays" as used here refers to X-rays with wavelength longer than those available from commercial X-ray tubes. These low-penetrating-power X-rays have been called "soft" since the early days of medical X-ray application to distinguish them from "hard" X-rays that could penetrate even the densest and thickest parts of the human body.

The distinction between "diffraction" and "scattering" is made to differentiate the superposition of waves scattered from periodic structures from the scattering on aperiodic electron density fluctuations (Figure 2). Long-wavelength X-rays are usually dismissed as "impractical" because technical X-ray tubes are not commercially available (for several good reasons), and because the heavy interaction of these long-wavelength X-rays with matter imposes serious limitations on tube window and sample volume. For polymers, however, some of these shortcomings can be virtues. Polymeric compounds consist almost exclusively of low-atomic-

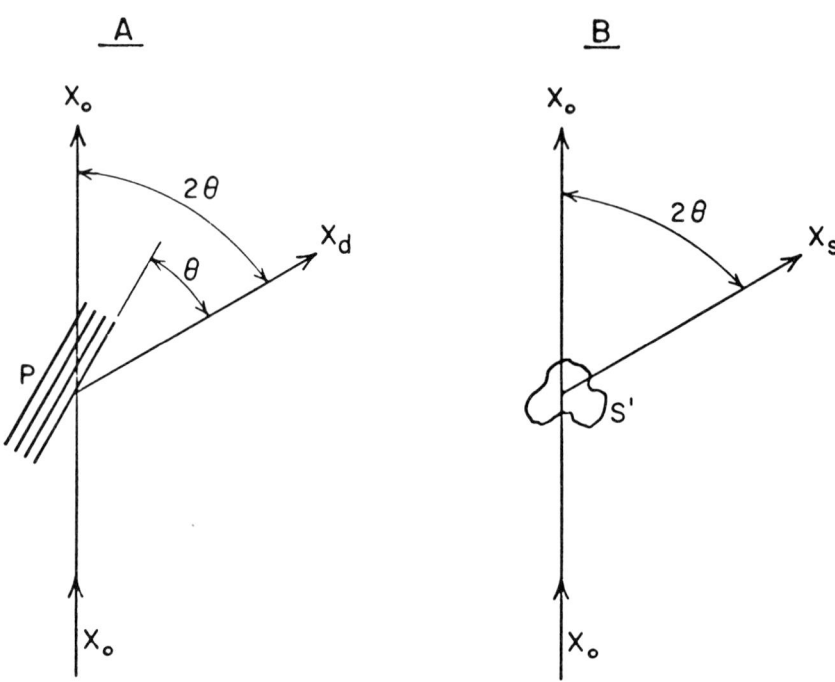

Figure 2. Bragg angle θ and scattering angle $2\theta = \epsilon$. x_o = primary X-ray direction; x_d = X-ray, diffracted on set of planes P; x_s = X-ray, scattered on volume S' without periodic electron density variations. **N.B.** Intensity is found only at x_d in case A; x_s is not unique in case B.

TABLE I
Cross Sections in $[cm^2\,g^{-1}]$ of Carbon [3]

Radiation	Photoelectric	Compton (incoherent)	Thomson (coherent)
Aluminum $K\alpha_{1,2}$ $\lambda = 8.34$ Å $E = 1.78$ keV	$7.08 \times 10^{2*}$	$3.00 \times 10^{-2\dagger}$	9.67×10^{-1}
Copper $K\alpha_{1,2}$ $\lambda = 1.54$ Å $E = 8.04$ keV	3.94^*	$1.28 \times 10^{-1\dagger}$	2.19×10^{-1}

*If compared with cross section of silver for copper radiation (2.18×10^2), the combination: aluminum radiation/polymers becomes quite comparable to copper radiation/heavy metals.
†The wavelength independent maximum Compton shift $\Delta\lambda = 0.0485$ Å at $\epsilon = 180°$, is 2.5% for $\lambda = 1.54$ Å, but only 0.4% for $\lambda = 8.34$ Å. At small angles it becomes insignificant.

number elements, where photoelectric absorption of large λ's is tolerable. Furthermore, undesirable Compton scattering is less significant both in intensity and wavelength shift, but coherent or Thomson scattering (the source of all diffraction phenomena) is delightfully large and sensitive to small changes in electron density. Table I and Figure 3 back up these claims. Figure 3 illustrates particularly the favorable effect of λ on the coherent scattering cross sections. Not only do the scattering cross sections themselves increase, but the differences between scatterers grow with λ. From this figure one can derive the higher sensitivity of aluminum

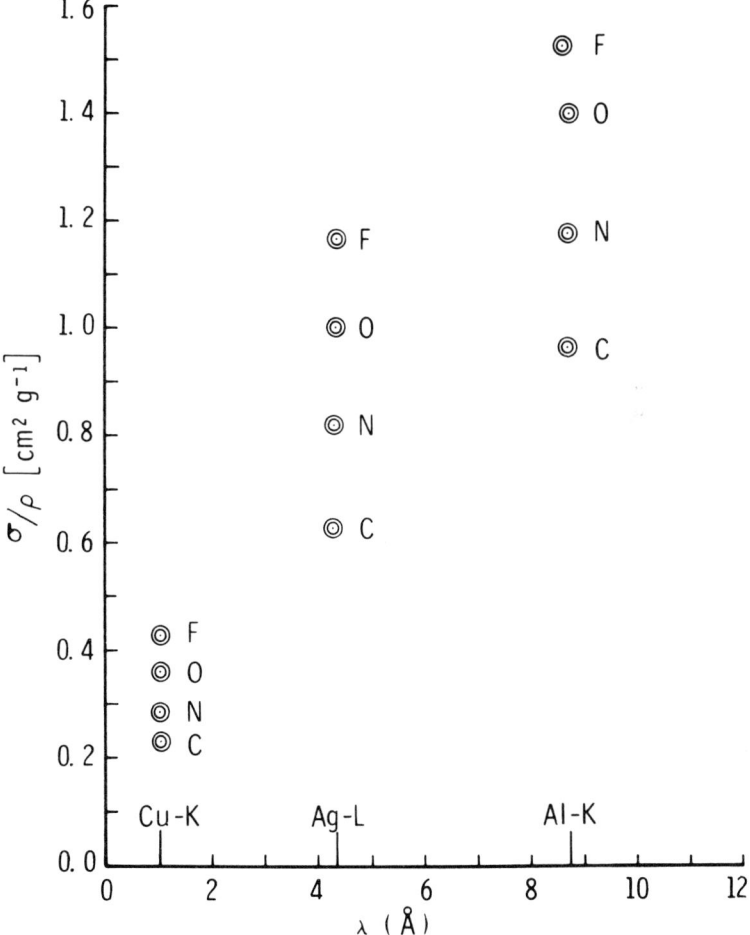

Figure 3. Cross sections of carbon [$cm^2 g^{-1}$] for coherent scattering of characteristic X-radiations from Cu K (λ = 1.54 Å), Ag L (λ = 4.15 Å), and Al K (λ = 8.34 Å) [3].

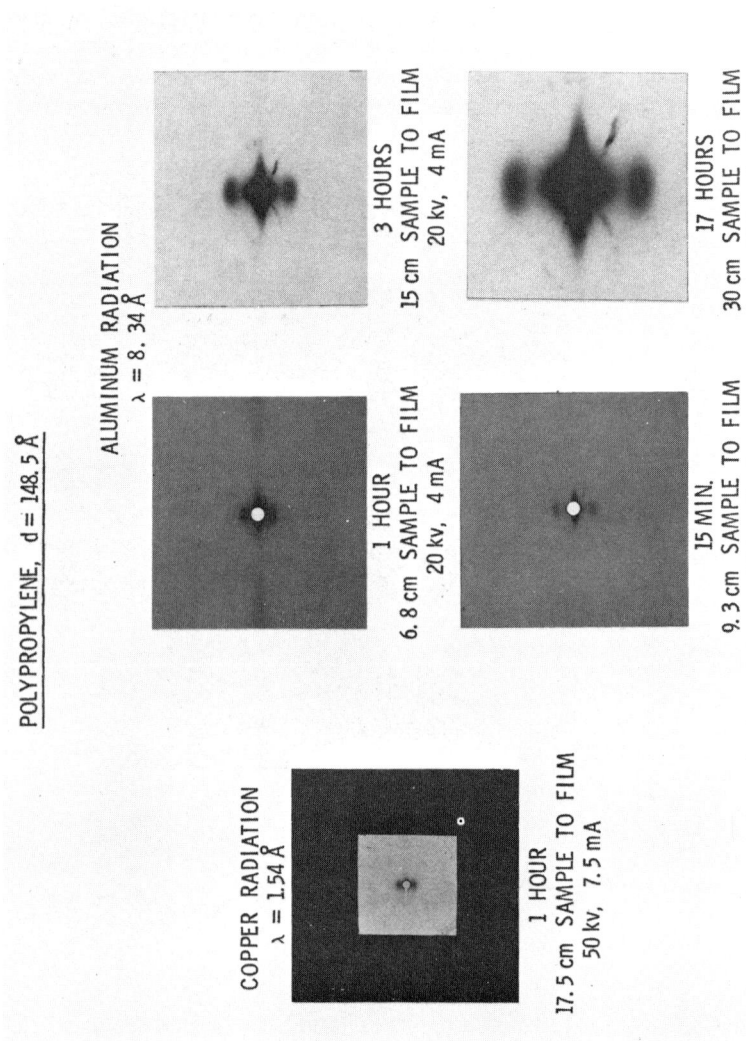

Figure 4. Polypropylene fiber, with a 148.5 Å meridional periodicity and equatorial scattering.

radiation to fluctuations of electron density common in polymers. Figure 4 gives evidence. It shows first of all that diffraction on the periodicity of 148.5 Å in this polypropylene fiber appears at more than 5× larger Bragg angle, as expected from the wavelength ratio 8.34:1.54. This in itself is a most desirable feature in polymers where such large repeat distances are common. For the 148.5 Å periodicity, the Bragg angle θ expands from 18' to the far better manageable 1° 37'. Figure 5 gives another example of Bragg angle expansion. Furthermore, the equatorial scattering streak on the aluminum pattern in Figure 4 contains information unavailable from a copper pattern expanded by the wavelength ratio. The higher sensitivity of aluminum radiation has recorded aperiodic lateral differences in electron density (perpendicular to the fiber axis), invisible to the copper ratio. In a first-order approximation, these aperiodic scatterers can be evaluated by a Guinier plot[4,5], where the logarithm of intensity is plotted against the square of the scattering angle ϵ^2. The superiority of Al radiation in Guinier plots is demonstrated in Figure 6.

The Al patterns of Figure 4 also show other unique phenomena in the form of reflections shaped in the form of a horizontal figure 8, arising from the arrangement of crystallites in the fiber. Details of this effect will be presented in a separate publication.

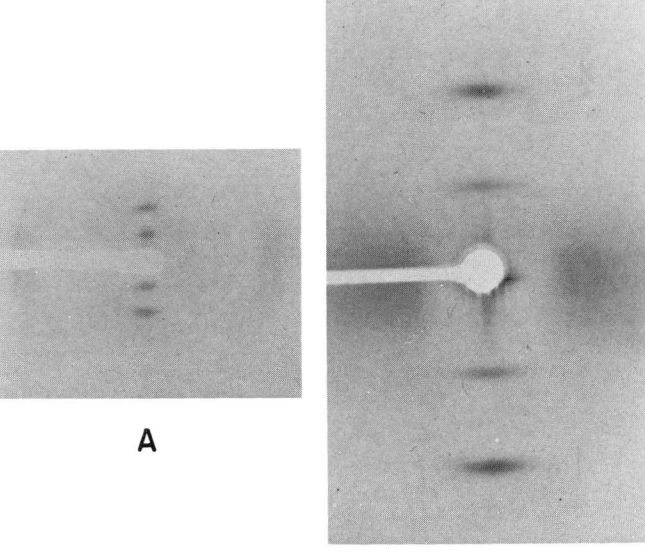

Figure 5. Wide-angle patterns of fiber with a 22 Å periodicity. (A) Copper radiation, flat film 5 cm sample-to-film distance; (B) aluminum radiation, cylindrical film, 4 cm radius.

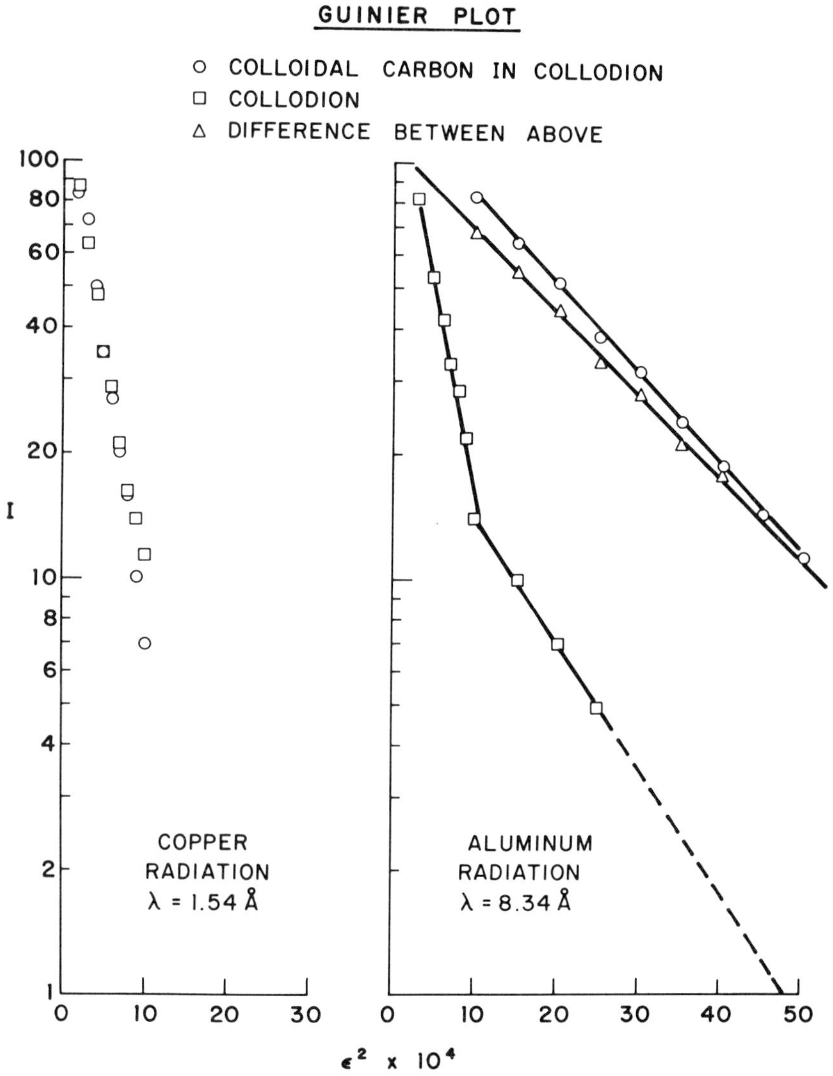

GUINIER PLOT

○ COLLOIDAL CARBON IN COLLODION
□ COLLODION
△ DIFFERENCE BETWEEN ABOVE

COPPER
RADIATION
$\lambda = 1.54$ Å

ALUMINUM
RADIATION
$\lambda = 8.34$ Å

$\epsilon^2 \times 10^4$

Figure 6.   Small-angle scattering by collodion films.

The high attenuation of aluminum radiation by matter imposes limitations on sample shape and form. The optimum sample thickness of the common polymers for $\lambda = 8.34$ Å is between 0.01 and 0.02 mm, closely met by many industrial polymeric fibers. One arrives at this thickness by simple calculation, assuming that the scattered or diffracted intensity is proportional to the mass, *i.e.,* thickness, of the sample, but

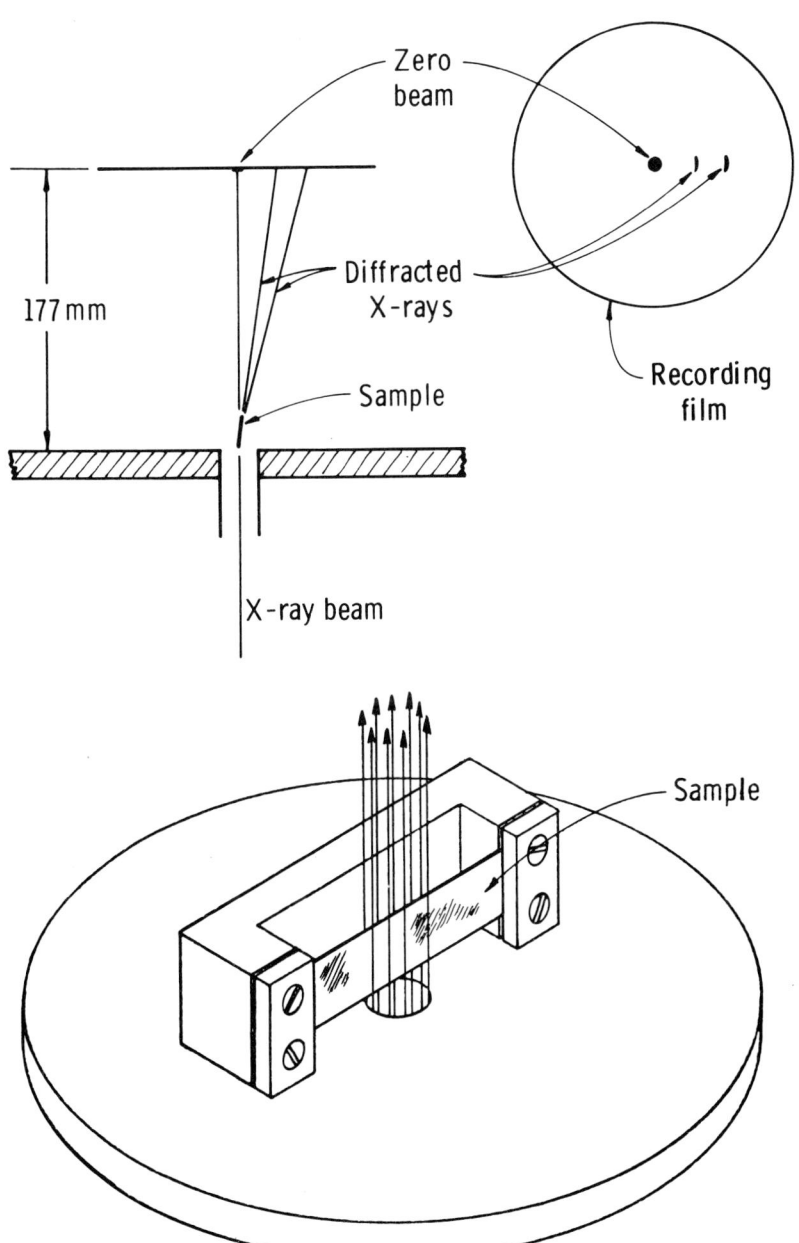

Figure 7.   Edge-on arrangement of film sample in X-ray beam.

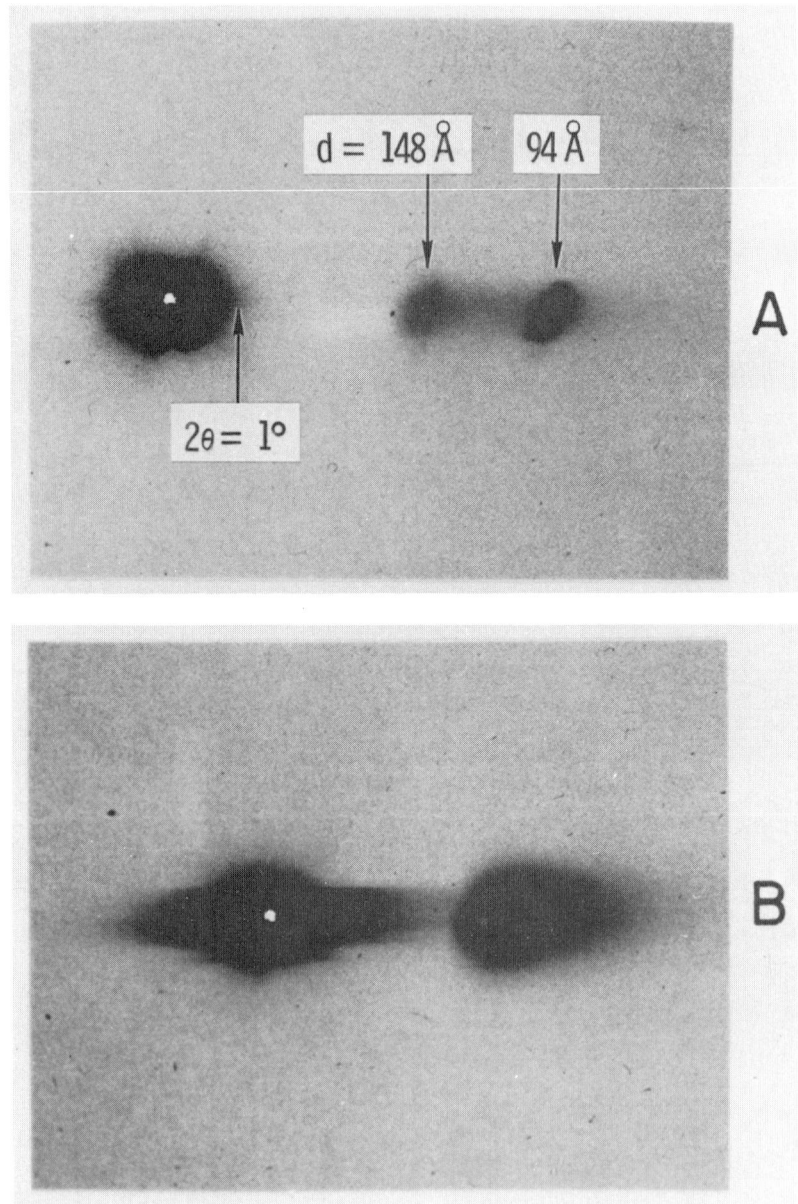

Figure 8.   Edge-on small-angle diffraction patterns of poly(ethylene terephthalate) with lamellar order (A), and disorder (B). Aluminum radiation, 20 kV, 4 mA, 30 min.

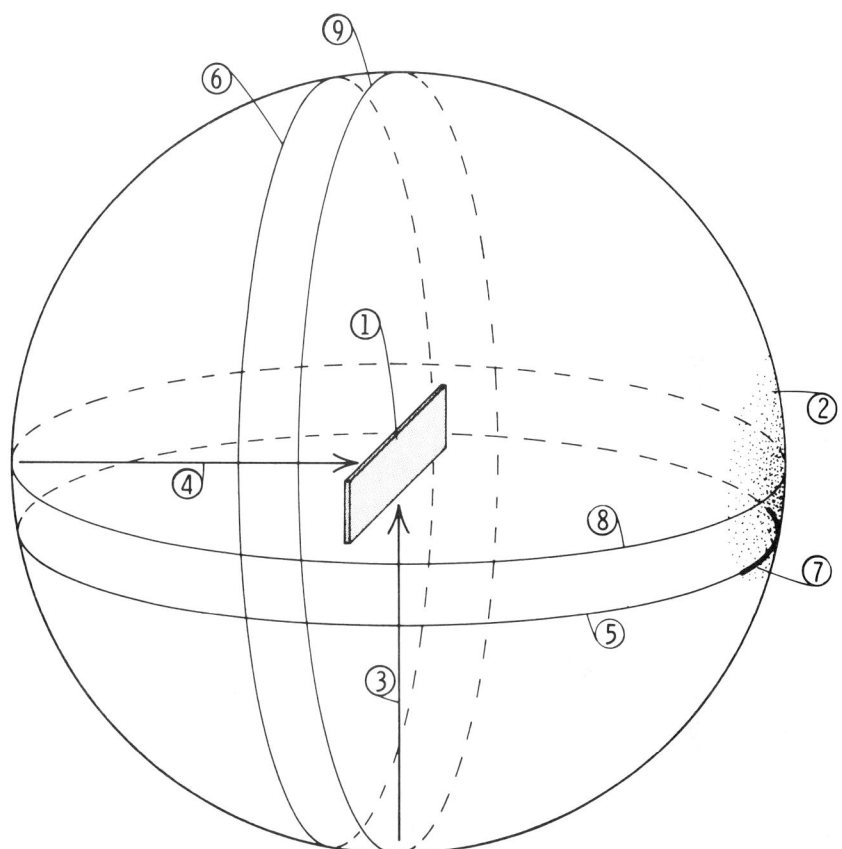

Figure 9. Spherical projection of lamellar plane vectors and diffraction circles
$(90 - \theta)$ for edge-on and transmission: (1) sample; (2) pole of lamellar plane vectors;
(3) X-ray beam edge-on; (4) in transmission; (5) diffraction circle $(90 - \theta)$ for "edge-on"
position; (6) for transmission; (7) reflecting planes; (8) equator; (9) meridian.

diminished by attenuation in the fiber. This leads to the equation

$$I(x) = I_o \sigma x e^{-\tau x} \tag{1}$$

τ = linear total attenuation coefficient
$I(x)$ = intensity of X-rays at distance x from surface
I_o = intensity in primary beam
σ = coherent scattering coefficient

Its derivative dI/dx equated to zero leads to the optimum thickness:

$$x_{\mathrm{opt}} = 1/\mu \tag{2}$$

Samples too thick to be penetrated by aluminum radiation can be microtomed, although the damage done can be noticeable. Polymeric films often yield valuable information if exposed "edge-on" (Figure 7). Slight tilt of the film or asymmetry in the x-ray beam usually renders the pattern asymmetric, as indicated in Figure 7. This is no bar to evaluation of the pattern. Likewise, thickness of the film larger than the collimator exit pinhole forces exposure of one film surface only, producing an asymmetric pattern. Figure 8 provides two examples of poly(ethylene terephthalate) films of different previous histories that exhibit high and low order in the lamellar chainfold structure of this material, with periodicities ranging from 80–150 Å, sometimes much larger. The edge-on sample technique necessitated by the nature of soft X-rays becomes a virtue here. Lamellae nearly parallel to the surface cannot be in diffracting position when analyzed in transmission, as demonstrated in spherical projection in Figure 9. Copper radiation produced no patterns in these two cases with the edge-on technique, although the requirements of Figure 9 are met. This is another manifestation of the additional information obtainable by soft X-rays.

Dual Wavelength X-Ray Diffraction

The information extracted from the classical crystallographic or "wide-angle" X-ray pattern (for which Figure 1 is but one of a great variety of examples), in combination with messages from the small-angle patterns exemplified by Figure 4, has resulted in models of the solid-state molecular structure. Figure 10 describes a representative case. Part A of the figure reminds us, for the case of polyethylene, that the same van der Waals forces between molecules generating organic crystals are at work in polymers and lead to a "crystallographic" order described by the unit cell. The strong covalent bonds within the polymeric chain interfere with the uninterrupted multiplication of these unit cells and prevent buildup to a macroscopic crystal. The model of Figure 10B emerges, representing many polymers after certain previous treatment. This and similar models are found quite frequently in the literature, but it should be remembered that they are an incomplete and imperfect representation of the actual polymeric structure. The two organizations of Figure 10 referred to here as the crystallographic and the supramolecular are, of course, interdependent, and thermal, mechanical, or chemical treatment of a polymer can change both organizations.

If one observes the interplanar or repeat distances in the two lattices, it becomes clear that copper radiation is well suited to generating diffraction maxima in the crystallographic lattice, while aluminum radiation is hardly ever diffracted by crystallographic interplanar distances d. This is immediately evident from Bragg's equation written in the form

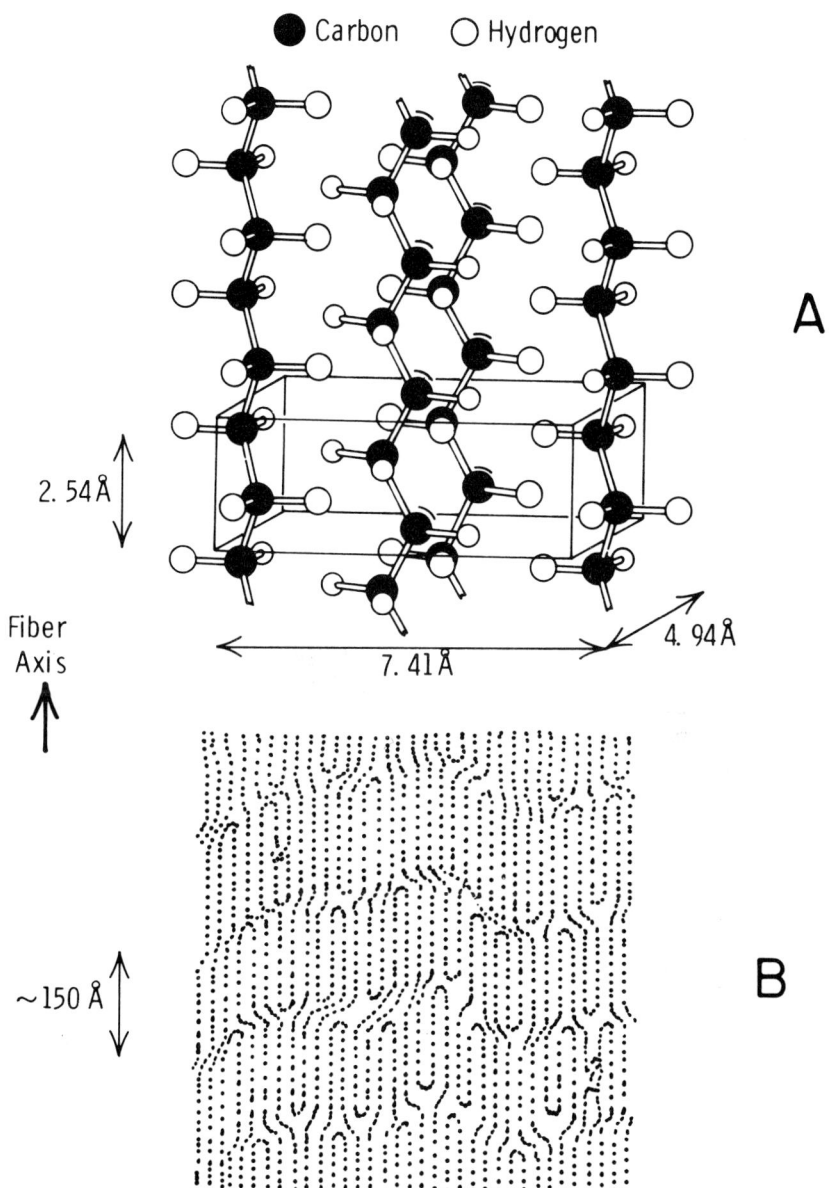

Figure 10.   Model of crystallographic and supramolecular structure of polyethylene. (A) crystallographic unit cell of polyethylene; (B) lamellar chainfold structure, found in many polymers.

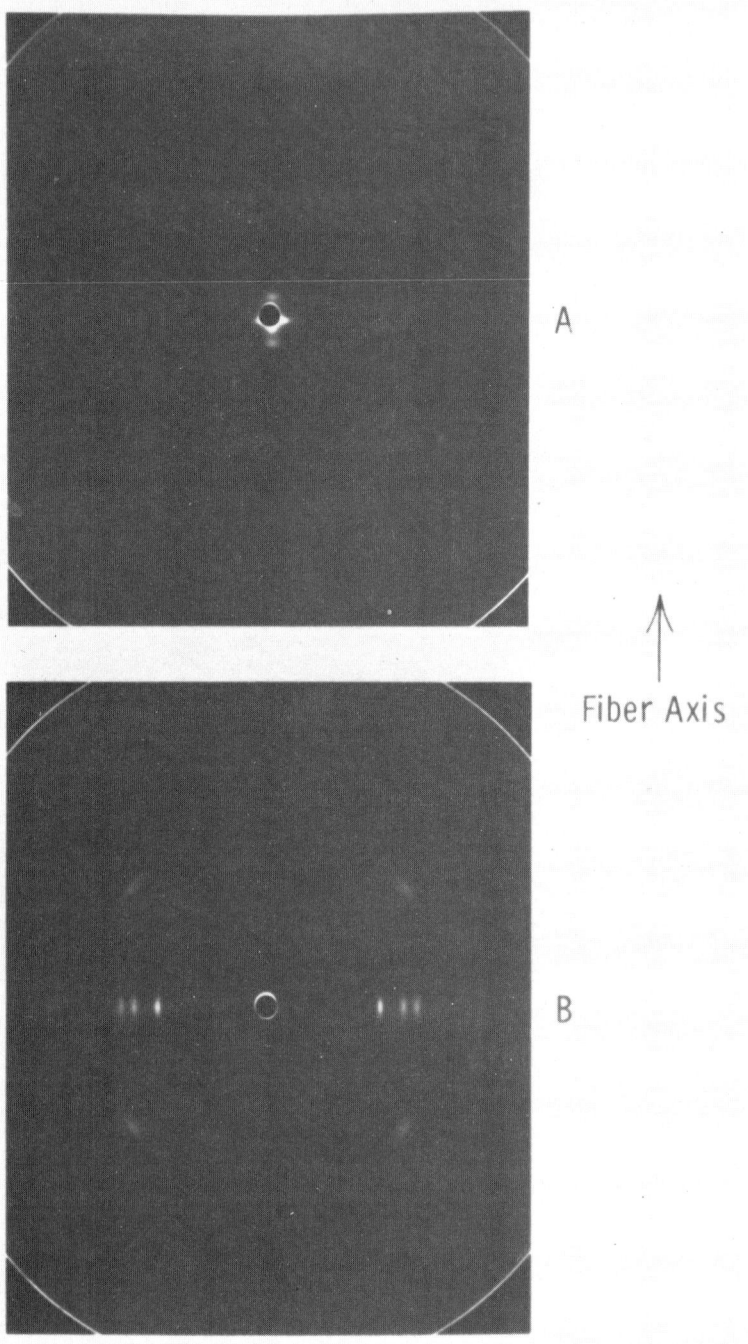

Figure 11. Dual-wavelength pattern of polypropylene fiber with copper and aluminum radiation. (A) Al radiation recorded on single-coated thin film; (B) Cu radiation on double-coated X-ray film; sample-to-film distance: 5 cm.

116

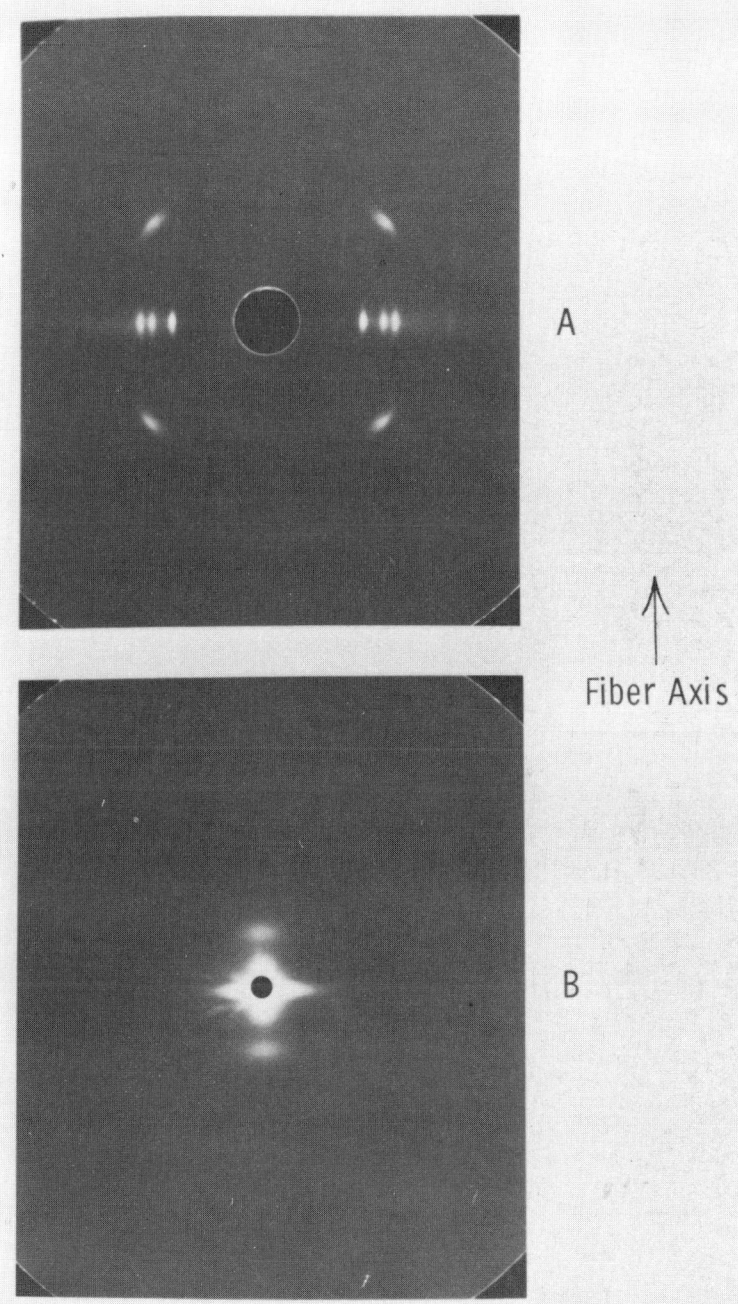

A

↑
Fiber Axis

B

Figure 12. Dual wavelength pattern of same fiber as in Figure 10. (A) Wide-angle section recorded on film with large central perforation, at 5 cm from sample; (B) Small-angle section at 17.5 cm sample-to-film distance, recording what passed through central perforation of (A).

117

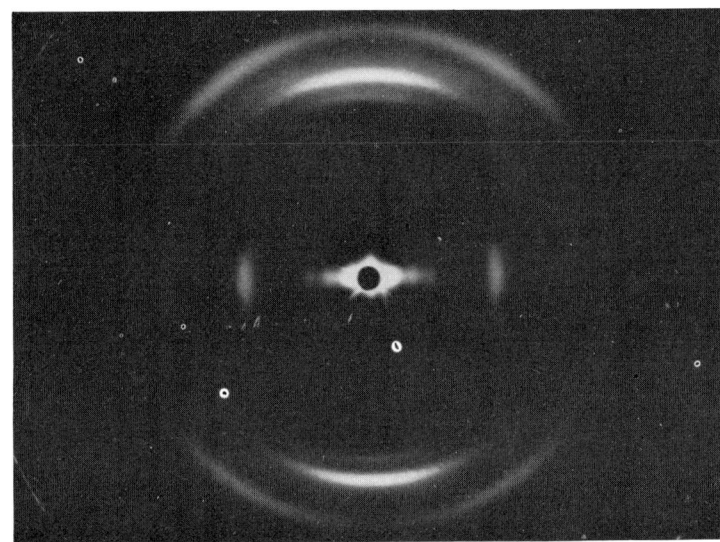

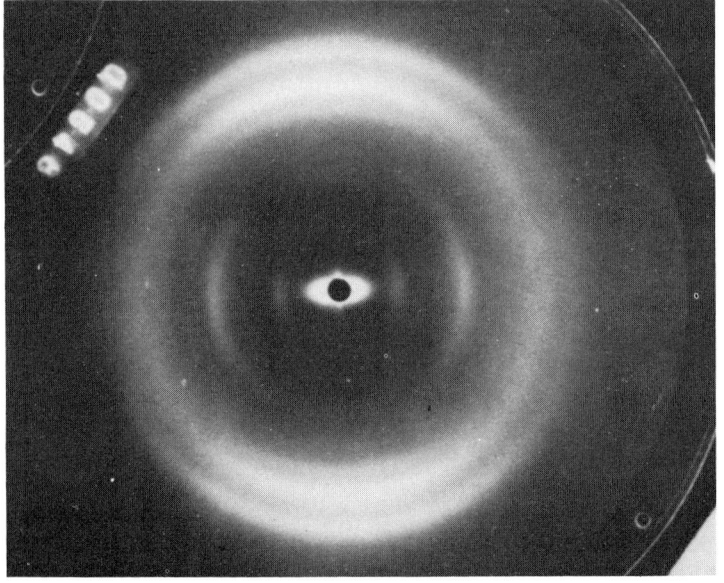

Figure 13. Dual-wavelength pattern of hollow nylon fiber before (A) and after formic acid leach (B).

$$d = \frac{n\lambda}{2\sin\theta} \qquad (3)$$

with $\sin\theta < 1$, and $n = 1$, we obtain the inequality

$$d > \frac{\lambda}{2} = 4.17\,\text{Å} \qquad (4)$$

Aluminum radiation, on the other hand, is far superior for large periodicities, $d > 100$ Å as we have seen in Figure 4. A composite X-ray tube target simultaneously emitting both radiations therefore probes both lattices and is capable of revealing interactions between the two organizations. Dual-wavelength diffraction is demonstrated in Figure 11, where the same polypropylene fiber sample as in Figure 4 is subjected to dual-wavelength diffraction. The dual pattern is recorded at 5 cm sample-to-film distance on a stack of two films. The first consists of a single thin emulsion on a thin base; the second is normal, double-coated X-ray film. Aluminum radiation is completely absorbed in the first film, where it records a pattern. The copper radiation passes through the first film with little interaction but leaves a strong pattern in the second, double-coated X-ray film. This helps to separate the two patterns, particularly if they overlap. To expand the small-angle part of the pattern, it is possible to have two films or pairs of film at different sample-to-film distances (Figure 12). Radiation passing through the large perforation on the "first floor" film is recorded on the second at a larger sample-to-film distance.

An application of the dual-wavelength method is shown in Figure 13, where patterns of hollow polyamid fibers were recorded before and after acid treatment. The effects of the treatment, evident in both crystallographic and supramolecular structure, are profound, as described elsewhere[6]. All results recorded here were obtained with pinhole geometry and photographic recording. Modern instrumentation favors slit geometry for increased intensity and electronic readout for faster and quantitative evaluation without microdensitometry. Pinhole geometry, however, is preferred because it preserves the shape of the diffraction pattern, which can be extracted from slit geometry only by "desmearing" operations. Photographic recording is superior to electronic scanning if unknown and unexpected features are to be recorded.

Instrumentation

One of the obstacles to routine utilization of long-wavelength X-rays is the necessity to operate demountable X-ray tubes. Continuous pumping, ultrathin windows, lack of stability, and low intensity are some of the difficulties cited against soft X-rays. We have developed instrumentation that eliminates these difficulties to a large extent and permits pushbutton

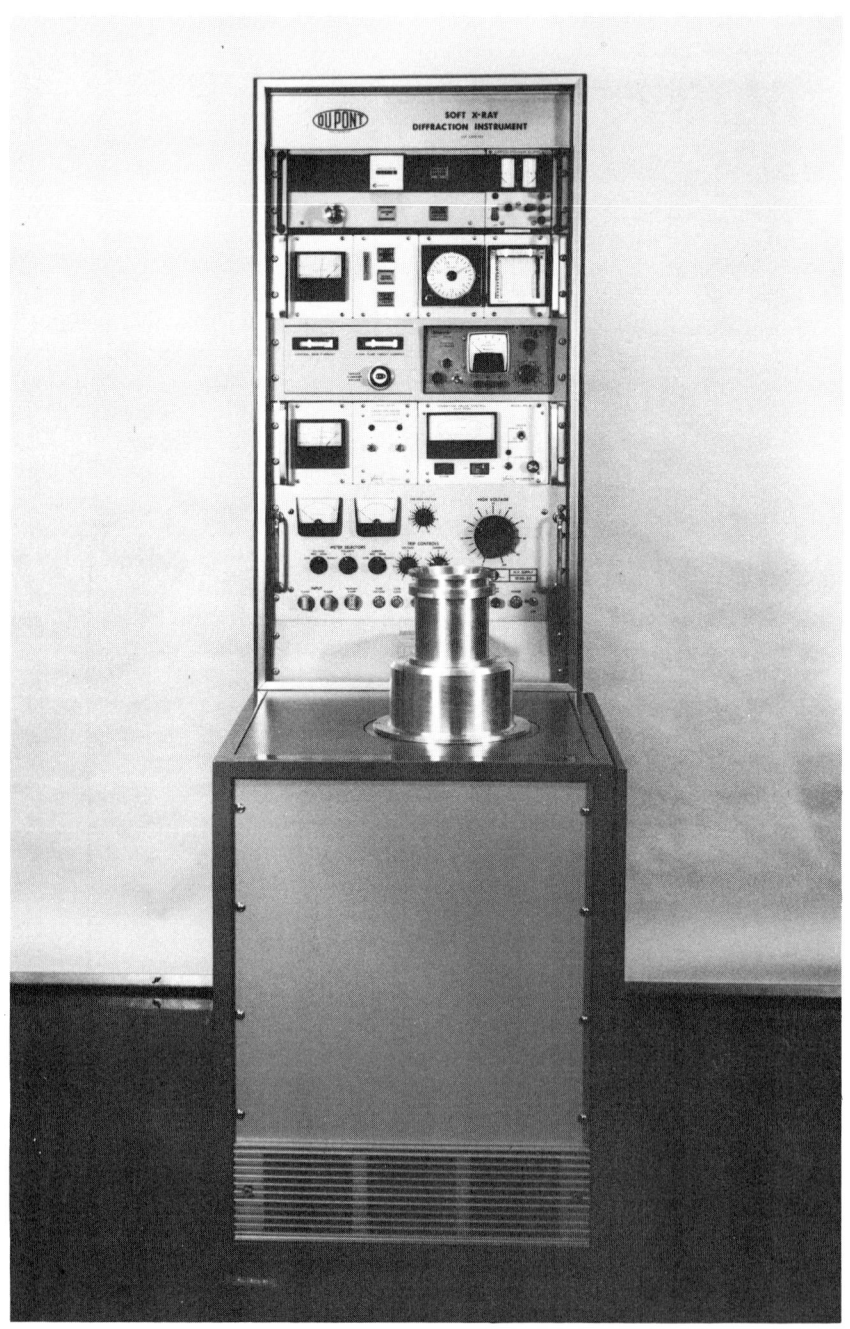

Figure 14.   Latest version of soft and dual-wavelength XRD instrument.

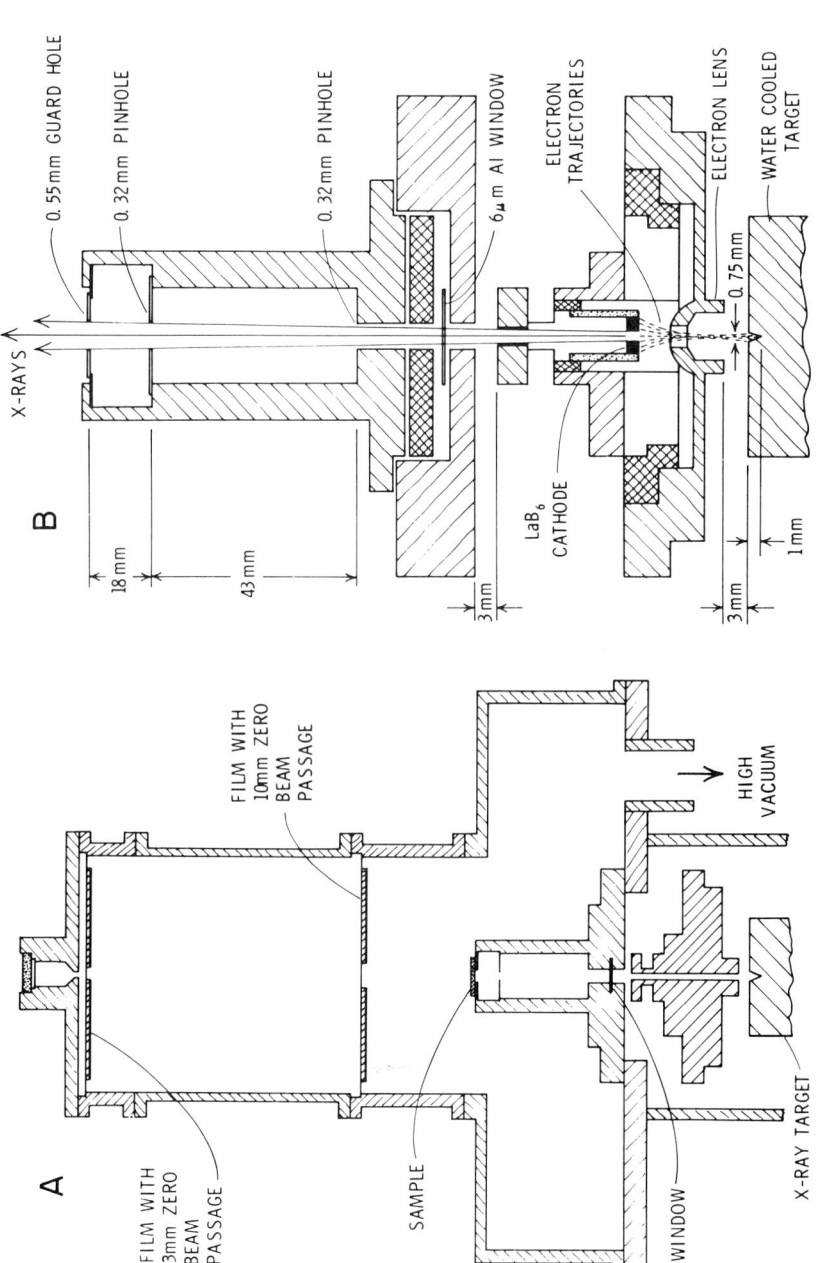

Figure 15. (A) Drawing of source and camera of instrument of Figure 14; (B) Detail of target/cathode/collimator arrangement.

start-up, unattended operation at high stability and considerable intensity. A small turbomolecular pump provides oil-free vacuum without the disadvantages of diffusion pumps. A cavity target and a feedback-controlled grid between cathode and target are responsible for high intensity and stability. The target is easily exchangeable, so that the aluminum target can be replaced by a composite target of copper–aluminum. Although other radiations and combinations have been tried, aluminum and copper are most advantageous. They come from targets resistant to erosion by electron bombardment, are therefore easily produced with high intensity, and experience little absorption in the 6 μm aluminum window of the X-ray source. Figure 14 illustrates the most recent version of the soft XRD instrument, while Figure 15 shows schematics of source and camera. More details are found in previous publications[5,7,8].

Electron Spectroscopy for Chemical Analysis (ESCA)

A few years ago, Du Pont's Engineering Physics Laboratory devoted considerable effort to the development of a soft X-ray spectrograph, based on the critical angle of total reflection, for the analysis of light elements below sodium in atomic number. The concept of this spectrograph was demonstrated in the laboratory and reported in the literature[9,10]. B. Henke[11] has drawn attention to the relative abundance of photoelectrons and Auger electrons, compared with fluorescent X-ray photons from light elements, and proposed their use in lieu of fluorescent X-ray photons for compositional analysis. When K. Siegbahn and his school at the University of Uppsala, Sweden, issued their comprehensive study of photoelectron and Auger electron spectroscopy[12], attractive possibilities for polymer characterization became apparent. This made us switch from soft X-ray spectroscopy to electron spectroscopy in pursuit of our goal of polymer characterization. When a sample of any material is bombarded by electrons, X-rays, or short-wavelength UV light, "secondary" (photo- and Auger) electrons and "secondary" (fluorescent) photons emerge (Figure 16). The processes involved are summarized in the simplified Rutherford–Bohr model of Figure 17. Absorption of an electron or photon by an atom causes core ionization, generating a photoelectron in each ionization act. Reconstitution of the atom is followed by either a fluorescent photon or one or several Auger electrons. The latter is the radiationless reconstitution involving two atomic levels. The Auger electrons carry away the excess energy of the ionized over the reconstructed atom. Information about the atom is found in the most straightforward manner from photoelectrons by the energy balance

$$h\nu = E_{\text{binding}} + E_{\text{kin}}$$

where $h\nu$ is the energy of the ionizing monochromatic photon, E_{binding} the

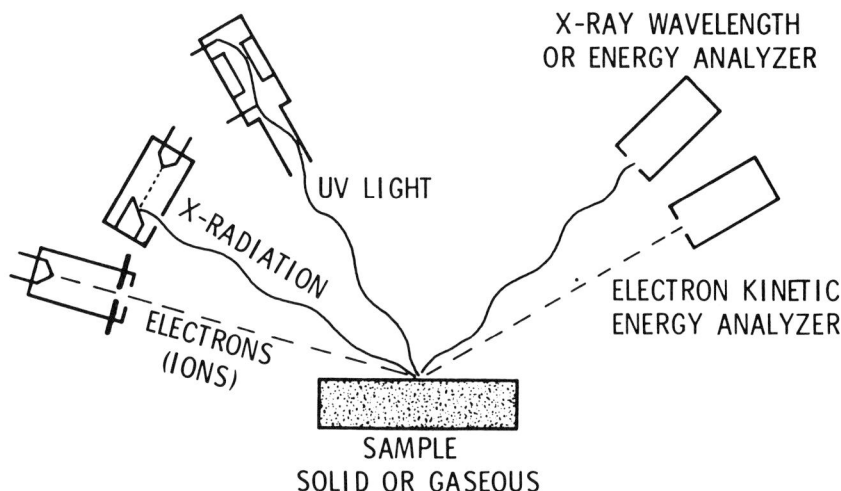

X-RAY WAVELENGTH
OR ENERGY ANALYZER

UV LIGHT

X-RADIATION

ELECTRON KINETIC
ENERGY ANALYZER

ELECTRONS
(IONS)

SAMPLE
SOLID OR GASEOUS

Figure 16. Excitation of a sample and recording of consequential events.

energy of ionization, and E_{kin} the excess of $h\nu$ over $E_{binding}$ carried away as kinetic energy of the photoelectron. A scan of number of electrons (count rate in counts/sec) versus kinetic energy reveals atomic energy levels.

Fluorescent photons and Auger electrons represent the difference between two (or three) atomic energy levels of the atom, respectively. Since atomic levels are well defined, measurement of kinetic energy of the emerging electrons or of the fluorescent photons can identify the atom from which they emerge and are therefore useful for analytical purposes. Both electrons and X-rays also carry messages about the chemical bonding in which the emitting atom finds itself. The chemical bonding aspect of X-rays has been summarized in the literature[13,14]. ESCA's superior bonding information has also been described[12,15]. While X-ray fluorescence analysis has been used routinely by the analytical chemist since the early 1950s, the same popularity came only recently to electron spectroscopy. K. Siegbahn, who deserves credit for making the method available to the chemist, coined the name ESCA, electron spectroscopy for chemical analysis. Both X-ray fluorescence and electron spectroscopy look back at a venerable age. Both were demonstrated before or around 1920. Figure 18 shows an instrument recording "corpuscular" spectra on a photographic plate by means of a magnetic field perpendicular to the plane of the figure after Rutherford et al.[16]. A spectrum obtained with this instrument is shown in Figure 19 and compared with that obtained by Siegbahn's spectrometer in Figure 20[12]. The improvement in resolution and signal strength is dramatic, making the method interesting to analytical chemists. It needed, however, a few other developments by Siegbahn's school and by others to assure its present popularity.

PHOTOELECTRON, FLUORESCENT X-RAYS, AUGER-ELECTRONS

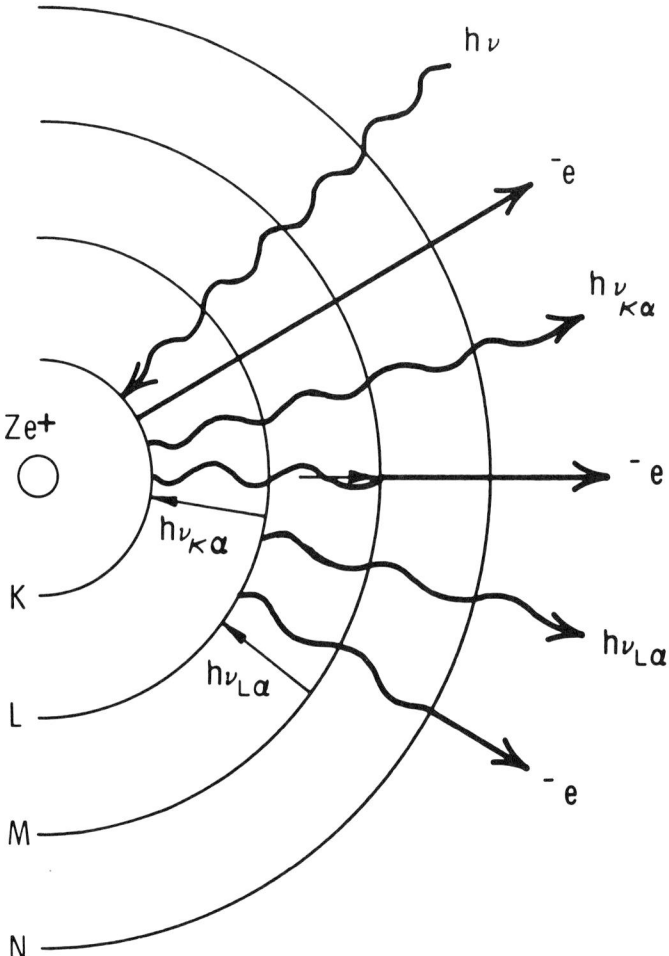

Figure 17. Core ionization and consequential events depicted in simplified Ruther-ford–Bohr model of the atom.

ESCA Instrumentation

The results of Siegbahn *et al.* pointed to the tremendous potential of ESCA, in its version of X-ray excited photoelectron spectroscopy, for polymer characterization because of:

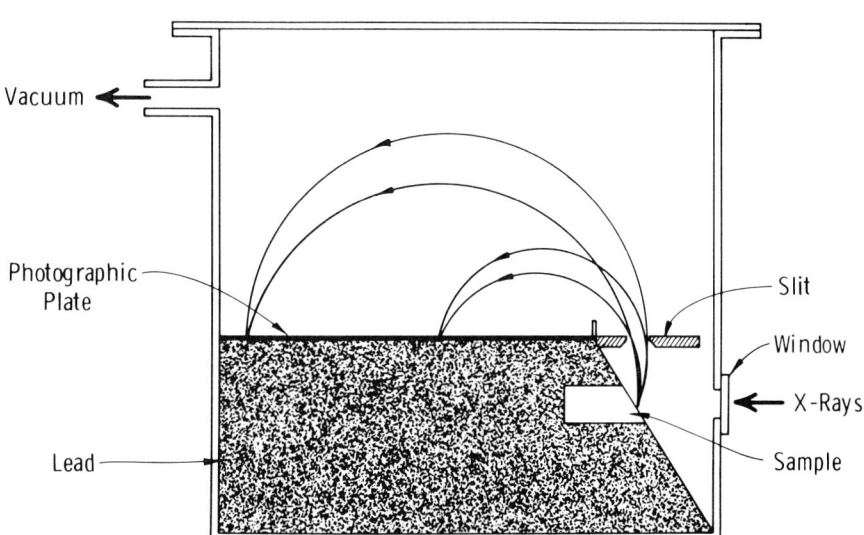

Figure 18. Apparatus for recording of corpuscular magnetic spectra in the 1920s (summarized in [16]).

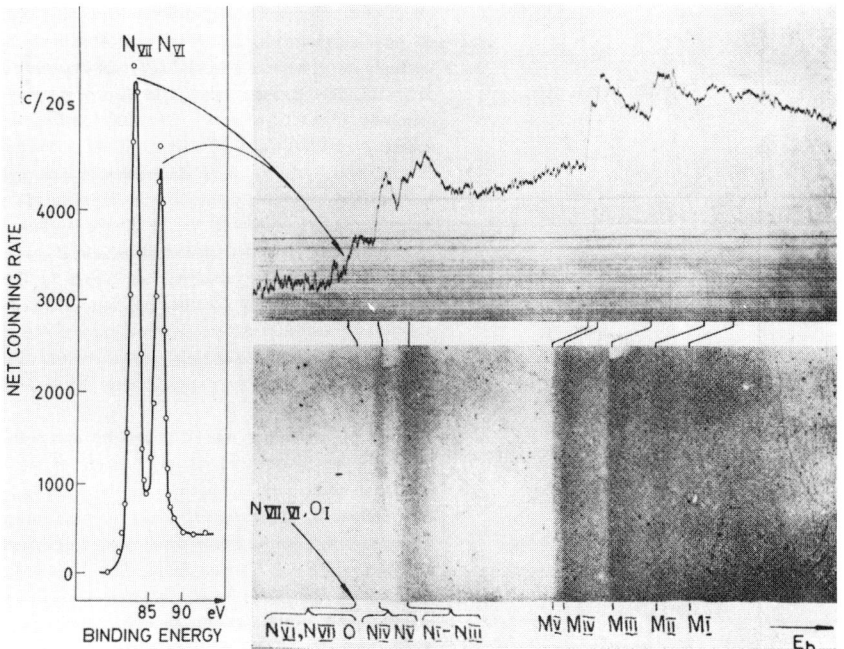

Figure 19. Spectra obtained with instrument of Figure 18 compared with those obtained by magnetic spectrometer of Figure 20 [12].

Figure 20. Magnetic spectrometer. Helmholtz coils are necessary to compensate external magnetic field [12].

1. mild, nondestructive excitation of organic material;
2. all-element sensitivity, particularly for light elements;
3. uncomplicated interpretation;
4. straightforward bonding information;
5. possibility of surface characterization apart from bulk.

The last benefit listed becomes obvious from Figure 21. Aluminum radiation penetrates deep into the polymer, but only photoelectrons from a shallow, less than 100 Å-layer, escape without energy loss, and contribute to the peaks of Figure 19; the rest is either reabsorbed or contributes to background. B. L. Henke has recently presented a quantitative treatment of the penetration of exciting radiation and the escape of photoelectrons[17]. This faculty of surface analysis can be either curse or blessing,

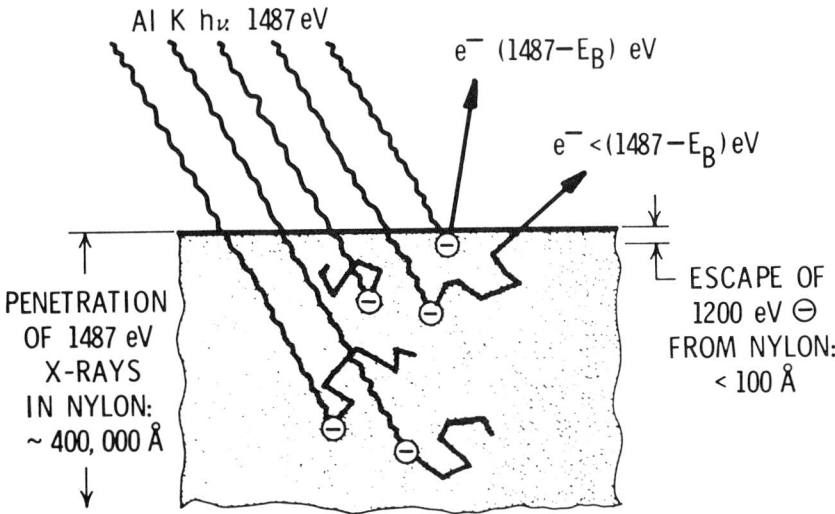

Al K hν 1487 eV

e$^-$ (1487−E$_B$) eV

e$^-$ < (1487−E$_B$) eV

PENETRATION
OF 1487 eV
X-RAYS
IN NYLON:
~ 400, 000 Å

ESCAPE OF
1200 eV ⊖
FROM NYLON:
< 100 Å

Figure 21. Penetration of aluminum radiation in nylon and escape depth of photo-electrons.

Figure 22. Laboratory instrument for ESCA and energy dispersive X-ray fluorescence spectroscopy.

127

COMBINED ESCA / XRS INSTRUMENT

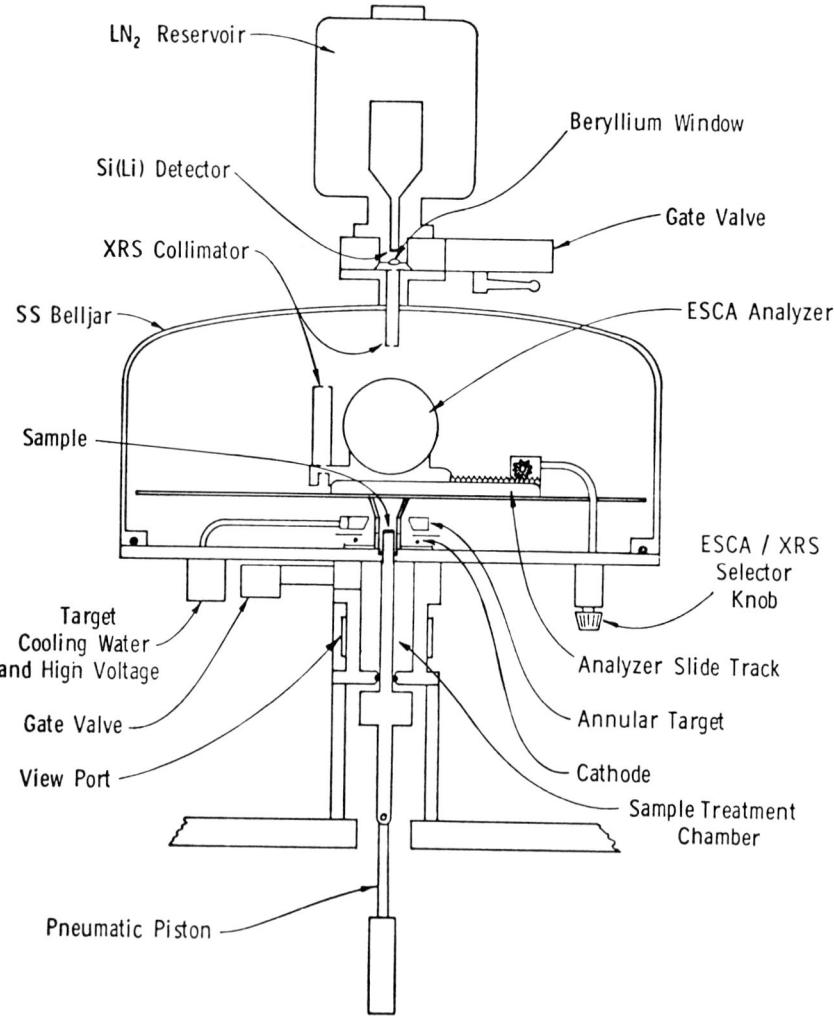

Figure 23. Scheme of combined ESCA/X-ray fluorescence instrument [21].

depending on the intentions of the analyst. Since the surface is a place where many things like adhesion, corrosion, etc., occur, it is of great interest. An analyst who wants to know bulk composition has to make sure that the surface he looks at represents the bulk.

With our soft X-ray fluorescence spectrograph and the incentives described above, we entered into the development of a high-sensitivity, high-resolution ESCA instrument. Our electron energy analyzers[18,19] and high-power, efficient excitation source[10] were incorporated in laboratory instruments having several unique features[20,21]. The instrument shown in Figure 22 is combined with an energy-dispersive X-ray analyzer [22] so that ESCA analysis for all elements in the surface can be combined with bulk analysis for elements above sodium on the same sample. Details of this setup are evident from Figure 23[21]. An example of application is seen in Figure 24. A galvanized steel coated by a fluoro polymer containing TiO_2 pigment is identified by the combination ESCA/ X-ray fluorescence.

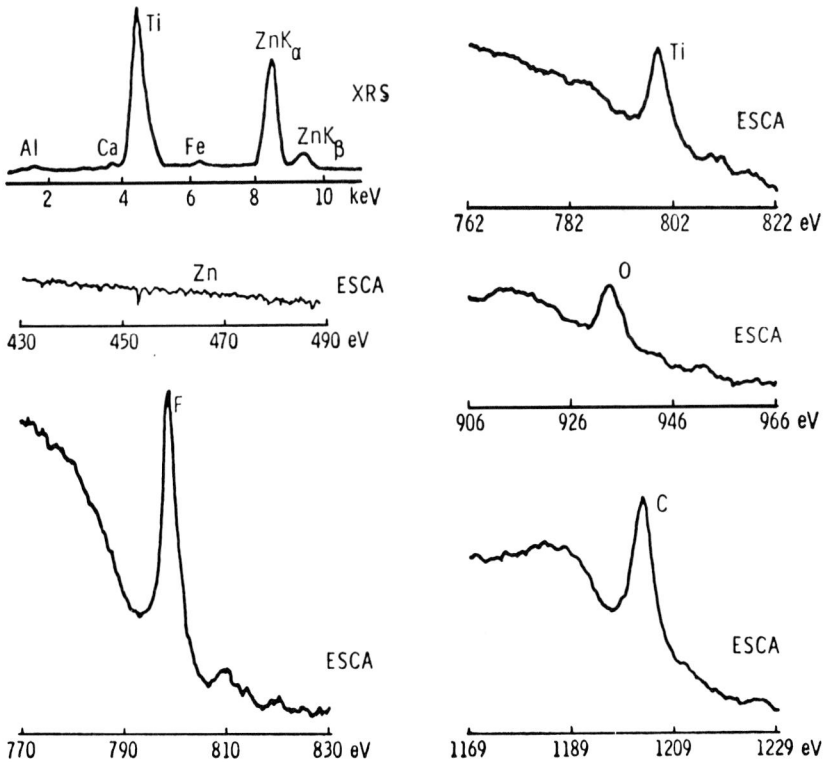

Figure 24. ESCA and X-ray fluorescence spectra of a coated metal.

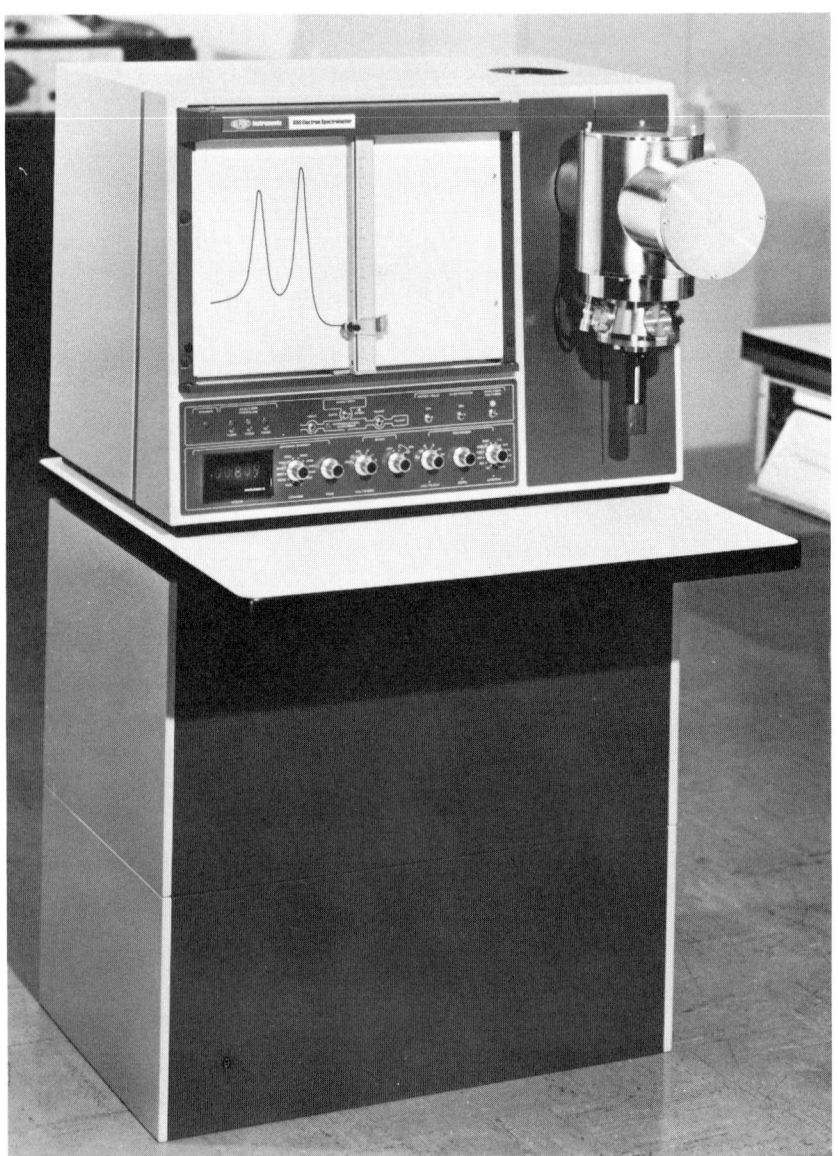

Figure 25.   Du Pont's low-cost, compact ESCA 650 instrument.

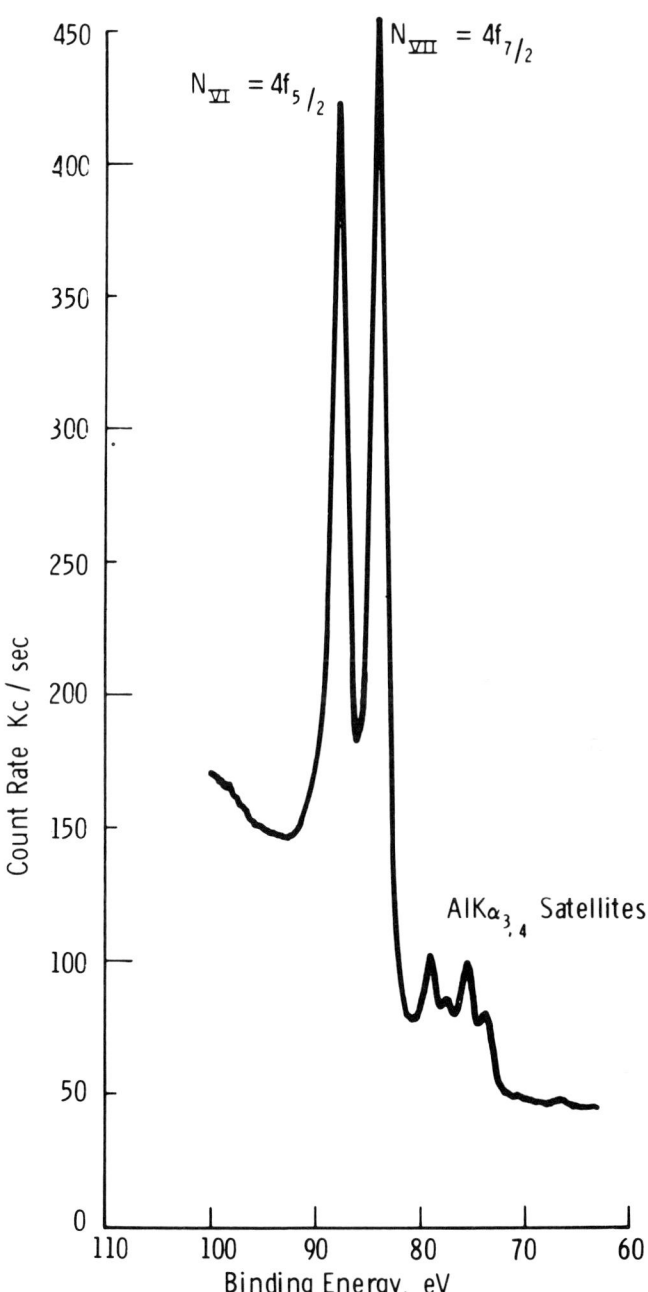

Figure 26. N_{VI-VII} ($4f_{5/2}$, $4f_{7/2}$) doublet of gold obtained with instrument of Figure 25. Peak count rate: 450,000 c/sec.

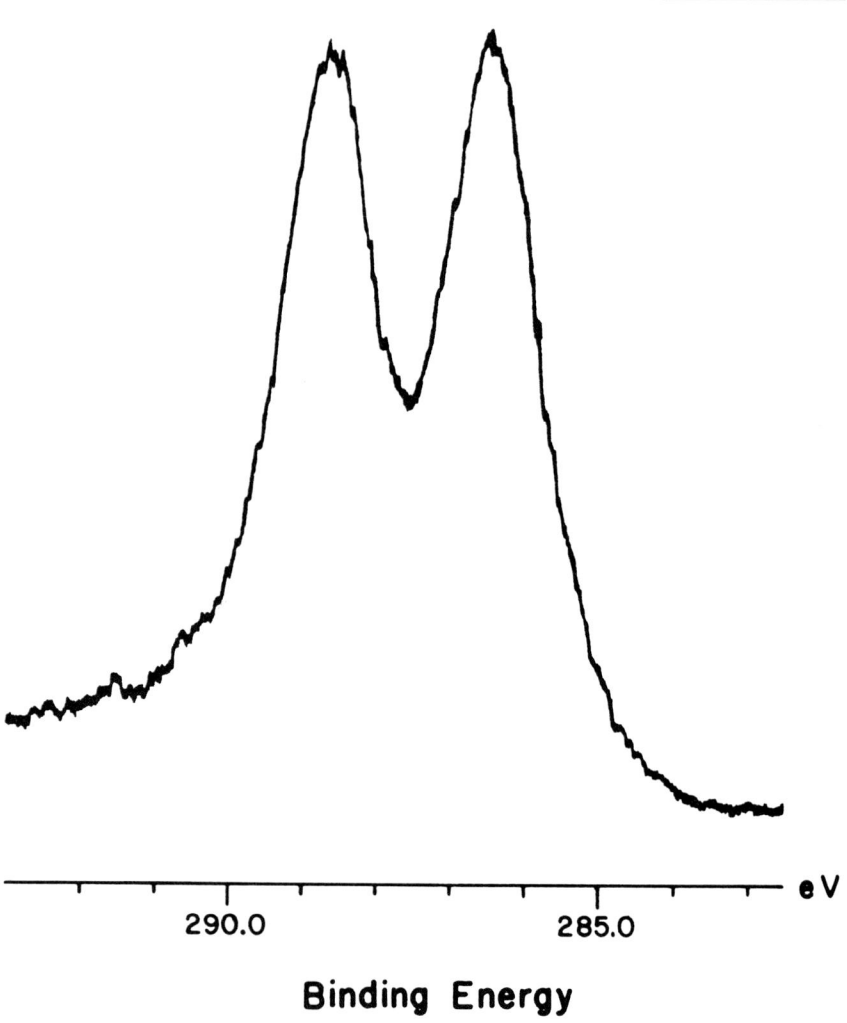

Binding Energy

Figure 27.   Carbon K (Cls$_{1/2}$) peaks in polyvinyl fluoride.

The final step in the instrument development is the compact, low-cost Du Pont ESCA 650 analyzer shown in Figure 25. Built around the same electron energy analyzer[19] as the instrument of Figure 22 and utilizing the same concept of X-ray source[10,20], it is streamlined for compactness, simplicity, and low cost at high performance. Figure 26 shows the same gold N_{VI-VII} doublet as Figure 19, but notice the difference in count rate (200 counts/sec vs. 450,000) at equivalent resolution; and the difference in the size of the two instruments with which these results were obtained. These differences result from a decade of development but in no

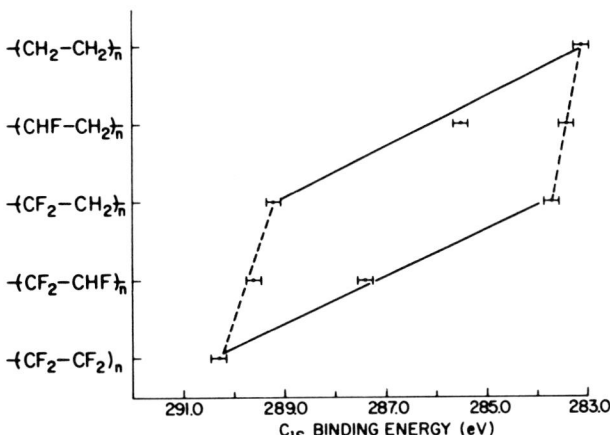

Figure 28. Observed $Cls_{1/2}$ binding energies of fluoropolymers vs. amount of fluorine substitution.

way diminish the merits of the instrument of Figure 20, which brought about the breakthrough.

A few results obtained with the instrument of Figure 25 should elucidate the value of both method and instrument. Figure 27 shows two carbon peaks in poly(vinyl fluoride) corresponding to different carbon atoms

$$-\left[\begin{array}{c} F \\ | \\ CH-CH_2 \end{array}\right]-$$
$$12$$

Figure 28 brings an example of dependence of binding energies of carbon peaks on binding partners[23]. This has delighted polymer chemists, since they can use it to analyze, *e.g.*, the concentration of blocks in block polymers or components in polymer mixtures. Figure 29 is an example of ESCA analysis in chemical surface treatment, which removed fluorine and introduced oxygen into a fluoropolymer[23].

Conclusion

The complexity of polymer structure with its almost unlimited variations makes it impossible fully to characterize a polymer by one or even a few methods. Any novel method furnishing additional, incremental informa-

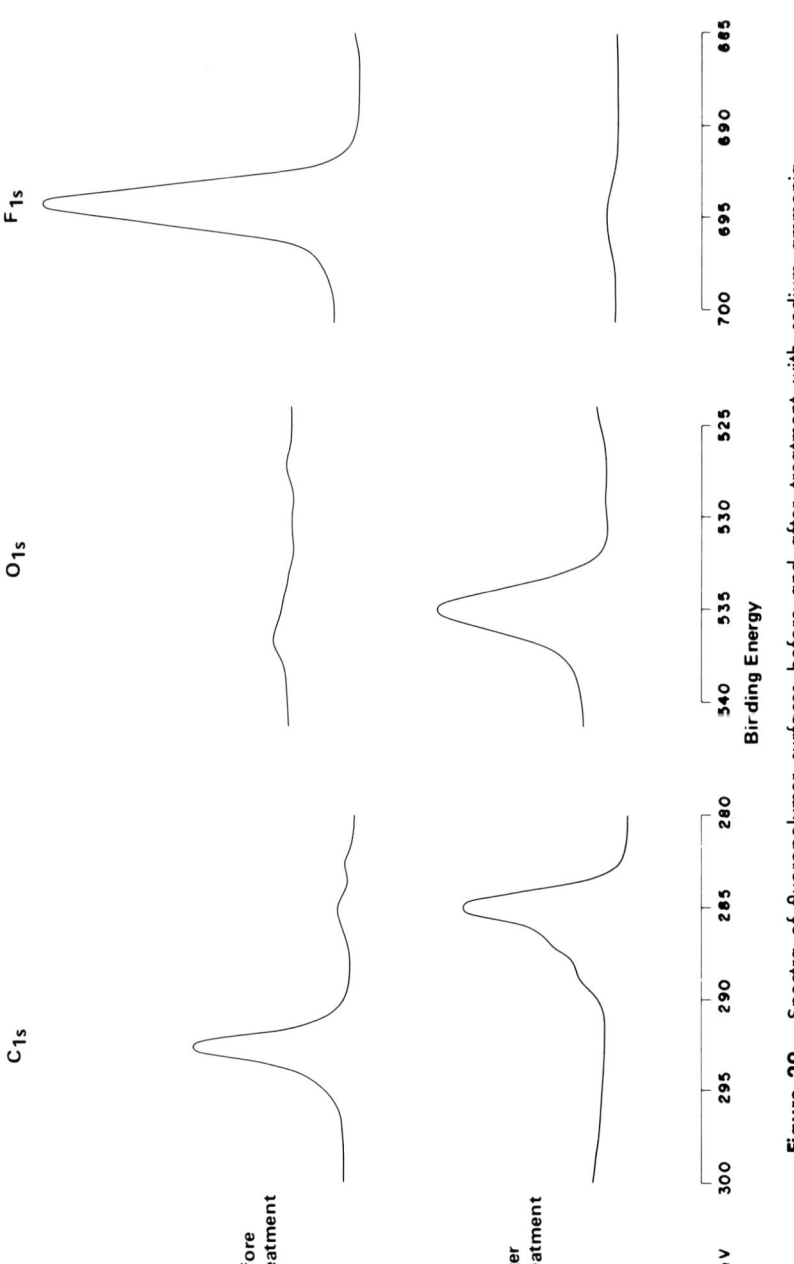

Figure 29. Spectra of fluoropolymer surfcces before and after treatment with sodium–ammonia.

tion, and any new instrument easing the analysts' burden are therefore welcome. If the infinite variability of polymeric features is discouraging at times to the researcher, consolation can be derived from the fact that this same variability is the base of the genetic code, the fundamental feature of life.

References

1. Alexander, L.E., *X-Ray Diffraction Methods in Polymer Science,* New York: John Wiley & Sons, 1969.

2. Kakudo, M. and Kasai, N., *X-Ray Diffraction by Polymers,* Tokyo: Kodansha, Ltd., and Amsterdam: Elsevier Publishing Company, 1972.

3. McMaster, W.H., Del Grande, K.N., Mallett, J.H. and Hubbell, J.H., "Compilation of X-Ray Cross Sections," University of California, Lawrence Radiation Laboratory, Livermore, U.S. Atomic Energy Commission Contract Report No. UCRL-50174, Sect. 2, Rev. 1, May 1969.

4. Guinier, A., *X-Ray Diffraction in Crystals, Imperfect Crystals, and Amorphous Bodies,* San Francisco and London: W.W. Freeman and Company, 1963.

5. Herglotz, H.K., "Study of Structural Detail by X-Rays of Wavelength Greater than 4 Å," in *Advances in X-Ray Analysis, Vol. 14,* C.S. Barrett, J.B. Newkirk and C.O. Ruud, eds., New York: Plenum Press (1971), 275–92.

6. U.S. Patent No. 3,551,331, December 29, 1970, *Reverse Osmosis Separations Using a Treated Polyamide Membrane,* L.A. Cescon and H.H. Hoehn to E.I. du Pont de Nemours and Company.

7. U.S. Patent No. 3,502,925, March 24, 1970, *High Intensity X-Ray Source,* H.K. Herglotz to E.I. du Pont de Nemours and Company.

8. U.S. Patent No. 3,743,841, July 3, 1973, *Method of Dual Wavelength X-Ray Analysis,* H.K. Herglotz to E.I. du Pont de Nemours and Company.

9. Herglotz, H.K., "Use of Total Reflection at the Critical Angle for Dispersion of Ultrasoft X-Rays," *Develop. Appl. Spectrosc.,* 7A (1969), 108–20.

10. Davies, R.D. and Herglotz, H.K., "A Total Reflection X-Ray Spectrograph for Fluorescent Analysis of Light Elements," in *Advances in X-Ray Analysis, Vol. 12,* C.S. Barrett, J.B. Newkirk and G.R. Mallett, eds., New York: Plenum Press (1969), 496–505; U.S. Patent No. 3,567,928, March 2, 1971, *Fluorescent Analytical Radiation Source for Producing Soft X-Rays and Secondary Electrons,* R.D. Davies and H.K. Herglotz to E.I. Du Pont de Nemours and Company; U.S. Patent No. 3,418,466, December 24, 1968, *X-Ray Spectrograph Apparatus Using Low Angle X-Ray Reflecting Units and Means to Vary the X-Ray Incidence Angle,* H.K. Herglotz to E.I. Du Pont de Nemours and Company.

11. Henke, B.L., "Production, Detection, and Application of Ultrasoft X-Rays," in *X-Ray Optics and X-Ray Microanalysis,* H.H. Pattee, V.E. Cosslett and A. Engstrom, eds., New York and London: Academic Press (1963), 157–72.

12. Siegbahn, K. *et al., ESCA: Atomic, Molecular and Solid State Structure Studied by Means of Electron Spectroscopy,* Uppsala: Almquist & Wiksells Boktrykeri AB, 1967.

13. Faessler, A., "Rontgenspektrum und Bindungszustand," in *Landolt Bornstein, Vol. 1,* 6th ed., Berlin: Springer-Verlag (1955), 769–868.

14. Herglotz, H., "Einflusse der Bindung auf das Rontgenspektrum," *Trans. Doc. Center Technol. Economy,* 13 (1952), 1–53.

15. Siegbahn, K. *et al.*, *ESCA Applied to Free Molecules*, Amsterdam: North Holland Publishing Company, 1971.

16. Siegbahn, M., *Spektroskopie der Röentgenstrahlen*, 2d ed., rev. Berlin: Springer-Verlag, 1931.

17. Henke, B.L. and Ebisu, E.S., "Low Energy X-Ray and Electron Absorption Within Solids (100–1500 eV Region)," in *Advances in X-Ray Analysis, Vol. 17* (1974), 150–213.

18. Lee, J.D., "A New Electrostatic Energy Analyzer for ESCA," *Rev. Sci. Instrum.*, 43 (1972), 1291–94.

19. Lee, J.D., "A Nondispersive Electron Energy Analyzer for ESCA," *Rev. Sci. Instrum.*, 44 (1973), 893–98.

20. Davies, R.D., Herglotz, H.K., Lee, J.D. and Suchan, H.L., "High Sensitivity ESCA Instrument," in *Advances in X-Ray Analysis, Vol. 16*, L.S. Birks, C.S. Barrett, J.B. Newkirk and C.O. Ruud, eds., New York: Plenum Press (1973), 90–101.

21. Herglotz, H.K. and Lynch, D.R., "A Combined Photoelectron/X-Ray Fluorescence Spectrometer," in *Advances in X-Ray Analysis, Vol. 17* (1974), 509–20.

22. Russ, J.C., *Energy Dispersion X-Ray Analysis: X-Ray and Electron Probe Analysis*, ASTM Special Technical Publication No. 485, Philadelphia: American Society for Testing and Materials, 1971.

23. Riggs, W.M. and Fedchenko, R.P., "Design and Application of a New Concept Electron Spectrometer," *Amer. Lab.*, 4 (1972), 65–66, 68, 70–73.

CERAMIC CHARACTERIZATION

Moderator: R. Nathan Katz

Army Materials and Mechanics Research Center
Watertown, Massachusetts

R. H. ARENDT, R. J. CHARLES,
C. D. GRESKOVICH, AND J. A. PALM
General Electric Company
Schenectady, New York

Chapter **6**

The Role of Powder Processing
in Developing Material Properties

ABSTRACT

Several powder-processing procedures are discussed to illustrate the critical role of the initial powder preparation in the development of preferred properties of dense, polycrystalline materials. While the materials developed from these procedures serve widely different applications (optical, structural, magnetic), it is shown that the powder-preparation processes for these materials have many features is common. For each application, however, there exist steps in the powder-preparation procedures which require special emphasis if optimum properties are to be achieved. These steps are identified and described in detail for ThO_2-doped Y_2O_3, $MFe_{12}O_{19}$ (M = Pb, Ba, Sr), Co-rare earths, and SiC.

Introduction

A large number of useful inorganic materials exhibit intrinsic properties which prevent effective use of forming processes based on mechanical deformation, solidification, or gas-phase reaction and deposition. Such intractable materials often respond, however, to fabrication procedures based on the consolidation of fine particles of these materials into dense structures by solid-state reaction processes at high temperatures. The overall process is termed sintering, and the mass transport required to achieve consolidation occurs by chemical diffusion of material components and may be assisted or accelerated by the application of mechanical pressure.

Since the process is basically a chemical one, its successful prosecution depends on environmental control at high temperatures and, most particularly, on the physical and chemical characteristics of the powder before the consolidation step. With respect to physical characteristics,

139

the state of subdivision of the material as reflected by particle size and distribution, particle morphology, and agglomeration are of dominant concern. In addition, crystal perfection, the presence or absence of microphase structures, and the deformation state of individual grains may markedly influence the development of final properties. With respect to chemical characteristics, two questions are generally of major importance: can adequate chemical and phase control of major components, contaminants, impurities, and solutes be exercised, and can the desired spatial distributions of all chemical components be achieved in the final product?

The following is an attempt to illustrate the flexibility and utility of the sintering process for the development of new and improved materials and, specifically, to illustrate the importance of the powder-preparation step in such processes.

The powder-preparation procedures for four materials have been selected for discussion. The intrinsic properties of all four of these materials are such that sintering from the powder state is currently the only practical method of fabrication. One of these materials requires the development of extraordinary optical properties since its intended use is in laser and IR window applications. The second is under development for use as high-temperature, load-bearing components in the hot sections of gas turbines. The last two materials are permanent magnet materials, one of which has been of long standing and the other is of recent development.

Ultratransparent ThO$_2$-Doped Y$_2$O$_3$ Ceramic

Introduction

Polycrystalline ceramics are promising materials for use in visible and infrared optical systems because they can be produced in large sizes with good optical quality and high strength by the sintering process. Although there is no intrinsic limitation in fabricating large, complex shapes by the cold pressing and sintering approach, the main problem with polycrystalline materials is the removal of light-scattering centers. The most common scattering defects in supposedly single-phase polycrystalline substances are residual porosity, which is often responsible for opacity, and small amounts of solid second phases. For particles much smaller than the wavelength of light in the transmitting medium, the scattering cross section is directly proportional to the sixth power of the particle radius, inversely proportional to the fourth power of the wavelength, and a function of the relative refractive index[1]. For particles much larger than the wavelength, the scattering cross section is simply equal to twice the geometrical cross section[1]. The scattering efficiency for particles residing in a transparent medium usually reaches a maximum when the particle size is of the order of the wavelength of incident light and decreases to a constant value at larger particle sizes. Another source of light scattering in cubic, optically isotropic solid solutions is a variation in refractive index, which probably arises from small concentration gradients throughout the sample. In the case of anisotropic materials such as aluminum oxide, birefringence can cause scattering at a grain boundary which represents a region of misorientation between two adjacent grains.

The processing of optical ceramics completely free from pores and secondary phases is the primary goal of the ceramic technologist. Although many physical and mechanical properties of ceramic materials could be enhanced by achieving a state of theoretical density and phase purity, relatively few materials closely approach this state. Lucalox® and Yttralox® ceramics (registered products of the General Electric Company), the IRTRAN series of hot pressed materials[2] and hot pressed PLZT ferroelectrics[3] fall into this category.

The present investigation was designed to develop powder-preparation and processing techniques which could produce ultratransparent Yttralox ceramic for lasers, IR windows, and domes, lenses, etc., by a sintering approach. This cubic oxide ceramic has the C-type rare-earth structure, and is composed of a solid solution of 0–10 mole % ThO_2 in Y_2O_3. (For laser applications, 1 cation % neodymium is introduced into the lattice.) After describing the importance of the grain-growth inhibitor (ThO_2), the interrelationships between powder preparation and processing, microstructure and some of the resulting optical properties will be discussed.

Grain-Growth Control Mechanism

Broad experience in the sintering of relatively pure ceramic powder shows that, in addition to the use of a fine powder for the attainment of nearly theoretical density, it is usually necessary to control the rate of grain growth during solid-state sintering by means of a grain-growth inhibitor. In particular, the inhibitor should prevent discontinuous grain growth so that pores remain on grain boundaries during sintering and essentially disappear by the high flow of matter from the grain boundaries (matter source) to the pore surface (matter sink). A clear example of the effectiveness of a solid-state grain-growth inhibitor is illustrated by comparing microstructures of Y_2O_3 [Figure 1(a)] and ThO_2-doped Y_2O_3 [Figure 1(b)] sintered under identical conditions. The sintering of pure Y_2O_3 powder yields an opaque ceramic caused by high residual porosity inside grains which have undergone discontinuous grain growth. The addition of 10 mole % ThO_2 to the Y_2O_3 powder produces a microstructure with a drastic reduction in average grain size and pore entrapment and yields a highly-dense, single-phase transparent material. Since the concentration of ThO_2 added is well below the solubility limit[4,5] at the sintering temperature, it was reasonably well demonstrated that the function of ThO_2 in Y_2O_3 is to reduce the average grain-boundary mobility—probably by a mecha-

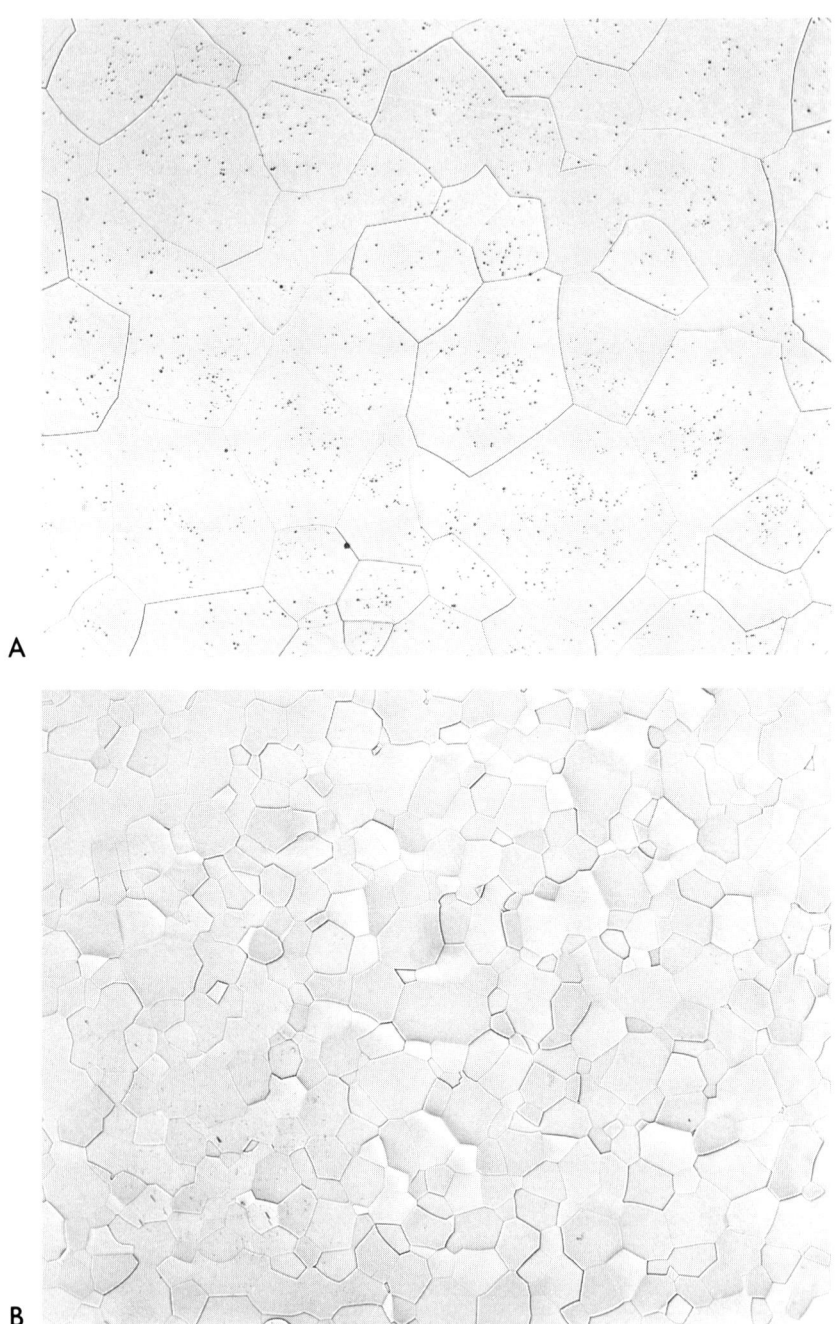

A

B

Figure 1. Microstructures of Y_2O_3 (a), and Y_2O_3 containing 10 mole % ThO_2 in solid solution (b), sintered at 2170°C for 10 hours. 100X.

142

nism of preferential solute segregation at or near grain boundaries. Theories[6,7] suggest that a solute profile across a migrating grain boundary exhibits a retarding force or solute drag on the boundary, thereby causing a reduction in the grain-boundary mobility rather than the driving force.

The method of preparing and processing a starting Y_2O_3 powder which contains the ThO_2-inhibitor for discontinuous grain growth often plays a dominant role in the development of the microstructure during sintering. It will be shown that the ThO_2-inhibitor should be present in sufficient amounts to be effective as a solid solution additive and should have a very small particle size or be directly incorporated into the Y_2O_3 lattice during powder preparation to enhance the rate of solid-state reaction by chemical diffusion. In addition, the inhibitor should be uniformly dispersed throughout the starting powder to insure chemical homogeneity and maximum control of grain-boundary mobility during the grain-growth process. Two methods developed to meet these requirements will now be discussed.

Characterization of Powder and Sintered Material Produced by the Sulfate Process

The sulfate process[8] consisted of adding an aqueous suspension of yttrium oxide powder (average particle size ~ 3 μm) into a filtered aqueous solution of thorium and neodymium sulfates. The batch composition was usually 89 mole % Y_2O_3, 10% ThO_2, and 1% Nd_2O_3; the use of the 10 mole % ThO_2 was found necessary to retard discontinuous grain growth effectively. The commercially designated purity of the Y_2O_3 and sulfate powders were 99.99 percent and 99.99 percent, respectively. While the suspension was continuously stirred, heat was applied to dry the slurry rapidly and to prevent salt-particle segregation. After drying, sieving through a nylon screen to improve particle size and compositional uniformity, and calcining in air at 1000°C, the calcined powder was pressed isostatically to minimize density gradients in compacts with relative green densities ~ 60 percent of theoretical. Specimens were then given a second calcination step at 1350°C for 15 hours in air to remove residual sulfur found to be responsible for specimen "slumping" during sintering. Finally, specimens were heated to temperatures between 2000°C and 2170°C for 20–125 hours in dry H_2 (dew point $\sim -70°C$), and finally cooled to room temperature in wet H_2 (dew point $\sim 25°C$) to obtain desirable oxygen stoichiometry conditions.

The morphology and particle size of the powder calcined for 7 hours at 1000°C in air are illustrated in Figure 2. Note that the Y_2O_3 particles (≈ 3 μm) are coated with fine particles (≈ 1000 Å or less) of ThO_2 and Nd_2O_3. This "coating" behavior enhances composition homogeneity and promotes inhibition of discontinuous grain growth.

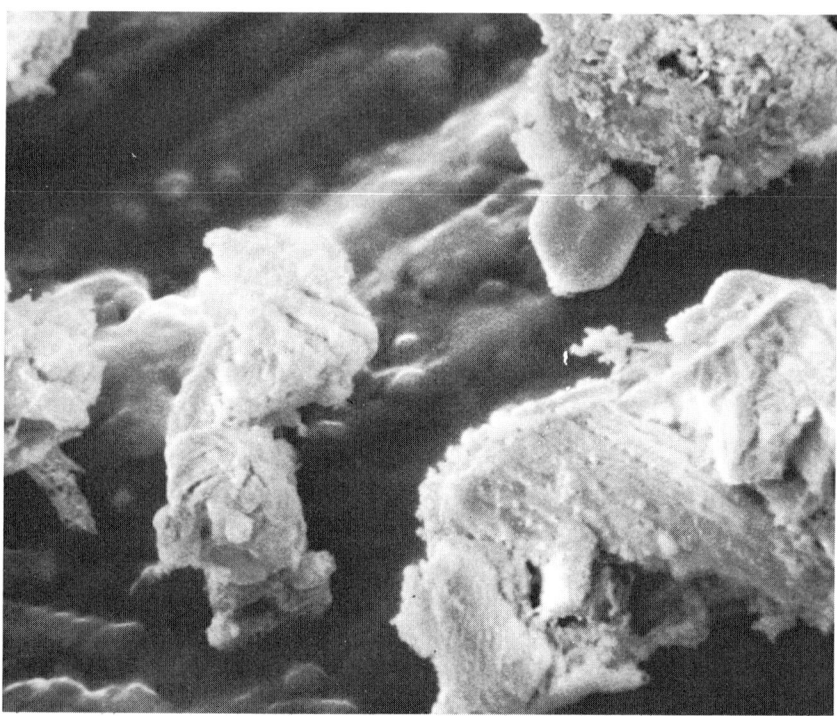

Figure 2. SEM photomicrograph of calcined powder prepared by the sulfate process[8]. 6,200X.

Sintered material[8] could be produced with good optical transparency and low porosity ($\sim 10^{-4} - 10^{-5}$) by the sulfate process, but deviations from optical perfection appear to originate primarily from impurities and agglomerates in the starting powder. Nonwhite impurity particles in the commercially available Y_2O_3 powder were discernable with the unaided eye and, subsequently, found to contain high levels of Al, Si, Fe, and Zn. These impurity particles, plus the presence of a small fraction of Y_2O_3 agglomerates > 10 μm in size, are most likely responsible for the small number density ($\sim 2/mm^3$) of large pores in the 20–50 μm range generally found in all sintered material. Particle comminution of the Y_2O_3 agglomerates and better dispersion of ThO_2 could be achieved by colloid milling the Y_2O_3-sulfate suspension. Although this processing step did give rise to highly transparent sintered material consistently having porosities between 10^{-5} and 5×10^{-6}, a visible haze in the transparent medium was always associated with colloid milling and was due to metallic contamination from the stainless steel wear surfaces of the mill. Finally, it was very difficult completely to eliminate sulfur in the powder compact before final sintering. This impurity caused the formation of an intergranular, sulfur-rich liquid phase which is an effective light scatterer and, consequently, creates high optical losses in the sintered product. In order to eliminate this phase, extended sintering times at high temperature in H_2 are required. It was finally concluded from these findings that the only method of reproducibly synthesizing Yttralox or Nd-doped Yttralox ceramic with extraordinary optical perfection is to work under higher purity conditions and, in particular, to use powders synthesized in our own laboratory instead of working directly with commercial oxide powders.

Characterization of Powder and Sintered Material Produced by the Oxalate Process

An oxalate precipitation method was developed to produce a starting powder with variable composition (96.5–89 mole % Y_2O_3, 2.5–10% ThO_2, and 1% Nd_2O_3) and with physical and chemical properties favorable for the cold pressing and sintering approach. This method has been described in detail[8]. Briefly, yttrium, thorium, and neodynimum nitrates are the starting materials and have purities of 99.99%, 99.9%, and 99.99%, respectively. Appropriate amounts of these nitrates are dissolved in demineralized water and the solution is filtered twice to remove undissolved foreign matter present in the nitrates. Similarly, an oxalic acid bath is prepared and filtered. The nitrate salt solution is then dripped or sprayed (by an aerosol method) into the oxalic acid solution to create an oxalate precipitate consisting of fine particles. During precipitation, the oxalic acid solution is vigorously mixed with a motor-driven stirring propeller to promote high reactivity at the nitrate droplet–oxalic acid solution interface. The precipitate is then washed with filtered demineralized water to remove residual acid, vacuum filtered, dried, and calcined for 4 hours between 800°C and 850°C to convert the oxalates to the oxides. The calcined powder plus 1 wt % stearic acid is then milled for 6 hours in a rubber-lined mill containing ThO_2-doped Y_2O_3 cylinders. Disks or rods are formed from the milled powder by isostatic pressing at 30,000 psi. Specimens having green densities $\approx 60\%$ of theoretical are then given an oxidizing treatment at 1100°C for 2 hours in air to remove organic materials introduced during the milling procedures. In several instances, the calcined powder is not ball milled but cold pressed directly into shape. Finally, the specimens are sintered under similar experimental conditions as those described under the sulfate process.

SEM photomicrographs of powder particles synthesized at various stages in the oxalate process are shown in Figure 3. Several photomicrographs of the oxalate particles (Figure 3a) show that the particles are in the approximate size range of 0.3–3 μm. High-purity oxalate particles with this narrow particle size distribution can be synthesized very reproducibly by this method. The calcined powder is composed of small

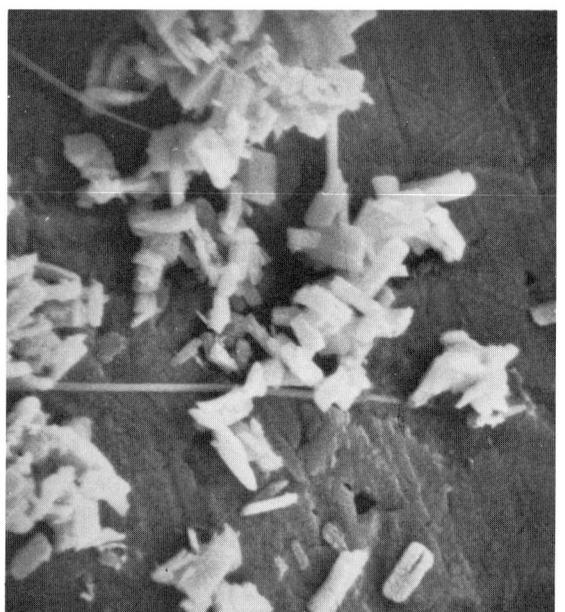

A

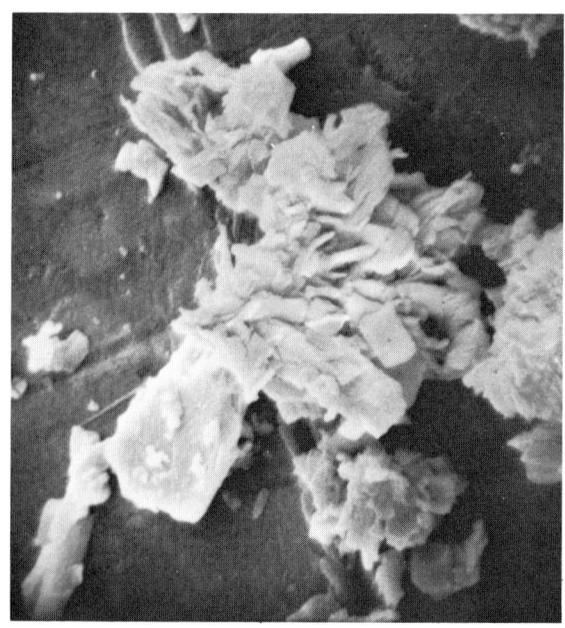

B

C

Figure 3. SEM photomicrographs of (a) powder particles synthesized by the oxalate process, 5,000X; (b) oxalate powder calcined for 4 hours at 800°C in air, 5,000X; and (c) ball-milled powder[8], 10,000X.

agglomerates (Figure 3b) which can be broken into particles smaller than 0.1 μm by dry ball milling (Figure 3c). The milled powder consists of very small "soft" agglomerates, which present no problems during specimen pressing or sintering. A small amount of sulfur is picked up from the vulcanized rubber lining during milling, but this impurity can be removed during the sintering step.

Residual porosity in sintered Nd-doped Yttralox (NDY) ceramic with high transparency can be measured by a microscopic method and is found to depend on powder preparation and processing parameters, composition, sintering time and temperature. The following discussion relates to a single change in the overall process, with other variables remaining essentially constant. Sintered material prepared from unmilled powder contains less porosity when oxalate powder derived by the aerosol technique is used instead of that prepared from oxalate powder formed by the drip method. The majority of pores in both of these samples are between 0.5 and 5 μm, but the pore density of $2 \times 10^4/cm^3$ is about an order of magnitude lower in sintered material prepared by the aerosol method. Since the average particle size is found to be essentially the same for both starting powders, and since nearly all pores are located inside grains of average size ~ 100 μm, it is believed that the aerosol technique produces a starting powder with greater chemical homogeneity, *i.e.,* a better dispersion of ThO_2 which reduces the average grain-boundary mobility during grain growth. It is also found that sintered material made from powder milled in a rubber-lined mill contains porosity as low as 3.5×10^{-7}, whereas that made from unmilled powder has a porosity $\sim 3 \times 10^{-5}$. This difference again suggests that the ball-milling operation enhances dispersion of the ThO_2-grain growth inhibitor which, during sintering, permits most pores to remain on grain boundaries for extended times and shrink. Powder milled in a polyurethane-lined mill consistently gives rise to sintered material containing large tubular pores originating from polyurethane particles formed by excessive abrasion of the lining and pressed into the powder compact.

Another important consideration in the milling step is to use fine-grain size ThO_2-doped Y_2O_3 cylinders to prevent impurity contamination and chipping or fracturing of the grinding media. It has been demonstrated[9] that large chips ($\sim 50 \ \mu m$ in size) can fragment off of the grinding media, mix into the fine ($\sim 0.1 \ \mu m$) powder, and act as nuclei for discontinuous grain growth which entrap pore clusters (Figure 4) inside rapidly growing grains. The elimination of pore clusters is essential for producing sintered material with low porosity and high optical quality because it has been established[10,11,12] that pores shrink much more slowly when they are located off of, rather than on, grain boundaries.

The influence of composition on residual porosity in NDY ceramics is generally found as expected. By increasing the ThO_2 content from 2.5 to 5 to 10 mole %, the residual porosity decreased from 5×10^{-6} to 1.1×10^{-6} to 0.33×10^{-6}, respectively. The pore size distributions for NDY ceramics containing 2.5 and 10 mole % ThO_2 are shown in Figure 5. Note that the peak in the size distribution occurs between 1 and 2 μm in both specimens and that there is just a decrease in the number density of pores in the specimen containing 10 mole % ThO_2. Because the average grain size is about 1.5 times larger in sintered specimens containing 2.5% ThO_2 than in material containing 10% ThO_2, these findings suggest that higher ThO_2 contents are more effective in reducing average grain-boundary mobility. The presence of 1 mole % Nd_2O_3 in these powders is found to play an insignificant role in the development of the microstructure. Finally, for the case when it is desirable to remove closed pores located primarily inside grains, it has been found[12] convenient to increase the sintering time or temperature so that migrating grain boundaries will intersect pores for a sufficient time to permit rapid pore shrinkage.

Another optical inhomogeneity in transparent NDY ceramic prepared from unmilled calcined powder is "orange peel," which is an optical waviness or distortion caused by refractive index variations within the single-phase solid solution matrix as detected by the unaided eye[8]. The degree of orange peel depends strongly on the method of powder preparation and processing and on the ThO_2 concentration. It is generally observed that, other things being equal, sintered material made from oxalate

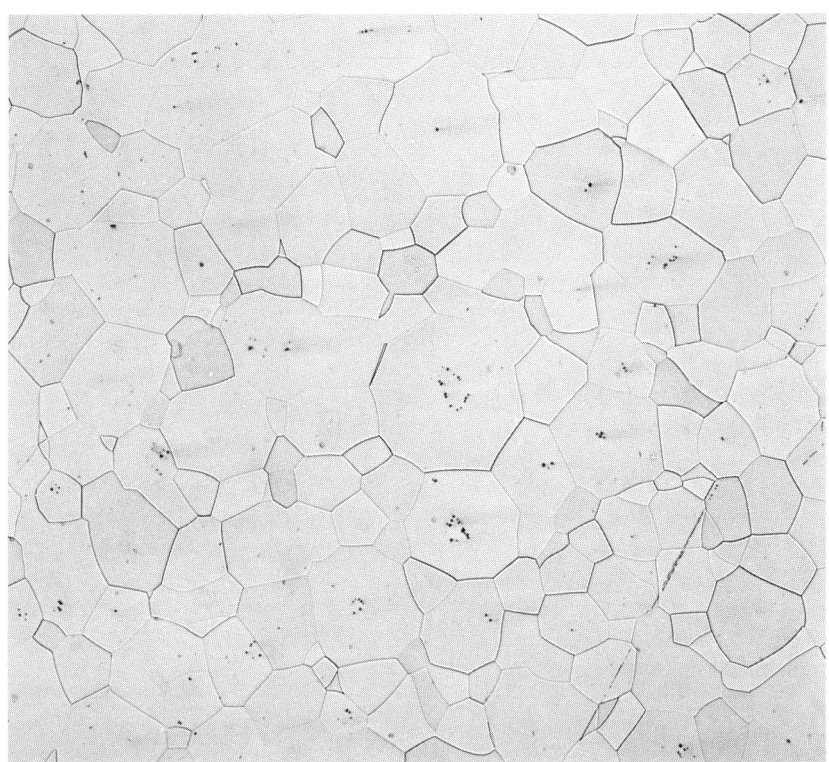

Figure 4. Pore clusters located inside grains. Specimen sintered for 16.5 hours at 2160°C. Reflected light. 100X.

powder produced by the aerosol method contains less orange peel than sintered material made from oxalate powder derived by the drip method. For a given method of powder preparation and without using a milling step, powders richer in thorium concentration generally yield sintered specimens having more orange peel. It is discovered[8], however, that orange peel can be reduced appreciably by ball milling any calcined oxalate powder. Evaluation of these experimental findings indicate that (1) small droplets of Y–Th–Nd nitrate favor a faster reaction rate with the oxalic acid and form particles with more uniform dispersion and/or composition; (2) chemical inhomogeneities apparently increase with increasing ThO_2 concentration; and (3) ball milling is a very efficient method of minimizing chemical inhomogeneities or chemical segregation in the starting powder. We currently believe, therefore, that orange peel is related to minute variations in chemical composition within the solid solution matrix and originates primarily from chemical segregation of yttrium and thorium oxalate during the precipitation step.

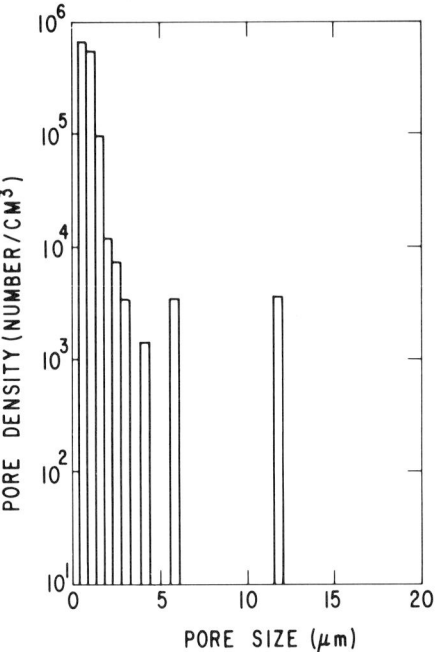

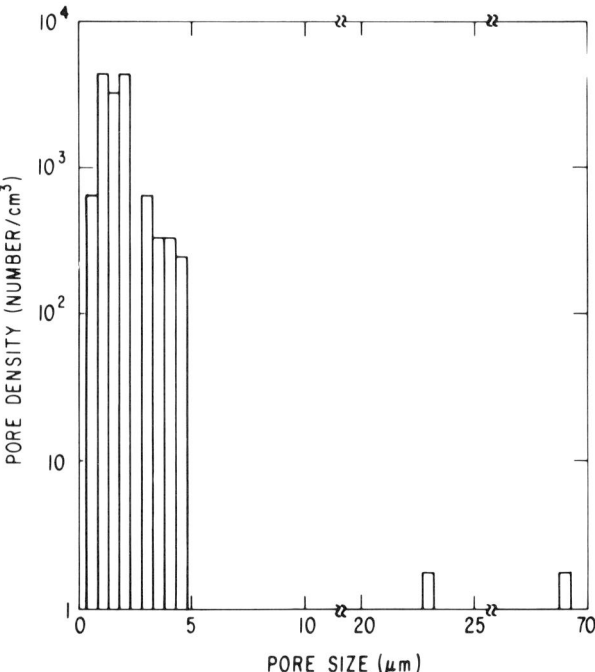

Figure 5. Pore density vs. pore size in Nd-doped Yttralox ceramic containing 2.5 mole % ThO$_2$ (a), and 10 mole % ThO$_2$ (b), identical heat treatments.

150

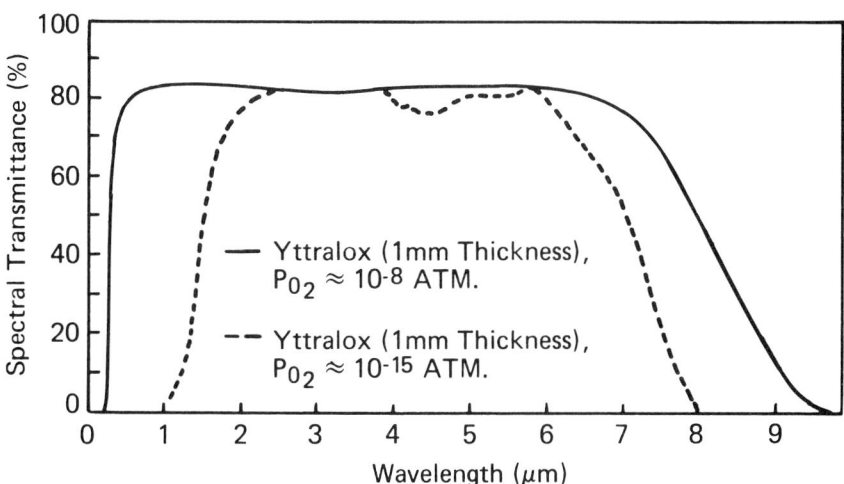

Figure 6. Spectral transmittance of Yttralox ceramic (90 mole % Y_2O_3 + 10% ThO_2) sintered in wet and dry H_2 at 2170°C.

Extraordinary optical properties can be achieved in Yttralox ceramic provided that extraordinary care is taken in powder preparation and processing and that the influence of atmosphere and thermal cycling conditions are understood and controlled. Controlled oxidation of Nd-free Yttralox ceramic at high temperature can generate a pronounced middle infrared pass filter or a visible-to-middle infrared pass filter (Figure 6). The color of Yttralox (90 mole % Y_2O_3 + 10% ThO_2) material sintered in wet H_2 ($P_{O_2} \sim 10^{-8}$ atm) at 2170°C is water-white, whereas that sintered in dry H_2 ($P_{O_2} \sim 10^{-15}$ atm) is black. Both materials are cubic single phases with nearly the same lattice parameters. The high spectral transmittance in the 1 μm region makes "clear" Yttralox ceramic a promising laser host for the Nd^{3+} ion. Nd-doped Yttralox rods 3″ × 1/4″ can be synthesized with excellent optical quality and exhibit intermediate-gain lasing behavior between that of two important commercial lasers, Nd:YAG single crystal and Nd:glasses[13]. Lasing efficiencies[14] as high as 94 percent of that of the best commercially available glass laser rods can be achieved under pulsed mode conditions if (1) residual porosity is $< 10^{-6}$, (2) orange peel is minimized by the ball milling step, and (3) the sintered rod is rapidly cooled from the sintering temperature to retard the nucleation and growth of coherent ordered zones or extended defects in the solid solution matrix during cooling. Laser output as a function of electrical input energy for NDY ceramic containing 90 mole % Y_2O_3, 5% ThO_2 and 1% Nd_2O_3 is shown along with laser glass (Owens-Illinois, ED-2) in Figure 7. The lasing efficiency at 40 J of input energy is 0.32%, and the optical attenuation coefficient is 2% cm^{-1} at the lasing wavelength of 1.074 μm. Scattering losses account for 80 percent of the total optical

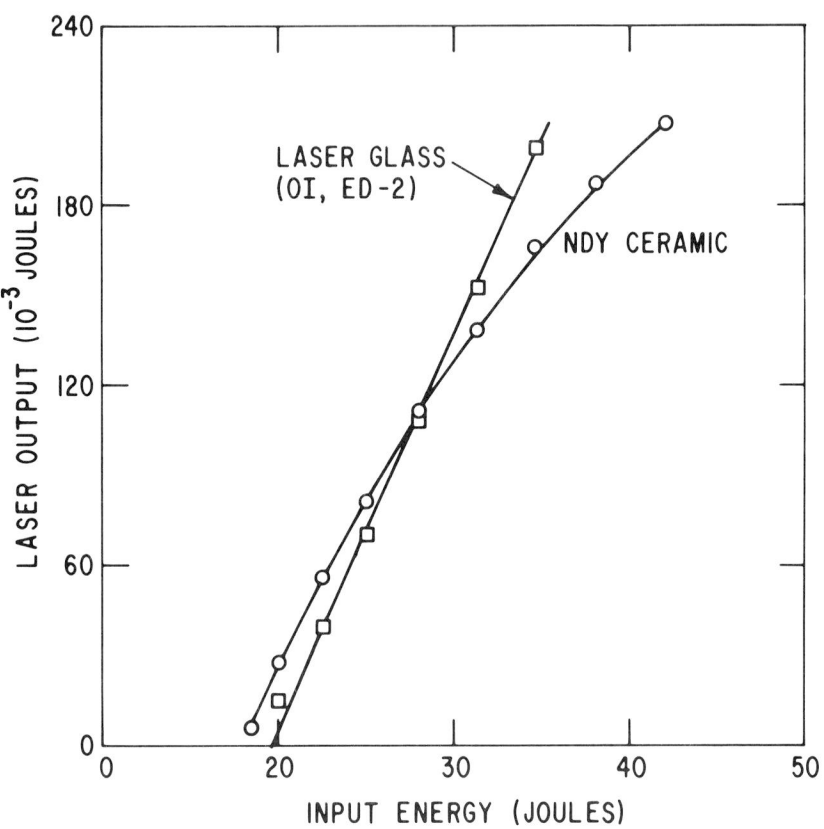

Figure 7. Pulsed mode laser output of Nd-doped Yttralox (NDY) rod (3″ × 1/4″) in comparison with that of the best commercially available Nd-doped laser glass rod. Output mirror reflectivity is 80%.

attenuation, and are believed to be caused primarily by orange peel and extended defect formation in these polycrystalline ceramic lasers[14].

SiC Powder Synthesis for High-Temperature Structural Ceramics

Introduction

Considerable emphasis has recently been placed on the fabrication of ceramic materials suitable for use as hot-section parts in high-temperature gas turbines. One of the few candidates which shows inherent properties suitable for such application is dense SiC. This material is characteristically prepared in dense form by processing sequences which involve hot pressing of powder compacts. Recent work has clearly shown that the success of the hot-pressing operation is sensitively dependent on both the physical and chemical properties of the initial SiC powder and whatever sintering aids are necessary for densification. For example, Nadeau[15] has shown that the hot-pressing conditions necessary to achieve full densification of pure SiC are extreme. This work indicated that densities of greater than 99 percent of theoretical could only be achieved at die pressures of the order of 50 Kb and temperatures well in excess of 2000°C. Other investigators[16,17] have shown, however, that minor additions (W, B, etc.) to SiC have a marked effect on lowering both the temperatures and pressures required to consolidate SiC by hot pressing to a dense form. These results show that the level of minor component additions and the physical characteristics of the starting powders are critical if the benefits available to the hot-pressing operation are to be achieved without sacrifice of the final properties of the densified SiC.

There is, therefore, considerable incentive in the processing of SiC by this latter route to have available powder-preparation procedures of SiC which allow close control of purity of product, dopant levels, grain morphology and particle size and distribution. The object of this discussion is to describe a powder-preparation process which generally meets the above criteria.

Sic Powder Process

The process under discussion is a simple one corresponding to the well known reduction of silica by carbon,

$$SiO_2 + 3C \rightarrow SiC + 2CO \uparrow \tag{1}$$

A flow chart of the process is shown in Figure 8. Because of the concern of achieving as small a SiC particle size as possible, silica "smokes" are employed having particle sizes of 10–15 mμ which is equivalent to a surface area of close to 200 m^2/gm. For the same reason, carbon powders having a particle size of about 50 mμ are used. An additional advantage gained by using extremely small particle sizes is the high surface contact between the reactants which results in high chemical reactivity. The second process control exercised is the reduction process itself. It is carried out with the intimately mixed reactants contained in graphite boxes which are batched, through a H$_2$ furnace operating at 1500–1600°C. The reaction time for each charge is about 4 hours. With a 2-foot-long hot zone in the furnace, 2 graphite boxes each containing about 1 Kg of the "black mix" of SiO$_2$ and carbon are reacted at one time.

After reaction, the product is a soft, easily handled powder containing a small excess of carbon, which is readily removed by oxidation in air

HYDROGEN PROCESS FOR MAKING SiC POWDER

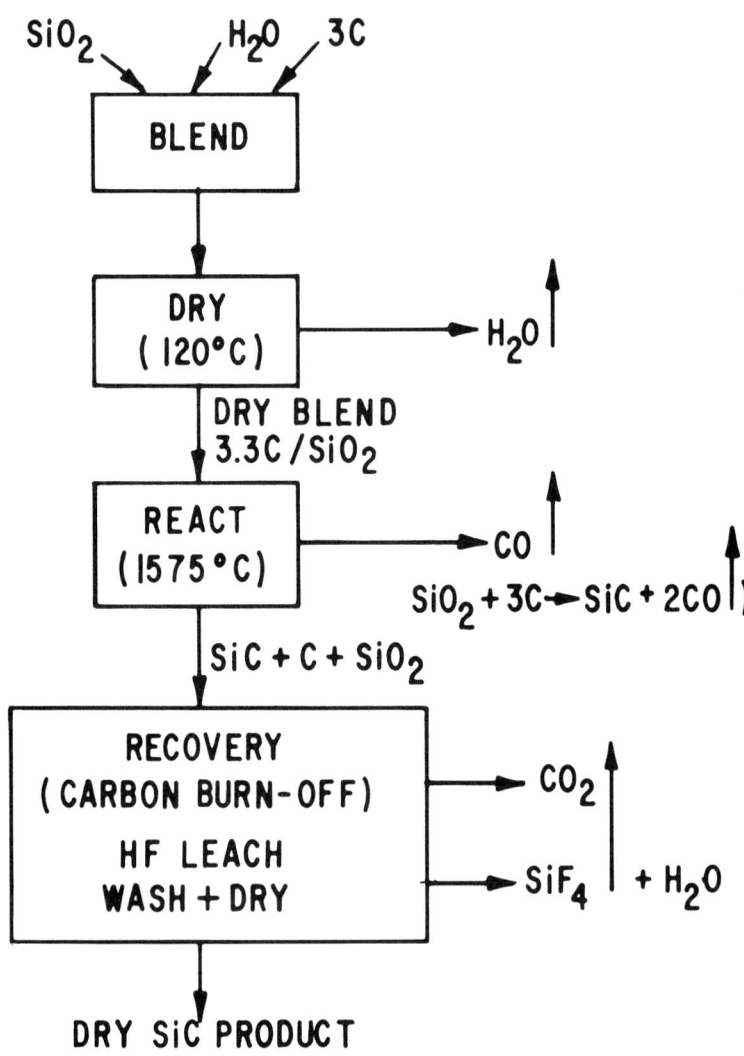

Figure 8. Flow chart of the hydrogen process for SiC powder synthesis.

	132 (B) BEFORE LEACH	132(B) AFTER LEACH
O	$0.4\% \approx 0.75\% \, SiO_2$	$0.058\% \approx 0.11\% \, SiO_2$
Al	0.1	0.02
Ca	0.1	0.01
B	0.05	0.02
Co	0.001	<0.001
W	0.005	<0.005
Cr	0.001	<0.001
Fe	0.01	0.003
Mg	0.001	0.001
Ti	0.005	0.002

Figure 9. Comparison of levels of contaminating elements before and after HF leaching.

at 700°C. This latter step does cause some oxidation of the SiC to take place, which in turn has led to the practice of leaching the SiO_2 out of the product with HF. Figure 9 shows a comparison of the contaminating elements before and after leaching, and shows achievement of a significant improvement in quality in terms of O, Al, and Ca removal from the product by the leaching operation.

The exact role, if any, of the H_2 atmosphere in the reaction between the SiO_2 and C is not known, and perhaps an inert gas atmosphere would work equally as well. It is known that, because of the CO evolution, reactions in vacuum of batches larger than a few hundred grams are difficult to contain. With the H_2 atmosphere, it is possible to plunge the reactants directly into the 1600°C atmosphere without any operational difficulties.

Reaction temperature makes a considerable difference in the powder structure obtained. Figure 10 shows a typical SEM of an HF leached SiC powder made by the H_2 process using a 1600°C reaction temperature. The mixture of platelets and elongated particles is completely different, as seen in Figure 11, from the mixture of elongated layered growths and spherical particles obtained when the reaction is run at 1500°C. Physically, the two powders respond markedly differently to powder packing; the bulk density of the 1600°C powder being 1.3 gm/cc as compared to 2.5 gm/cc

Figure 10. SEM of SiC powder formed at 1600°C reaction temperature.

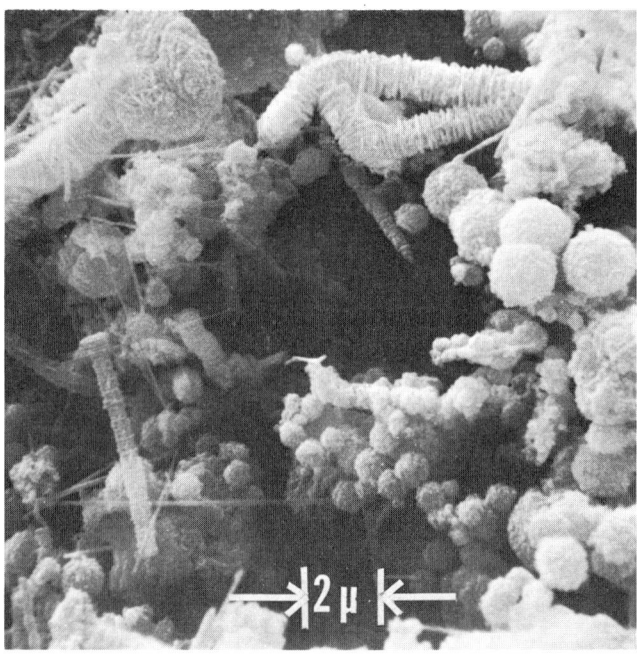

Figure 11. SEM of SiC powder formed at 1500°C reaction temperature.

156

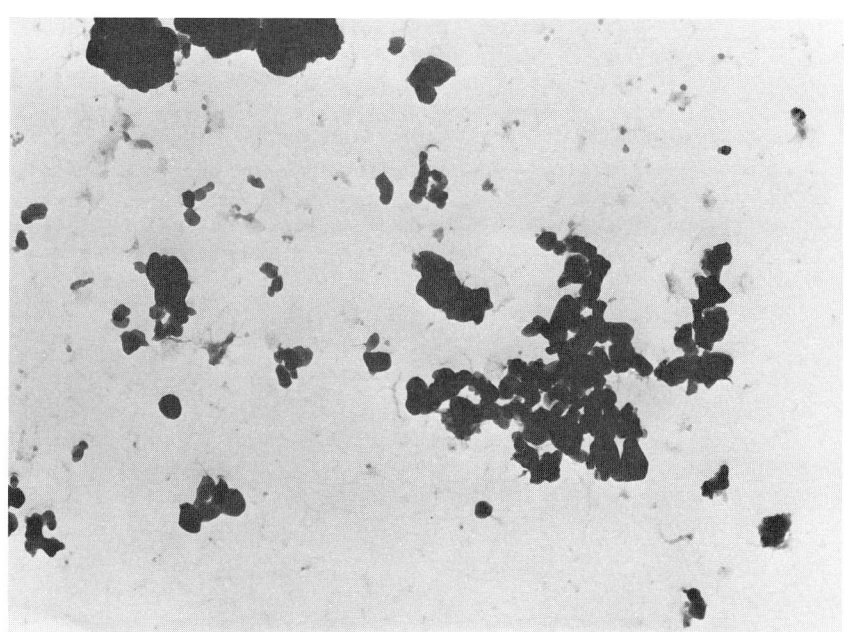

Figure 12.   Transmission electron micrograph of SiC powder. 25,000X.

being for the 1500°C powder.   Hot-pressed samples of the above two powders have been made to ascertain any detectable differences in ceramic microstructure and integrity between the two.   The morphology of SiC powder as produced, is of the β modification and it is largely submicron in particle size.   A transmission electron micrograph (Figure 12) shows the smallest individual particles to be about 0.2 μ.   The corresponding ceramic made from these powders is shown in Figure 13, which reveals the general excellence of its microstructure, featuring high integrity and submicron grain size.

An important observation that distinguishes SiC ceramic made with the H_2-process powder from several others is its apparent resistance to exaggerated grain growth.   The following figures show the microstructures of two hot-pressed SiC ceramics which had been subjected to 2000°C for an hour.   Before heat treatment, each was fine grained with no evidence of large grains.   In Figure 14 one sees the fine-grained microstructure and absence of grain growth found in the ceramic made from hydrogen processed powder.   Figure 15 shows the microstructure of another ceramic made with a commercial SiC powder.   It clearly shows the exaggerated grain growth occurring after a one-hour heat treatment at 2000°C.

In addition to being especially amenable to hot pressing and consolidation into structural ceramic bodies, these powders can be further processed to form masses of relatively large crystals of the α modification.

Figure 13. Photomicrograph of hot-pressed SiC ceramic made from H_2 process SiC powder. 40,000X.

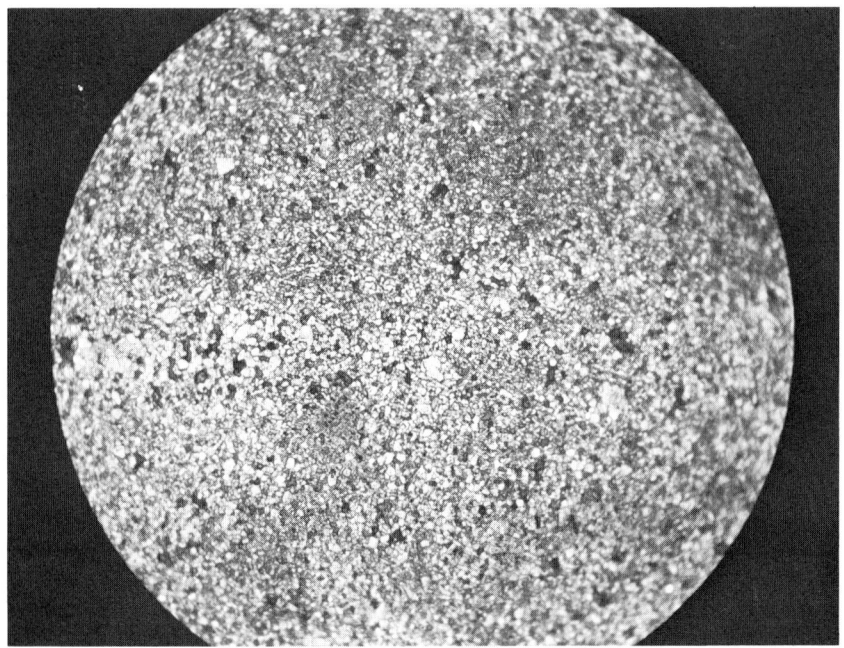

Figure 14. Photomicrograph of hot-pressed SiC ceramic made from H_2-process SiC powder after 1 hour at 2000°C. 500X.

158

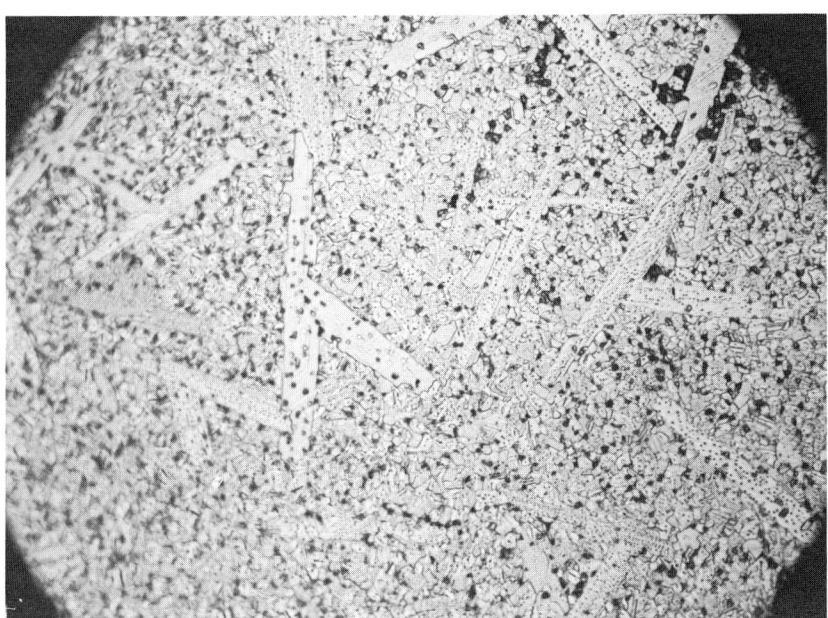

Figure 15. Photomicrograph of hot-pressed SiC ceramic made from commercial grade SiC powder after one hour at 2000°C. 500X.

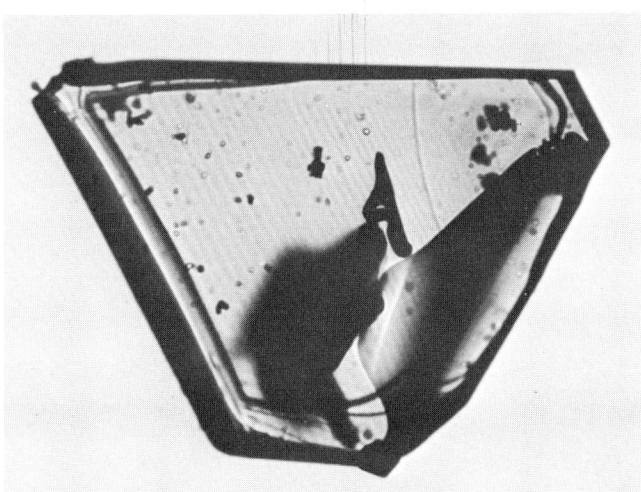

Figure 16. Photograph of single-crystal α-SiC formed at 2300°C in argon from submicron β-SiC powder. 150X.

Such materials are of value to the manufacturers of varistors for the protection of electrical equipment. A one- to two-hour heat treatment of the undoped, H_2-processed powder at 2300°C in argon will yield a 50 percent transformation of the β to α form. X-ray analyses indicate that doping the starting powder with aluminum will increase the α-form product yield to about 75 percent, and is comprised largely of the 6H polytype, with lesser amounts of the 15R and 4H polytypes. Well-formed crystals as shown in Figure 16 are typical of the transformation product. During the 2-hour heat treatment the submicron powder undergoes crystal growth to the extent that 65 percent of it is in the 200 μm particle size. About 20 percent of it will be 300 μm and larger, and 15 percent will be less than 100 μm.

Preparation and Characterization of $MFe_{12}O_{19}$ Powders

Introduction

The hard ferrites possessing the chemical formula $MFe_{12}O_{19}$, M = Pb, Sr, Ba, represent a class of permanent-magnet materials of increasing technological importance. The initial investigation of these materials was made by J. J. Went et al.[18] in 1952, and a considerable volume of literature on the subject now exists[19]. Present applications utilize $BaFe_{12}O_{19}$ and $SrFe_{12}O_{19}$ predominantly in the form of sintered bodies. However, a significant effort has been made to develop methods for the preparation and utilization of bodies composed of magnetically oriented ferrite crystallites bound in an organic polymer matrix. In this case, the powder magnetic properties determine the properties of the finished body. Although the present work was directed toward the polymer matrixing approach, the methods and results derived are equally relevant for sintered magnets and the powders used in their preparation.

Conventional Powder Synthesis

$MFe_{12}O_{19}$ powders are conventionally prepared by the high-temperature (1000–1300°C) solid-state reaction of an MO-yielding compound (usually the carbonate) and an iron oxide. The reactant stoichiometry employed is generally the ratio $MO:Fe_2O_3$ equal to 1:4.5–5.6, compared with the crystallochemical stoichiometry of 1:6. This is done so as to compensate for either vaporization of MO (PbO) during reaction or iron oxide pick-up during subsequent processing. The chemical homogeneity of the product is improved by subjecting the reactants to a prereaction mixing procedure. An extreme example is to coprecipitate the cations from an aqueous solution as the hydroxides, carbonates, or salts of organic anions. In general, the completeness of reaction, and hence the properties of the product, are proportional to the efforts made to homogenize the reactant mixture.

The product of the solid-state reaction is in the form of crystallites, or aggregates of crystallites, ranging in size from microns to millimeters. Permanent-magnet applications require the crystallites to have dimensions on the order of the magnetic domain size, $\sim 1.0 - 1.5$ μm for $MFe_{12}O_{19}$ [20]. The product is therefore milled to give crystallites in this size range. Milling, however, introduces numerous imperfections (vacancies, dislocations, irregular crystal surfaces) into the crystallites which adversely affect the magnetic properties [21]. Subsequent processing procedures such as high-temperature anneals [21,22] and chemical etching with mineral acids [23] are required to minimize the deleterious affect of milling. It is apparent that the sintering process similarly results in a significant reduction in the number of imperfections.

Molten Salt Powder Synthesis

An alternative synthetic process for $MFe_{12}O_{19}$ powder of high magnetic quality has been developed in this laboratory. It is based on the use of molten salts, NaCl–KCl in particular, as solvents for the chemical reaction [24]. Briefly, the process consists of mixing together MCO_3 and Fe_2O_3 (1.1:6 on a molar basis) with the molten salt solvent components (NaCl–KCl in any ratio) such that $MFe_{12}O_{19}$ constitutes 20–50 wt % of the reaction mixture. The mixture is heated to 1000–1050°C in a Pt, dense α–Al_2O_3, or dense-stabilized ZrO_2 crucible for 30–60 minutes, then air cooled to room temperature. The salt and excess MO are water-leached from the solidified reaction mixture to yield a stoichiometric 1.0 μm-particle-sized $MFe_{12}O_{19}$ product, the properties of which will be discussed below.

Powder Characterization

The most important characteristics of $MFe_{12}O_{19}$ powders are those which govern its magnetic properties, $i.e.$, σ_s, the saturation magnetization, and H_{ci}, the intrinsic coercive field. The saturation magnetization at 298°K is 74.3 e.m.u.g.$^{-1}$ for pure $SrFe_{12}O_{19}$ and 72.0 e.m.u.g.$^{-1}$ for pure $BaFe_{12}O_{19}$ [25]. H_{ci} for coherent rotation of magnetization in single magnetic domain particles was treated theoretically by Stoner and Wohlfarth [26]. The values of H_{ci} at 298°K are 6,700 Oe for $SrFe_{12}O_{19}$ and 6,000 Oe for $BaFe_{12}O_{19}$ [25]. The values of these parameters in turn are directly related to chemical purity, crystallite size, and crystallite perfection.

A convenient method to measure H_{ci} is to orient the crystallites in molten beeswax, using a magnetic field and solidifying the wax while still under the influence of the field. The hysteresis loop of the compact is then measured, using a vibrating sample magnetometer, with the applied field axis parallel to that of the orienting magnetic field. H_{ci} is that value of the reverse applied field where the sample magnetization is zero. The magnetometer can also be used to measure the value of σ_s by plotting the magnetization of a randomly oriented powder sample of known mass versus the inverse of the applied field, H, or H^{-2}. The intercept at H^{-1} or $H^{-2} = 0$ gives the saturation magnetization.

The values of σ_s for solid-state-reacted $BaFe_{12}O_{19}$, the compound which has received the greatest attention technologically, range from the intrinsic value quoted above to as little as 50 percent of that value in commercial preparations where highly sophisticated preparative procedures are not practical. Generally, values within 10–15 percent of the intrinsic value are commonly attained. In the case of molten-salt-synthesized $BaFe_{12}O_{19}$ and $SrFe_{12}O_{19}$, the intrinsic values of σ_s are attained consistantly with the process as outlined above as well as with an improved modification given in [24].

The H_{ci} of solid-state-reacted $BaFe_{12}O_{19}$ typically ranges from ~ 1,000 Oe for the milled product to ~ 3,000 Oe for specially treated, milled product. This value can be increased to ~ 5,400 Oe for acid-etched powder, but acid etching is currently not considered to be a viable commercial process. The modified molten salt synthesis [24] will yield powders with H_{ci} = 5,400 Oe for $BaFe_{12}O_{19}$ and H_{ci} = 6,000 Oe for $SrFe_{12}O_{19}$ without post-synthetic treatment of the powder. This is accomplished by control of the reaction chemistry, in particular the Fe(II) content, coupled with the inherent characteristics of flux-grown crystals.

Chemical Purity

The chemical purity of the reactants affects the magnetic properties of the product in several ways. Cations which will substitute for ions in the $MFe_{12}O_{19}$ lattice disturb the magnetic interactions and result in decreased values of both σ_s and H_{ci}. Generally, trivalent ions such as Al^{+3}, Ga^{+3}, Cr^{+3} will substitute for Fe^{+3}. Direct substitutions for the divalent cation are uncommon and, generally, divalent impurities result in distinct impurity phases. Ferrous ion is a notable exception in that it apparently substitutes for Fe^{+3} and causes extreme reductions in H_{ci}, even when present at levels below detection [24].

Cations which do not substitute for ions in the $MFe_{12}O_{19}$ lattice generally result in distinct impurity phases. Divalent impurities such as Ca^{+2}, Mg^{+2}, Ni^{+2}, Co^{+2}, etc., result in spinel phases (MFe_2O_4) among others. Tetravalent impurities, such as Si, Ti, Zr, etc., can, under certain circumstances, result in lattice substitution, but most often result in distinct impurity phases, *i.e.,* silicates, titanates, etc. Impurity phases, along with unreacted MO and Fe_2O_3, act as magnetic diluents which result in lower than intrinsic values of σ_s. Since it is generally impossible to remove these diluents from solid-state-synthesized powder, their concentration must be minimized through control of reactant purity and completeness of reaction. In the case of the molten salt synthesis, nonmagnetic impurity phases are present as distinct particles. These particles can be separated from the $MFe_{12}O_{19}$ by magnetic sedimentation of the latter in a aqueous slurry. The impurity particles remain suspended in the fluid and are removed by decanting off the liquid phase. This behavior is one of the virtues of the molten salt synthesis.

Crystal Size, Morphology, and Lattice Perfection

Crystallite size and morphology are most conveniently determined using the scanning electron microscope. In general, special techniques are required to prepare specimens for examination such that the crystallite features are not obscured by the mounting medium. Procedures which accomplish this task have been developed, as shown in Figure 17. The crystallites shown were randomly selected from samples of molten-salt-

Figure 17. SEM photomicrograph of $BaFe_{12}O_{19}$ synthesized by molten salt technique. 15,000X.

synthesized $BaFe_{12}O_{19}$. The crystallite size and morphology are clearly shown in this example.

The crystallite size distribution can be measured by one of several conventional techniques. Coulter counters, under specific conditions, can be used with these materials. Other inorganic powder classification systems can also function with some degree of success. However, the problem is complicated by the fact that the magnetic powder tends to agglomerate under the influence of the relatively large attractive magnetic forces. Hence, the scanning electron microscope has been found more advantageous for the examination of these magnetic powders.

Crystal lattice perfection is determined by precision lattice parameter measurements coupled with X-ray structure factor analysis. An appropriate reference standard material such as a flux-grown single crystal, or the published literature[27], is required for comparison against the powder under consideration. These measurements will give a relative measure of lattice distortions produced by impurities, vacancies, dislocations, etc., which may then be correlated with other physical data on the sample.

An extremely convenient measure of the cumulative effect of all these parameters is H_{ci}, the measurement of which was considered above. It is possible to draw correlations between H_{ci} and any one of these parameters, if all other parameters are maintained constant.

Powder Consolidation

Although the previously discussed work was aimed at preparing a powder for use in an organic polymer matrix environment, the techniques and results are directly applicable to the sintered bodies which represent the major useful form for hard ferrite applications. Efforts have been made to understand the mechanism of densification of isotropic and anisotropic ferrite compacts, and the effects of this process on the magnetic properties of the finished product. The literature on this subject is voluminous, and no attempt will be made here to review it fully. Rather, a brief discussion will be made of the work done to sinter $SrFe_{12}O_{19}$ and $BaFe_{12}O_{19}$ prepared by the molten salt synthesis into dense bodies with desirable magnetic properties.

It is evident from the literature that $MFe_{12}O_{19}$ compacts will only yield sintered products of desirable magnetic properties in the presence of trace sintering additives such as SiO_2, Al_2O_3, Bi_2O_3, PbO, $SrSO_4$, etc. The work of Tokar[28] indicates that $PbFe_{12}O_{19}$ requires the presence of a liquid-phase sintering agent. It was demonstrated that a similar situation existed with respect to $BaFe_{12}O_{19}$ and $SrFe_{12}O_{19}$.

It was found that the pure molten-salt-synthesized $MFe_{12}O_{19}$ would not densify in the normal sintering temperature range of $1100-1300°C$[29]. However, it was found that additive phases chosen from the chemical system $MO-SiO_2$, $MO-SiO_2-Al_2O_3$, $MO-B_2O_3-SiO_2$ and $MO-B_2O_3-$

$SiO_2-Al_2O_3$ would accomplish this task. The additives were prepared separately to the desired homogeneous chemical composition. The additive constituted 1–6 wt % of the total sample mass. It was found that uniform dispersion of the additive was essential to its effectiveness. Unfortunately, milling was found to be the only effective additive dispersion method. This process resulted in significant physical damage to the ferrite crystallites, and therefore a much lower value of H_{ci}, than results from the molten salt synthesis. The sintering process did, however, remove some, but not all, of this damage during the course of densification. The sintered product based on the molten-salt-synthesized $MFe_{12}O_{19}$ was magnetically indistinguishable from that obtained with solid-state-reacted $MFe_{12}O_{19}$, indicating that the liquid-phase sintering additives are required for magnetic property development. Further evidence is provided by the fact that solid-state-reacted (commercial) $MFe_{12}O_{19}$ generally contains excess MO and has SiO_2, Al_2O_3, etc., present either as an intentional addition or as a reactant impurity (particularly in the iron sources). It is likely that the solid-state-reacted $MFe_{12}O_{19}$ has the appropriate liquid-phase sintering agent generated during the ferrite synthesis. In those cases where SiO_2, Al_2O_3, etc., are added during the ferrite milling, the initial step in the sintering process is the formation of this liquid-phase sintering agent.

Powder Processing of Advanced Permanent Magnets

Introduction

Remarkable advances in the properties attainable by permanent magnets have been shown recently by intermetallics derived from cobalt and a number of rare earths. While the remanences shown by these materials are often surpassed by alloys based on aluminum, nickel, cobalt, and iron, the coercive forces generated by these newer materials may be several times greater than those of their predecessors. It is the combination of high coercive force and high remanence which permits a magnetic energy density in these magnets which exceeds that in conventional materials by a factor varying between two and eight. In addition, the coercive force property, by itself, allows the new materials to be used in conditions of extreme demagnetization resulting from either adverse shape factors or reverse fields without change in the initial, pre-set magnetic properties. These properties promise many new device applications, and particularly those which employ variable-gap or field-reversing conditions.

Magnetic Processing Procedures

The generation of high coercive force in a magnetic material is generally difficult since many processes, dependent on local structures and chemistry, tend to diminish coercivity. In 1966 Becker[30] reported measurements of exceptionally high intrinsic coercive forces of single- and multi-crystal grains of Co_5Sm many microns in size. In the same size range, hard ferrites, magnetic materials based on alloys of Al, Ni, and Co, and other 5–1 cobalt–rare earth compounds characteristically showed coercive forces of at the most only a few kOe, whereas the Co_5Sm particles exhibited coercivities of several tens of kOe. This finding suggested that practical permanent magnets of exceptional properties should be possible if such high-coercivity particles or grains of Co_5Sm could be magnetically aligned and assembled in dense form.

While the 5–1 cobalt–rare earth compounds develop coercivity by mechanisms based on crystalline rather than shape anisotropy, it had generally been expected—by analogy with other materials—that the particle size necessary for the development of high coercivity in cobalt–rare earths by single magnetic-domain behavior would be well into the submicron size range. Recently, Livingston and McConnell[31] have shown that of the members of the 5–1 cobalt–rare earth family only Co_5Sm, and perhaps Co_5Gd, are unique in this respect. These authors show that due to a particularly high domain-wall energy the characteristic domain-wall spacing in Co_5Sm is larger, and consequently single domain behavior begins to be approximated at particle sizes well in excess of 1 μm.

From the above it is apparent that sintering of aligned intermetallic powders would be a preferred method to develop dense magnets of high energy, and it is this process which has been recently used to advantage. The processing requires strict atmosphere, temperature, and chemistry control but is amenable to large-scale manufacturing, as is currently being demonstrated. Specific details of processing procedures have been reported many times and are available elsewhere[32,33,34].

A major factor which, without significant change, may have limited the commercial applicability of cobalt–rare earth magnet materials results from the commercial availability of the necessary raw materials. Because of its limited industrial use, the metal samarium was available, a few years ago, only in small amounts and at prices exceeding \$100/lb. Oxide powder products, containing a large fraction of Sm_2O_3, were available, however, at substantially reduced cost and promised a further lowering in price should volume of use increase. This fact indicated that there could be major advantage to a magnet producer by processing intermetallic materials directly from oxides or from compounds easily converted to the powdered oxide form (*i.e.,* halides, carbonates, etc.) and it is this process that will be described.

Powder-Preparation Processes

The rare-earth oxides are the most stable oxides, except one, in nature, and this exception is CaO. In principle, one could therefore reduce rare-earth oxides by calcium metal through the proper choice of temperature and atmosphere, but, unfortunately, the affinity for alloying between calcium and rare earths is high and the desired reduction process to a pure metal cannot be achieved. In addition, calcium alloys significantly with Co and one would normally expect that cobalt–rare earth intermetallics would also alloy significantly with calcium metal. Cech[35] was the first to show that such is not the case, and on this finding was able to develop a process of powder manufacture ideally suited to the preparation of new magnet materials. As shown in the following equation,

$$5Co + \frac{1}{2} Sm_2O_3 + \frac{3}{2} CaH_2 \rightarrow Co_5Sm + \frac{3}{2} CaO + \frac{3}{2} H_2 \uparrow \quad (2)$$

the basis of the process is that the presence of cobalt in the reduction process allows the reaction to proceed, forming an intermetallic with very low calcium content (<0.05 w/o). The reaction is basically a solid-state reaction yielding directly the intermetallic and CaO in powder form. Separation of the intermetallic from the reaction product CaO may be

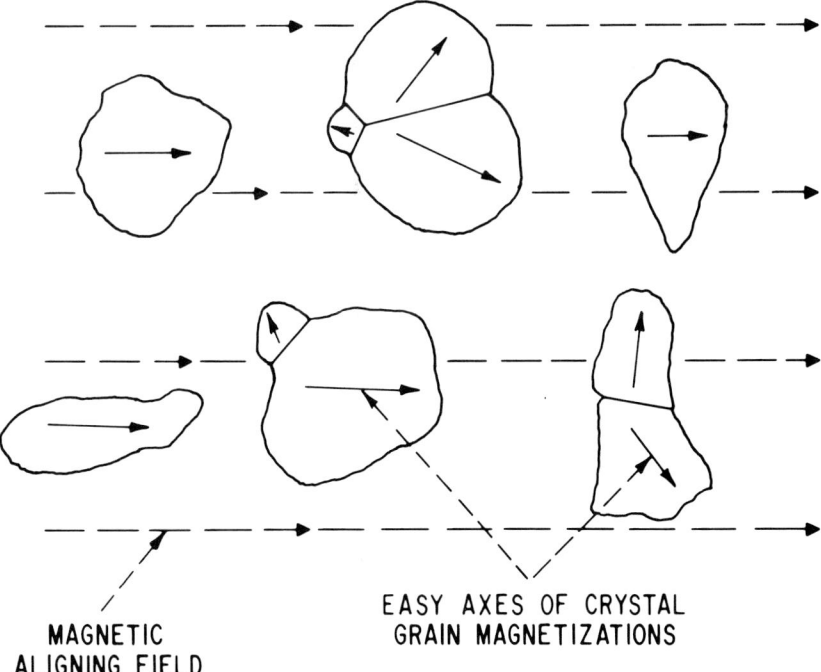

MAGNETIC
ALIGNING FIELD

EASY AXES OF CRYSTAL
GRAIN MAGNETIZATIONS

Figure 18. Schematic diagram showing the effect of aligning magnetic fields on the aligning orientations of polycrystalline and monodispersed particles of permanent-magnet materials.

easily achieved by many procedures, but most often those based on partial hydration, magnetic separation, and washing. In some cases, the use of dilute acids followed by alcohol rinse is utilized in the washing.

From the magnet producer's standpoint, it is important to note that, since the reaction is essentially solid-state, the resultant size characteristics of the intermetallic powder are largely determined by the initial particle-size distribution of the cobalt (or cobalt oxide) powder used in the reaction. The process is, in fact, capable of producing particles sizes ($\approx 10 \mu$) which are properly suited to the sintering of cobalt–rare earth magnets with high coercivity. There is, however, one important aspect in the production of solid-state reduction—diffusion powders for the preparation of high-energy magnets. In the reduction process, it is clear that some sintering and formation of grain boundaries between initially separate particles will occur. Such aggregates cannot achieve the degree of alignment in the powder-pressing field possible with individual, separate particles, and, consequently, the use of partially aggregated powders

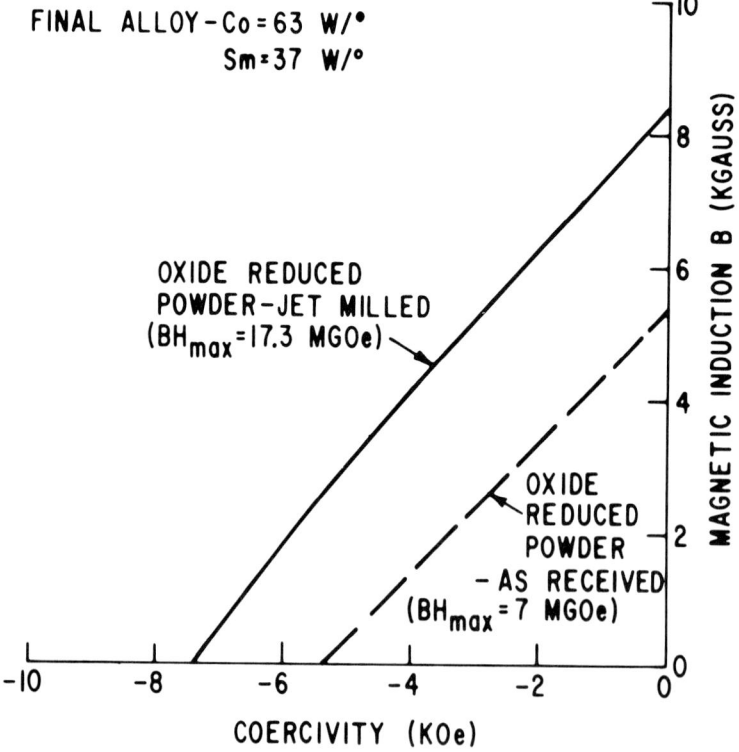

Figure 19. B-H curves for Co_5Sm powder-sintered magnets showing the effect of milling on properties of oxide-reduced powders before pressing and sintering.

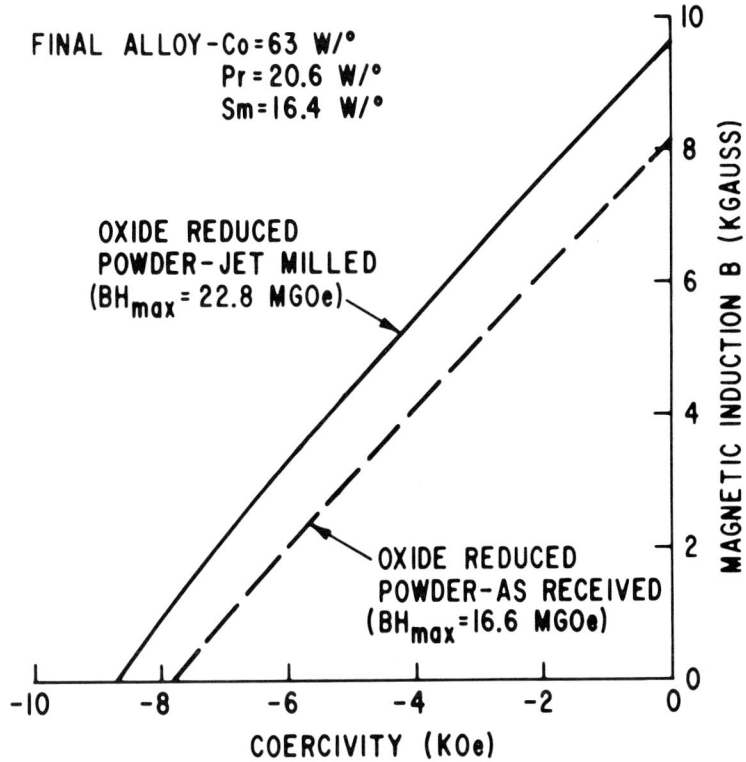

FINAL ALLOY-Co=63 W/°
Pr=20.6 W/°
Sm=16.4 W/°

OXIDE REDUCED
POWDER-JET MILLED
(BH_{max} = 22.8 MGOe)

OXIDE REDUCED
POWDER-AS RECEIVED
(BH_{max} = 16.6 MGOe)

COERCIVITY (KOe)

MAGNETIC INDUCTION B (KGAUSS)

Figure 20. B-H curves for a CoPrSm powder-sintered magnet showing the effect of milling on properties of oxide-reduced powders before pressing and sintering.

necessarily reduces the remanence and energy-product values otherwise possible. Figure 18 schematically shows the effect of aligning field on aggregated and single particles and Figures 19, 20, and 21 illustrate the effect of such powders on the properties of typical magnets prepared from cobalt and combinations of samarium mischmetal and praseodymium oxides. The materials in these latter figures are the initial, as-received powders from the reduction processes and the same powders jet milled to produce separate grains, which are single crystals.

Figure 22 shows the demagnetization curve for a particular $Co_5 \cdot$ (Sm,Pr) magnet material in which the base powder was produced by the oxide-reduction process. The powder was liquid-phase sintered by using a small amount of additive powder produced from a melt corresponding to 60% Sm and 40% Co. The magnetic properties of this magnet exceed those heretofore reported, and, additionally, this magnet demonstrated for the first time[36] that coercive forces in excess of 10 kOe are realizable in particle magnets.

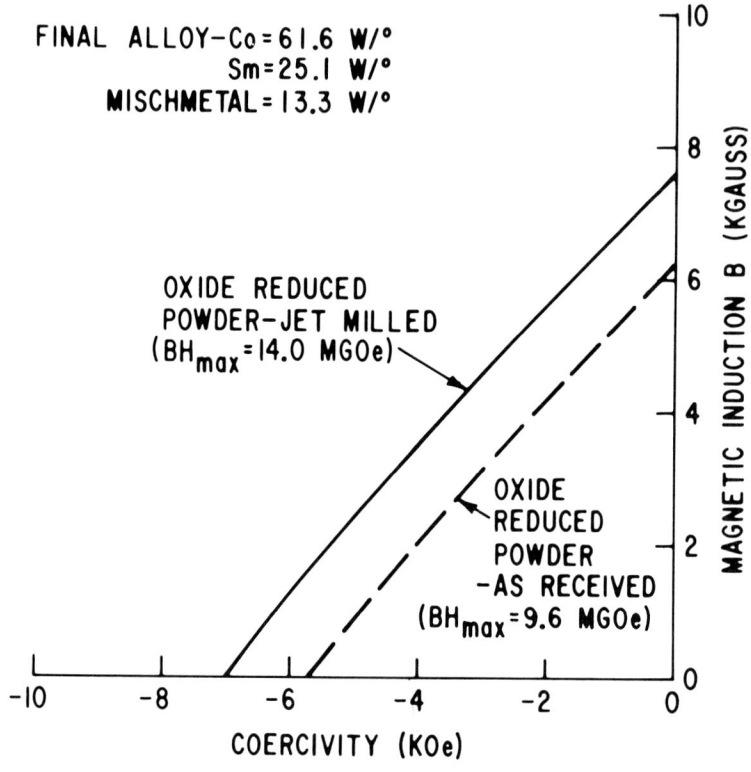

Figure 21. B-H curves for a Co-Sm-mischmetal powder-sintered magnet showing the effect of milling on properties of oxide-reduced powders before pressing and sintering.

Summary

Polycrystalline nonmetallic and intermetallic materials can be developed with extraordinary properties (optical, mechanical, magnetic) by powder-sintering processes. Careful preparation and adjustment of powders, synthesized in the laboratory, is a key factor in the development of controlled microstructure, reproducibility, and properties. Fine powders ($\lesssim 1$ μm) are desirable because they give rise to high densification rates and chemical reactivity, leading to sintered materials with low porosity and high chemical homogeneity. Impurities are shown to have a deleterious effect on the microstructure and resulting physical properties of the optical, structural, and magnetic materials selected for discussion. Such impurities may promote discontinuous grain growth, develop second-phase, act as absorption or scattering centers, and as magnetic diluents. Intentional additives, such as ThO_2 in Y_2O_3 and $MO-SiO_2$ in $MFe_{12}O_{19}$, are valuable as sintering aids and to a large extent are empirically discovered. A care-

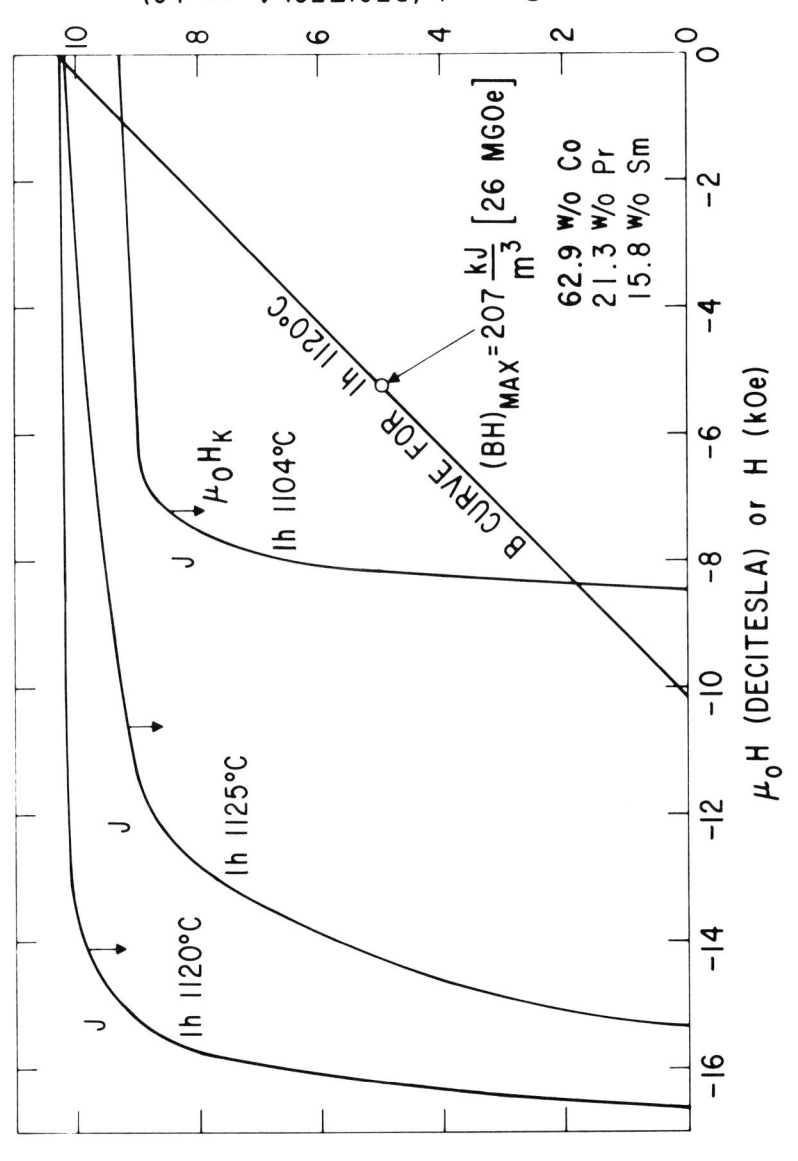

Figure 22. Induction and magnetization curves for a CoPrSm powder-sintered magnet showing an induction coercivity in excess of 10 kOe.

171

fully selected milling treatment of even fine powders helps to reduce the average particle size for favorable sintering and magnetic results and to chemically homogenize the starting reactants. On the other hand, the pick-up of impurities and the introduction of lattice imperfections in individual particles during milling may lead to inferior properties in the final material. In this case, further processing steps are usually required to produce powders with the desired characteristics. For the four advanced types of materials discussed herein, it is shown that an understanding of the role of compositional and physical changes, and particularly the use of additives, is of invaluable assistance in achieving solutions to new sintering problems.

Acknowledgments

The authors gratefully express their appreciation to L. Valentine and D. L. Martin, who collaborated in developing data on the magnetic materials reported herein; W. J. Dondalski, who contributed to the development of the SiC powder synthesis process; and C. O'Clair for preparing powders used in the investigation on optical materials. The research work on optical materials was partially supported by the Advanced Research Projects Agency of the Department of Defense and was monitored by the Office of Naval Research under contract N00014-70-C-0360.

References

1. Van De Hulst, H.C., *Light Scattering by Small Particles,* New York: John Wiley & Sons, 1957.

2. Eastman Kodak Company, Rochester, New York, Publication No. U-72, 1971.

3. Haertling, G.H., "Improved Hot-Pressed Electrooptic Ceramics in the (Pb,La) (Zr,Ti)O$_3$ System," *J. Amer. Ceram. Soc.,* 54 (1970), 303–309.

4. Subarrao, E.C., Sutter, P.H. and Hrizo, J., "Defect Structure and Electrical Conductivity of ThO$_2$-Y$_2$O$_3$ Solid Solutions, *J. Amer. Ceram. Soc.,* 48 (1965), 443–46.

5. Jorgensen, P.J. and Anderson, R.C., "Grain-Boundary Segregation and Final-Stage Sintering of Y$_2$O$_3$," *J. Amer. Ceram. Soc.,* 50 (1967), 553–58.

6. Lucke, K. and Detert, K., "A Quantitative Theory of Grain-Boundary Motion and Recrystallization in Metals in the Presence of Impurities," *Acta Met.,* 5 (1957), 628–37.

7. Cahn, J.W., "The Impurity-Drag Effect in Grain Boundary Motion," *Acta Met.,* 10 (1962), 789–98.

8. Greskovich, C. and Woods, K.N., "Fabrication of Transparent ThO$_2$-Doped Y$_2$O$_3$," *Bull. Amer. Ceram. Soc.,* 52 (1973), 473–78.

9. Greskovich, C.D., "Oxide Ceramic Laser." Annual Technical Report, 1 June 1971–

31 May 1972. General Electric Corporate Research and Development, Schenectady, N.Y. Office of Naval Research Contract Report No. SRD-72-105, July 1972 (AD 746 521).

10. Alexander, B.H. and Balluffi, R.W., "Mechanism of Sintering of Copper," *Acta Met.*, 5 (1957), 666–77.

11. Burke, J.E., "Role of Grain Boundaries in Sintering," *J. Amer. Ceram. Soc.*, 40 (1957), 80–85.

12. Rosolowski, J.H. and Greskovich, C., "Analysis of Pore Shrinkage by Volume Diffusion During Final Stage Sintering," *J. Appl. Phys.*, 44 (1973), 1441–50.

13. Greskovich, C. and Chernoch, J.P., "Polycrystalline Ceramic Lasers," *J. Appl. Phys.*, 44 (1973), 4599–4606.

14. Greskovich, C., "Oxide Ceramic Laser," Final Technical Report, 1 June 1970–31 May 1973. General Electric Corporate Research and Development, Schenectady, N.Y., Office of Naval Research Contract Report No. SRD-73-108, July 1973 (AD 765 345/4).

15. Nadeau, J.S., "Very High Pressure Hot Pressing of Silicon Carbide," *Bull. Amer. Ceram. Soc.*, 52 (1973), 170–74.

16. Prochazka, S., "Investigation of Ceramics for High Temperature Turbine Vanes," Final Report, 1 April 1971–31 March 1972. General Electric Corporate Research and Development, Schenectady, N.Y., Office of Naval Research Contract Report No. SRD-72-035, March 1972 (AD 742 857).

17. Alliegro, R.A., Coffin, L.B. and Tinklepaugh, J., "Pressure-Sintered Silicon Carbide," *J. Amer. Ceram. Soc.*, 39 (1956), 386–89.

18. Went, J.J., Rathenau, G.W., Gorter, E.W. and van Oosterhout, G.W., "Ferroxdure, A Class of New Permanent Magnet Materials," *Philips Tech. Rev.*, 13 (1952), 194–208.

19. Cochardt, A., "Recent Ferrite Magnet Developments," *J. Appl. Phys.*, 37 (1966), 1112–15.

20. Tenzer, R.K., "Influence of Particle Size on the Coercive Force of Barium Ferrite Powders," *J. Appl. Phys.*, 34 (1963), 1267–68.

21. Heimke, G., "The Saturation Magnetization of Barium Ferrite Powders with Different Grinding and Annealing Processes," *Z. Angew. Phys.*, 17 (1964), 181–83.

22. Fahlenbrach, H., "Koerzitivfeldstarke von Bariumferritpulver nach Verschiedenen Warmbehandlungen," *Tech. Mitt. Krupp Forschungsber.*, 23 (1965), 26–35.

23. German Patent No. 1,925,056, November 27, 1969, *Increasing the Coercivity of Ferrite Powder*, J.J. Becker to the General Electric Company.

24. Arendt, R.H., "The Molten Salt Synthesis of Single Magnetic Domain $BaFe_{12}O_{19}$ and $SrFe_{12}O_{19}$ Crystals," *J. Solid State Chem.*, 8 (1973), 339–47.

25. Shirk, B.T. and Buessem, W.R., "Temperature Dependence of M_s and K_1 of $BaFe_{12}O_{19}$ and $SrFe_{12}O_{19}$ Single Crystals," *J. Appl. Phys.*, 40 (1969), 1194–96.

26. Stoner, E.C. and Wohlfarth, E.P., "A Mechanism of Magnetic Hysteresis in Heterogeneous Alloys," *Phil. Trans. Roy. Soc., London, Ser. A*, A240 (1948), 599–644.

27. Braun, P.B., "The Crystal Structures of a New Group of Ferromagnetic Compounds," *Philips Res. Rep.*, 12 (1957), 491–548.

28. Tokar, M., "Microstructure and Magnetic Properties of Lead Ferrite," *J. Amer. Ceram. Soc.*, 52 (1969), 302–306.

29. Arendt, R.H., "Liquid-Phase Sintering of Magnetically Isotropic and Anisotropic Compacts of $BaFe_{12}O_{19}$ and $SrFe_{12}O_{19}$," *J. Appl. Phys.*, 44 (1973), 3300–05.

30. Becker, J.J., "Research to Investigate the Microstructure of the Internal Magnetic Field in Selected Magnetic Materials," General Electric Corporate Research and

Development, Schenectady, N.Y., Air Force Materials Laboratory Contract Reports, Quarterly Report No. 8, October 1966 and Final Report No. AFML-TR-67-28, March 1967 (AD 801 541).

31. Livingstone, J.D. and McConnell, M.D., "Domain-Wall Energy in Cobalt-Rare Earth Compounds," *J. Appl. Phys.*, 43 (1972), 4756–62.

32. Cech, R.E., "Sintering of Die-Pressed Co_5Sm Magnets," *J. Appl. Phys.*, 41 (1970), 5247–49.

33. Das, D.K., "Temperature on Magnetic Properties of Samarium-Cobalt Magnets," *IEEE Trans. Magn.*, MAG-7 (1971), 432–35.

34. Benz, M.G. and Martin, D.L., "Cobalt-Samarium Permanent Magnets Prepared by Liquid Phase Sintering," *Appl. Phys. Letters*, 17 (1970), 176–77, and "Mechanism of Sintering in Cobalt-Rare Earth Permanent-Magnet Alloys," *J. Appl. Phys.*, 43 (1972), 3165–70.

35. Cech, R.E., "Cobalt-Rare Earth Intermetallic Compounds Produced by Calcium Hydride Reduction of Oxides," *J. Metals*, 26 (1974), 32–35.

36. Charles, R.J., Martin, D.L., Valentine, L. and Cech, R.E., "A 10,000 Oe B-Coercive Force Magnet," in *Magnetism and Magnetic Materials*, AIP Conference Proceedings No. 5, C.D. Graham and J.J. Rhyne, eds., New York: American Institute of Physics (1972), 1072–76.

JAMES W. McCAULEY
Army Materials and Mechanics Research Center
Chapter 7 Watertown, Massachusetts

Structural and Chemical Characterization of Processed Crystalline Ceramic Materials

ABSTRACT

Research on new materials demands systematic characterization, not only
to optimize fabrication parameters and to insure future quality control,
but also to enable optimum engineering properties to be achieved. Any
processed ceramic material should be uniquely defined by a necessary and
sufficient set of parameters including composition, grain size, shape,
orientation, and packing. Quantitative relationships derived among these
parameters, engineering properties, and utilization functions can be used
to control and optimize their properties and use.

Selection of the optimum characterization procedures requires a
careful preliminary analysis of the material and what is needed. Reviews
and examples are presented for visible light observation (macroscopic and
microscopic), X-ray analysis (single crystal and powder diffraction, energy
dispersive analysis, fluorescence, topography) and electron interaction in
solids (X-ray fluorescence, SEM, TEM, ESCA, and cathodoluminescence).

Introduction

Research and development of new materials demands systematic characterization, not only to optimize fabrication parameters and to insure
future quality control but also to enable the achievement of optimum engineering properties. This is especially true with brittle materials since
much more restricted tolerance on properties is necessary to insure failsafe
operation in structural applications. With primarily ductile materials,
much larger tolerances can be accepted by design engineers.

The area of characterization has been treated by many people, *e.g.*
[1,2], and was the subject of a comprehensive report by the Committee on
Characterization of Materials of the Materials Advisory Board of the National Academy of Sciences. Their report [3] establishes that characterization should be treated as a critical part of any materials program and not,

as so often is the case, as a "bootleg" operation. This report is clearly a definitive treatment of characterization, containing many references, and should be consulted much more than it has been for characterization problems. Many new methods and equipment have come along since the publication of the report, but it still can serve as a starting point in many areas.

Characterization is becoming even more critical. Modern technology is demanding that materials scientists push the absolute limit in the obtainable properties of materials. The energy crisis before us further demands that machinery be lighter and stronger so that more useful work can be performed for less energy input. However, the route to these exotic materials is not a simple one. Minute changes in composition or structure of a material are sufficient to decrease required engineering properties by factors of 2 or 3 and sometimes by orders of magnitude. Procurement groups must have precise characterization data available or obtainable so that meaningful specifications can be written for required material.

Figure 1 diagramatically depicts a possible scheme for effective materials research and development. Characterization is indicated in three areas on the chart, illustrating its overall importance. Not only is characterization information critical for the initial conceptualization stages (Phase I), but it is also required during the preliminary stages of constituent component preparation (Phase II) and finally to optimize the critical processing parameters (Phase III). This paper is concerned with Phase III characterization.

An intimate relationship must exist between characterization and processing so that processing variables can be rapidly modified to optimize fabrication. Further, careful characterization work in this part of the scheme will insure that the material can be reproduced to yield the required engineering properties. Another very important linkage is the optimization loop tying together processing, characterization, engineering property evaluation and testing. This is an iterative sequence of steps, especially between engineering property evaluation and testing. If the engineering properties have been unambiguously linked to the composition and microstructure (meaning atomic level, micro- or macrostructure), then these material parameters can be modified to opitmize the materials for particular uses. As pointed out by the Committee report (and as will be discussed later), it is more correct and practical to relate engineering properties to characterization parameters rather than processing variables.

Finally, materials scientists should be aware of flow charts of this nature so that the relationship between the function they are carrying out and all of the other pertinent stages is fully appreciated. It would be highly desirable for a materials scientist to conduct work in every phase of this sequence during a learning period so that the importance and interrelationship between the various stages will be fully appreciated.

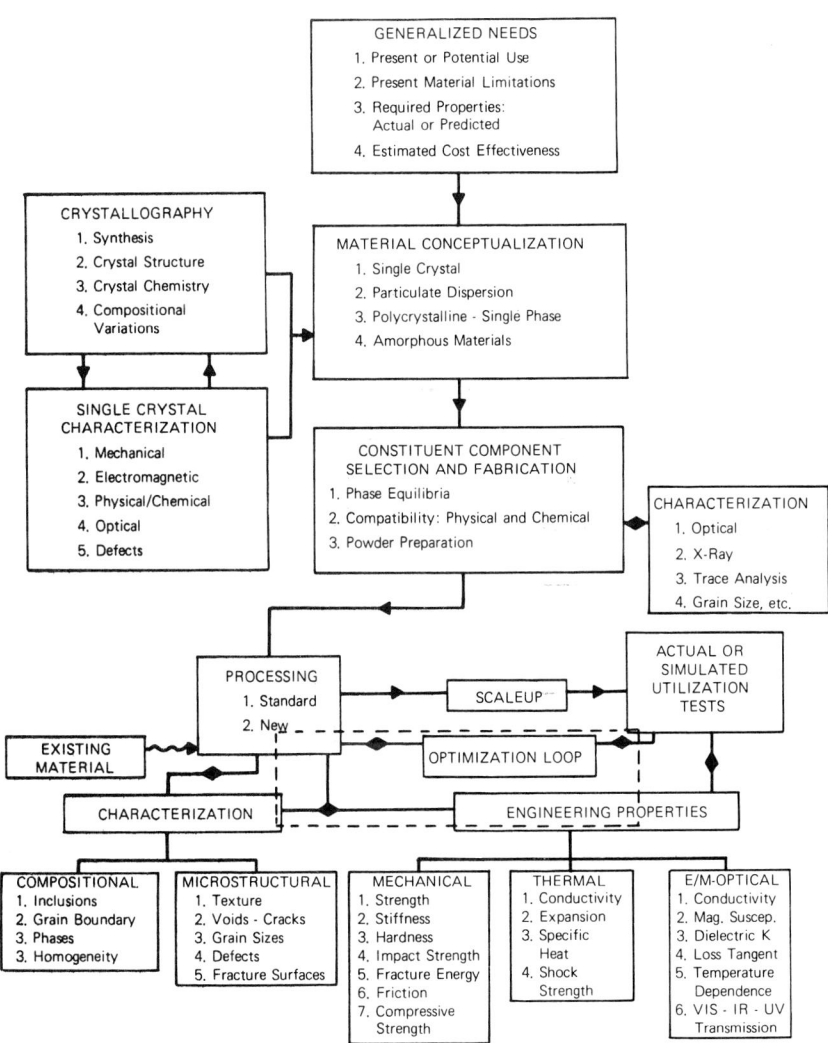

Figure 1. Flow chart for materials research and development.

Guidelines for Ceramic Characterization

Material Classification

A new solid-state material may be any one of several different types.

A. Single-crystal
B. Noncrystalline solid
   1. Monolithic material (glass)
   2. Loose or Coherent Aggregate (gels, hot-pressed fused silica)
C. Single-phase polycrystalline aggregate
D. Multiphase mixture—including all crystalline or partly crystalline and partly noncrystalline
   1. Equilibrium mixture (eutectic)
   2. Nonequilibrium mixture—composite

A different set of characterization parameters is required for each type of material. Particle size, for example, would have no meaning in a single crystal, but is extremely important in a polycrystalline ceramic. Thus, the most important aspect of characterization is to have a general plan as to what characterization data are required for a new material. This chapter will be concerned with processed single-crystal, and single and multiphase polycrystalline ceramic materials.

Definition of Population

According to Griffiths [4] (a sedimentary petrologist) any population including rocks or ceramic materials can be uniquely described by the following necessary and sufficient set of properties:

$$P = f(m, s, sh, o, p) \qquad (1)$$

where

   P = necessary and sufficient set of properties for unique description
   m = composition
   s = size of constituents
   sh = shape of constituents
   o = orientation—shape position of constituents
   p = packing—relative position of each constituent

This general formula will satisfy for any population. Considering a single crystal as a population described by the above definition, it can be fully and unambiguously represented by its composition, the size and shape of the atoms, and the orientation of the atoms with respect to each other and how far apart they all are. A chemical analysis and crystal-structure determination will provide this information for the fundamental unit cell of the crystal. Actually, most single crystals consist of mosaic blocks with various forms of defects. Hence, the definition of the single crystal must include the "microstructure" of mosaic blocks and defects.

For ceramic materials, it is convenient to have an equation that is a function of more familiar parameters:

$$PC = f(c, p_t, s, sh, Op, Or, d_t, d_G, g) \qquad (2)$$

where

c = composition of all constituents including grain boundaries
P_t = directional (tensor) properties of individual constituents (P_e = elastic constants, etc.)
s = grain sizes
sh = grain shapes
Op = orientation of grains with respect to processing medium
Or = orientation of grains with respect to each other
d_t = intergranular density
d_G = granular density
g = flaws

This equation is not meant to be an all-inclusive (necessary and sufficient) definition of a processed ceramic (PC), but a guideline to be improved on.

The directional properties (P_t) of the individual constituents [5] are oftentimes neglected in an analysis of the bulk properties of processed materials. These include the optical properties, elastic properties, and thermal expansion properties, as well as many others. Alumina is optically birefringent, but would be almost as transparent as window glass if all the crystal grains were perfectly aligned. Another example is thermal expansion anisotropy—which can be quite appreciable, as in alumina [6]. Figure 2(c) shows a gradient-furnace-solidified sapphire [7] disk which cracked on cooling, apparently because of the nucleation and growth of grains with large crystallographic misorientations. One of the grains in the disk exhibited a rumpled appearance, seemingly originating at the high-angle grain boundary. This is attributed to shear forces between the two grains resulting from an appreciable mismatch in thermal expansion (or contraction) between the grains [8]. Thermally etched slip bands also resulting from this interaction can be observed in 2(d). This disk was photographed in crossed polarized light in a modified polariscope.

Use of Defining Equation

Actually, Equation (2) can be much simplified since all of the defining parameters are related to either composition or microstructure (microstructure is used to include all structural features, including atomic,

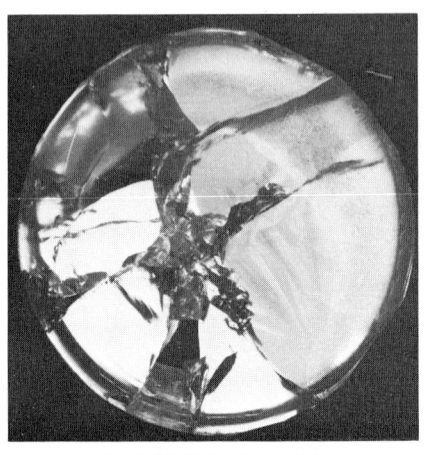

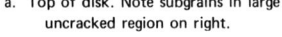

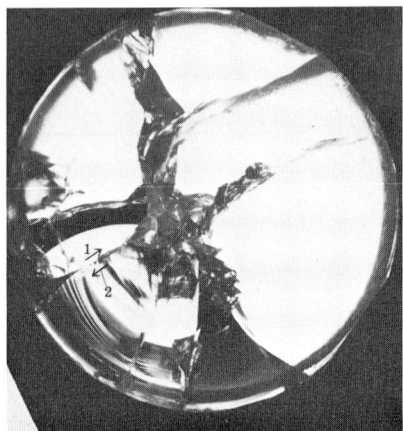

a. Top of disk. Note subgrains in large uncracked region on right.

b. Top of disk.

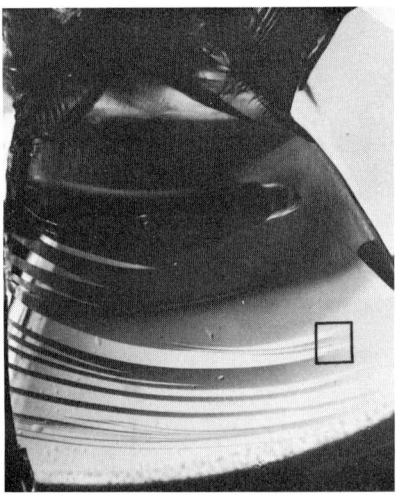

d. Magnified view of inclosed area in (c) showing thermally etched slip bands.

c. Magnified view of grain showing rumpling and thermally etched slip bands.

Figure 2.   Cracked sapphire disk in crossed polarized light.

micro- or macrostructures):

$$PC = f(c, M) \tag{3}$$

where

$$c = \text{composition}$$
$$M = \text{microstructure.}$$

The processing variables including thermal history, pressure (atmosphere) history, time, and starting materials are *not* fundamental parameters and thus should *not* be used in a defining equation. In this regard then, any property of a processed ceramic may be *quantitatively* represented by the defining equation. Thus,

$$\text{Strength} = f(c, M) \tag{4}$$

and multiple regression can be used to isolate source of variation in strength:

$$\textit{Strength} = a_1 X_1 + a_2 X_2 + a_n X_n + a_n X_n \tag{5}$$

where

a_1 = coefficients to be obtained in multiple regression

X_1 = fundamental parameters $(c, p_t, s, \text{etc.})$, which are mathematical functions (theoretical or experimental) of strength.

Since the fundamental parameters are controlled by the processing variables, the variation in strength may be quantitatively linked to processing *through* the fundamental parameters.

Some authors have opted to use engineering properties as part of the characterization scheme without specifying, completely, the composition and microstructure. Engineering properties can be successfully used as part of the characterization scheme only if composition and microstructure data are used also, so that the properties can be related to definable characteristics of the material. Actually, careful engineering-property evaluation sometimes reveals subtle variants that have not been observed in the initial characterization. For example, McCauley and Acquaviva [9] collected a full set of engineering-property data on well-characterized Mica–Al_2O_3 composites and found clear indications from thermal conductivity measurements that the Mica–85% Al_2O_3 composites must have an undetected characterization variable, *i.e.,* glass, grain-boundary phase, etc. All other properties closely followed known composite-material models.

In the general case, solution of the defining equations for multiple regression coefficients, when a random set of characterization parameters are available for a material, should allow one to optimize fabrication and predict properties. Chay, Palmour, and Kriegel[10] used this approach on hot-pressed spinel ($MgAl_2O_4$) to relate density and grain size to the mechanical properties.

In special cases, one can control the composition and/or microstructure to establish direct functional relationships and sources of variation. Gupta [11] has unambiguously shown that the change in strength after critical thermal shock is related to the grain size of Al_2O_3. In this case, the composition was kept constant while the microstructure was varied. However, both the composition and microstructure can be varied simultaneously to establish certain functional relationships. This has been clearly demonstrated by McCauley and Acquaviva[9], who showed linear relationships between mica content in an Al_2O_3 matrix and elastic modulus, but a definite nonlinear relationship between mica content and strength and fracture energy. These data can be seen in Figure 3. Obviously, microstructure has a pronounced effect on some properties like strength but not on other properties like elastic modulus. Working on isolated compositions would not have revealed this relationship. Hence, effort can now be undertaken to understand the mechanism so that the effect can be optimized.

Moreover, by systematic control of composition and microstructure, various utilization functions can be solved to predict the optimum engineering properties for specific applications. A utilization function is a mathematical combination of critical properties which can quantitatively predict how well a material will perform in a certain use or environment. An example of these utilization functions are the thermal shock resistance parameters that have been compiled by Hasselman[12]. Other functions could also be established for bearing applications, ballistic armor, turbine engine components, etc. For example, Seaton and McCauley[13] have demonstrated that by the systematic control of strength, elastic modulus, and thermal expansion, the direct functional relationship between the critical thermal shock temperature (T_c) and the thermal shock resistance parameter $(R = \sigma(1 - \nu)/\alpha E)$ can be obtained. Using this relationship, the critical thermal shock temperature can be predicted for most ceramic materials. Materials scientists can utilize these data, then, to fabricate materials with optimum properties.

The technique of establishing fundamental parameters and using multiple regression is not limited to microscopic observations. McCauley and Newnham[14] used multiple regression to predict crystal structural variations in micas from chemical composition data. Actual use of the calculated equations has proved to be quite useful in several cases, and allows a rough structure to be calculated without having to go through a tedious crystal structure refinement. Two structural parameters, Δ and α, represent the deviation of real mica structures from an ideal model. Both of these can now be calculated from the following equations:

$$\alpha(°) = 218 \, (b_t/b_0) - 1.5 \, (\text{I.F.S.}) - 221.5$$
$$\Delta(\overset{\circ}{A}) = 0.47 \, \alpha° \tag{6}$$

The coefficients 218, -1.5, and -22.15 were derived by multiple regression calculations for all known mica structures. The parameters, b_t/b_0 and I.F.S., are respectively the tetrahedral–octahedral layer misfit parameter and interlayer cation field strength; they can be routinely calculated from the chemical composition.

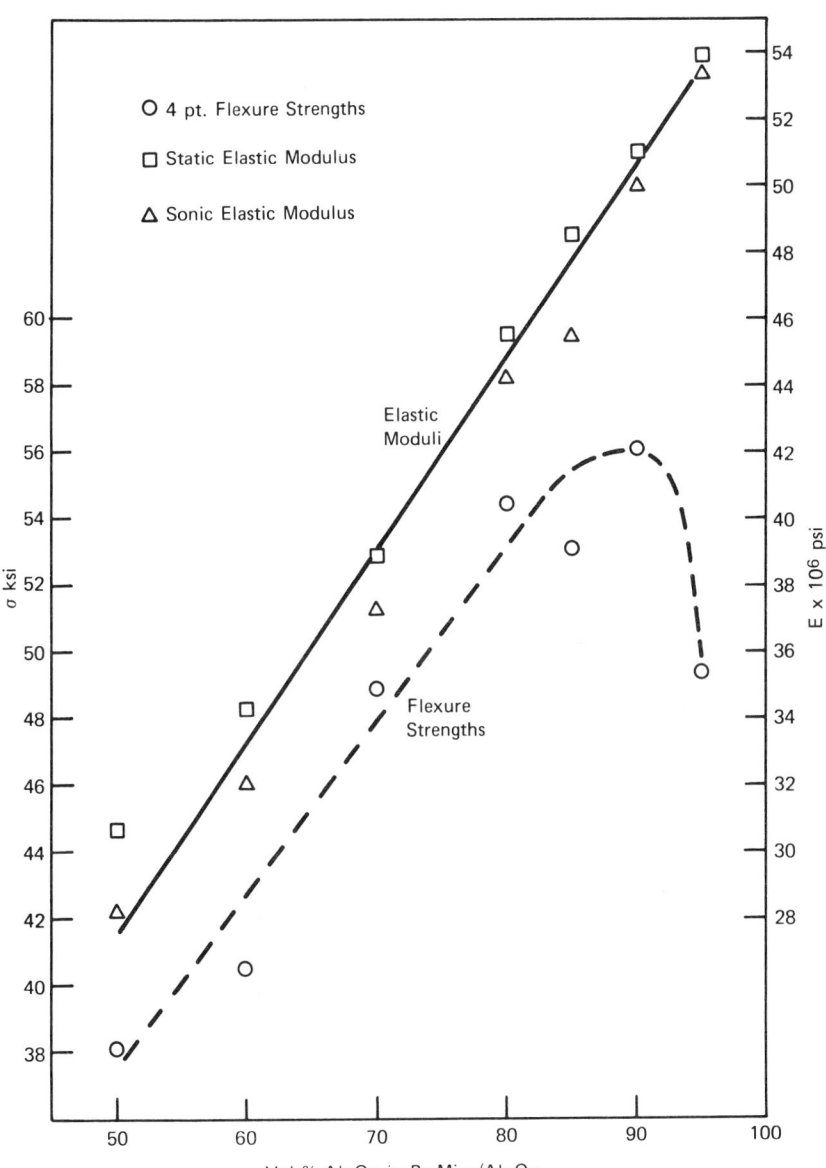

Figure 3. Four-point flexure strengths and elastic moduli of Ba-mica (30 μm)/ Al_2O_3 composites.

Specific Methods of Characterization

Initial Steps

Before attempting to characterize a material, there are several questions that should be evaluated. Some of these might be as follows:

1. What type of equipment is available?
2. How much sample is available?
3. Can testing be destructive or must it be nondestructive?
4. Are macro and/or micro gradients and/or inhomogeneities possible?
5. Are bulk, various components, or surface composition required?
6. Are trace, minor or major compositions required?
7. Is there a possibility of a noncrystalline phase being present?
8. What is known about the crystallography and physical properties of the individual components and is it necessary to determine them?

Often an investigator working in Phase I characterization (*i.e.,* crystallography and synthesis) carries out both the synthesis and characterization. But in Phase II and Phase III characterization, more often than not, the powder preparation and processing is carried out independent of the characterization step. This can lead to enormous problems if one is not careful. The role of characterization of powders has been discussed by Arendt *et al.,* in another part of this book. Suffice it to say that all powders obtained from sources other than one's own should be well characterized preliminary to utilization. Rapid examination by optical microscopy, X-ray powder diffraction, and emission spectroscopy may be adequate for some uses, whereas much more detailed procedures may be needed for others.

It is desirable to have one individual or a close working team carry out both processing and characterization. In many cases, however, this is not the case and processing people must rely on outside sources or independent characterization groups to evaluate the processed material. This may be beneficial because of the objectivity involved, but there are also pitfalls. In the latter situation, close interfacing between both groups is critical. The characterization group should be made aware of the potential importance of the material and what kind of data is needed and how fast. For example, a detailed crystal-structure determination should not be carried out on a material of minimal importance. During the early stages of a processing program, rapid characterization and very close interfacing is important, whereas in the late stages much more detailed work and less contact may be required.

There are numerous so-called "quick-and-dirty" measurements that suffice for determining preliminary data to establish how sophisticated the characterization should be and form the guidelines for future work.

Further, these types of methods may be adequate in the early stages of a program when only rough ball-park numbers are required to establish correct fabrication procedures.

During the initial stages of characterization, many routine procedures are expediently overlooked. The surface appearance should be noted, and especially its relationship to the processing environment. A good idea is to take a photograph of the as-processed body, noting carefully its relationship to the processing environment. Certain inhomogeneities noted in the characterization may eventually be linked to an unsymmetrical processing environment. Color and transparency of a sample may also provide valuable clues to further testing and processing modification. Finally, and perhaps of most importance, is a good density measurement. This can be carried out quickly and inexpensively and yields invaluable information in many cases, especially single-phase or multiphase mixtures. Precise density measurements on single crystals can also provide information in regard to defects and atomic substitutions.

In the succeeding sections, an attempt will be made to review various chemical and structural characterization procedures for processed crystalline ceramic materials. The review is not meant to be all-inclusive, but merely as an attempt to provide insights into certain procedures and stimuli for their use. Further, the material is presented more from a user's point of view, rather than that of an expert in the various fields.

Visible Light Observation

Visible light can be utilized in many ways. As pointed out previously, one can merely look at a sample carefully, noting any obvious peculiarities. For large transparent samples, it has been found very profitable to use a modified polariscope on both isotropic and anisotropic materials[8]. Figure 4(a) shows a schematic illustration of this technique; it is simply a macroscopic petrographic microscope. Variously oriented grains of Al_2O_3 in a polished sapphire disk are clearly discernible in crossed polarized light, as seen in Figure 4(c), but absent when viewed in ordinary light in 4(b). If properly used and related to the processing environment and history, these data on the spatial distribution of variously oriented Al_2O_3 grains are valuable for modification of the processing parameters and as a quality control check on sapphire for precise optical applications. It is also a quick way to monitor the thickness uniformity of a polished disk; Figure 4(d) illustrates the fringes formed in the subgrain-free area of the disk due to its slight wedge shape. This polariscopic technique was also used to obtain the photograph of the cracked sapphire disk shown in Figure 2.

This method can also be beneficially employed on isotropic materials to reveal areas of residual strain and noncubic phases. Figures 5(a), (b), (c) show various polariscopic views of a spinel ingot[15] fabricated by a

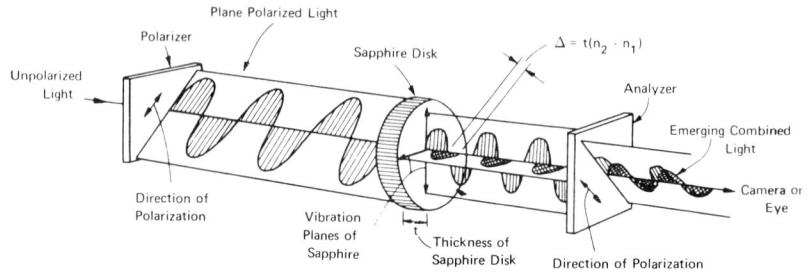

a. Schematic drawing of interaction and analysis of
polarized light in sapphire disk. (Adapted from Crites, et al.,[4])

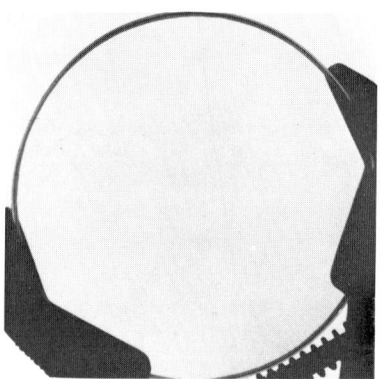

b. Polished; plane polarized light

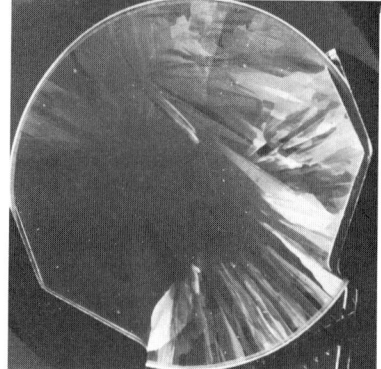

c. Polished; crossed polarized light

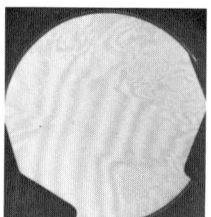

d. Isochromatic fringes
in polished disk

Figure 4.   Various views of sapphire disk before and after machining.

gradient furnace technique. The cross-sectional slice shown in Figure 5(c)
demonstrates that the {110}-type strain patterns seen in 5(b) are restricted
to the top corners, edges, and in the region of the seed (bottom center of
slice). The interior of the ingot is relatively strain-free and shows no non-
isotropic phases. Obviously, when selecting a small sample for detailed
study one must be aware of features such as these so that representative
samples are obtained.

a. Plane light photograph

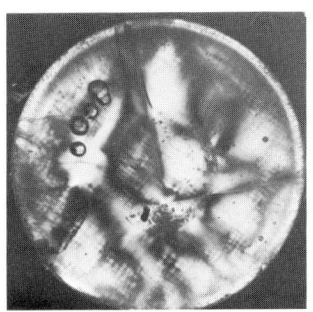

b. Same ingot viewed in
crossed polarized light

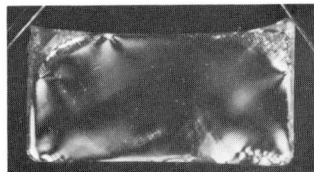

c. Slice through spinel ingot viewed
in crossed polarized light.

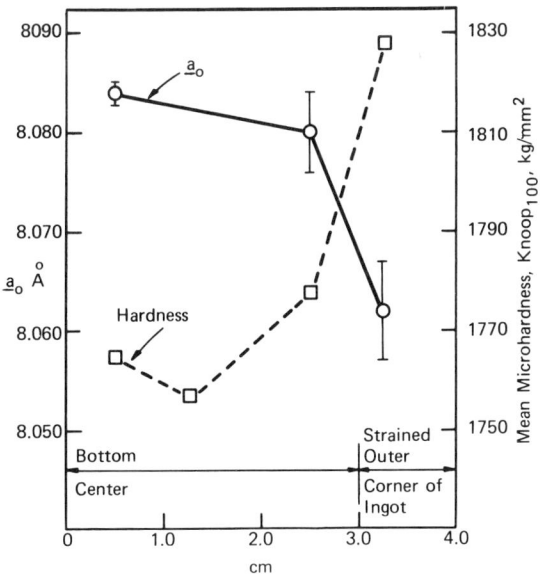

d. Spatial distribution of hardness and lattice parameter in spinel ingot

Figure 5.   Spinel $(MgAl_2O_4)$ ingot.

The optical microscope should be the most widely used instrument for characterization of the microstructure of materials, although recently it seems that the scanning electron microscope is being utilized even more. Much information can be acquired with both reflected and transmitted visible light microscopy. For opaque and extremely fine-grained transparent materials, the reflected light microscope provides the best data. With proper sample preparation procedures, magnifications up to 1,000X can be routinely achieved without distortion. Microstructures—including grain shapes, sizes, and interrelationships—may be readily obtained, as well as the substructure in individual grains; for moderately and highly reflective materials, use of a polarizer is often very advantageous for observing substructure and intergrowths. Another important use of reflected light microscopy is the observation of etch pits—mineralogists[16] have been using these for years to determine the symmetry of crystalline materials. Techniques are even available for chemical analysis under the microscope[17]. Figure 6 is a reflected light photomicrograph of a mica–Al_2O_3 composite illustrating the thermal crack energy absorption (deflection, generation of new cracks) by the alligned mica flakes. This sample was relief polished on a lead lap using a Cr_2O_3 grinding media.

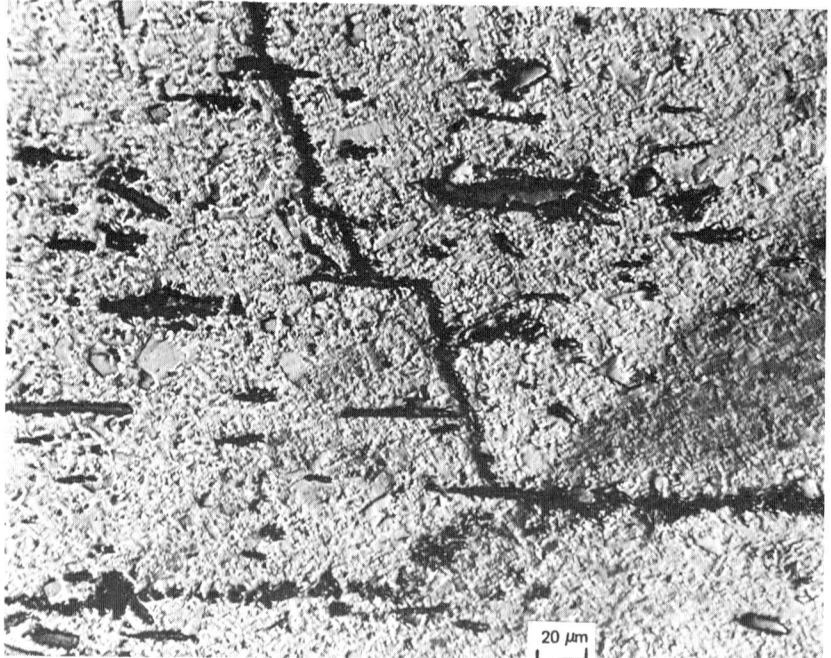

Figure 6. Photomicrographs of thermal cracks in 10 volume % Ba-mica (30 μm)/ Al_2O_3 resulting from quench from 700° C into water; relief polish.

Use of transmitted light in a polarizing (petrographic) microscope greatly expands the microscopist's capabilities[18,19]. Determination of the optical properties of single crystals, including refractive indices and crystal systems, can provide much information on crystal structure. Structural disorder (stacking faults) has also been determined by optical techniques[20]. Crystals can be rapidly oriented and aligned for single-crystal X-ray investigations or property measurements. Further, it affords a simple way of quickly checking for substructure (twinning and/or domains) in single crystals before pursuing an elaborate structure determination. Figure 7(a) and (b) illustrates a silica-gel-grown[21] single crystal of $Nd_2(CO_3)_3 \cdot 8H_2O$ in two different orientations—one inclined to extinction (a), and one at "extinction" (b). This crystal never went to extinction in total, but exhibited still-undetermined domain structures seen in (b). Single-crystal X-ray precession photographs taken perpendicular to (c) and parallel to (d) the plate in (a) and (b) did not indicate any suggestion of this domain structure.

McCauley and Roy[22,23] have studied the crystal growth and doping of various slightly soluble materials by the silica-gel technique. In one investigation, the amount of Nd^{3+} incorporated into $SrSO_4$ single crystals was monitored by refractive index measurements and emission spectroscopy.

$\gamma(Na_D$ light)	Wt % Nd_2O_3
1.631 (0.001)	0.1
1.604 (0.001)	0.7
1.602 (0.001)	0.9

As can be seen, the change in refractive index is very sensitive to the amount of Nd incorporated.

The usefulness of the petrographic microscope on thin sections of polycrystalline material is illustrated in Figure 8. These photomicrographs depict various materials solidified from the melt in the system $MgAl_2O_4-Al_2O_3$ by a gradient furnace technique[24]. It is apparent that much microstructural detail can be observed. Approximate orientations of the eutectic (about 9.5 mole % MgO) phases can be ascertained much more easily and rapidly than by reflected light or X-ray methods. Some three-dimensional texture is also obtainable by focusing the microscope into the sample rather than on the surfaces. Complicated crystallographically controlled decomposition or solid-state reaction are also readily observable in (a) and (d), complementing any X-ray diffraction analyses of these samples. X-ray diffraction analysis of 8(a) showed only monoclinic Al_2O_3; however, microscopic examination exhibited at least two phases with different optical properties. The X-ray powder diffraction pattern of spinel can be superimposed on a pattern for monoclinic Al_2O_3, making detection of spinel very difficult by X-ray methods when monoclinic Al_2O_3 is present.

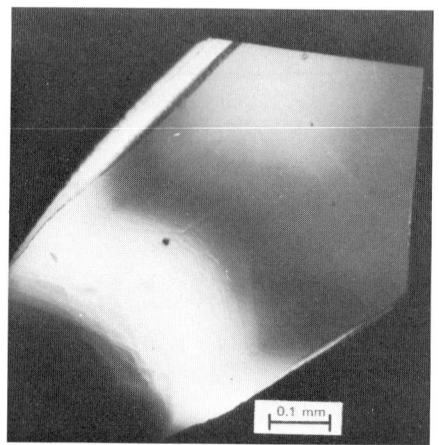

a. Inclined to Extinction

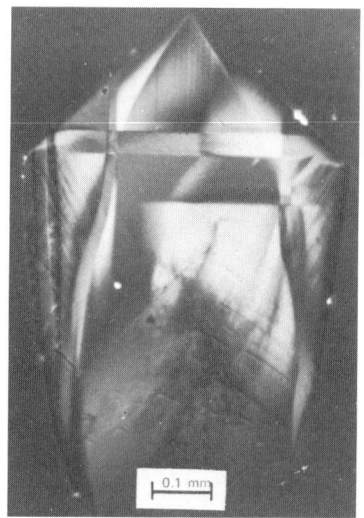

b. At Extinction

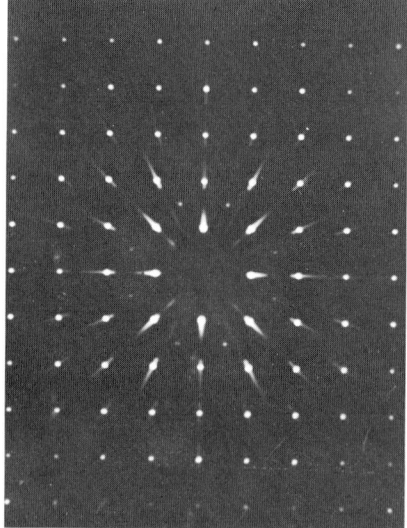

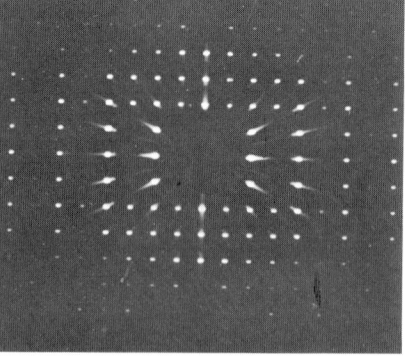

c. $\underline{C}$ - Axis X-Ray Precession Photograph   d. $\underline{b}$ - Axis X-Ray Precession Photograph

Figure 7.   Transmitted crossed polarized light photomicrographs [(a) and (b)] and X-ray precession photographs of $Nd_2(CO_3)_3 \cdot 8H_2O$.

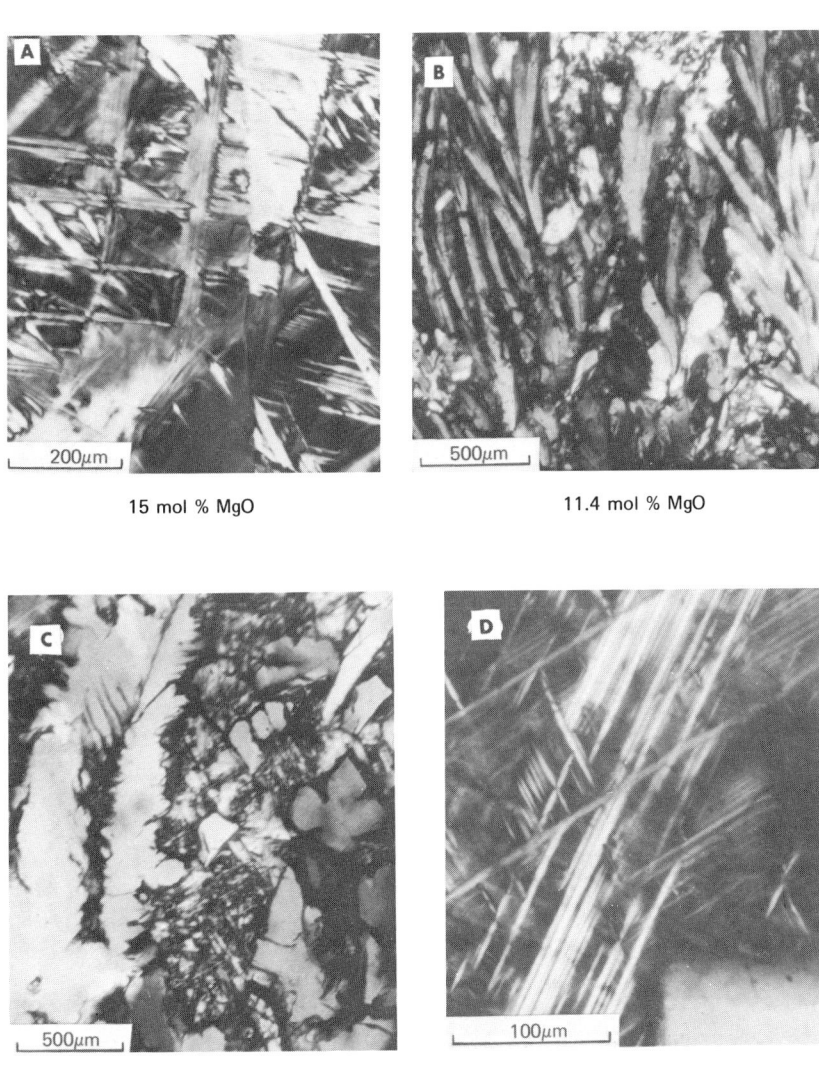

15 mol % MgO   11.4 mol % MgO

9.5 mol % MgO - Divorced Eutectic   9.5 mol % MgO - Enlargement of
                                             Spinel Phase in C

Figure 8. Microstructures of gradient-furnace-solidified materials in the $MgAl_2O_4$–Al_2O_3 System.

X-Ray Analysis

X-ray diffraction analysis is probably the most common method used for phase identification. Occasionally, investigators have shown tendencies to rely almost exclusively on this technique without studying the microstructure, texture, or actual arrangement of various phases within a material. Total reliance on X-ray methods may lead one to miss some very important morphologic features or the presence of noncrystalline material.

The two most important equations in X-ray diffraction analysis are the Bragg equation

$$n\lambda = 2d \sin\theta \qquad (7)$$

which determines the spatial location of peaks, and the structure factor formula

$$F_c = \Sigma_n f_n \exp[2\Pi i(hx_n + ky_n + \ell z_n)] \qquad (8)$$

which determines the magnitude of the diffracted intensity[25–27]. Actually, the intensity of diffracted radiation for most real crystals is porportional to the square of the structure factor. The International Tables for X-ray Crystallography[28] are very valuable reference work in this area. From Equation 7 the geometric measurements (lattice parameters) of the smallest orderly atomic unit in the solid can be obtained, while the arrangement of the atoms and their vibrational characteristics can be obtained with the use of Equation 8. Suitable electronic or photographic equipment can be placed around the radiated material to record the various positions and intensities of the reflections.

Phase identification can be carried out by measuring the d-spacings and intensities of the five most intense peaks on a powder diffraction pattern and comparing either manually or by computer to the powder diffraction file; programs and tapes are available for this purpose. (Powder Diffraction File, Joint Committee on Powder Diffraction Standards, 1601 Park Lane, Swarthmore PA 19081.) Precise measurement of lattice parameters can give a true measure of the amount of crystalline solution— in contrast to other analytical methods which yield only a bulk analysis, including impurities not in solution. As with microscopic examination, there are also X-ray methods for determining the texture and grain size of polycrystalline materials[26]. It is now possible to automate the above procedures completely so that all of them can be carried out without ever touching a diffraction pattern. It is a complex and expensive process to automate a system completely, including a linkage to a large computer, but the eventual benefits should justify the expense and time, especially in quality-control situations.

Crystallographers have paved the way as far as computerization of equipment and use of large computers is concerned. In a fully operational

X-ray crystallography laboratory it is now possible automatically to solve crystal structures of moderate size in a matter of days that fifteen years ago would have taken up to a year or two. Crystal-structure analysis is still very expensive, however, and is often hard to justify on materials with limited apparent applications. However, the atomic structure of crystals is the foundation of materials science, for without these kinds of data, interpretation of results is difficult-to-impossible in many cases. Knowledge of the crystal structure of materials allows for determinations of possible structural modification and ensuing property control and optimization by appropriate atomic substitutions. Jack[27] has recently shown the extent to which this type of atomic "engineering" is possible with the new class of oxynitride (SIALONS) materials. Roy and colleagues at the Pennsylvania State University have been using crystal chemistry for years to create new materials.

Routine crystal structure refinement results in precise atomic positions and their apparent thermal parameters. If the atom is assumed to be vibrating isotropically, then the scattering factor can be written as follows:

$$f = f_o \exp[-2\Pi^2 \bar{\mu}^2 \, | \, d^* \, |^2] \tag{9}$$

or

$$f = f_o \exp[B(\sin^2\theta)/\lambda^2] \tag{10}$$

where

$\bar{\mu}$ = average perpendicular displacement from mean position of plane of atoms,

and

$B = 8\Pi^2 \bar{\mu}^2$ = isotropic temperature factor

However, in anisotropic crystals the atoms may be vibrating nonisotropically. For this case, the scattering factor can be written as follows:

since

$$d^* = ha^* + kb^* + \ell c^*$$

and

$$| \, d^* \, |^2 = d^* \cdot d^* = h^2 a^{*2} + k^2 b^{*2} + \ell^2 c^{*2} + 2hka^*b^* \cos\gamma^*$$
$$+ 2h\ell a^*c^* \cos\beta^* + 2k\ell b^*c^* \cos\alpha^*$$

then

$$f = f_o \exp\left[-\frac{B}{4}(|\, d^* \,|^2)\right]$$

and

$$f = f_o \exp[\beta_{11}h^2 + \beta_{22}k^2 + \beta_{33}\ell^2 + 2\beta_{12}hk + 2\beta_{13}h\ell + 2\beta_{23}k\ell] \quad (11)$$

where

$$\beta_{11} = \frac{B}{4} a^{*2}, \text{etc.}$$

The β_{ij} terms are the so-called anisotropic temperature factors and describe the magnitude and position of a triaxial ellipsoid which describes the apparent vibrational features of the atoms.

More than one kind of atom may occupy an equivalent position in a crystal. Attempts are now being made, with some degree of success, to determine the position of the two species and the amount of each present; the latter has been termed site-occupancy refinement. It involves making an assumption that the total scattering factor for the site is a linear combination of two scattering factors[30].

$$f = xf_1 + (1 - x)f_2 \quad (12)$$

For this case, using least squares refinement, the optimum value of x can be determined, thus resulting in the amount of each atomic species in that site. However, one must assume or measure the approximate composition since there is no unambiguous way, at present, to assign specific scattering factors to the site. Eventually, with extremely precise data, this may be possible. When the two atoms are very close in atomic number—say, for example, Al and Si—site-occupancy refinement is almost impossible and measurement of average bond lengths becomes more meaningful since their ionic radii are quite different.

Determination of the position of two different atomic species in the same site is more difficult. Moss and Newnham[31] and McCauley and Gibbs[32] have made attempts at obtaining the relative positions of Cr and Al in ruby $((Al,Cr)_2O_3)$. A drawing of the structure is shown in Figure 9; the large circles are oxygen and the small ones are either Al or Cr and are denoted R. R_3 and R_6, for example, share an oxygen (O_4,O_5,O_6) octahedral face. In both aforementioned studies it was found that Cr shifts about 0.03 Å *toward* the nearest cation along c, contrary to theoretical calculations and spectroscopic evidence. In the figure, this would mean that if Cr was at site R_3, then it would be shifted 0.03 Å from the position that Al would have at this site toward the cation at R_6.

Anisotropic temperature factors were also obtained on this ruby[33]. Scaled thermal ellipsoids of the oxygen O_5 in Figure 9 is shown in Figure 10. The largest apparent vibration of the oxygen is into an area of least occupancy by other cations. The thermal parameters are referred to as "apparent" because positional disorder (atoms at same site occupying more than one mean position) and chemical disorder (two different atomic species at the same site) can lead to parameters which reflect only these types of disorders and not true thermal motion.

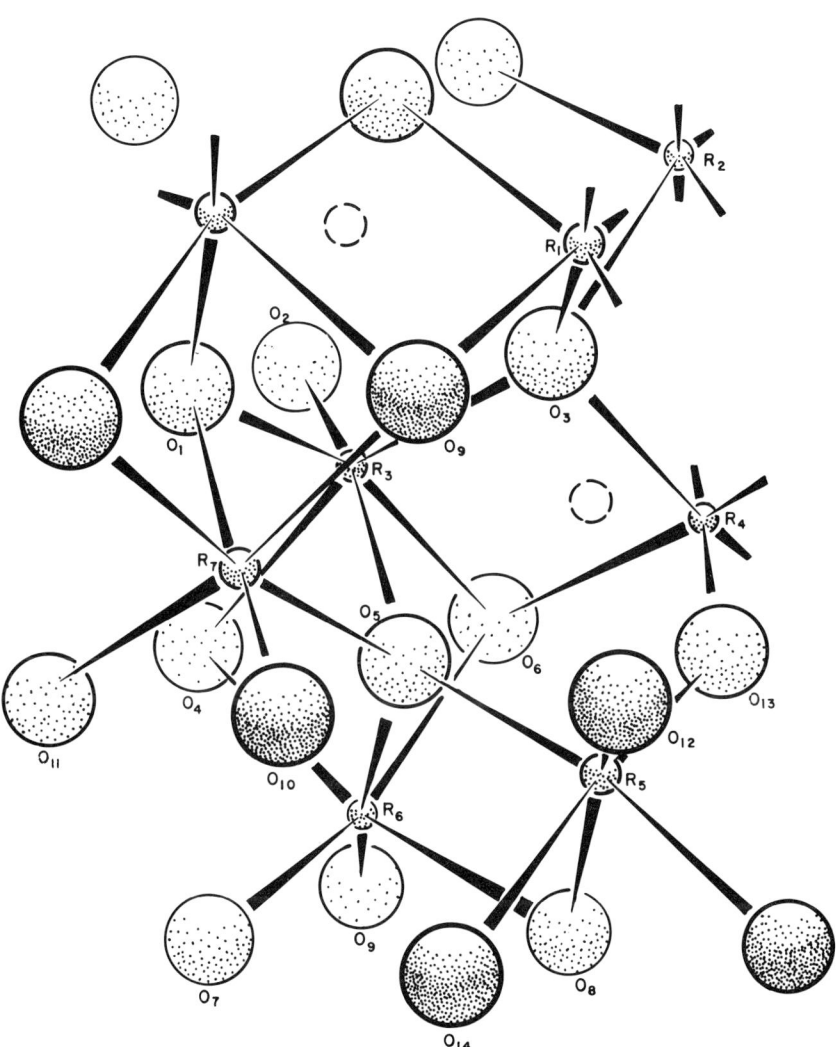

Figure 9. Drawing of the corundum structure.

The usefulness of a combination of X-ray and optical methods is illustrated in Figure 5. Careful lattice parameter measurement of various parts of the spinel ingot showed a significant decrease in a at the corner of the ingot, indicating volatilization of MgO with a concomitant relative increase of Al_2O_3 in crystalline solution.

Chemical analysis by X-ray methods can be carried out by various fluorescence techniques. The most commonly used method involves impinging an X-ray beam onto a sample and then using a known crystal to

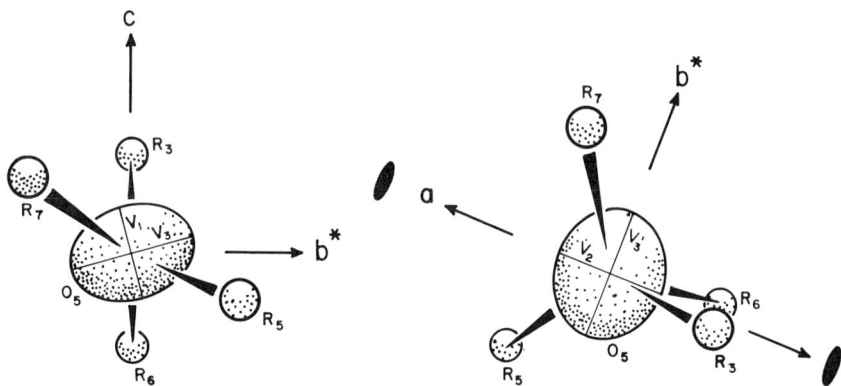

Figure 10. Pictorial representation of magnitude of vibration directions of O_5 with respect to the coordinating cations.

analyze the fluoresced radiation. X-rays can be excited from a sample by using a highly energetic polychromatic source[34]. By measuring the wavelength and intensity of the emitted X-rays, good chemical analysis for major and minor constitutents can be obtained rapidly and inexpensively.

X-ray fluorescence can also be stimulated by electron bombardment of a sample in an electron microprobe. Use of a crystal spectrometer (as described in the previous paragraph) allows for a rapid chemical analysis which can be quantitative[25,35,36]. The electron beam can be focused to a spot size of about 2 μm, but an area up to 10 μm is excited in this case, thus restricting the minimum area of analysis. The real beauty of this technique, however, is that the spatial distribution of composition can be readily determined, as can be seen in Figure 11. $CaCO_3$, in the high-

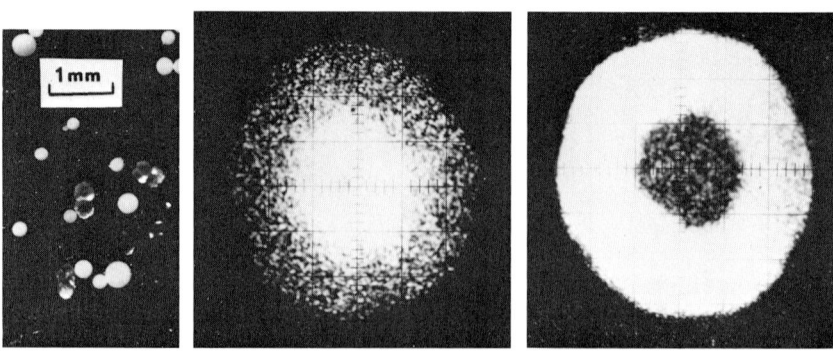

a. Reflected Visible Light   b. Sr X-Ray Scan   c. Ca X-Ray Scan

Figure 11. Photomicrograph and electron probe X-ray scan photographs of aragonite ($CaCO_3$) spherulites.

pressure aragonite structure, can be crystallized from solution by the addition of Sr to the precipitation medium[37]. Electron-probe analysis of halved aragonite spherulites shown in 11(a) unambiguously demonstrated that $CaCO_3$ in the aragonite form can be epitaxially overgrown on $SrCO_3$, which has the aragonite structure at 1 atm pressure. Finally, precise measurement of emitted X-ray wavelengths can give structural information for crystalline and noncrystalline solids[38].

In the last few years, X-ray energy dispersion (in contrast to X-ray wavelength dispersion using crystal spectrometers) has come into much increased usage, primarily because of advances in solid-state detectors and the increased usage of scanning electron microscopy. Both chemical analysis and X-ray diffraction are possible by this technique[39,40].

In energy dispersive X-ray analysis, the energies of fluoresced or diffracted X-rays are energetically dispersed in a multichannel analyzer, resulting in an energy pattern. The wavelength and d-spacings can be calculated from the following equations:

$$E = \frac{hc}{\lambda} = \frac{12.398}{\lambda} \tag{13}$$

and since

$$\lambda = \frac{2d}{\sin\theta}$$

$$d(hk\ell) = \frac{6.199}{\sin\theta} \cdot \frac{1}{E} \tag{14}$$

where

λ is expressed in agnstroms

and

E in kiloelectronvolts

The angle θ is chosen to maximize the number of powder lines and optimize the resolution.

Figure 12 is from Giessen and Gordon[39], and shows the type of information available by this technique. Potentially, one can get powder-diffraction patterns and chemical analysis almost instantaneously on the same pattern. Lattice-parameter refinement is also possible, but not to the precision of wavelength-dispersion techniques because of the lower resolution that is obtainable—at the moment, it is about an order of magnitude less precise. Giessen and Gordon have suggested the terms "Spectrometric Powder Diffractometry" for this type of "energy" diffraction, while the chemical analysis procedure is called "X-ray Energy Spectroscopy." A high-power X-ray source must be used to obtain the energy powder patterns, but radioactive materials such as Fe[55], Cd[109], and Am[241] can be used as the X-ray source in chemical analysis. The technique is more suit-

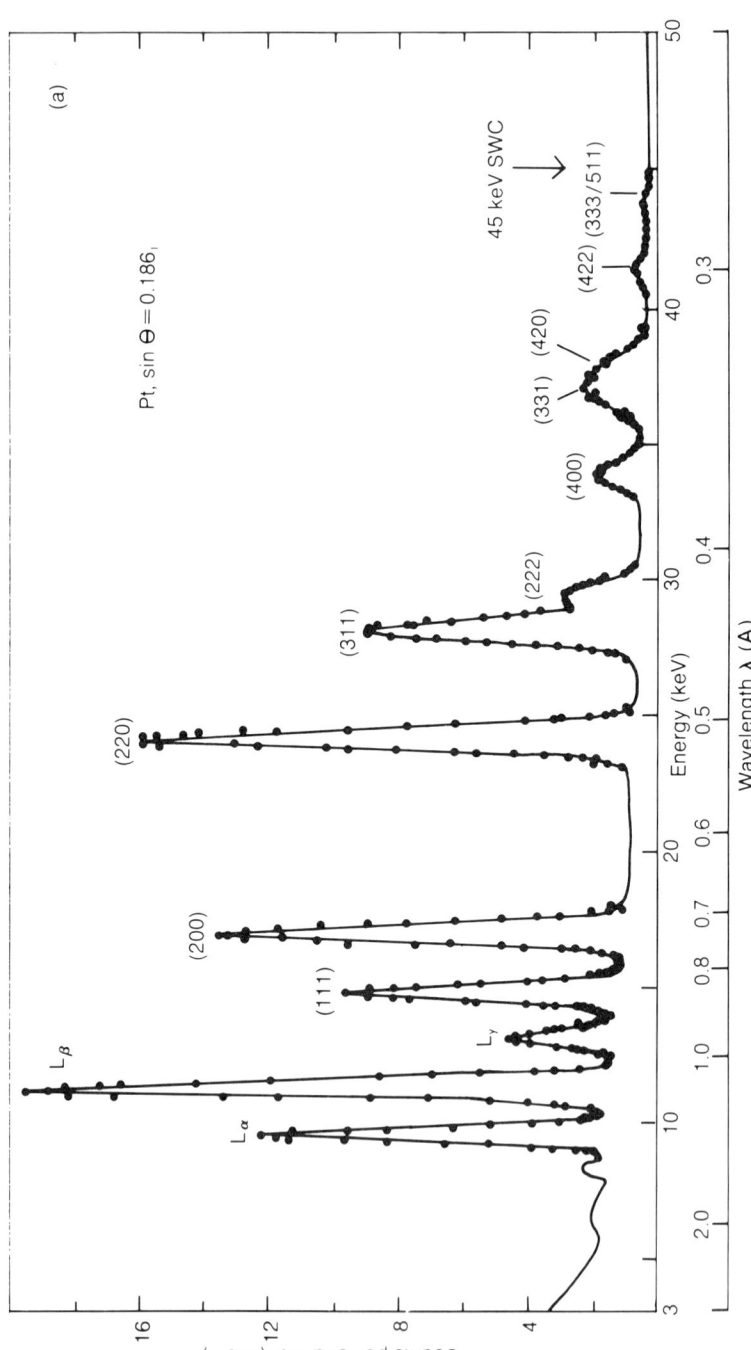

Figure 12. Diffraction pattern and L-fluorescence spectrum of platinum sheet, obtained by X-ray spectroscopy, SWC, shortwave cutoff of X-ray beam [39].

able for automation than other X-ray techniques because no moving parts need be involved and data gathering is much quicker. In-situ measurements of diffraction patterns of materials subjected to high temperature or high pressure is also more convenient. Further, since large generators are not necessary for chemical analysis, the procedure lends itself to mobilization.

Finally, X-rays have been utilized in the diffraction mode to examine the perfection (defect content) of materials. This procedure has been referred to as X-ray topography, since it essentially examines the intensity distribution (profile or topography) on a single X-ray diffraction peak[41–43]. Imperfections affect the diffracted intensity in various ways and these can be studied very conveniently by these methods. There are essentially two different techniques: the Berg–Barrett method, which operates in the back-reflection region, and the Lang arrangement, which utilizes transmitted reflections. The former method is primarily for surface characterization to depths about 10–100 μm, depending on the radiation used and the material surveyed. Grain boundaries, inclusions, voids, and other types of imperfections can be revealed by this procedure. In the transmission mode, X-ray topography can reveal volume imperfections and identify the Burgers vectors of dislocations. Domain structures can be studied as well. The resolution to about 1 μm is limited by the grain size of the recording film. A careful study of dislocations in gradient-furnace sapphire crystals has recently been completed by Caslavsky et al.[44,45], using X-ray transmission topography. X-ray topography is even being used for quality control of semiconductor materials.

Electron Interaction in Solids

Besides the generation of X-rays by bombardment of a material with electrons, other processes occur—depending on the energy of the electron beam. Visible light is sometimes produced, and is referred to as cathodoluminescence[46]. The intensity and wavelength of light emitted is a function of the amount and structural configuration of certain activators present as impurities or dopants in a material. Recently, the writer has been utilizing this technique to monitor the diffusion bonding of Mo-wire to sapphire single crystals and studying remnant liquid–solid interfaces in gradient furnace–produced sapphire single crystals.

Figure 13 is a mosaic optical photograph of four individual large-area electron-beam-irradiated Mo-sapphire interface areas, taken in an electron microprobe. The dark center portion is a Mo-wire, 0.0007" in diameter, completely embedded in a sapphire single crystal. As can be seen, the luminescent halo immediately surrounding the wire is different than that farther from the wire. The color of this halo is a light pink, whereas farther from the wire it is a deep red—characteristic of these sapphires grown by the gradient-furnace technique. This seems to indi-

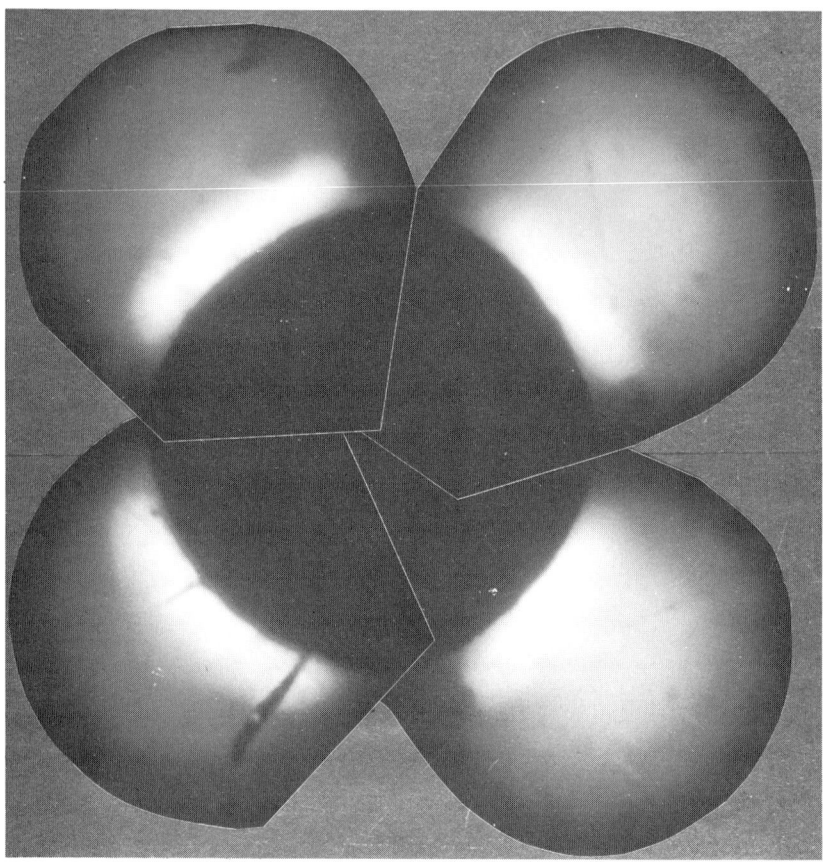

Figure 13. Mosaic photograph of cathodoluminescent halo around molybdenum wire in sapphire, excited in four areas by electron beam in electron microprobe; wire (dark center portion) is 0.007″ in diameter.

cate a diffusion gradient of Mo into sapphire, forming a good chemical (no reaction phase) bond.

This technique is also being used to examine remnant liquid–solid interfaces in gradient furnace sapphire. Figure 14(a) shows a cross-sectional slice through a 2″-diameter circular cylindrical ingot of sapphire, containing a two-dimensional Mo-wire screen. The cross section of the screen can be observed running from the top left of the section to the right center. Section 1 is the end of the screen attached to large, elliptically shaped W-wire used to hold the Mo screen in place. Area 2 is the top corner (near Mo crucible wall) of the ingot, and section 3 is the seed area; the slice has been cut in half, showing only one-half of the cross-sectional slice. Small changes in vacuum or temperature in the gradient furnace

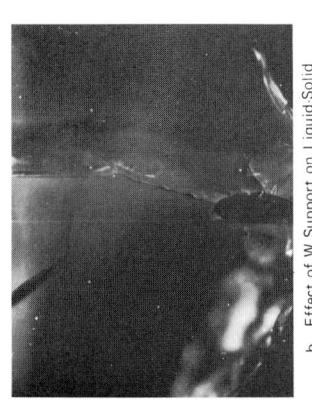

b. Effect of W Support on Liquid-Solid Interface; Area 1

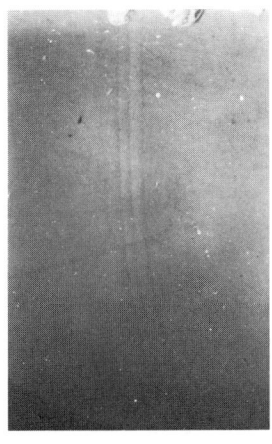

d. Remnant Liquid Solid Interface, Near Seed, Back of Area 3

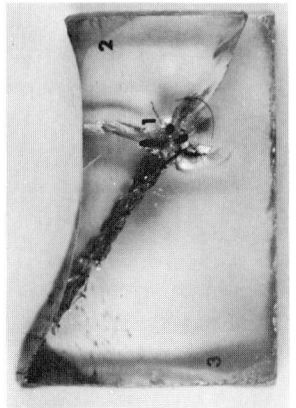

a. Cross-Sectional Slice Through an Mo/Sapphire Cylindrical Ingot; Bottom Dimension ≥ .94″

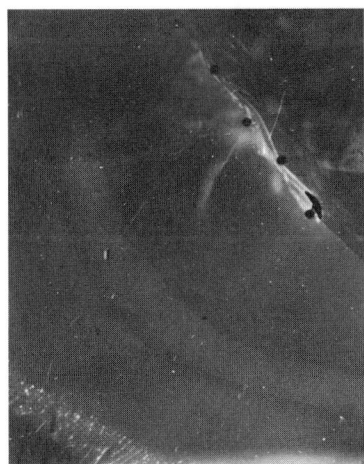

c. Remnant Liquid Solid Interfaces and Halos Around Mo Wire; Back of Area 2

Figure 14. Cathodoluminescence in gradient-furnace fabricated Mo/sapphire composites.

during solidification will cause a change in the growth processes. If this effect results in a valence change of the incorporated Mo (from wire or crucible walls) or in the amount of Mo incorporated, this should result in luminescent rings or bands where the anomaly occurred. Indeed, these bands are observed in the solidified ingot, as can be seen in 14(b), (c), and (d). In these photographs, the darker bands are purple in color, while the lighter areas are red. All these photographs were taken in a relatively inexpensive cathodoluminescent device. Color pictures were initially taken which have been rephotographed in black and white for reproduction purposes. Figure 14(b) illustrates that the liquid–solid interface was pinned by the W-wire (the large, dark elliptical object); 14(c) and (d) show remnant interfaces at the top corner and seed area, respectively. Much more detail should be obtainable using more carefully polished thin sections. Growth kinetics studies can now be carried out on gradient-furnace material through purposely putting in bands, either by changing the vacuum or the temperature. Further, and perhaps of more importance, the distribution (homogeneity) of impurities in the crystals can be monitored routinely—distribution of dopants (e.g., Nd in YAG) could also be followed routinely. The technique has even been used successfully in differentiating various natural $CaCO_3$ polymorphs[47].

  Electron spectroscopy for chemical analysis (ESCA) has recently become a popular analytical technique due to improved vacuums and the availability of commercial machines[48,49]. Essentially, a beam of electrons or photons are focused onto a sample, resulting in the emitting of electrons in varying amounts and energies. When all of the energy of the photon or electron is transferred for the ejection of an electron, then the binding energy of the electron can be calculated from the following formula:

$$E_b = hv - E_k \qquad (15)$$

where

E_b = binding energy of ejected electron
hv = energy of photon or electron focused on sample
E_k = kinetic energy of emitted electron

This is called photoelectron spectroscopy. Two types are now being carried out: (1) photoelectron spectroscopy of the outer shells (PESOS) using low-energy photon bombardment with the helium-I resonance line at 21.22 eV or the helium-II line at 40.8 eV; (2) using more energetic bombardment with, for example, $AlK\alpha$ at 1487 eV and $MgK\alpha$ at 1254 eV, deeper penetration into inner shells (photoelectron spectroscopy of inner shells—PESIS) is possible. These techniques are being used to study the bonding characteristics of materials.

  Secondary electrons (Auger electrons) are also produced, due to the formation of a hole in an inner shell, forming small maxima in the energy profile which are characteristic of the elements involved[50]. Electronic differentiation of the intensity versus energy curve $(dN(E)/dE)$ sharpens these small bumps for unambiguous identification. A high-energy beam of electrons is used, but kept less than 10 KeV to minimize X-ray fluorescence. This procedure lends itself to light-element analysis for surface layers up to about 20 Å deep. By using the electron beam to evaporate material from the surface, elemental profiles can be obtained down through a sample[51]. Techniques have also been worked to minimize charging effects in ceramic insulators[51]. Auger spectroscopy has already provided results in regard to grain-boundary contaminants of polycrystalline ceramics (Al_2O_3, Si_3N_4), modifying previously accepted conclusions.

The secondary electrons being ejected from a sample can also be used in the scanning electron microscopy (SEM) mode to provide electronic images of the surface down to a resolution of about 100 Å [52–55]. The depth of focus is remarkable for this technique: 1 cm at 100× to about 1 μm at 10,000×. This is demonstrated in Figure 15, which is a series of SEM photographs of the fracture surface of a Mo-sapphire bend bar. Figure 15(b) and (c) are at a magnification of about 400×, while 15(d) is at 1,700×. Used in conjunction with an X-ray energy-dispersive system, chemical analyses can also be carried out in an SEM unit. For quantitative evaluation of surfaces or powders, either stereoscopic photographs or computerized SEM images are being used[55].

Before the advent of the scanning electron microscope, transmission electron microscopy (TEM) was the only tool available for observation of surfaces (by replica techniques) and very thin samples at magnifications higher than those obtainable in optical microscopes[56]. The SEM has significantly decreased the use of transmission microscopy in recent years. However, ultrahigh-energy beams (up to 5,000 KeV) and new ion thinning devices are providing new incentives for its use. Ceramics can be thinned by this technique to around 2000 Å, so that they can be viewed directly in transmission. Heuer and co-workers, at Case Western Reserve University, have carried out very careful work on dislocations in oxides by these procedures[57]. X-ray topography does not have the resolving power to compete with electron microscopy for examining dislocations in finely crystalline materials or when the dislocation densities are high.

Transmission electron microscopy can also be used for electron diffraction as well. For an electron beam of 40 KeV, the wavelength of the electrons is about 0.01 Å. The diffraction pattern from a sample using electrons of this wavelength is almost an undistorted reciprocal lattice because of the extremely large radius ($r = 1/\lambda$) of the reflection sphere. Hence, no moving parts are necessary to produce an undistorted diffraction photograph. Since penetration is so small—about 3000 Å for a 100 KeV source—the sample must be very thin. Lattice parameters can be obtained to a precision of about 1 percent, but crystal-structure refinement is not possible; only general structure data can be obtained. Zvyagin

c. SEM of Mo Wire (#2) Fracture

d. High Magnification SEM of Mo/Sapphire Interface in (c)

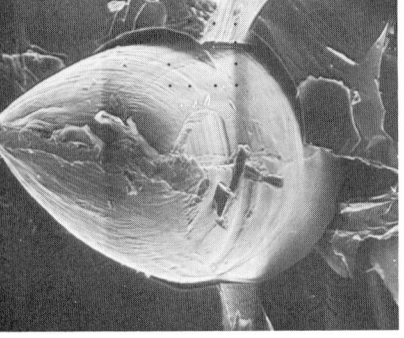

a. Normal Reflected Light Photograph of Mo/Sapphire Bend Bar Fracture Surface; Sides = 0.191" X 0.191"

b. SEM of Mo Wire (#1) Fracture; Wire = 0.007" Diameter

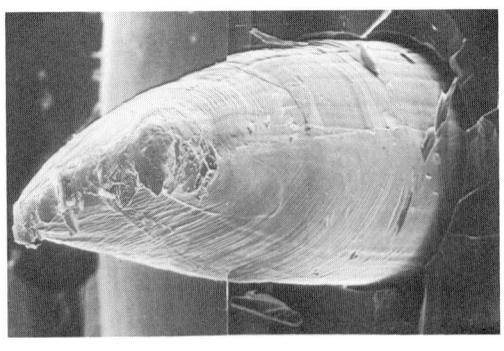

has been very active in the area of structure determinations by electron diffraction[58].

Images are obtained in transmission microscopy by absorption contrast, using electron optical procedures and by diffraction contrast. As discussed previously, conventional diffraction patterns can be routinely obtained from crystalline materials. Moreover, Bragg-diffracted electrons can also be brought to focus in an image plane, resulting in diffraction-produced images. This procedure is analogous to X-ray transmission topography. Recent work by Steeds *et al.*[59] has demonstrated that much crystal structure information can be obtained by proper diffraction-image plane analysis. Finally, very exciting work is being carried out using lattice-imaging techniques. By this procedure, using very thin specimens, a slightly underfocused image is representative of charge distributions in a sample[60]. Direct observation of the domains in disordered materials becomes possible at the 5–10 Å level of resolution. Once-complex structural relationships in phases such as found in the WO_3–Nb_2O_5 ($x = 0.2$–0.35) systems are being unraveled in terms of so-called Wadsley defects—defined as coherent intergrowths, often only half a unit cell wide, of one structural entity in a matrix of another. The surfaces separating the structural entities are called crystallographic shear planes. Materials once thought to be solid solutions in many Nb_2O_5-based systems are now known to be a series of closely related structures forming intergrowth or domains at the unit cell level. Haggerty and Peters discuss this further in another section of this book. Materials like ruby[32] may eventually be found to be very intimate intergrowths of two distinct structures.

Other Techniques

Diffraction can also be obtained using a neutron source; however, the sample size required is much larger than that for X-rays, and the equipment needed is much more elaborate[61]. In this case, the wavelength at $25°$ C is about 1 Å. Neutron diffraction has been used for the solution of magnetic structures, where X-ray diffraction fails. In X-ray diffraction structure determinations, the location of hydrogen atoms is carried out mainly by inference and bond-length information. However, the neutron-scattering factor for hydrogen is large, thus allowing for its direct location. Recently, Coppens[62] has used a combination of X-ray and neutron-diffraction data to determine the distribution of bonding electrons. Since X-rays are diffracted by the electron cloud and neutrons by the nuclei, a subtraction of the two sets of data will result in a map (X–N map) which successfully shows the distribution of non-bonding electrons.

Many other techniques can be utilized to characterize the structure and chemistry of ceramic materials[63]. For example, very powerful techniques which can be carried out quickly and cheaply are infrared absorption analysis and emission spectroscopy. A "quick and dirty" chemical analysis can be obtained from an emission spectroscopic analysis, whereas an IR spectra can provide much structural and also some chemical information. Both can also provide precise quantitative information. Hence, the techniques described in this chapter are not the only ones that can be utilized in a full characterization program and should not be the only ones. However, any further discussion is beyond the scope of this chapter, and though one dealing with chemical characterization would be incomplete without mentioning wet chemical analysis, a discussion of these procedures could fill many books such as this; it remains a tried-and-true procedure for obtaining a precise chemical analysis of a bulk sample.

Summary

The importance of characterization in ceramics research and development has been discussed. Guidelines have been suggested for the adequate characterization of processed crystalline ceramic materials which have included the use of composition, grain size, grain shape, relative orientation, and packing (spatial distances) in their full and unambiguous definition. Utilization of these concepts could provide a way for the routine engineering of materials for specific applications. A review of commonly used ceramic characterization procedures has been presented, along with highlights from relatively new techniques, including site-occupancy refinement in crystal-structure analysis, cathodoluminescence, and ESCA.

Acknowledgment

Various parts of the work discussed in this chapter were carried out in the Materials Research Laboratory at the Pennsylvania State University under Advanced Research Projects Agency Contracts SD-132 and DA-49-083 OSA-3140, Army Materials and Mechanics Research Center Contract DA-19-0660 AMC-325(X), and in the Ceramics Research Division of the Army Materials and Mechanics Research Center.

References

1. Roy, R., "Parameters Necessary for the Adequate Characterization of a Solid-State Material," *Phys. Today*, 18 (1965), 71–73.
2. Fulrath, R.M. and Pask, J.A., eds., *Ceramic Microstructures*, New York: Wiley, 2968.
3. "Characterization of Materials," Materials Advisory Board, Washington, D.C., Report No. MAB-229-M, March 1967 (AD649 941).

4. Griffiths, J.C., "Measurement of Properties of Sediments," *J. Geol.,* 69 (1961), 487–98.

5. Nye, J.F., *Physical Properties of Crystals,* London: Oxford University Press, 1964.

6. Campbell, W.J., and Grain, C., "Thermal Expansion of Alpha Alumina," U.S. Bureau of Mines Report No. 5757, 1961.

7. Schmid, F. and Viechnicki, D., "Growth of Sapphire Disks from the Melt by a Gradient Furnace Technique," *J. Amer. Ceram. Soc.,* 53 (1970), 528–29.

8. McCauley, J.W., "Polariscopic Characterization of Sapphire and Spinel," Army Materials and Mechanics Research Center, Watertown, Mass., Technical Report No. AMMRC TR 72-1, January 1972 (AD 741 810).

9. McCauley, J.W., and Acquaviva, S.J., "Property Characterization of Ba-Mica/Al_2O_3 Composites," paper presented at the 75th Annual Meeting of the American Ceramic Society, Cincinnati, Ohio, 29 April—3 May 1973 (Abstract: *Bull. Amer. Ceram. Soc.,* 52 (1973), 364).

10. Chay, D.M., Palmour, H. III and Kriegel, W.W., "Microstructure and Room-Temperature Mechanical Properties of Hot-Pressed Magnesium Aluminate as Described by Quadratic Multivariate Analysis," *J. Amer. Ceram. Soc.,* 51 (1968), 10–16.

11. Gupta, T.K., "Strength Degradation and Crack Propagation in Thermally Shocked Al_2O_3," *J. Amer. Ceram. Soc.,* 55 (1972), 249–53.

12. Hasselman, D.P.H., "Thermal Stress Resistance Parameters for Brittle Ceramics: A Compendium," *Bull. Amer. Ceram. Soc.,* 49 (1970), 1033–37.

13. Seaton, C.C. and McCauley, J.W., "Thermal Shock Characteristics of Ba-Mica/Al_2O_3 Composites," paper presented at the 26th Pacific Coast Meeting of the American Ceramic Society, San Francisco, CA, 30 October—2 November 1973 (Abstract: *Bull. Amer. Ceram. Soc.,* 52 [1973], 708–709).

14. McCauley, J.W., and Newnham, R.E., "Origin and Prediction of Ditrigonal Distortions in Micas," *Amer. Mineral,* 56 (1971), 1626–38.

15. Viechnicki, D., Schmid, F. and McCauley, J.W., "Growth of Nearly Stoichiometric $MgAl_2O_4$ Spinel Single Crystals by a Gradient Furnace Technique," *J. Appl. Phys.,* 43 (1972), 4508–12.

16. Honess, A.P., *The Nature, Origin and Interpretation of the Etch Figures on Crystals,* New York: John Wiley & Sons, 1927.

17. Short, M.N., *The Microscopic Determination of the Ore Minerals,* 2d ed., U.S. Department of the Interior, Geological Survey Bulletin 914, Washington, D.C.: U.S. Government Printing Office, 1948.

18. Wahlstrom, E.E., *Optical Crystallography,* New York: John Wiley & Sons, 1960.

19. Larsen, E.S. and Berman, H., *The Microscopic Determination of the Nonopaque Minerals,* 2d ed., U.S. Department of the Interior, Geological Survey Bulletin 848, Washington, D.C., U.S. Government Printing Office, 1934.

20. Bloss, F.D., Gibbs, G.V. and Cummings, D., "Polymorphism and Twinning in Synthetic Fluorophologopite," *J. Geol.,* 71 (1963), 537–47.

21. McCauley, J.W. and Gehrhardt, H.M., "Crystal Growth of $CaWO_4$ and $Nd_2(CO_3)_3$ $8H_2O$ by the Gel Technique." Army Materials and Mechanics Research Center, Watertown, Mass., Technical Report No. AMMRC TR 70-13 (AD 710 236).

22. McCauley, J.W. and Roy, R., "Gel-Growth of Pure and Doped $CaCO_3$ (Calcite) and $SrSO_4$ Single Crystals," paper presented at the American Ceramic Society, Basic Science Division Meeting, Pittsburgh, Pa., 1965 (Abstract: *Bull. Amer. Ceram. Soc.,* 44 [1965], 635–37).

23. McCauley, J.W., "Control of Nucleation, Crystal Growth and Doping of Various Calcium Carbonate Phases by the Gel Technique," unpublished M.S. dissertation, Pennsylvania State University, 1965.

24. Viechnicki, D., Schmid, F. and McCauley, J.W., "Liquidus-Solidus Determinations in the System MgAl$_2$O$_4$-Al$_2$O$_3$," *J. Amer. Ceram. Soc.*, 57 (1974), 47-48.

25. Cullity, B.D., *Elements of X-Ray Diffraction*, Reading, Mass.: Addison-Wesley Publishing Company, Inc., 1959.

26. Olphen, H. van and Parrish, W., eds., *X-Ray and Electron Methods of Analysis*, New York: Plenum, 1968.

27. Stout, G.H. and Jensen, L.H., *X-Ray Structure Determination*, New York: The MacMillan Company, 1968.

28. *International Tables for X-Ray Crystallography, Vols I-III*, Birmingham, England: The Kynoch Press, 1952-1962.

29. Jack, K.H., "Nitrogen Ceramics," *Trans. J. Brit. Ceram. Soc.*, 72 (1973), 376-84.

30. Finger, L.W., "Determination of Cation Distributions by Least-Squares Refinement of Single-Crystal X-Ray Data," in Annual Report of the Director, Geophysical Laboratory, Carnegie Institution (1967-1968), 216-17.

31. Moss, S.C. and Newnham, R.E., "The Chromium Position in Ruby," *Z. Kristallogr.*, 120 (1964), 359-63.

32. McCauley, J.W. and Gibbs, G.V., "Redetermination of the Chromium Position in Ruby," *Z. Kristallogr.*, 135 (1972), 453-55.

33. McCauley, J.W. and Gibbs, G.V., "Significance of Weighting Schemes in an Anisotropic Least-Squares Refinement of Ruby," in *Molecular Dynamics and Structure of Solids*, NBS Special Publication 301, R.S. Carter and J.J. Rush, eds., Washington, D.C.: U.S. Government Printing Office (1969), 277-82.

34. Jenkins, R. and deVries, J.L., *Practical X-Ray Spectrometry*, New York: Springer-Verlag, 1969.

35. Anderson, C.A., ed., *Microprobe Analysis*, New York: John Wiley & Sons, 1973.

36. Beaman, D.R. and Isasi, J.A., *Electron Beam Microanalysis*, ASTM Special Technical Publication No. 506, Philadelphia: American Society for Testing and Materials, 1972.

37. McCauley, J.W. and Roy, R., "Evidence for Epitaxial Control of CaCO$_3$ Phase Transformation as the Mechanism of the Influence of Impurity Ions," paper presented at the American Geological Union Meeting, Washington, D.C., 1966 (Abstract: *Trans. Amer. Geophys. Union*, 47 [1966], 202-03).

38. White, E.W., "Applications of Soft X-Ray Spectroscopy to Chemical Bonding Studies with Electron Microprobe," in *Microprobe Analysis*, C.A. Anderson, ed., New York: John Wiley & Sons (1973), 349-69.

39. Giesson, B.C. and Gordon, G.E., "X-Ray Diffraction: New High-Speed Technique Based on X-Ray Spectrography," *Science*, 159 (1968), 973-75.

40. Waldseth, R., *All Your Ever Wanted to Know About X-Ray Energy Spectrometry*, Burlingame, Calif.: Kevex Corp., 1973.

41. Weissmann, S., "Recent Advances in X-Ray Diffraction Topography," in *Fifty Years of Progress in Metallographic Techniques*, ASTM Special Technical Publication No. 430, Philadelphia: American Society for Testing and Materials (1968), 141-91.

42. Authier, A., "Contrast of Images in X-Ray Topography," in *Modern Diffraction and Imaging Techniques in Material Science*, Edited by S. Amelinckx, R. Gevers, G. Remaut, and J. Van Landuyt, New York: American Elsevier Publishing Company (1970), 481-520.

43. Newkirk, J.B., "The Observation of Dislocation and Other Imperfections by X-Ray Extinction Contrast," *Trans. Met Soc. AIME*, 215 (1959), 483-97.

44. Caslavsky, J.L., Gazzara, C.P. and Middleton, R.M., "The Study of Basal Dislocations in Sapphire," *Phil. Mag.*, 25 (1972), 35-44.

45. Caslavsky, J.L. and Gazzara, C.P., "Dislocation Behavior in Sapphire Single Crystals," *Phil. Mag.*, 26 (1972), 961–75.

46. Kniseley, R.N. and Laabs, F.C., "Applications of Cathodoluminescence in Electron Microprobe Analysis," in *Microprobe Analysis*, C.A. Anderson, ed., New York: John Wiley & Sons (1973), 371–82.

47. Sommer, S.E., "Cathodoluminescence of Carbonates, 1. Characterization of Cathodoluminescence from Carbonate Solid Solutions," *Chem. Geol.*, 9 (1972), 257–73.

48. Carlson, T.A., "Electron Spectroscopy for Chemical Analysis," *Phys. Today*, 25 (1972), 30–39.

49. Ignatiev, A. and Rhodin, T.N., "Secondary Electron Spectroscopy, A Surface Sensitive Tool," *Amer. Lab.*, 4 (1972), 8–17.

50. Chang, C.C., "Auger Electron Spectroscopy," *Surface Sci.*, 25 (1971), 53–79.

51. Marcus, H.L., Harris, J.M. and Szalkowski, F.J., "Auger Spectroscopy of Fracture Surfaces of Ceramics," in *Fracture Mechanics of Ceramics, Vol I, Concepts, Flows, and Fractography*, R.C. Bradt, B.P.H. Hasselman, and F.F. Lange, eds., New York: Plenum Press (1973), 387–98.

52. Thorton, P.R., *Scanning Electron Microscopy*, London: Chapman and Hall, 1968.

53. Johari, O., "Total Materials Characterization with the Scanning Electron Microscope," *Res. Devel.*, 22, No. 7 (1971), 12–20 (Reference: *Res. Devel.* 22, No. 8 [1971], 46–47).

54. Kammlott, G.W., "Some Aspects of Scanning Electron Microscopy," *Surface Sci.*, 25 (1971), 120–46.

55. White, E.W., McKinstry, H.A. and Diness, A., "Quantitative Surface Finish Characterization by CESEMI," in *The Science of Ceramic Machining and Surface Finishing*, NBS Special Publication 348, S.J. Schneider and R.W. Rice, eds., Washington, D.C.: U.S. Government Printing Office (1972), 306–16.

56. Thomas, G., Fulrath, R.M., and Fisher, R.M., eds., *Electron Microscopy and Structure of Materials*, Berkeley, Calif.: University of California Press, 1971.

57. Heuer, A.H., "Exploitation of Transmission Electron Microscopy in Ceramics and Mineralogy," paper presented at the 74th Annual Meeting of the American Ceramic Society, Washington, D.C., 8–10 May 1972 (Abstract: *Bull. Amer. Ceram. Soc.*, 51 [1972], 437).

58. Zvyagin, B.B., *Electron-Diffraction Analysis of Clay Mineral Structure*, New York: Consultants Bureau, 1967.

59. Steeds, J.W., Tatlock, G.J. and Hampson, J., "Real Space Crystallography," *Nature*, 241 (1973), 435–39.

60. Allpress, J.G., "The Application of Electron Optical Techniques to High-Temperature Materials," in *Solid State Chemistry*, NBS Special Publication 364, R.S. Roth and S.J. Schneider, Jr., eds., Washington, D.C.: U.S. Government Printing Office (1972), 87–111.

61. Bacon, G.E., *Neutron Diffraction*, London: Oxford University Press, 1955.

62. Coppens, P., Sabine, T.M., Delaplane, R.G. and Ibers, J.A., "An Experimental Determination of the Asphericity of the Atomic Charge Distribution in Oxalic Acid Dihydrate," *Acta Cryst.*, B25 (1969), 2451–58.

63. Newnham, R.E. and Roy, R., "Structural Characterization of Solids," in *The Chemical Structure of Solids, Vol. I, Treatise on Solid Chemistry*, N.B. Hannay, ed., New York: Plenum Press (1973), 437–533.

L. L. HENCH
United of Florida
Gainesville, Florida

Chapter **8**

Characterization of Glass

ABSTRACT

Research techniques suitable for determining the chemical, structural, microstructural, and surface characterization features of glass are reviewed. Basic principles of operation of the instrumental tools are presented. Recent applications of the methods to a variety of glasses are made and limitations of the techniques are discussed. These results are used as a basis for discussing the selection of a suitable characterization program for glass.

Introduction and Objectives

The materials advisory board of the National Research Council [1] has defined characterization as: "Characterization describes those features of the composition and structure (including defects) of a material that are significant for a particular preparation, study of properties, or use, and suffice for the reproduction of the material." Discussion of the objectives and philosophy of characterization of materials occurs in several additional sources [2–4]. Consequently, in order to characterize a glass it is necessary to evaluate the composition, atomic structure, microstructure, and surface sufficiently that the glass and its properties can be reproduced. It is quantitative analysis of these characterization features that is of primary concern in present glass science and technology. The objective of this chapter is to review the various tools appropriate for characterization of each of these primary characterization features. In addition, recent applications of the characterization tools to representative glassy systems are discussed. Finally factors involved in the selection of a suitable characterization program for a particular glass product are discussed.

211

Chemical Composition Characterization

Quantitative chemical analysis of the glass composition is nearly always a requisite characterization step. The analytical problem generally can be considered as two separate topics: bulk compositional analysis and trace element analysis. The number of methods developed to obtain such analyses encompass nearly the entire field of chemistry, as recently reviewed [5]. Over 30 different analytical techniques must be considered, falling into 5 major catagories: (1) separation methods, (2) optical methods, (3) electrochemical methods, (4) thermal methods, and (5) nuclear radiation and other miscellaneous methods. Detailed descriptions of these various chemical analytical methods appear in a number of sources [6–8], including several recent summaries of chemical analyses applications to glasses [9–11].

Instrumental methods are gradually gaining wider and wider acceptance as the primary tools for chemical compositional analyses of glasses. X-ray spectrochemical analysis appears to be the primary choice. This is indicated by the success of recent symposia on such topics[12]. The reasons for the popularity of X-ray spectrochemical analysis is speed of sample preparation, speed of analysis, and accuracy of results and availability of automated equipment [13]. All of these factors add up to a single overriding advantage—low cost. Details, both as simplified references and complicated references, of X-ray spectroscopy techniques are available from numerous sources [14, 15].

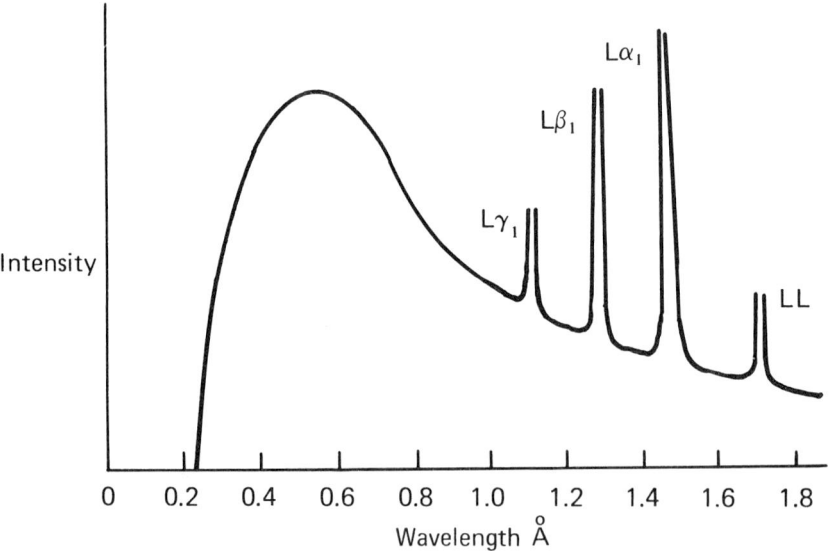

Figure 1. Schematic of an X-ray spectrometer.

X-ray spectrographs usually operate with a continuous X-ray spectrum from a tungsten-target spectrograph tube, such as shown in Figure 1. When high-energy X-rays are incident upon an atom of a solid such as glass, electrons can be excited from inner shells of the atom. Vacancies in inner electron shells are not stable and become filled instantly by transitions of electron from the outer shells. The transition involves a loss of energy, which appears in the form of an X-ray photon (Figure 2). The wavelength of the X-ray photon emitted from the atom is directly proportional to the energy difference between the shells in which the transition occurs. Because each atom has a characteristic distribution of electron shells, the X-ray spectrum produced by a high-energy incident X-ray beam is characteristic of the atom. The intensities of the fluorescent X-rays produced by the electron transitions from the atoms are additive, and consequently the intensity of X-rays emitted from the solid at a given wavelength is proportional to the concentration of atoms present in the solid being exposed to the incident exciting X-ray beam.

A problem that has limited the accuracy, applicability, and detection limits of X-ray spectroscopy is the fact that neighboring atoms in a solid can absorb the fluorescing X-rays emitted. This characteristic of the solid, called the mass absorption coefficient, has required theoretical corrections or development of analytical correction factors. The accuracy of these methods has generally been limited for many elements, and, as a consequence, X-ray spectrographic analysis has generally been limited to measurement of bulk concentrations of chemical species (0.01 percent or greater). The accuracy demonstrated by this method of rapid routine bulk analyses has become outstanding. Table I shows the small variation in the fourth decimal place of the X-ray spectrographic analysis of a large series of standard commercial glasses.

A recent major advance in X-ray spectroscopy has occurred through the development and use of a monochromated X-ray source by R. L.

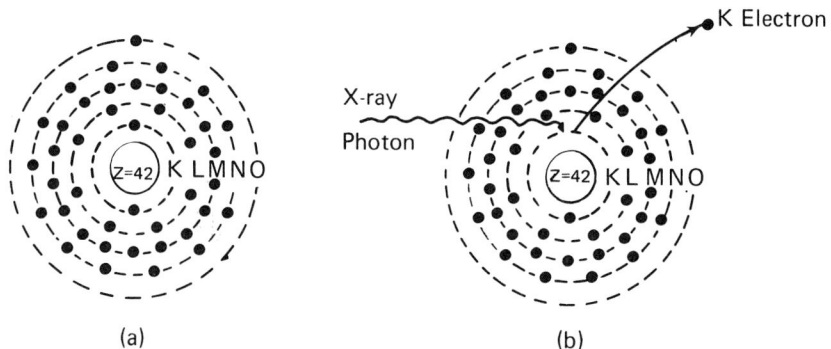

(a)   (b)

Figure 2. Electron transition resulting in production of characteristic X-rays used in spectrochemical analysis.

TABLE I

Range of Compositional Analyses Obtained for Typical Commercial
Glasses Using an Automated X-Ray Spectrometer

	Borosilicate Glass		
	Average %	High %	Low %
SiO_2	76.66	76.86	76.50
Al_2O_3	2.05	2.08	2.02
CaO	0.51	0.51	0.50
Na_2O	5.05	5.07	5.04
As_2O_3	0.24	0.24	0.24
Fe_2O_3	0.018	0.019	0.018
B_2O_3	remainder		

	Soda–Lime–Silicate Glass		
	Average %	High %	Low %
SiO_2	76.52	71.55	71.45
BaO	0.57	0.57	0.56
Al_2O_3	1.16	1.17	1.16
MgO	0.57	0.58	0.57
CaO	10.40	10.47	10.33
K_2O	0.16	0.16	0.16
Na_2O	15.20	15.29	15.16
Fe_2O_3	0.210	0.211	0.209
SO_3	0.14	0.15	0.12

Sparks[16]. The use of the monochromated X-ray source now makes it possible to apply the rapid automated X-ray spectrographic analysis method to trace-element analyses in solids such as glasses. Analysis in the range of parts per million have been reported, as shown in Table II, and continued development of this technique shows promise for obtaining trace-element analyses in the parts-per-billion range on a routine basis.

Since a thorough characterization of trace-element effects on the properties of glass has seldom been attempted because of the expense and complexity of such analyses, the recent development of this instrumental method should herald a major advance in the chemical aspects of glass characterization.

One of the requirements for accurate application of X-ray spectroscopic instrumental chemical analysis is that of homogeneous distribution of the elements being analyzed. Lack of a homogeneous density of the sample under X-radiation results in significant errors which can reach as large as a factor of 50 percent or more [17]. The scale of homogeneity of most glass articles makes such a requirement easily satisfied. In fact, it is possible to employ glass technology as a great aid in achieving homo-

TABLE II
Compositional Analysis of Coal Using Sparks Monochromated
X-Ray Fluorescence Method[16]

NBS 1632 Reference Standard (ppm)	XRF (ppm)
(8600)	7150 ± 800
18 ± 2	22.6 ± 3
37 ± 4	36.6 ± 1.4
5.9 ± 0.4	3 ± 2ᵃ
2.8 ± 0.2	5.5 ± 0.5
	23.7 ± 3.2
	28.6 ± 3.2
	155 ± 15
30 ± 9	23 ± 0.9

geneous distribution of elements for other materials which must undergo analyses. Minerals, ceramics, metals, particulate solids, effluents, etc., can be reduced to a state where X-ray spectrochemical analysis is possible by dissolving the analysate in a glass. This process, called the glass-fusion technique, has undergone recent advancements by the application of glass science to the dissolution process [18,19]. However, additional studies need to be conducted in order to further optimize this important new method for general automated chemical analyses.

Another exciting advantage of X-ray spectrographic technique from a glass-characterization standpoint is the portability of the system when radioactive isotopes are used for the incident energy source [13]. Approximately a 1 mC source is sufficient for obtaining spectra. High-energy resolution, solid-state detector diodes in multichannel pulse height analyzers are used as the detector systems. The isotope spectrographic source makes a small portable item possible, which can permit the use of chemical analyses of large manufactured glass objects already in service. The analysis is nondestructive and therefore can be used to monitor alterations in the chemical compositions of the material, providing a chemical characterization base for evaluating maintenance or changes in performance.

Structural Characterization

Diffraction techniques [20] are the primary means for characterization of the atomic or ionic structure of glasses. However, the lack of long-range order in glasses produces an additional complexity in the interpretation of diffraction analyses of glass structure that results in ambiguity. The ambiguity arises as a result of the necessity of interpreting the nearly featureless glass-diffraction spectra with the use of a structural model.

The recent advances in obtaining a statistical structural model interpretation of diffraction data by Mozzi and Warren [21] is a major advance in glass-structure characterization. Two significant improvements were achieved in their study. First, the use of a diffracted beam detector system, which eliminates Compton-radiation contributions to the experimental diffraction data, produced considerable additional accuracy in the diffraction intensity data at high $s = 2 \sin\theta/\lambda$ values. A somewhat simpler system than that used by Warren and Mozzi is shown in Figure 3 [22]. Second, the diffraction data was analyzed by Mozzi and Warren with the use of pair-function distributions instead of radial distribution functions characteristic of the earlier glass-diffraction studies of the 1940s and 1950s [23].

The net effect of the Mozzi and Warren improvements was a resolution of the pair-function distribution peaks which were characteristic of the second, third, and fourth coordination shells of the ions in the glass structure. When the improved resolution was combined with the detailed pair-function distribution analysis, a statistical description of the extent of disorder–order in the glass structure became possible. As shown in Figure 4, a distribution of Si–O–Si bond angles (α) and bond distances (r_{ij}) was obtained. The half-breadth of the distribution peak can be considered as a characterization feature of the extent of disorder of the glass structure [22]. This is because a completely crystalline material will exhibit a bond-angle distribution curve function which is nearly a delta function having a very narrow width.

Computer automation of the statistical distribution-function interpretation of the diffraction data developed by the author and colleagues [22] has made it possible to obtain a series of disordered distribution functions from X-ray diffraction data which characterize the structural disorder in vitreous SiO_2. Examples of the distributions of Si–2nd Si and oxygen–2nd oxygen bond distances for vitreous SiO_2 obtained in this

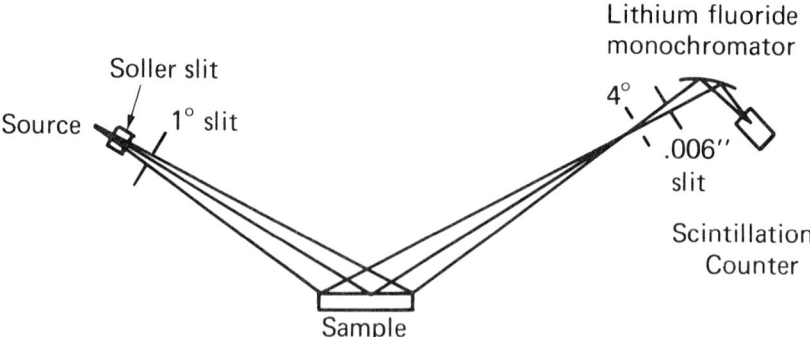

Figure 3. Schematic of an X-ray diffracted-beam monochromated system which experimentally eliminates much of the Compton scattering.

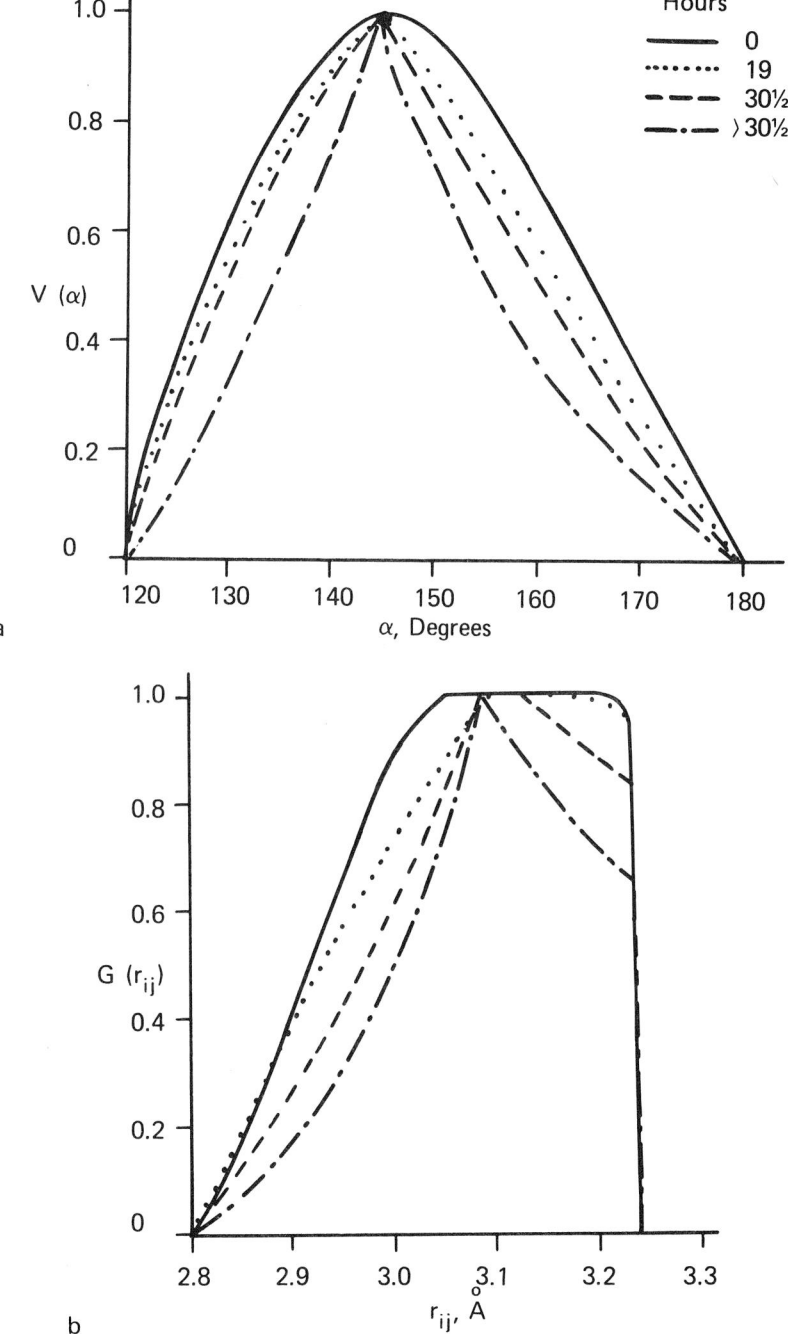

Figure 4. The distribution of Si-O-Si bond angles (α) and distance of separation R_{ij} of vitreous SiO_2 are shown as a function of thermal treatment at 1450°C. The changes in distributions shown are due to progressive formation of crystals in the glass[22].

217

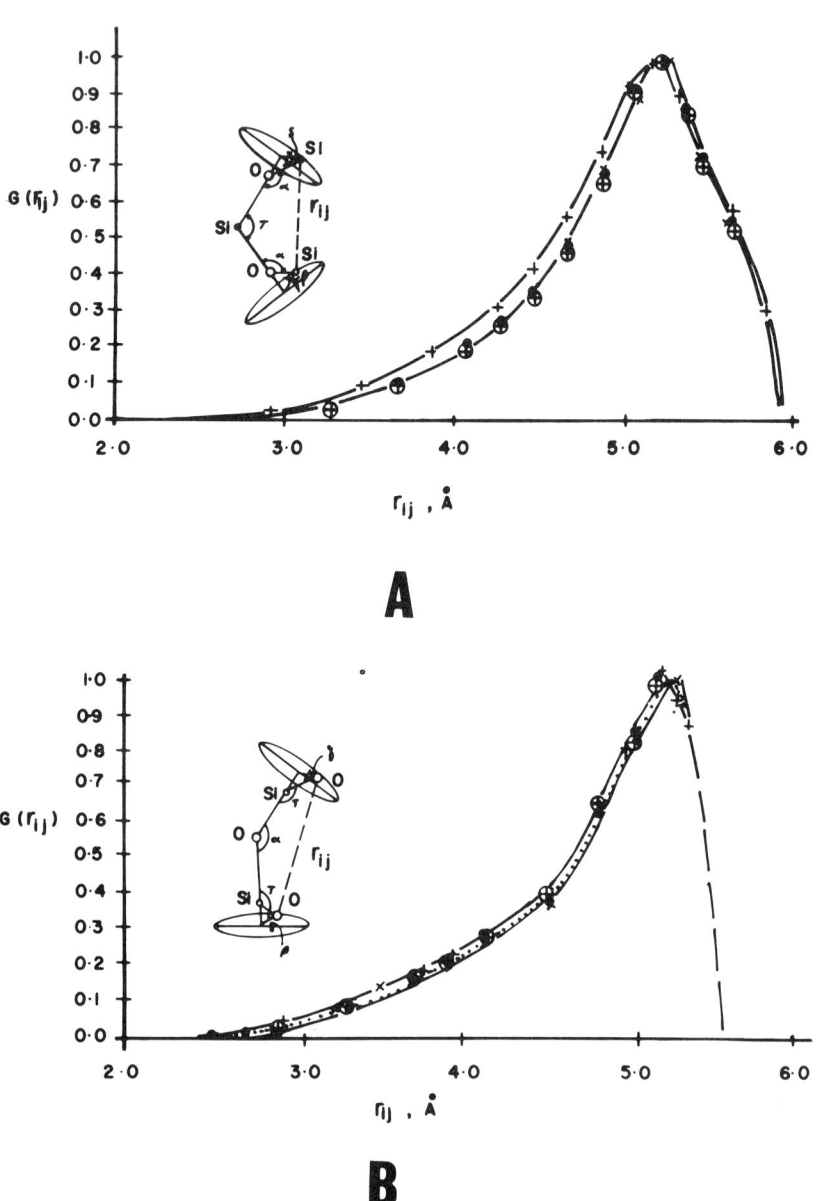

Figure 5. A. Distribution of Si-2nd Si bond distributions in vitreous SiO$_2$[22]. B. Distribution of oxygen-2nd oxygen bond distributions in vitreous SiO$_2$. Changes in distributions shown are due to the onset of crystallization at 1450°C[22].

manner are shown in Figures 5(a) and 5(b). Application of this method to the interpretation of fast neutron irradiation damage in fused silica has shown the advantages of the disorder distribution function for glass structural characterization [24]. An effort to follow thermal processing by use of this characterization method, however, proved to be of no avail in the fused-silica system as a result of heterogeneous nucleation and crystallization domination of the thermal processing (Figures 4 and 5). The method has been shown to be useful, however, in identifying structural differences between vitreous silica glasses prepared by different processing methods [24].

Electron diffraction techniques are particularly useful in characterizing the structure of glassy thin films [25]. Because amorphous thin films can be prepared by vapor-deposition processes that are compatible with integrated circuit technology, there has been considerable interest in the characterization of local order of such amorphous films during the last few years. An electron diffraction pattern is similar to that described above for X-ray diffraction in that it yields only very diffuse rings. The intensity profile across the diffuse rings must be measured with suitable elimination of effects due to incoherent and multiple scattering. A numerical Fourier transform is carried out on the resulting intensity profile, yielding, in effect, a spectrum of the predominant interatomic distances in the specimen, termed the radial distribution function.

A notable advance in obtaining high-precision intensity measurements has recently been developed through the use of scanning or direct recording of the electron diffraction intensity profile [26]. In this type of instrument (Figure 6), the scattered electron beams are reflected back and forth across a very small aperture by a pair of magnetic coils situated beneath the specimen. Electrons that have lost more than a few volts in energy are rejected by an electrostatic filter. Transmitted electrons are collected by a Faraday cage or other electron detector. As a consequence, the diffraction profile obtained (such as shown in Figure 7 for amorphous Ge) has incoherent, inelastic scattering contributions experimentally eliminated. Excellent comparisons of processing parameters such as time, temperature, or composition and substrate effects can be achieved. An important feature is the extension of the measurements to a greater angular range. The improved resolution of the diffracted intensity over larger angular range produces a much more detailed radial distribution function. Consequently, the interpretation of various models which can account for the distribution of atoms in the amorphous thin films is more easily achieved. A number of reviews on the use of this and other electron-diffraction methods for a variety of amorphous thin films are presented by several authors [25,27].

The scattering factors of atoms are different for neutrons and electrons and X-rays. Consequently, structural analysis of glasses containing two or more atoms can be improved by comparing neutron-diffraction spectra with those obtained by X-ray diffraction or electron-diffraction

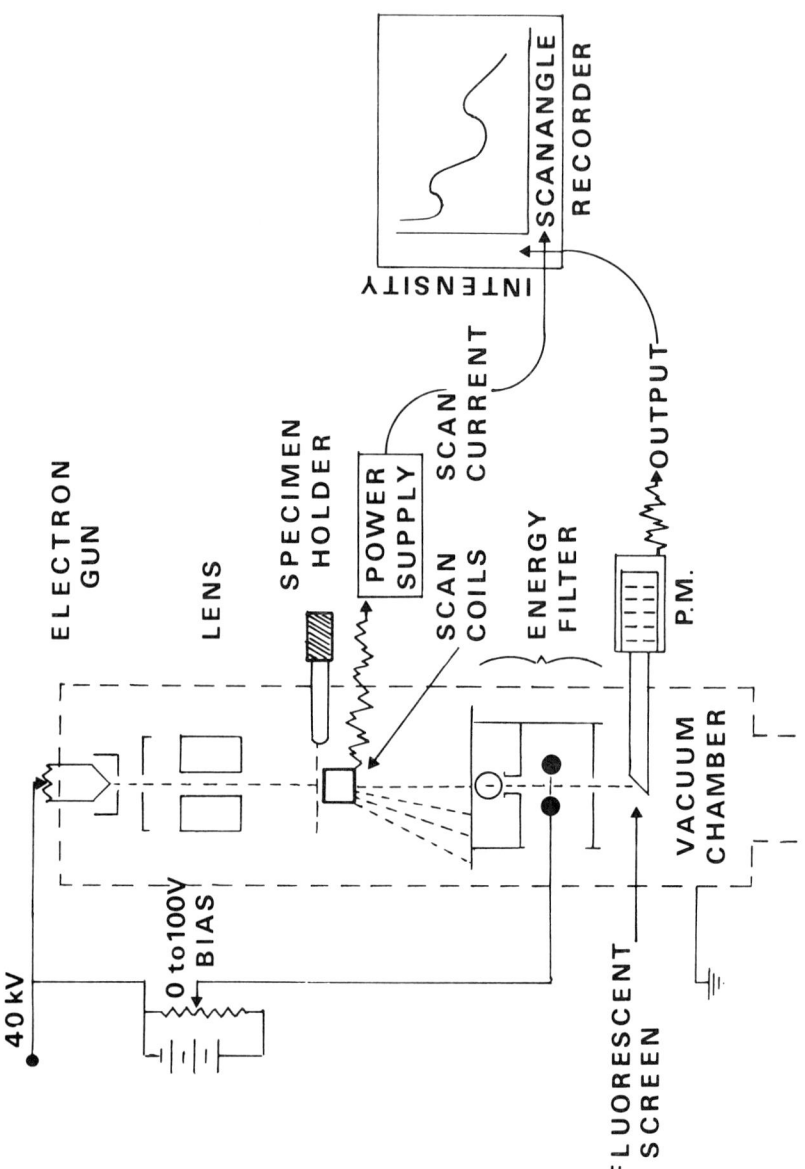

Figure 6. Schematic of a direct-recording scanning electron diffraction system for studying glassy thin films [25].

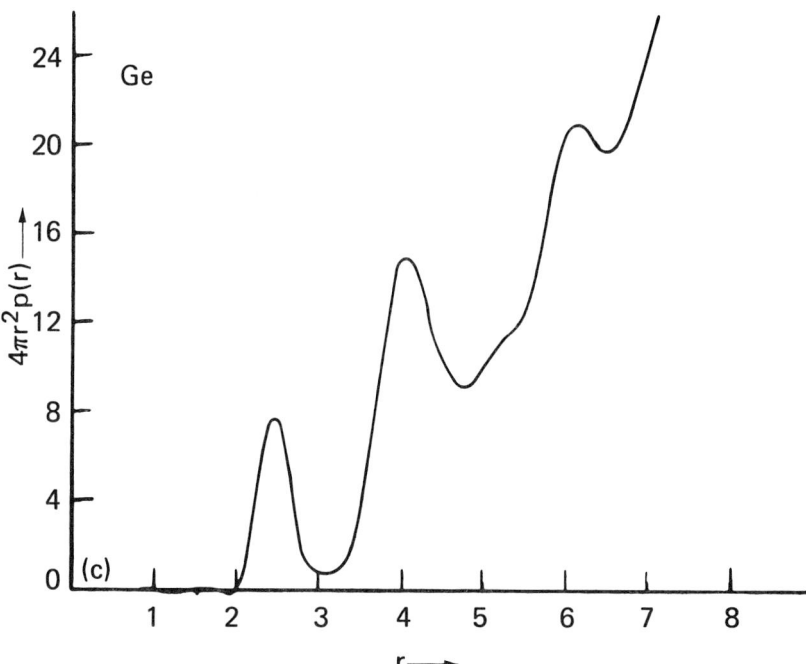

Figure 7. Radial distribution function of Ge atoms in amorphous Ge obtained using scanning electron diffraction [25].

methods. However, the availability of neutron-diffraction facilities severely limits this method as a widely employed tool for structural analysis. Significant results obtained by this method are the separation of the overlapping radial distribution function peaks of oxygen and metallic elements [28].

Specific structural information on the chemical coordination of various ions in glass is obtained using nuclear magnetic resonance (NMR) techniques [29]. Understanding of changes in B^{3+}–O^{2-}–coordinations in borate glasses from threefold to fourfold symmetry with NMR has confirmed X-ray interpretations [30]. Application to anomalous behavior in glasses such as the mixed-alkali effect have also been achieved [31].

A powerful new technique for structural determination of noncrystalline materials is the Fourier inversion of extended X-ray absorption fine structure (EXAFS) [32]. The radial distribution function obtained in this manner contains information on the number, distance, and distribution of atoms surrounding the absorbing atom. It is based upon the sensitivity of the fine structure observed on the high-energy side of an X-ray absorption edge to the short-range order around the absorbing atom even in complex glasses. Sayers, Lytle, and Stern have applied the technique to amorphous Ge, GeO_2, and GeSe, and reveal considerable additional detail over X-ray and electron-diffraction methods [32].

Microstructural Characterization

It is now generally agreed that all glasses cannot be considered as homogeneous isotropic solids on a scale of 10 Å or more, and possess various microstructural features. The microstructural character of the glasses may be as minimal as small density fluctuations on a scale of 10–100 Å, as shown for single-phase glasses [33]. However, the microstructure may consist of glass and glass separation of phases on a scale of 20–10,000 Å or greater [34,35]. Bulk microstructural defects, such as crystals, seeds, stones, cords, and striae may also be present, and, if they are, may dominate the physical and chemical behavior of the material. Consequently, microstructural characterization is one of the dominant themes in current glass research. New instrumental experimental tools has made the analysis and interpretation of microstructural features possible on a scale heretofore unachievable. As a result the concept of what constitutes a glass has broadened considerably to include a much wider range of materials. The definition of glass has also been extended to exclude a material which contains a large volume fraction of crystals of a controlled size, termed glass-ceramics. The analytical techniques to be discussed in this section are applicable to both glasses and glass-ceramic materials.

X-ray small-angle scattering is one of the most powerful tools for evaluating structural heterogeneities in the size range of 10–10,000 Å, as reviewed by various authors [36,37]. The requirements for obtaining X-ray small-angle scattering in a solid is the presence of heterogeneous regions of sufficient electron-density difference from the matrix that they scatter the incident X-ray diffraction beam, and that the volume fraction of the particles is sufficiently large that the scattered intensity is measurable using available detector systems. A typical X-ray small-angle scattering configuration is shown in Figure 8, and a typical pattern obtained from the scattering from spherical particles is shown in Figure 9. The particle parameters that are obtainable using the small-angle scattering method include: radius of gyration, integrated intensity (Q_o), volume per particle, number of particles per volume, total surface area of particles, and size distribution of particles. The mathematical relationships required to obtain the microstructural parameters are too involved to discuss in this review chapter, but are well reviewed in the literature[36,37]. The small-angle X-ray method has been used to determine fluctuations in nearly homogeneous, simple glasses, phase separation in glasses, development of nuclei, radiation damage, and crystallization.

An example of data obtained from the characterization of 33 mole percent lithia–silica glasses that have undergone a nucleation treatment are shown in Figure 10. Measurement of the sizes of the nuclei particles using this technique, shown in Table III, were essential in interpreting the nucleation state of this material. The relationship of physical properties of the nucleated material and subsequent processing steps was made possible by the use of this characterization method[38].

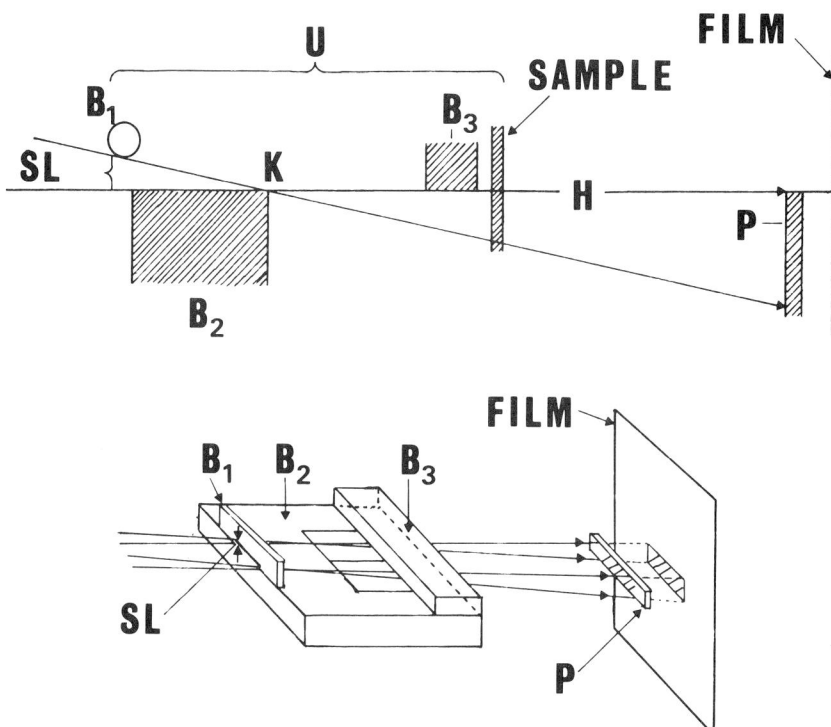

Figure 8. Schematic of a Kratky X-ray small-angle scattering system.

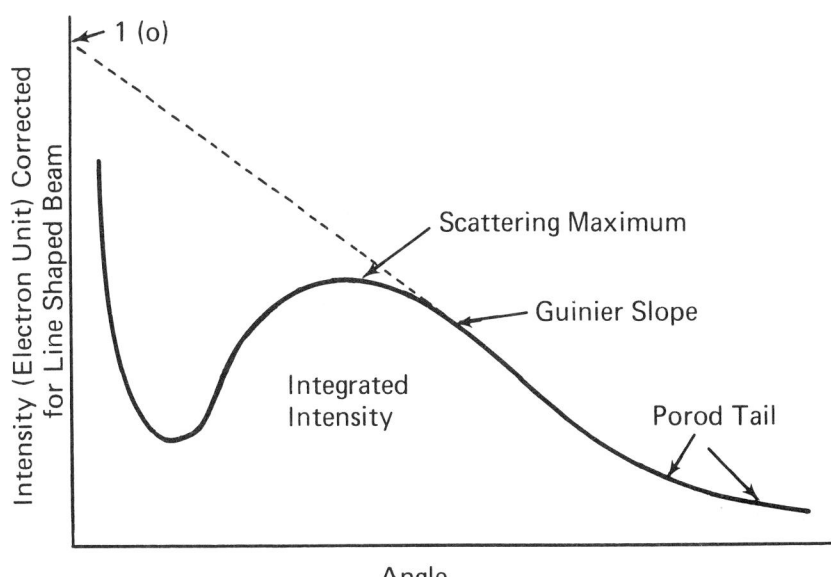

Figure 9. A typical X-ray small-angle scattering pattern obtained from spherical particles[36].

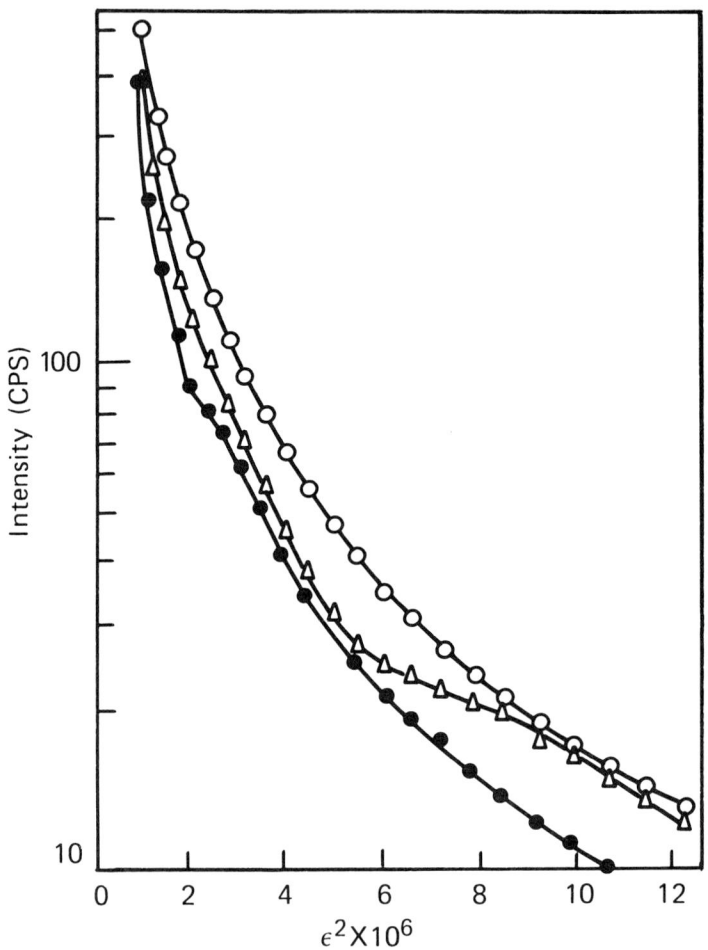

Figure 10. X-ray small-angle scattering intensity as a function of a scattering angle for 33 mole % Li_2O–SiO_2 glass. The changes in scattering observed are due to formation of nuclei in the glass[38].

When the heterogeneities present in the glass structure are of sufficient composition difference that they possess a higher electrical conductivity than the bulk glass matrix, quantitative dielectric relaxation techniques can also be used as a characterization tool. This method is largely based upon the development of Maxwell–Wagner–Sillers dielectric losses at the interfaces of the heterogeneities at a characteristic electrical frequency[39]. The size of the particles, aspect ratio, and volume fraction can be established by measuring the frequency dependence and magnitude of the dielectric loss. Application of this technique to characterizing the

TABLE III
Particle Sizes Obtained from X-ray Small-Angle Scattering[38]

Heat-treatment time (hours)	Guinier radius (Å) in glass containing		Porod radius (Å) in glass containing	
	30 mol % Li$_2$O	33 mol % Li$_2$O	30 mol % Li$_2$O	33 mol % Li$_2$O
2	265–302	—	159	—
3	—	256	—	162
5	228–272	233	149	196
10	237–310	242	143	194

nucleated state of a lithia–silica glass is shown in Figure 11. The change in dielectric loss angle (tan δ) frequency dependence indicates a change in aspect ratio of nuclei. The decrease in magnitude of the loss peak indicates a diffusional change in composition of the nuclei[40].

Electron microscopy is a standard analytical tool for microstructural characterization[41,42]. Electron microscopy can be applied through the use of replicas of various types or through electron transmission. Replica electron microscopy has been a standard technique for analysis of glass

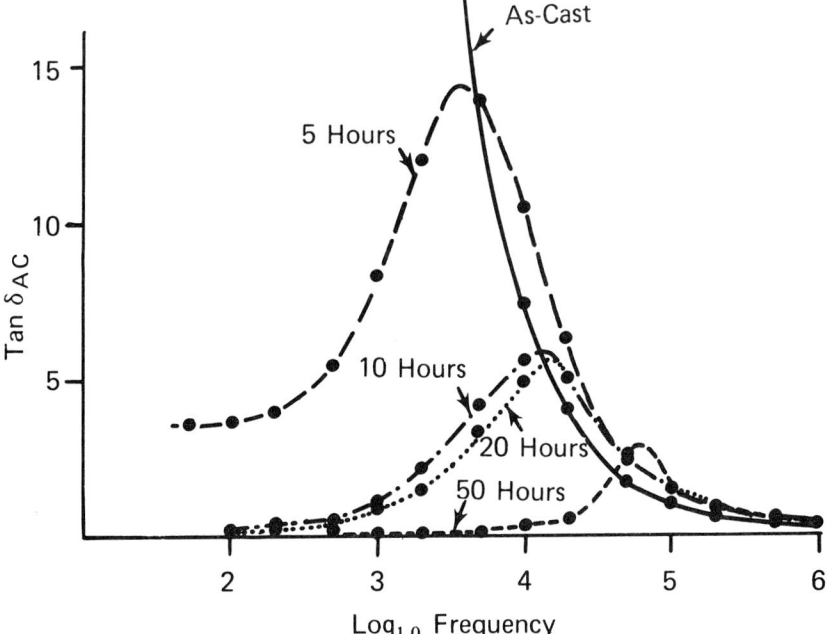

Figure 11.   Changes in the heterogeneous dielectric relaxation loss spectra of a 33 mole % Li$_2$O–SiO$_2$ glass due to growth and resorption of metastable nuclei[40].

microstructural features for many years. A shadowed replication micro-graph of the fracture surface of a partially crystallized Li_2O–SiO_2 glass is shown in Figure 12. Resolution limits for the replication method gen-erally are in the range of 20 Å for single-stage or two-stage carbon replicas shadowed with platinum metal. It is the particle size of the platinum shadow that limits resolution of the replica to about 20 Å. Unshadowed carbon replicas may have resolutions <10 Å, but the image contrast usually is low and phase-contrast effects can contribute to difficulty in image interpretation. Although, as discussed below, scanning electron microscopy (SEM) has tended to dominate recent topographical char-acterization studies of microstructural heterogeneities, the replication method has an advantage in that its surface topography can be observed at a resolution better than the 100 Å limit of most SEM instruments.

Direct observation of microstructural heterogeneities in glasses can be obtained with transmission electron microscopy (TEM). If the hetero-geneity is crystalline, identification can also be achieved using selected area electron diffraction. The difficulty in applying transmission electron microscopy techniques to glass characterization is twofold. First, the sampling volume is extremely small. Because of the very high resolution—less than 10 Å—the disadvantage of observing only a very small portion of a bulk sample may be overcome by the detail in the observation. The

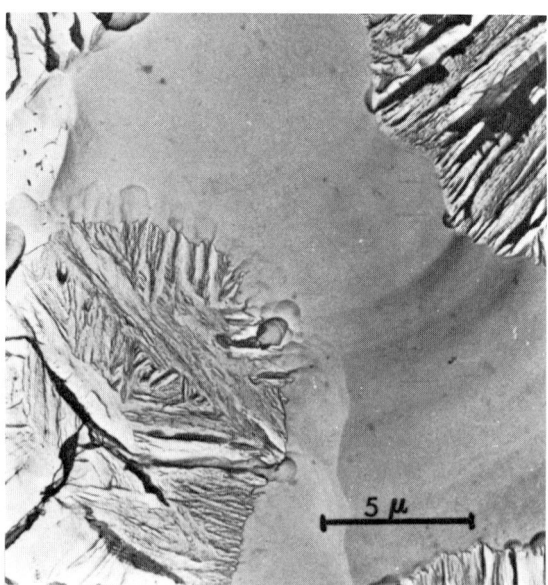

Figure 12.   Electron micrograph replica of a fracture surface of a partially crystal-lized 33 mole % Li_2O–SiO_2 glass[65].

second problem encountered is that of obtaining a suitably thin sample for transmission of electrons.

Three methods are used for preparing samples for transmission electron microscopy. The simplest is a fracturing technique whereby one simply powders the samples and searches for a feather edge on one of the glass fragments that is thin enough for electron transmission. The fracture method has the advantage that it is quick and simple but has the disadvantage that preferential fracture at heterogeneities may occur, yielding poor statistical basis for interpreting the data. In addition, the selection process of locating suitably thin specimens is quite tedious. A transmission electron micrograph specimen from the fracture technique is shown in Figure 13.

The most standard method employed has been that of chemical etching[43]. The appropriate etch is usually a mixture of hydrofluoric acid with acetic and hydrochloric acid addition[43]. Recent studies have shown that such an etched edge does not produce a chemical change in silicate glass specimens[44]. The hole in the center of a small specimen is masked off and the acid is allowed to eat through the hole. The feather edge of the glass around a hole is usually thin enough for observation in the electron microscope.

A third technique now being widely used is that of ion-beam milling.

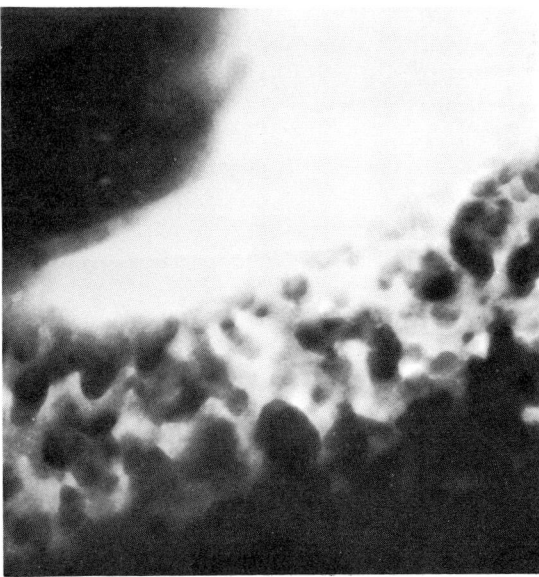

Figure 13. Transmission electron micrograph of a thin fracture edge of a soda-lime-silica glass with a phosphate addition showing structural fluctuations. 200,000X.

A high-energy beam of ions, usually argon, is allowed to impinge on the sample surface. The bombardment of the energetic ions on the glass surface produces an erosion of the material, uniformly eating its way through until a hole is produced. Observation of the thin glass around the hole is made in the electron microscope.

The scanning electron microscope has rapidly become one of the most widely used tools for microstructural characterization of glasses[45,46]. The ease of sample preparation, large-sized samples, wide range of magnification from 10x to 25,000x and extreme depth of field are responsible for the popularity of SEM. One minor drawback to the technique is the necessity of applying a thin conductive layer on the samples before observation in order to avoid charging effects. However, this is a simple technique in any electron microscopy laboratory. Another limitation of SEM is the difficulty of obtaining quantitative microscopy data. This is because the large depth of field produces a 3-D effect to the image which does not satisfy the planar section requirements of quantitative microscopy relations. The three-dimensional imaging characteristic of this

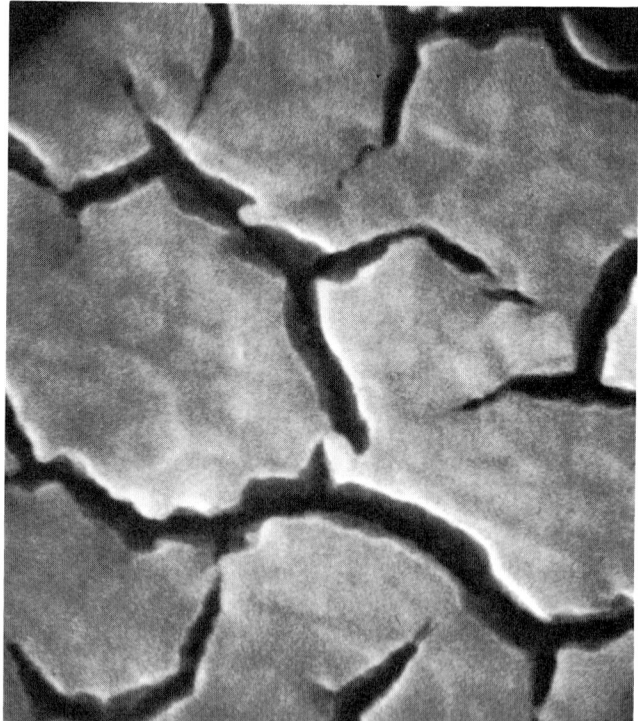

Figure 14.   Scanning electron micrograph of surface cracks in a chemically etched soda-lime-silica glass with phosphate additions. 5,000X.

tool can be seen in Figure 14, which is a scanning electron micrograph of the crack pattern on a glass surface due to thermal stressing.

The use of X-ray nondispersive analysis accessories in the SEM makes compositional analysis of the observed surface also possible[47]. An X-ray spectra from a glass sample taken with the nondispersive analysis unit in an SEM is shown in Figure 15[48]. The time required to take such spectra is only minutes, and the data can be made quantitative through the use of comparative standards of known composition at the same time. The combination of high resolution, near 3-dimensional imaging, and compositional analysis of the microstructural features observed make the SEM unit perhaps the most powerful general characterization tool currently available.

Electron microprobe analysis has also proven to be a useful tool for the chemical analysis of microstructural features of glass[49,50]. The electron microprobe also produces an analysis of the intensity of X-rays produced by the bombardment of the sample surface by an incident electron beam. The width of a standard microprobe beam is in the range of 0.5 μm, and thus chemical analysis of microstructural features in the range of several microns are possible. An example of the use of a microprobe to obtain a compositional profile of a corroded glass surface is shown in Figure 16[51]. Both intensity measurements of elements and image displays of the distribution of the elements are possible.

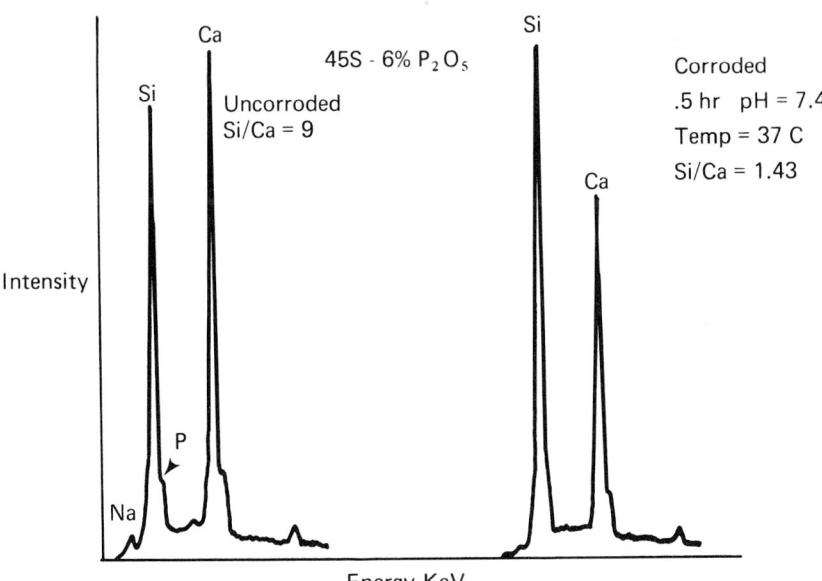

Figure 15. X-ray nondispersive analysis spectra of a soda-lime-silica glass with phosphate additions before and after aqueous surface corrosion[48].

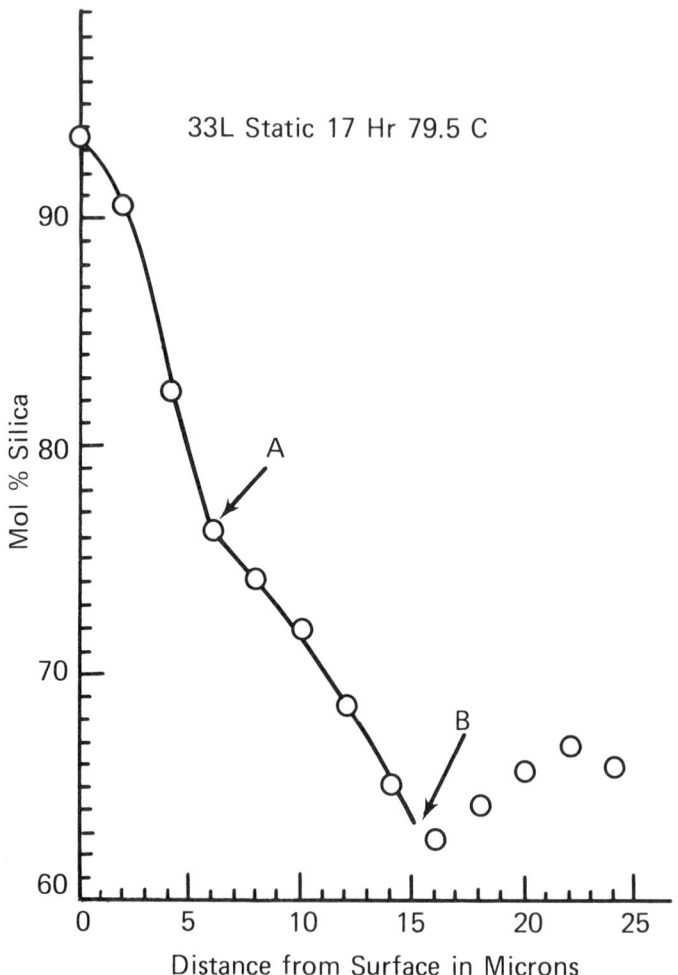

Figure 16.   Surface profile of increased SiO_2 content due to aqueous water attack of a 33 mole % Li_2O–SiO_2 glass obtained with an electron microprobe. Point counting of a cross sectioned surface was employed[51].

Optical microscopy using polarized light has been the classical tool for microstructural analysis of glasses[52,53]. The ease of operation, inexpensiveness, and large background of literature in this area still makes optical microscopy one of the first techniques that should be considered in the observation and interpretation of large-scale defects in glass. Identification of crystalline products or refractory stones is possible using optical mineralogy methods[54]. An example of optical microscopy of a partially crystallized 33 mole % Li_2O–SiO_2 glass viewed under polarized light is shown in Figure 17[55]. Use of polarized light and stress optical data

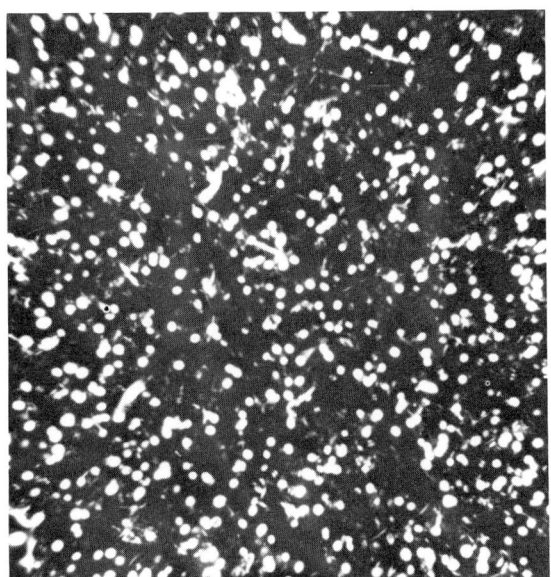

Figure 17. Optical photomicrograph taken with polarized light showing ~ 5 μm crystals grown in a 33 mole % Li_2O–SiO_2 glass[55].

makes the interpretation of cords and striae in glass also possible. A combination of mass spectroscopy and optical microscopy can be used for analysis of the gas content in bubbles in glasses[56].

Optical scattering of light can be used for characterization of microstructural features in the size range of 0.5–1.5 μm[57]. There are several fine reviews of the development and theory of light scattering[58,59]. The principles are similar to those involved in X-ray small-angle scattering discussed above. However, since the wavelength of light is in the range of 0.5–1.5 μm, light scattering is sensitive to microstructural features in this range rather than the 10–10,000 Å range characteristic of XSAS.

Scattering by particles which are small with respect to the wavelength of the scattered light can be described simply with Rayleigh scattering. However, the more general Mie theory[57] must be used for larger particles. The theory assumes that the scattering particle is an optically isotropic sphere with a constant refractive index. Computer programs are available from the National Bureau of Standards for calculation of various Mie scattering functions covering a wide range of particle size and relative refractive index. This calculational basis, plus the availability of lasers which provide the monochromatic high-intensity light source, make light scattering a most useful tool. Use of appropriate mathematical approximations will yield both particle size and size distributions with this method.

The study by Hammel of light scattering of phase-separation nucleation in glasses illustrates the applicability of this method for microstructural analysis[60]. Recent studies by Macedo and colleagues following the kinetics of structural change in phase-separating systems have employed line-shape analyses of light-scattering curves[61].

Quantitative analysis of microstructural features is best achieved by combining one of the scattering techniques—XSAS, or light scattering—with one of the direct observational techniques. In order to achieve quantitative interpretation of TEM, SEM, or optical micrographs, it is necessary to employ quantitative stereological analyses as reviewed by Freiman[62] and discussed in detail in DeHoff and Rhines[63]. Application of these quantitative methods is only beginning to be applied to glass microstructural features[64-66].

Surface Characterization

Understanding the character of a glass surface is one of the most important features of a characterization program[62]. This is because the flaws in the glass surface are responsible for a decrease in strength greater than 100 x. Also, the glass surface interacts with the environment and the change in structure and composition at the surface may drastically alter bulk-glass properties. Many new and powerful surface analytical tools are now available for glass-surface characterization and are being applied in research programs.

For example, Auger spectroscopy is a relatively new nondestructive analytical technique for surface chemical analysis[68,69]. It is based upon the analysis of energies of secondary electrons ejected from a solid. The phenomenon is initiated by bombarding the material with a primary beam of electrons. The mechanism involved is shown in Figure 18. The initial beam of electrons ionizes one of the electron levels, such as the K-level shown. As the system tries to return to equilibrium, an e^- from a level such as the M-level drops in to fill the vacancy in the K-shell. When this happens, energy is transferred by the Auger process to an electron such as one occupying the N-level, which can then have enough energy to eject from the solid. This Auger electron is characteristic of the atom from which it is released. Only a few transitions occur among a wide number of possibilities for each element. Thus, by sorting out electrons according to their energies, identification of elements is relatively simple.

Auger electrons produce peaks in the secondary-electron energy distribution. This is shown in the lower curves of Figure 19. These Auger peaks are enhanced in practice by recording the derivative of this secondary distribution, as shown in the upper curve of Figure 19. Thus, in standard application, a plot of $N(E)/dE$ vs. E is referred to as the "Auger spectrum," and elements are identified by the energies at which these peaks occur[68,69].

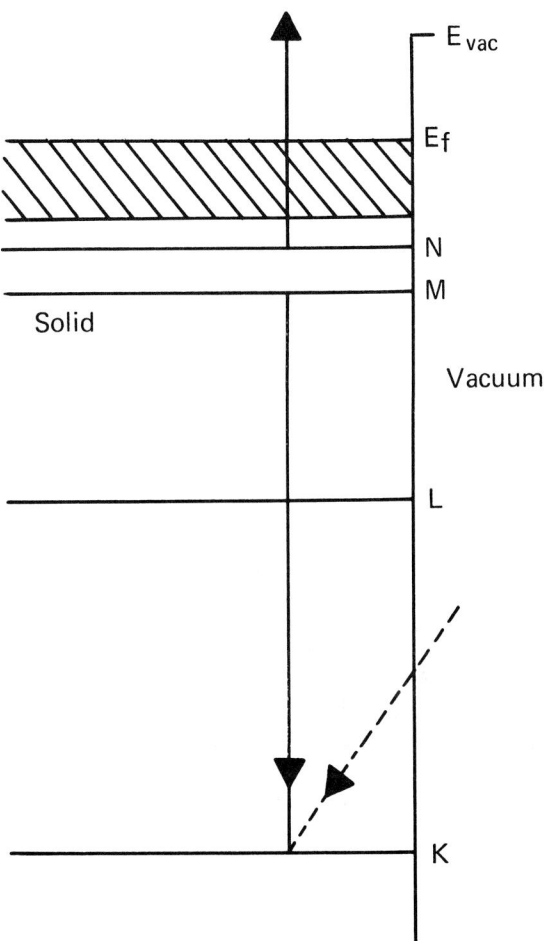

Figure 18. Mechanism of emission of an Auger electron from a glass due to incidence of a primary electron beam (dotted arrow).

Only if the Auger electron is created within a few atomic layers of the sample surface will it be able to migrate to the surface and escape, without the loss of energy. Because of this, its use is limited to examination of material surfaces within a few monolayers. It is a very sensitive technique, with detection and identification of much less than 1 percent of a layer of atoms possible. All elements except H and He are identifiable by this method. Although Auger spectroscopy is primarily a qualitative tool at present, quantitative measurements can be made by utilizing peak heights, which are proportional to the number of atoms producing the peak, and standard means of calibration.

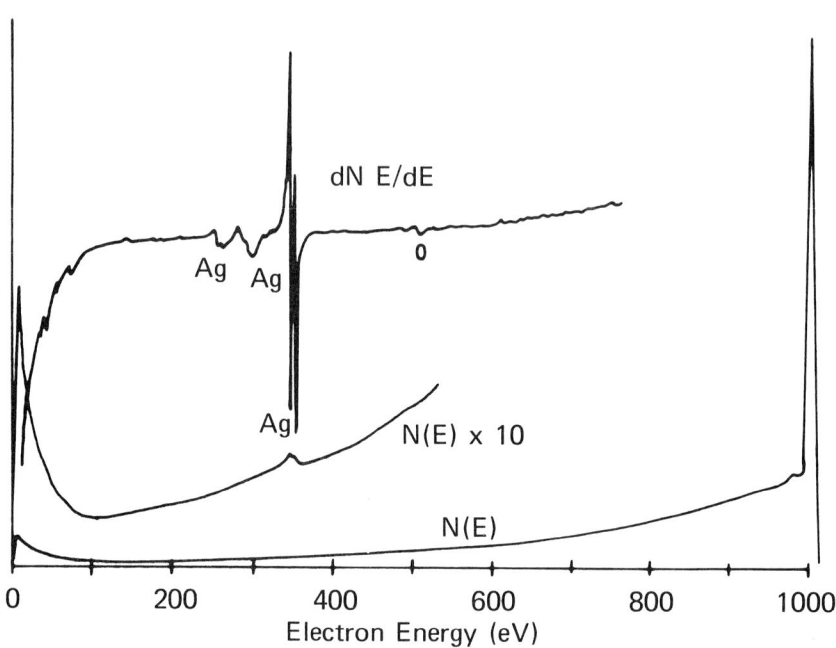

Figure 19. A typical Auger electron spectra showing Ag peaks. Both raw (lower) and differentiated data (upper) curves are shown.

Figure 20 shows a schematic arrangement for the electron optics and the basic electronics involved in the Auger-electron spectroscopy system[70]. The system consists of an electron gun used for excitation, a coaxial cylindrical analyzer for analyzing the electron energies being emitted, an electron multiplier for amplification of the Auger-electron signal, and a system of standard modulation and phase-sensitive detection electronics.

Auger spectroscopy can be used widely in the study of a variety of materials, including metals, semiconductors, and insulators, and including glasses. The capability of ion-sputtering a surface coupled with simultaneous Auger analyses has opened up a wide range of research opportunities in the area of compositional profiling.

Figure 21 shows some representative data which can be obtained using an Auger facility[70]. These curves represent some recent work which has been undertaken to evaluate the properties and function of very thin metallic oxide and organic coatings which are applied to commercial glass containers to increase their durability and mechanical strength. The lower curve shows a representative Auger spectrum of a plain uncoated glass container immediately after the forming operation. Peaks 1, 2, 3, and 4 are, respectively, silicon, carbon, calcium, and oxygen. The upper

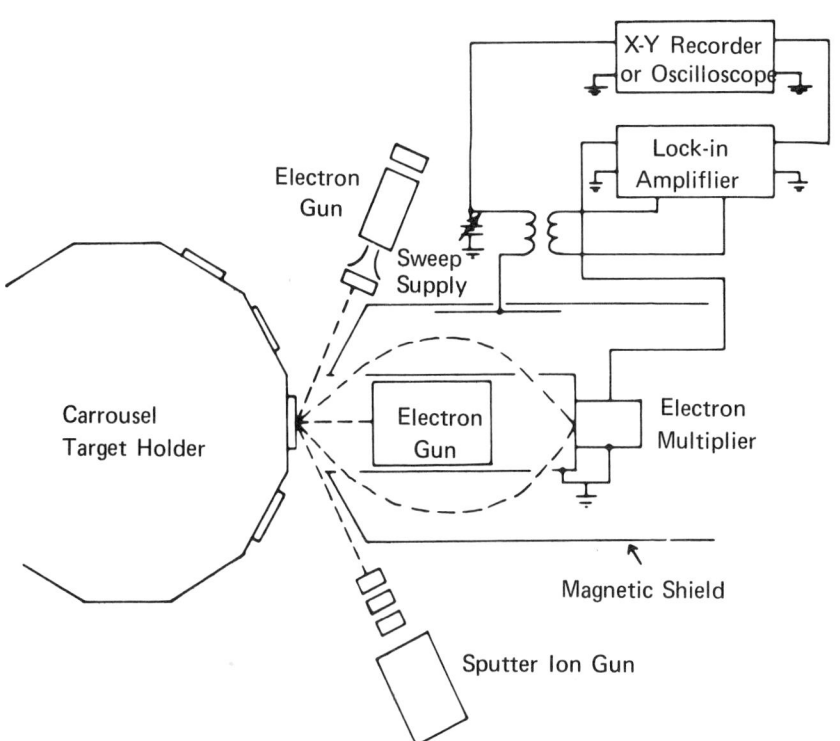

Figure 20. Schematic of an Auger spectrometer.

curve is the surface analysis of a glass container which has been coated with tin oxide. Peaks 1, 2, 3, 4, and 5 in this spectrum represent, respectively, silicon, carbon, calcium, tin, and oxygen. Considering that these surface coatings are approximately 50 Å in thickness, it is apparent from the data in Figure 21 that Auger spectroscopy is a very valuable technique in glass-surface analysis.

Electron spectroscopy for chemical analysis (ESCA) has been developed in the last few years and has been proven to be a valuable technique for structural and material analysis of surfaces[71]. It has been applied with much success to the study of electron binding energies in different atomic environments. The results provide a microscopic fingerprint which corresponds to the chemical combinations of the various elements.

Basically, the technique involves the measurement of the energy distribution of photoelectrons emitted from a material under bombardment by characteristic X-rays (Figure 22). The resultant spectra gives the kinetic energies of the emitted photoelectrons, which are then related to the specific binding energies of the electrons in the atom itself, such as in

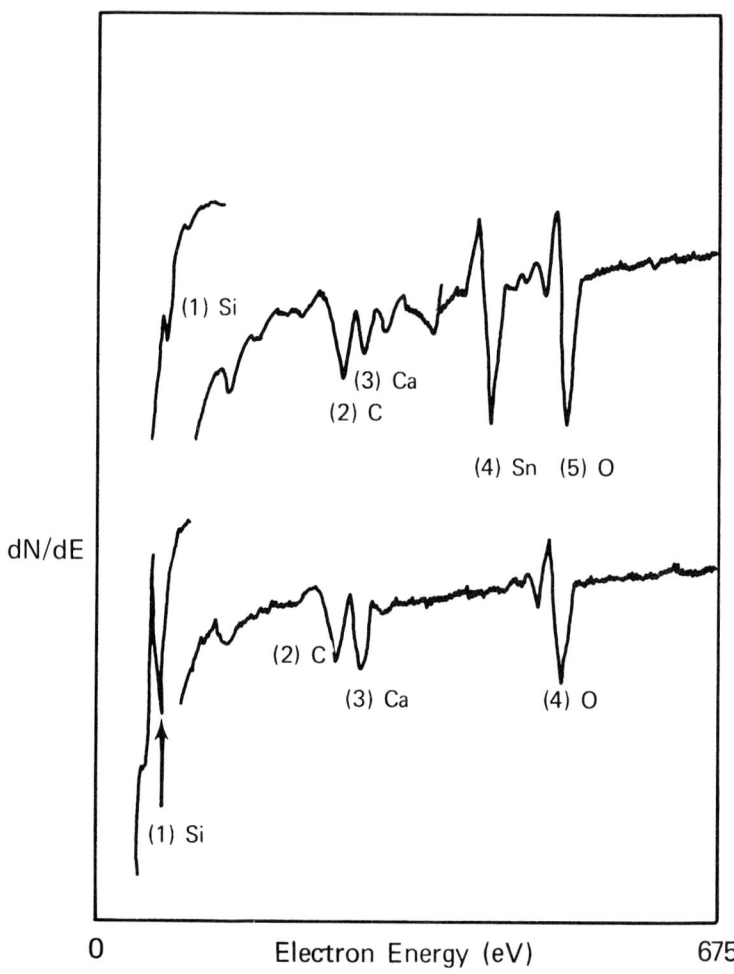

0   Electron Energy (eV)   675

Figure 21. Auger spectra of an uncoated glass container surface (lower curve) and a SnO coated surface (upper curve)[70].

the 30 mole % $BaO-V_2O_5$ glass shown in Figure 23[72]. The photoelectrons emitted from the sample originate within a relatively small depth below the sample surface, according to the penetration of the incident radiation; hence, the technique is essentially one suited to the study of surface phenomena and thin films.

Since ESCA is a relatively new technique, a review of its applications to date is limited, particularly in the glass field. Nonetheless, the potential of the method for problems in glass research and technology are relatively obvious. Perhaps the most attractive facet of this technique is its de-

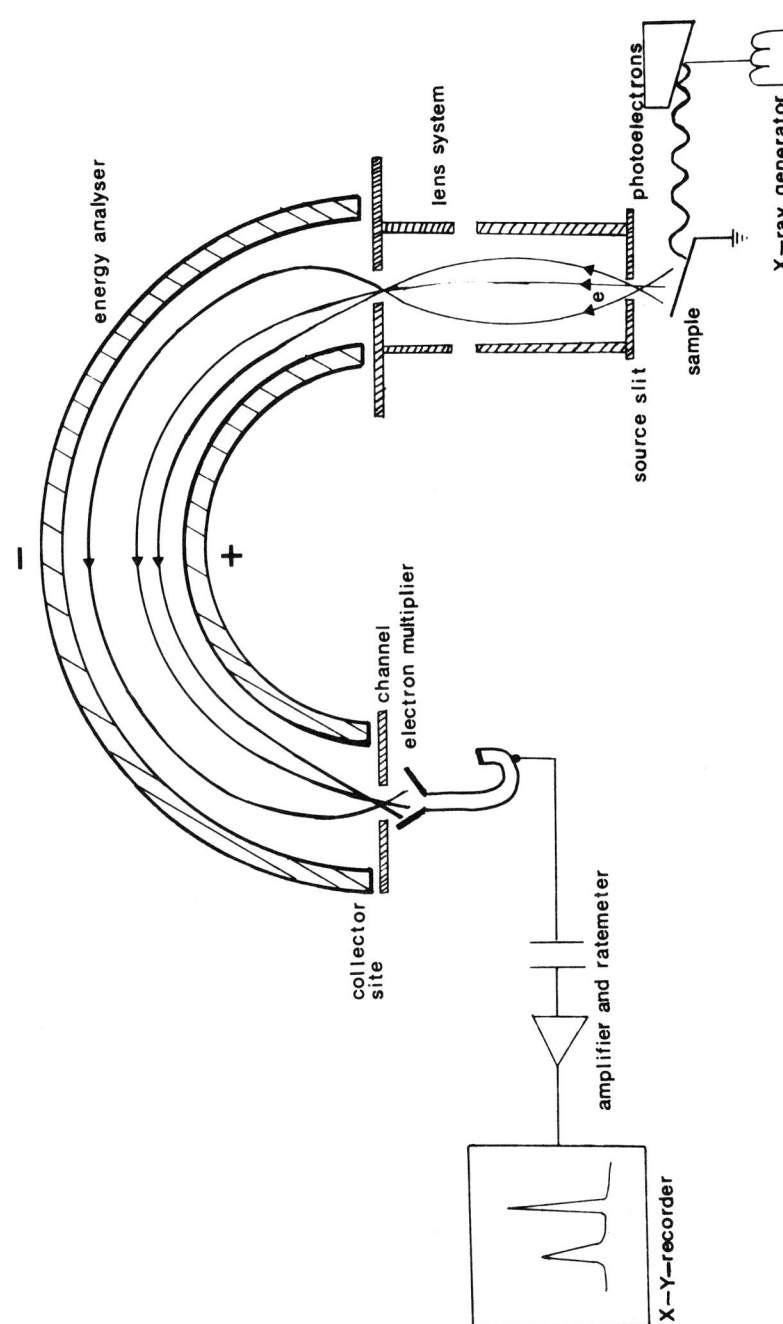

Figure 22.   Schematic of an electron spectrometer for chemical analysis (ESCA).

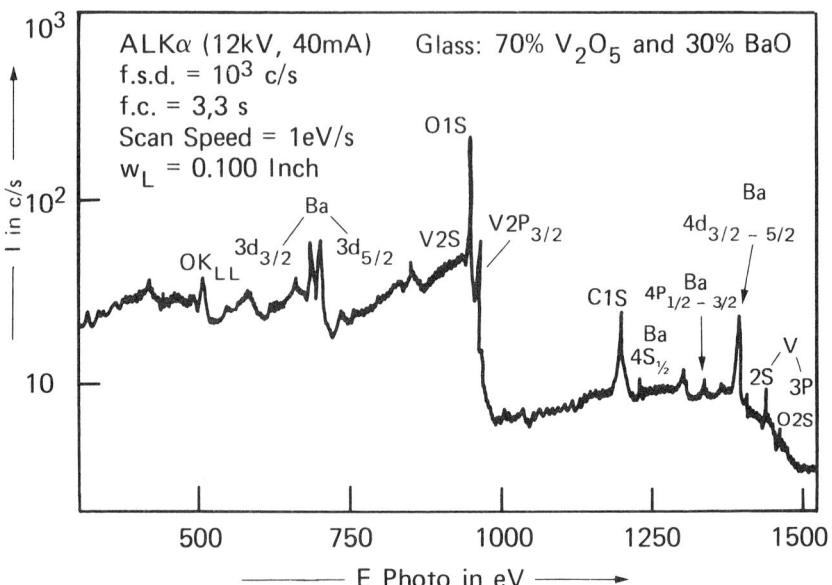

Figure 23.   ESCA spectra of a 30 mole % BaO-70 mole % V_2O_5 glass surface[72].

pendence upon "chemical structure." In the case of glass, it certainly has the potential for clarifying many of the structural features of the vitreous state. Most obvious is its ability to clearly determine the oxidation state(s) of the elemental components[72]. Carrying such a study to its limit, one can monitor the bonding character (percent ionic or covalent) as a function of composition. Therefore, ESCA provides a means of studying such structural variations at the surface as a function of thermal history. Its surface sensitivity can be further exploited for the study of adsorbed monomolecular layers and thin films on the glass surface. It should prove particularly useful for studying the nature of corrosion films (gels) which form on the glass surface as a result of certain environments.

Secondary ion mass spectroscopy (SIMS) involves the bombarding of a sample with a 1–25 keV beam of primary ions such as argon[73]. This process sputters off the surface layers of the sample, producing a variety of secondary species including neutral atoms and molecules, secondary electrons, photons, and positive and negative ions. Mass spectrometric analysis of the secondary positive and negative ions provides a microchemical characterization of the surface. SIMS analysis provides full elemental coverage from hydrogen to uranium, including the capability of isotopic characterization.

SIMS systems range from the simple to the very complex. A simple combination of an ion-milling source and mass spectrometer is useful for general surface analysis and depth profiling. An example of an elaborate system is the "probe, scanning imaging, or direct-imaging" ion micro-probe mass spectrometers which provide less than 2 mm lateral image resolution, point-to-point analysis, and in-depth profile analysis.

An example of the application of SIMS to a $Li_2O–Al_2O_3–SiO_2$ glass is shown in Figure 24[74]. The lower curve in the figure shows a Li de-pleted region at the surface and the upper curve shows an aluminum-enriched surface region. The depth of the total compositional profile extends to nearly 500 Å.

An advantage of SIMS profiling over AES is the detectability of H^+ and the ability to monitor volatile components such as the alkali ions, which in AES are often problems because of the influence of the electron beam. However, coupling a mass spectrometer with an ion-milling gun in an AES system provides much of the capability of the SIMS system.

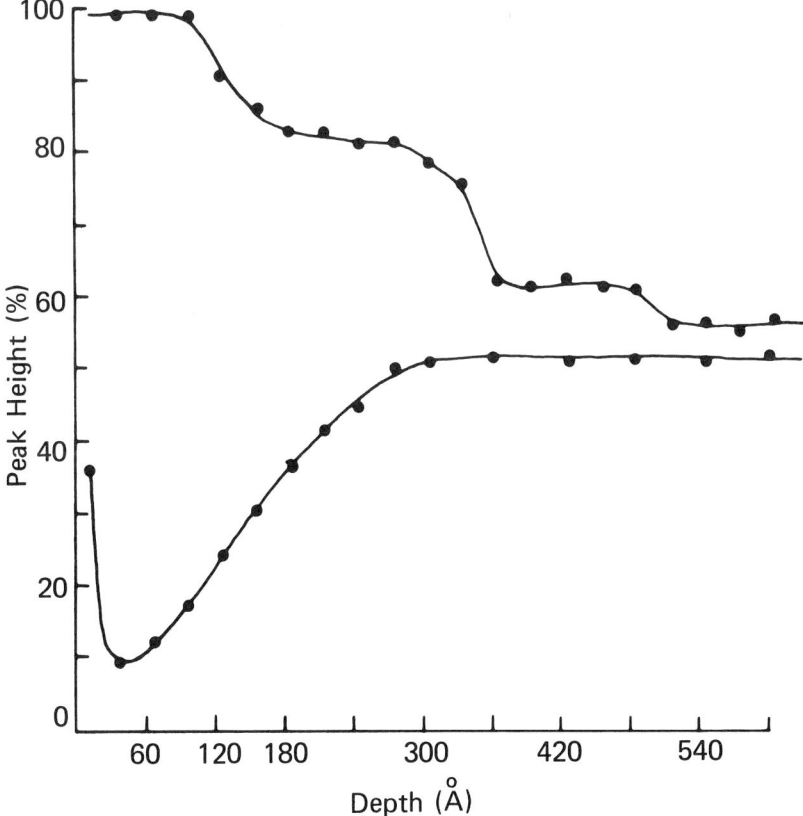

Figure 24. A secondary-ion mass spectrometer profile of the surface of a lithia-alumina-silica glass. Upper curve Al^{3+}, lower curve Li^+. (Courtesy of Commonwealth Scientific, Corp.)

Infrared reflection spectroscopy (IRRS) has recently been developed as a quantitative tool for characterizing the surface structure and composition of glasses[44,51,75]. There are several important advantages of the technique such as: not requiring vacuum and energetic bombardment; thus IRRS does not alter the surface of the glass as may AES, ESCA, and SIMS. Also, it is applicable to in-situ glass surfaces of nearly any configuration; analysis of large or small areas; it is rapid and inexpensive, and requires only standard infrared spectrometers. The IR reflection method can be used as an automated analytical tool and can also be coupled with other solution-analysis techniques, making it especially suitable for characterization of surface environmental interactions[44,51].

The IRRS method can use either single- or double-beam reflection installations, as shown in Figure 25[75]. For a double-beam setup, it is possible to reference a standard surface to the unknown surface and measure difference spectra to improve detection of small structural or compositional changes if necessary. A typical single-beam infrared reflection spectra of a 33 mole % Li_2O–SiO_2 glass surface is shown in Figure 26[51]. The spectra shown is sampling approximately a 0.1 μm thickness of the surface of the sample, using a total reflection angle of 25°. Silicon–oxygen–silicon stretching (~ 1100 cm^{-1}) and rocking peaks (~ 500 cm^{-1})

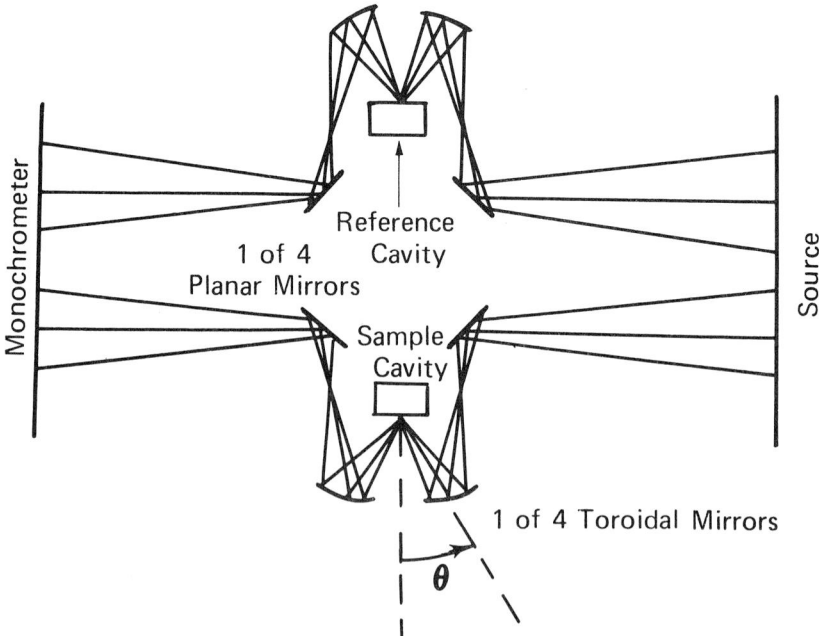

Figure 25. Schematic of double-beam infrared reflection spectrometer for glass-surface characterization[75].

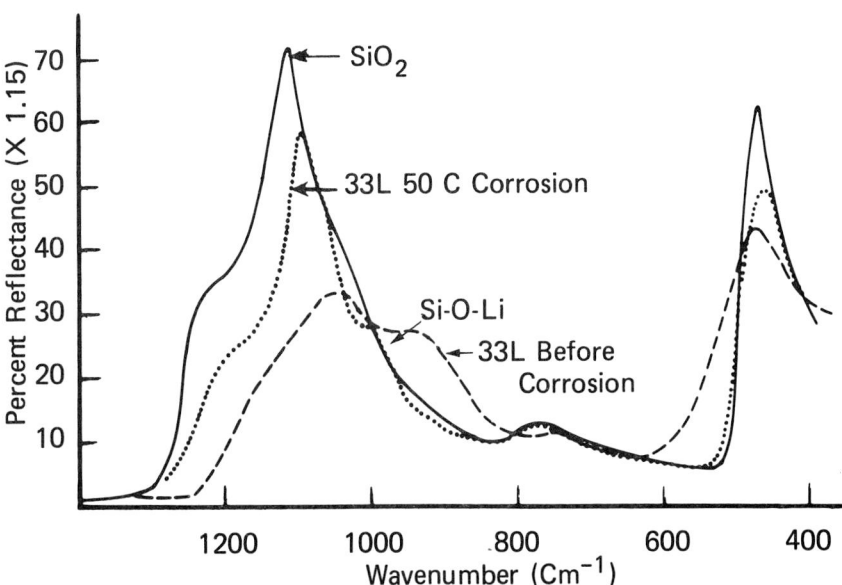

Figure 26. Infrared reflection spectra of a noncorroded 33 mole % Li_2O-SiO_2 glass surface (dashed curve) compared with vitreous silica (solid curve). A vitreous silicalike surface is formed during corrosion (dotted curve)[51].

are shown on the spectra giving the SiO_2 concentration of a glass. The silicon–oxygen–alkali (Si–O–Li) stretching peak at ~900 cm^{-1} is also shown, which yields the modifier concentration of the glass surface. Exposure of the glass surface to a chemical environment alters the relative concentration of both the silica and the alkali species, due to preferential leaching of the alkali ions. This produces a change in the infrared reflection spectra as shown in the figure. The surface film developed on the glass is similar to vitreous SiO_2. Thus, in-situ observations of structural or compositional differences of bulk-glass surfaces can be made in time to detect the effect of variations in processing or applications on the glass surfaces.

Recent results show that corrosion of glass surfaces produces a heterogeneous attack which can alter subsequent mechanical properties[76]. Consequently, the infrared reflection characterization of the glass surface may provide a nondestructive means of testing whether or not strength changes could occur. Additional studies to establish quantitative correlations in this area must be done.

Profilometers resolving microinches can be used to measure surface roughness and profiles of scratches, corrosion attack, and other surface heteorgeneities[76]. Correlation of profilometer analysis of specific regions on a glass surface with the large areal composition analysis using tools such as infrared reflectence spectrometry provide a powerful characterization combination.

Glass surfaces that have reacted with the environment undergo a structural change which requires characterization to understand the behavior of the glass. Recent studies by McGee and colleagues[78] have shown that adsorption isotherms of nitrogen and water vapor are useful in obtaining an understanding of the microporosity developed on glass surfaces. They have found that surface porosities in the range of 20 Å are present through the use of the adsorption methods. The process is one of exposing the vacuum-desorbed glass surface to a vapor as a function of increasing vapor pressure. The quantity of vapor adsorbed is measured quantitatively by precision gravimetric means. After saturation, the vapor pressure of the gas specie is progressively decreased and the weight change measured producing an adsorption isotherm. Extensive details of the calculation of volume fraction, shape, and size distribution of surface porosity is possible using these methods, as discussed by Linsen and others[79].

Ellipsometry provides a nondestructive method for the measurement of both thickness and refractive index of transparent coatings on glasses[80,81]. In ellipsometry, monochromatic, collimated, polarized light is reflected from a sample surface. The state of polarization of the light is defined by the phase and amplitude relationships between the component of the light electric field in the plane of incidence p and the component normal to the place of incidence s. A difference in phase between the two components other than 180° corresponds to elliptical polarization. Reflection of the light beam from a surface (coated or uncoated) results in a change in the phase difference between the two components $\Delta p - \Delta s$ and a change in the ratio of the amplitudes $\rho/\rho s$, which is dependent upon the optical constants of the coating and substrate, the angle of incidence, and the coating thickness. A program has been written by McCrackin (NBS Tech. Note 479) which allows one to determine the thickness of a coating and its refractive index, given the optical constants of the substrate and the range in which the coating refractive index lies.

A schematic of a typical ellipsometer is shown in Figure 27. The collimated incident beam is first linearly polarized by the polarizer and then elliptically polarized by the compensator. The light reflected from the sample is transmitted through a second polarizer (the analyzer), after which its intensity is determined by a photomultiplier detector. The compensator is normally set so that its fast axis is ±45° to the plane of incidence. The polarizer and analyzer are rotated until extinction is obtained. Under these conditions, the ellipticity caused by the polarizer–compensator–analyzer combination is the opposite of that produced by reflection from the film and substrate. From the extinction settings of the polarizer, compensator, and analyzer, $\Delta p - \Delta s$ and $\rho p/\rho s$ can be determined.

An application of ellipsometry to the study of polished glass surfaces was made by H. Yokato et al.[82]. They determined the refractive index

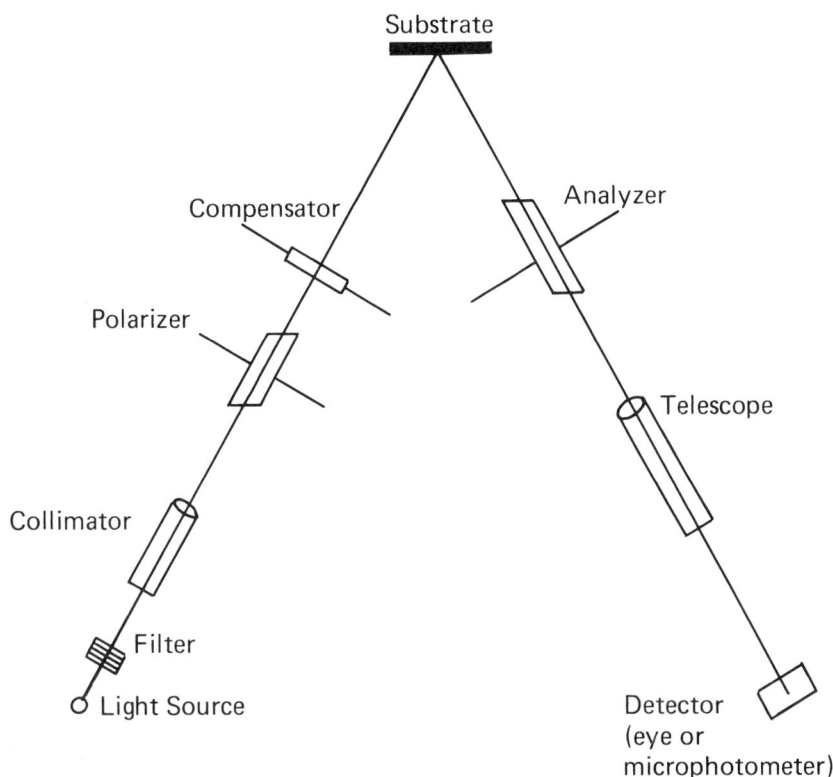

Figure 27. Schematic of an ellipsometer used to characterize glass surface films.

n_f and thickness d_f of the "polished layer" on the surfaces of various kinds of glass polished under standard conditions. It was found that $n_f < n_b$ (n_b = bulk index of glass) with glasses which are susceptible to weathering, and $n_f > n_b$ with chemically durable glasses such as Vycor®, Pyrex®, and silica glass. Other applications are discussed in this volume by Vedam.

In other studies in progress, ellipsometry is being used to measure the thickness of ultrathin oxide and organic films on glass[83]. It is particularly useful for calibrating ion-milling rates for AES or SIMS profiling studies.

Electronic Defects

Characterization of the electronic defects in glasses which give rise to variations in optical absorption spectra are worthy of a review chapter alone. The interested reader is referred to the excellent review by

Kreidl[84]. The application of electron spin resonance (ESR), electron paramagnetic resonance (EPR), Mössbauer, and photoluminescence techniques to characterizing impurity-charge centers in glasses are discussed by Kreidl and related to the optical spectra of various silicate and non-silicate glasses.

Stress Chracterization

Measurement of residual stresses within a glass or at a glass surface, due to differential rates of cooling, must also be considered in a characterization program. Stress optical measurements using a polariscope is the standard method. The technique is well established, and is capable of resolving 100 psi or less[85].

Stress effects can also be used to detect and analyse flaws in a glass surface. As developed by Ernsberger, exchange of different sized alkali ions from a salt solution with alkali ions in the glass stresses flaws already present in the glass surface[86]. The flaws are opened by the induced stresses, making them visible. Once visible, the size and distribution of flaws can be calculated. Whether the flaws so revealed are the critical flaws in the glass is still open to speculation.

An instrument termed the "differential surface refractometer" has also been proposed for analysis of stressed surface layers. Such surface compressive layers cannot be viewed in cross section by cutting since failure results. The instrument operates on the principle of measuring the difference in index of refraction between the surface and the bulk glass. Results have been difficult to reproduce[87].

Indirect Characterization Methods

A couple of characterization methods are traditionally employed for glasses although they are indirect in the pure sense of characterization definitions[1,2]. Analysis of "water" in glass is more typically done by the intensity of infrared absorption of the glass at three bands, 2.75–2.95 μm, 3.35–3.8 μm, and 4.25 μm, rather than by other chemical analysis techniques. This is because of the extensive quantitative studies conducted by Scholze relating H_2O content and type of bonding in the glass to the infrared spectra[88,89].

The second method is the use of density in a characterization description of a glass. Density measurements are a valid and necessary technique because glass is a nonequilibrium solid state of matter. Consequently, the volume of a glass, and therefore its density, is a function of its thermal history. As shown in Figure 28, the quench temperature and quenching rate of a glass influences the density of the glass.

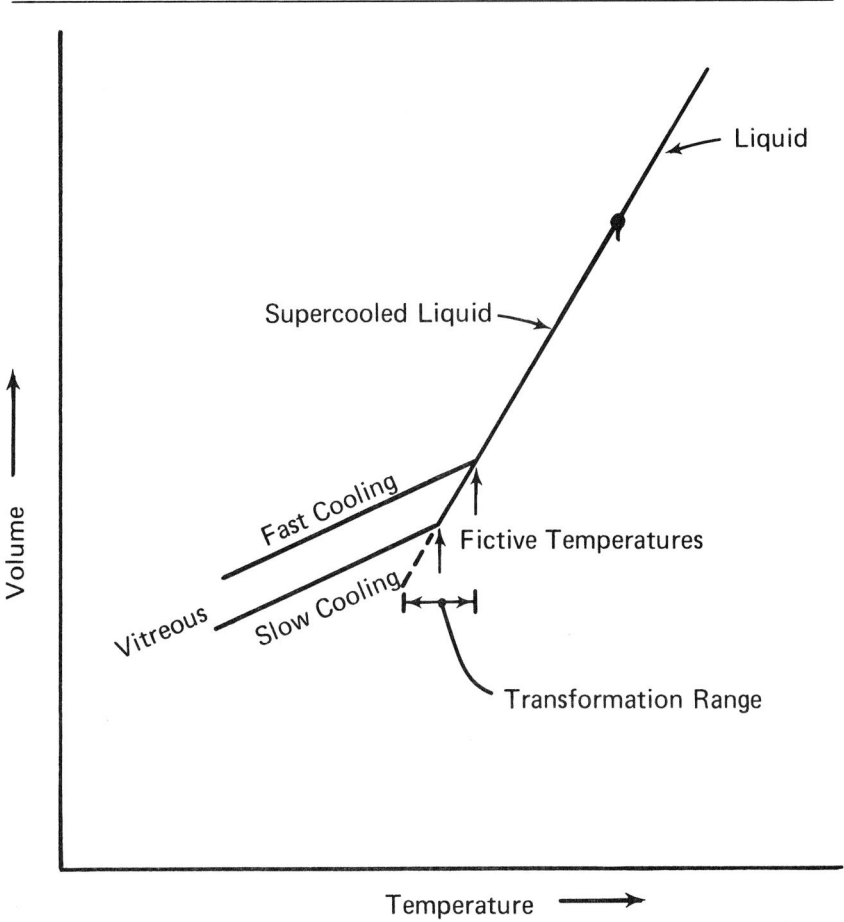

Figure 28. Thermal history dependence of the volume of a glass.

If a complete 3-D description of the atom positions in a glass could be established by structural characterization and the composition known, as a crystal, then density measurements would not be a necessary and valid characterized tool. However, limitations on the structural description of a glass imposed by the lack of long-range order require another characterization method that is sensitive to thermal history. Specific volume, or density, is generally accepted as the most direct method. In addition, it is simple, rapid, inexpensive, and highly accurate using modern gravimetric methods. Because of surface-corrosion reactions of glass with water and other solutions, care must be taken to make the measurement in a nonreactive manner.

The dependence of index of refraction n on both composition and density prevents it from being used as an independent characterization tool. However, if constant-composition glasses are being investigated, measurements of n can also be used to determine thermal history effects.

Selection of a Characterization Program

Selection of a characterization program for a particular product is most difficult, as can be seen from the large selection of characterization tools discussed above. With the many characterizations paths possible, establishing a characterization area to evaluate is perhaps the most critical and most difficult step in a characterization program. For a commercial product, a single criterion for establishing a suitable characterization program should be kept in mind, *e.g.,* to select those methods which provide the property control required for application or sales at least cost. Since characterization costs are additive to production costs, they must therefore be economically justified from terms of increased sales, higher reliability, or lower waste, etc.

The characterization effort required to understand the properties of exploratory products or devices obtained from a research and development effort may be much more complex. However, it is essential to establish the most suitable characterization activities even in a research and development effort, since the expense and time involved can seriously hamper rate of R&D activities if the characterization is unduly time consuming and expensive.

Because of the economic and time factors involved, it is usually wise to consider the least expensive and least time-consuming characterization techniques applicable to the product or device of concern. For example, characterization of the surface of a glass before or after application could involve simple observation under the optical microscope, polariscope examination, and an infrared reflection spectra. Such a rapid nondestructive characterization might well prove suitable for property control, and the extraordinary length and expense of an Auger or ESCA analysis may well be extraneous. If the more simple characterization methods prove unsatisfactory for obtaining an understanding of the reproduceable properties of the material, then the more elaborate instrumental methods described above must be examined one by one.

The literature in each of the characterization areas described above is increasingly voluminous. Consequently, before adopting a characterization step in even a research and development effort, it is essential that a thorough understanding be achieved of the alternative means of evaluating the chemical, structural, microstructural, or surface features that might be involved. Often, major savings in time and expense can be made by studying the various alternative methods available from previous applications before embarking on an experimental or production characterization program.

Acknowledgments

The author gratefully acknowledges most helpful contributions and suggestions by G. Y. Onoda, S. W. Freiman, B. Molnar, C. Pantano, M. Dilmore, and R. Nastasi, and financial support of ONR Contract, No. N00014-68-A-1073-0006.

References

1. "Characterization of Materials," Materials Advisory Board, Washington, D.C., Report No. MAB-229-M, March 1967 (AD 649 941).

2. "Ceramic Processing." Materials Advisory Board, Washington, D.C., National Academy of Sciences Publication NAS-NRC-Pub-1576, 1968 (AD 679 886).

3. Stover, E.R., "A Critical Survey of Characterization of Particulate Ceramic Raw Materials," University of California, Berkeley, Air Force Materials Laboratory Contract Report No. AFML TR-67-56, May 1967 (AD 816 014).

4. Hench, L.L., "Introduction to the Characterization of Ceramics," in *Characterization of Ceramics,* L.L. Hench and R.W. Gould, eds., New York: Marcel Dekker, Inc. (1971), 1–5.

5. Rankin, P., "General Analytical Chemistry," in *Characterization of Ceramics,* L.L. Hench and R.W. Gould, eds., New York: Marcel Dekker, Inc. (1971), 9–38.

6. Krugers, J. and Keulmans, A.M., eds., *Practical Instrumental Analysis,* London: Elsevier Publishing Company, 1965.

7. Kolthoff, I.M., Elving, P.J. and Sandell, E.B., *Treatise on Analytical Chemistry,* New York: John Wiley & Sons, 1952–1974.

8. Day, R.A., Jr. and Underwood, A.L., *Quantitative Analysis,* Englewood Cliffs, N.J.: Prentice-Hall, 1967.

9. Chemical Analysis Committee of the Society, "The Chemical Analysis of Colourless Soda-Lime-Silica Glasses," *Glass Technol.,* 14 (1973), 5–13.

10. Subcommittee AII of the International Commission on Glass, "The Chemical Durability of Glass; The Determination of Boric Oxide in Durability Extract Solutions," *Glass Technol.,* 14 (1973), 14–19.

11. Chemical Analysis Committee of the Society, "Limestone for Making Colourless Glasses: Revised Specification 1971," *Glass Technol.,* 13 (1972), 104–09.

12. "Glass Analysis Methods," papers presented at 74th Annual Meeting of the American Ceramic Society, Glass Division, Washington, D.C., 8–10 May 1972 (Abstracts: *Bull. Amer. Ceram. Soc.,* 51 [1972], 367, 370).

13. Gould, R.W., "X-Ray Spectroscopy," in *Characterization of Ceramics,* L.L. Hench and R.W. Gould, eds., New York: Marcel Dekker (1971), 39–87.

14. Liebhafsky, H.A., Pfeiffer, H.G., Winslow, E.H. and Zemany, W.D., *X-Ray Absorption and Emission in Analytical Chemistry,* New York: John Wiley & Sons, 1960.

15. Bertin, E.P., *Principles and Practice of X-Ray Spectrochemical Analysis,* New York: Plenum Press, 1970.

16. Sparks, C.J. Jr., Cavin, O.B., Harris, L.A. and Ogle, J.C., "Simple Quantitative X-Ray Fluorescent Analysis for Trace Elements," paper presented at the 7th Annual Conference on Trace Substances in Environmental Health, University of Missouri, June 1973.

17. "Powder Sample Preparation Errors in X-Ray Spectrochemical Analysis," *Norelco Reporter,* 17 (December 1970), 25-29.

18. Claisse, F., "Analyse quantitative precise par fluoresence X." Ministere de Mines, Quebec, Report No. 327, 1956.

19. Stephenson, D.A., "An Improved Flux-Fusion Technique for X-Ray Emission Analysis," *Anal. Chem.,* 41 (1969), 966-67.

20. Gould, R.W., "X-Ray Diffraction," in *Characterization of Ceramics,* L.L. Hench and R.W. Gould, eds., New York: Marcel Dekker, Inc. (1971), 135-76.

21. Mozzi, R.L. and Warren, B.E., "The Structure of Vitreous Silica," *J. Appl. Crystallogr.,* 2 (1969), 164-72.

22. Gokularathnam, C.V., Gould, R.W. and Hench, L.L., "X-Ray Diffraction Analysis of Structural Changes in Glass," in *Advances in Nucleation and Crystallization in Glasses,* Special Publication No. 5, L.L. Hench and S.W. Freiman, eds., Columbus, Ohio: American Ceramic Society (1971), 61-70.

23. Warren, B.E., "Summary of Work on Atomic Arrangement in Glass," *J. Amer. Ceram. Soc.,* 24 (1941), 256-61.

24. Gokularathnam, C.V., "X-Ray Diffraction Analysis of Structural Changes in Vitreous Silica," unpublished Ph.D. dissertation, University of Florida, 1971.

25. Dove, D.B., "The Determination of Local Order in Amorphous Semiconducting Films," in *Physics of Electronic Ceramics, Part A,* L.L. Hench and D.B. Dove, eds., New York: Marcel Dekker, Inc. (1971), 227-67.

26. Grigson, C.W.B. and Dove, D.B., "Scanning Electron Diffraction of Film Growth," *J. Vac. Sci. Technol.,* 3 (1966), 120-32.

27. Chopra, K.L., *Thin Film Phenomena,* New York: McGraw-Hill Book Publishing Company, 1969.

28. Henninger, E.H., Buschert, R.C. and Heaton, L., "Atomic Structure and Correlation in Vitreous Silica by X-Ray and Neutron Diffraction," *J. Phys. Chem. Solids,* 28 (1967), 423-32.

29. Weeks, R.A., "The Uses of Electron and Nuclear Magnetic Resonance Fluorescence in Studies of Glass," in *Introduction to Glass Science,* L.D. Pye, H.J. Stevens, and W.A. La Course, eds., New York: Plenum Press (1972), 137-71.

30. Bray, P.J., "NMR Studies of Glasses and Related Crystalline Solids," in *Magnetic Resonance,* C.K. Coogan, N.S. Ham, S.N. Stuart, J.R. Pilbrow, and G.V.H. Wilson, eds., New York and London: Plenum Press (1970), 11-40.

31. Hendrickson, J.R. and Bray, P.J., "A Theory for the Mixed Alkali Effect in Glass. Part I," *Phys. Chem. Glasses,* 13 (1972), 43-49 and "Part 2," *Phys. Chem. Glasses,* 13 (1972), 107-15.

32. Sayers, D.E., Lytle, F.W., and Stern, E.A., "Structure Determination of Amorphous Ge, GeO_2, and GeSe by Fourier Analysis of Extended X-Ray Absorption Fine Structure (EXAFS)," *J. Non-Cryst. Solids,* 8-10 (1972), 401-07.

33. Goldstein, M., "Depolarized Components of Light Scattered by Glasses. I. Measurements of Twelve Optical Glasses," *J. Appl. Phys.,* 30 (1959), 493-500.

34. Hammel, J., "Nucleation in Glass—A Review," in *Advances in Nucleation and Crystallization in Glasses,* Special Publication No. 5, L.L. Hench and S.W. Freiman, eds., Columbus, Ohio: American Ceramic Society (1971), 1-10.

35. Vogel, W. and Gerth, K., "Catalyzed Crystallization in Glass," in *Symposium on Nucleation and Crystallization in Glasses and Melts,* M.K. Reser, G. Smith, and H. Insley, eds., Columbus, Ohio: American Ceramic Society (1962), 11-22.

36. Gould, R.W., "Small Angle X-Ray Scattering," in *Characterization of Ceramics,* L.L. Hench and R.W. Gould, eds., New York: Marcel Dekker, Inc. (1971), 371-94.

37. Gerold, V.K., *Small Angle X-Ray Scattering*, New York: Gordon and Breach Science Publishers, Inc., 1965.

38. Hench, L.L., Freiman, S.W. and Kinser, D.L., "The Early Stages of Crystallisation in a Li_2O-$2SiO_2$ Glass," *Phys. Chem. Glasses*, 12 (1971), 58–63.

39. Hench, L.L. and Schaake, H.F., "Electrical Properties of Glass," in *Introduction to Glass Science*, D. Pye and H. Simpson, eds., New York: Plenum Press (1972), 586–659.

40. Kinser, D.L. and Hench, L.L., "Effects of Metastable Precipitate on the Electrical Properties of an LiO_2-SiO_2 Glass," *J. Amer. Ceram. Soc.*, 51 (1968), 445–48.

41. Tufts, C.F., "Transmission Electron Microscopy and Electron Diffraction," in *Characterization of Ceramics*, L.L. Hench and R.W. Gould, eds., New York: Marcel Dekker, Inc. (1971), 177–218.

42. Wischnitzer, S., *Introduction to Electron Microscopy*, New York: Pergamon Press, 1972.

43. Kinser, D.L. and Hench, L.L., "Hot Stage Transmission Electron Microscopy of Crystallisation in a Lithia-Silica Glass," *J. Mater. Sci.*, 5 (1970), 369–73.

44. Sanders, D.M. and Hench, L.L., "Environmental Effects on Glass Corrosion Kinetics," *Bull. Amer. Ceram. Soc.*, 52 (1973), 662–65, 669.

45. Bates, S.R., "Scanning Electron Microscopy," in *Characterization of Ceramics*, L.L. Hench and R.W. Gould, eds., New York: Marcel Dekker, Inc. (1971), 419–34.

46. Freiman, S.W., Onoda, G.Y. Jr. and Pincus, A.G., "Spherulitic Crystallization in Glasses," in *Advances in Nucleation and Crystallization in Glasses*, Special Publication No. 5, L.L. Hench and S.I. Freiman, eds., Columbis, Ohio: American Ceramic Society (1971), 141–50.

47. Russ, J.C., "ORTEC Application Note (AN36)," Ortec, Inc., Oak Ridge, Tennessee.

48. Clark, A.E. and Hench, L.L., "Effect of P^{5+}, B^{3+} and F^- Additions on the Surface Reactions of Soda-Lime-Silica Glass," to be published.

49. Lewis, G., "Electron Microprobe," in *Characterization of Ceramics*, L.L. Hench and R.W. Gould, eds., New York: Marcel Dekker, Inc. (1971), 505–28.

50. Birks, L.S., *Electron Probe Microanalysis*, New York: John Wiley & Sons, 1963.

51. Sanders, D.M. and Hench, L.L., "Mechanisms of Glass Corrosion," *J. Amer. Ceram. Soc.*, 56 (1973), 373–78.

52. Frechette, V.D., "Petrographic Analysis," in *Characterization of Ceramics*, L.L. Hench and R.W. Gould, eds., New York: Marcel Dekker, Inc. (1971), 257–72.

53. Insley, H. and Frechette, V.D., *Microscopy of Ceramics and Cements*, New York: Academic Press, 1955.

54. Kerr, P.F., *Optical Mineralogy*, New York: McGraw-Hill Book Publishing Company, 1959.

55. Gokularathnam, C.V., Freiman, S.W., DeHoff, R.T. and Hench, L.L., "Thickness Error in Quantitative Transmission Microscopy of Ceramics," *J. Amer. Ceram. Soc.*, 52 (1969), 327–31.

56. Fowkes, A.J. and Parkinson, R.T., "The Analysis of Gaseous Inclusions in Glass Using a Quadrupole Mass Spectrometer," *Glass Technol.*, 13 (1972), 126–32.

57. Boughton, J.H., "Light Scattering," in *Characterization of Ceramics*, L.L. Hench and R.W. Gould, eds., New York: Marcel Dekker, Inc. (1971), 435–56.

58. Van de Hulst, H.C., *Light Scattering by Small Particles*, New York: John Wiley & Sons, 1957.

59. Kerker, M., *The Scattering of Light: and Other Electromagnetic Radiation*, New York: Academic Press, 1969.

60. Hammel, J.J., "Light Scattering from Glass," in *Physics of Electronic Ceramics, Part B,* L.L. Hench and D.B. Dove, eds., New York: Marcel Dekker, Inc. (1971), 963–86.

61. Macedo, P., personal communication.

62. Freiman, S.W., "Applied Stereology," in *Characterization of Ceramics,* L.L. Hench and R.W. Gould, eds., New York: Marcel Dekker, Inc. (1971), 555–79.

63. DeHoff, R.T. and Rhines, F.N., eds., *Quantitative Microscopy,* New York: McGraw-Hill Book Publishing Company, 1968.

64. Freiman, S.W. and Hench, L.L., "Crystallization Kinetics of Lithia-Silica Glass," *J. Amer. Ceram. Soc.,* 51 (1968), 382–87.

65. Freiman, S.W. and Hench, L.L., "Effect of Crystallization on the Mechanical Properties of Li_2O–SiO_2 Glass-Ceramics," *J. Amer. Ceram. Soc.,* 55 (1972), 86–90.

66. McMillan, P.W., "The Constitution, Microstructure and Properties of Glass-Ceramics," in *Advances in Nucleation and Crystallization in Glasses,* Special Publication No. 5, Columbus, Ohio: American Ceramic Society (1971), 224–50.

67. Berrin, L. and Sundahl, R.C., "Characterization of Ceramic Surfaces," in *Characterization of Ceramics,* L.L. Hench and R.W. Gould, eds., New York: Marcel Dekker, Inc. (1971), 583–624.

68. Chang, C.C., "Auger Electron Spectroscopy," *Surface Sci.,* 25 (1971), 53–79.

69. Burhop, E.H.S., *The Auger Effect and Other Radiationless Effects,* New York: Cambridge University Press, 1952.

70. Dove, D.B., Onoda, G.Y. and Pantano, C., "Auger Spectroscopic Analysis of Coated Glass Surfaces," paper presented at the Conference on Surfaces and Interfaces of Glasses and Ceramics, Alfred University, 27–29 August 1973, to be published.

71. Siegbahn, K. *et al., ESCA: Atomic, Molecular and Solid State Structures Studied by Means of Electron Spectroscopy,* Uppsala: Almquist & Wiksells Boktryckeri Ab, 1967.

72. Von Hickson, K., "Electron Spectroscopy," *Glastech. Ber.,* 44 (1971), 537–42.

73. Evans, C.A. Jr., "Secondary Ion Mass Analysis: A Technique for Three-Dimensional Characterization," *Anal. Chem.,* 44, No. 13 (1972), 67A–80A.

74. "Mass Spectrum No. 73–185," 1973, Commonwealth Scientific Corporation.

75. Sanders, D.M., Person, W.D. and Hench, L.L., "New Methods for Studying Glass Corrosion Kinetics," *Appl. Spectrosc.,* 26 (1972), 530–36.

76. Semenov, N.I., Paplauskas, A.B. and Ryabov, V.A., "Effect of Surface Microstructure on the Strength of Glass," *Glass Technol.,* 13 (1972), 171–75.

77. Sanders, D.M. and Hench, L.L., "Surface Roughness and Glass Corrosion," *Bull. Amer. Ceram. Soc.,* 52 (1973), 666–69.

78. Huang, R.J., Demirel, T. and McGee, T.D., "Calculation and Interpretation of Surface Free Energy of Wetting of E-Glass by Vapors," *J. Amer. Ceram. Soc.,* 56 (1973), 87–91.

79. Linsen, B.G., ed., *Physical and Chemical Aspects of Adsorbents and Catalysts,* New York: Academic Press, 1970.

80. Vedam, K. and White, E.W., "Characterization of Real Surfaces by Ellipsometry and Soft X-Ray Spectroscopy," paper presented at Office of Naval Research Glass Environmental Workshop, University of Florida, Gainesville, 6–7 September 1973.

81. Pliskin, W.A. and Zamin, S.J., "Film Thickness and Composition," in *Handbook of Thin Film Technology,* L.I. Maissel and R. Glang, eds., New York: McGraw-Hill Book Publishing Company (1970), 11–1, 11–54.

82. Yokota, H., Sokata, H., Nichibori, M., and Kinosita, K., "Ellipsometric Study of Polished Glass Surfaces," *Surface Sci.,* 16 (1969), 265–74.

83. Molnar, B. and Dove, D.B., personal communication.

84. Kreidl, N.J., "The Optical Absorption of Glasses," in *Physics of Electronic Ceramics, Part B*, L.L. Hench and D.B. Dove, eds., New York: Marcel Dekker, Inc. (1971), 915–61.

85. Rood, J.L., "Dispersion, Stress-Optical Effects in Glass, Optical Glasses," in *Introduction to Glass Science*, L.D. Pye, H.J. Stevens, and W.C. LaCourse, eds., New York: Plenum Press (1972), 373–89.

86. Ernsberger, F.M., "Strength and Strengthening of Glass. I. Strength of Glass," *Glass Ind.*, 47 (1966), 422–27, 481–83; and "II. Strengthening of Glass," *Glass Ind.*, 47 (1966), 483–87.

87. Freiman, S.W., personal communication.

88. Scholze, H., "Water Content of Glasses. II. Infrared Measurements of Systematically Varied Glass Compositions. Interpretation of Hydroxy Bands in Silicate Glasses," *Glastech. Ber.*, 32 (1959), 142–52; "III. Infrared Measurements of Additional Glasses," *Glastech. Ber.*, 32 (1959), 278–81; "IV. Effect of Temperature," *Glastech. Ber.*, 32 (1959), 314–20; "V. Diffusion of Water in Glasses at High Temperatures," *Glastech. Ber.*, 32 (1959), 381–86.

89. Scholze, H., "Glasses and Water in Glass," *Glass Ind.*, 47 (1966), 546–61, 622–28, 670–75.

MOLECULAR CHARACTERIZATION
OF POLYMERS

Moderator: Fred W. Billmeyer, Jr.

Rensselaer Polytechnic Institute
Troy, New York

FRED W. BILLMEYER, JR.
Rensselaer Polytechnic Institute
Troy, New York

Chapter 9

Introductory Remarks to Polymer Characterization Chapters

The discussion following Dr. Promisel's keynote address on "The Concept of Characterization" left no doubt as to the importance and difficulty of the characterization and specification of polymers as materials. It seems both appropriate and necessary to add a few remarks on this subject by way of introducing the chapters concerned with the molecular characterization of polymers and their characterization in bulk.

By now many people are aware of the importance of polymers as a class of materials. More than half of all chemists and chemical engineers in the U.S. work with polymers or in polymer-related fields; the annual production of polymers, by volume, exceeds that of aluminum and copper combined and is rapidly approaching that of steel; accounting for all elastomers, all adhesives, and all fibers (save asbestos and glass), polymers are essential to man's present technological accomplishments[1,2]. It is most appropriate, therefore, that polymeric materials be considered side by side with the other major classes of materials, metals, and ceramics, at the Sagamore Conferences.

It is my belief, however, that the characterization of polymers is inherently more difficult than that of other materials, for the following reasons:

1. At almost every level of organization, from microscopic to macroscopic, the structural features of polymeric materials are roughly equivalent in complexity to those of other materials, if not worse. For example, polymer crystal structures are closely analogous to those in other crystalline materials, but the combined crystal-amorphous nature of polymers has no direct counterpart that I am aware of in metals or ceramics. The nature of crystal defects appears to be similar in the polymeric and nonpolymeric cases. The larger structural entities of spherulites (not limited to polymeric systems, of course) are at least of the same order of size as grains in metals; and the segregation of elements at grain boundaries in metals has its counterpart in the segregation of low-molecular-weight

255

material at the boundaries of crystals in polymers. The structures of amorphous regions are at least similar in polymers and ceramics, and polymer blends and alloys are well known. No doubt other examples of correspondence could be cited.

2. In addition to all these, the structure of polymers is complicated at the molecular level in ways which, I submit, have no counterpart in other materials. Instead of the relatively small and simple groups of atoms forming the structural units of metals and ceramics, the building block of polymeric materials is the long-chain macromolecule consisting of thousands, if not hundreds of thousands, of atoms. Of course, there is not an infinite variety of ways in which these atoms are arranged, but the deviations from perfectly regular chain structure are both manifold and important. In the matter of structural isomerism alone, it has been calculated that during the entire production history of branched polyethylene since 1937, not enough molecules have been made to produce one of each possible geometric isomer, if all molecules were different from one another. In addition, there is a variety of stereoisomeric conditions and other chain irregularities to consider.

Moreover, in solution and presumably in the amorphous state, the typical polymer chain takes up a vast number of conformations, resulting in changes of size and shape, so that these parameters are only statistically defined. And a final degree of complexity is added by virtue of the fact that, except possibly for a few macromolecules of biological significance, every polymer sample prduced consists of molecules with a wide distribution of chain lengths.

3. As a result of the long-chain nature of polymeric materials, these substances exhibit a wide range of time-dependent elastic and viscoelastic behavior which, again, has no counterpart in other materials. Thus the bulk properties of polymers are dependent upon and exhibit many phenomena unknown prior to their availability.

In view of all this, it is a formidable task to describe the characterization of polymers in a small group of chapters. In attempting to do so, we have divided the subject in two parts:

The first three chapters consider the nature of the polymer molecule and its characterization as to mass, size, sequence of atoms, geometry, and stereoregularity, and the corresponding distributions of all these properties. In the first chapter Jack B. Kinsinger treats the characterization of molecular weight and its distribution. The geometrical and stereoregular aspects of the chemical makeup of the polymer chain are then considered by James Harwood, while James E. Mark's chapter on the characterization of polymer molecular size and conformation by statistical methods closes the section.

Three chapters on thᴇ characterization of polymers in bulk follow. Some aspects of the morphology of semicrystalline polymers are discussed by Leo Mandelkern. (The morphology of crystalline polymers has been studied intensively by various groups in recent years. Interpretations other than those given in Dr. Mandelkern's review exist[3].) Two of several possible areas related to the characterization of polymers in bulk are then covered by Stuart B. Clough (optical methods applied to bulk polymers) and William W. Graessley (rheological characterization of polymers).

Since it is not possible to describe all aspects of polymer characterization, either molecular or bulk, within the confines of a small group of chapters, those interested in pursuing the topic are referred to a recent book [4] covering some experimental aspects of polymer characterization.

Since the structural variability in a polymer material system is so great, it is important to consider how one might write a specification insuring the attainment of identical properties and performance from successive batches of such a material. This is a difficult task, and one which is rarely carried out successfully. Three arguments are pertinent.

One point of view is that if a given material from a specific supplier proves satisfactory, then just specify this material and supplier in the future. This has obvious practical advantages, and if the supplier is responsible and knowledgeable, it has much to recommend it. The disadvantages are just as obvious, however, as this approach simply begs the question.

Essentially the opposite point of view holds that one ought to be able to specify all aspects of polymer structure with our present state of knowledge, and therefore in principle could write a complete specification of polymer material performance from first principles. I seriously doubt that this can ever be accomplished, and if it were possible, it is almost certain that it would not be economically feasible.

A middle viewpoint appears to offer the best hope for sensible compromise. It is argued that performance is based largely on physical rather than chemical properties in a polymer materials system, and that it is possible to specify performance by means of a limited number of meaningful tests of physical properties. If properly selected, these go to the heart of the matter and guarantee the desired performance in a most direct way. The counterargument is that one is never sure that all the tests necessary to insure performance have been included unless the basic molecular and structural parameters are fixed. I agree in principle, but it is here that compromise is necessary, and it is my feeling that this is our best current approach to the specifications problem.

Those responsible for the polymer-related chapters in this book (E. A. Collins, K. J. Smith, and myself) hope that the following chapters provide some insight into the goals and rewards as well as the pitfalls in the characterization of polymer materials.

REFERENCES

1. Billmeyer, F.W. Jr., *Textbook of Polymer Science*, 2nd ed., New York: John Wiley & Sons, 1971.

2. Billmeyer, F.W. Jr., *Synthetic Polymers,* New York: Doubleday & Company, Inc., 1972.

3. Blackadder, D.A., "Ten Years of Polymer Single Crystals," *J. Macromol. Sci., Rev. Macromol. Chem.,* C1 (1967), 297–326; and Keller, A., "Polymer Crystals," *Rep. Prog. Phys.,* 31 (1968), 623–704.

4. Collins, E.A., Bareš, J. and Billmeyer, F.W. Jr., *Experiments in Polymer Science,* New York: John Wiley & Sons, 1973.

JACK B. KINSINGER
Michigan State University
East Lansing, Michigan

Chapter **10**

Macromolecules: Their Masses, Sizes, and Related Distributions

ABSTRACT

A qualitative description of experimental methods for measuring the molecular weight (MW) and molecular weight distribution (MWD) of macromolecules is given. The basic principles of these methods are reviewed. Use of these methods to investigate indices related to molecular architecture such as branching and copolymer composition are reviewed. Some new measurements based on laser light scattering spectroscopy as applied to the MW of macromolecules are given. Some recent advances in the new experimental methods are reviewed.

Introduction

We search for the detailed molecular structure of molecules with thousands, even tens of thousands of atoms; yet we are struck by our childlike questions; how large are they, how much do they weigh, and are they all alike? That so simple a set of questions could be posed and a review of procedure used to determine the answers, could find a place in a symposium on macromolecular characterization in 1973, nearly 40 years after the birth of macromolecular science, merely points to the experimental difficulties we suffer in measuring simple parameters on complex substances.

In chemical education we dwell on the relationship between chemical composition and physical properties of small molecules; however, when simple substances are connected with covalent bonds to form extended molecules of chainlike nature, unique properties arise which depend less on the chemical nature of the constituent units than on the length of the chain and collectively on the distribution or heterogeneity associated with the length. The initial task in characterizing a macromolecule is simple conceptually; to determine its molecular weight (MW) and its related distribution (MWD). Even this information becomes less meaningful if the macromolecules are fixed into an insoluble, indivisible network. Characterization of these later substances requires alternate methods and attention is focused on other structural attributes.

If we restrict our consideration to only homogeneous polymeric systems, that is, systems where there is a uniformity of repeating chemical units in the polymer chain, then after 30 years of research effort in this field of study we can summarize the status as follows: it is possible to obtain independent absolute molecular weights, that is, 3 different moments of a distribution within approximately 10 percent error each and at least 3 methods exist which can provide reasonable knowledge of the molecular weight distribution[1,2]. There are, of course, many limitations within such a general statement and these will be dealt with separately. The most disconcerting aspect, however, is that the information mentioned above cannot be obtained easily, at a low cost in either equipment or manpower, to the extent that many researchers in macromolecular science bypass searching for the absolute values available and rely heavily on related "measures" of these quantities. Therefore at this juncture in time we find our characterization methods have relatively low precision (in the pure analytical sense) and require considerable manhours and technical talent to complete. As we pass to relative "measures" of these absolute quantities, the uncertainty rises sharply. Notwithstanding, polymer science has progressed rapidly despite our limited analytical procedures, in part, because the physical and mechanical properties in many cases are not sharply dependent on these parameters so that relative information with high uncertainty could be tolerated and made useful.

The Distribution

Every investor understands the hidden agony in a single "average" which reflects an investment market. Likewise, no complex, heterogeneous macromolecular system can be adequately described by a single average value. Even if we restrict our considerations to chemically homogeneous, strictly linear macromolecular systems, we will note that the characterization will be complicated by a distribution of chain lengths or molecular weights. In the technological, scientific domain, most macromolecular samples do not have the pristine molecular weight distributions so neatly calculated in texts under highly restricted boundary conditions (for example, the "most probable distribution" of condensation polymerization and certain steady-state addition polymerizations, the Poisson distribution of the "living polymer" mechanism or other simple distributions associated with other steady-state kinetic steps), nor are many experimental samples graced with beautiful symmetrical distributions so often found in mathematical physics such as "The Gaussian." Rather, most

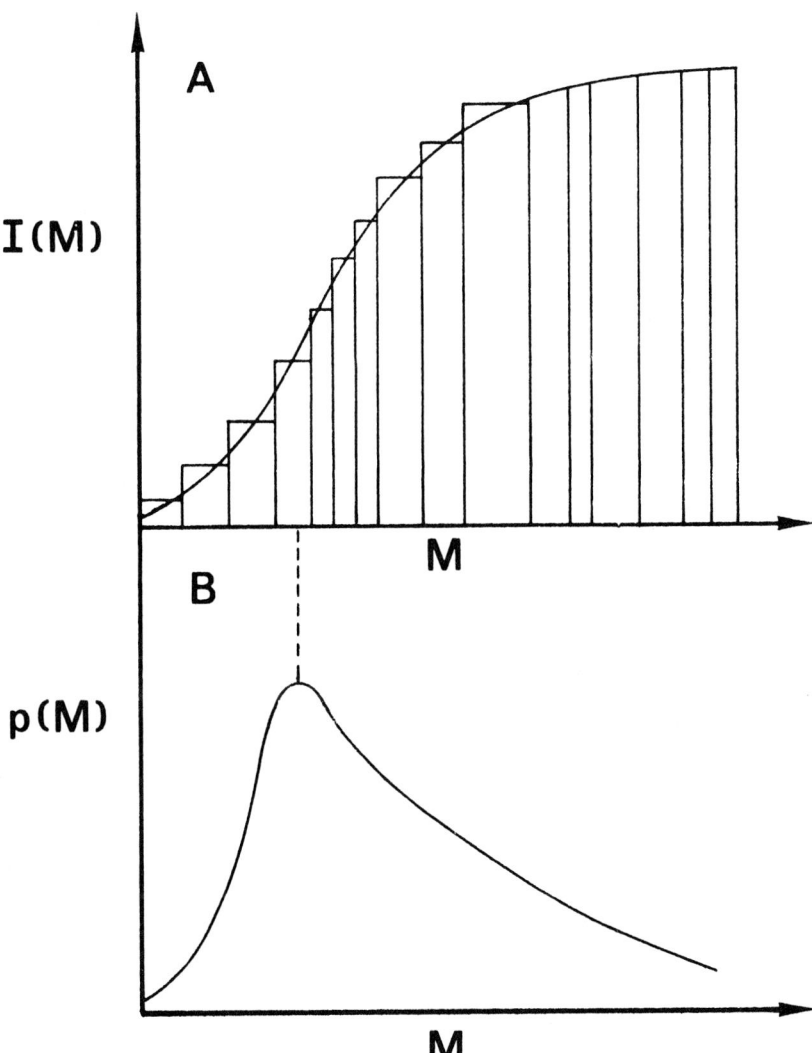

Figure 1. (A) Integral distribution curve, bars signify experimental data on discrete samples; (B) Differential distribution curve.

often a very unsymmetrical (skewed) distribution is found experimentally. A typical illustration of both an integral and a differential distribution curve overlaying experimental data bars is noted in Figure 1. The integral distribution curve is represented by the equation

$$I(M) = \int_0^M p(M)\,dM \qquad (1)$$

where $p(M)$ is subject to the normalization condition

$$\int_0^\infty p(M)dM = 1 \qquad (2)$$

Here M stands for the molecular weight of a chemically homogeneous linear chain and $p(M)$ is related to the probability that a sample will have a molecular weight between M and $M + dM$. The two curves shown in Figure 1 also illustrate graphically the integral and differential distribution curves and a visual representation of the experimental problem. A particular physical or mechanical property of such a sample may depend primarily on different components of the distribution. For example, lubricity of the sample might depend heavily on the proportion of short chains, tensile strength on the molecular weight associated with the largest weight proportion of the sample and moldability on the proportion of very large molecules in the system. Qualitative and quantitative interpretation of a distribution (MWD) usually proceeds along the flowing paths if the experimental data is displayed in the two forms shown (*i.e.*, A, cumulative data on discrete samples; B, continuous differential curve).

Path I

Qualitatively, one simply compares curves between various samples. Figure 1(B) is preferred in this sense, and the important features are the value of M at the maximum in $p(M)$; the length of the tail at large M, the general breadth of the peak, and the steepness of the curve as the peak is approached. This analysis is often used; it is largely visual, but various experimental parameters can be assigned to these features.

Path II

Attempt to fit the continuous curve to a mathematical distribution function. If a "good fit" is found, the moments of the function will provide various averages and the parameters of the function provide a quantitative measure of the shape analysis (note Table I). Or, digitalize the data and calculate the various averages and descriptive parameters directly with a summation procedure[6].

Path III

Experimentally determine various averages by measuring the molecular weight of each discrete sample and its proportion of the total, then:

$$\overline{M} = \sum_i p_i \overline{M_i}$$

TABLE I
Molecular Weight Distribution Functions

Probability Density	1 $\bar{M}_N$	2 $\bar{M}_W$	3 $\bar{M}_Z$	4 $\psi = \left(\dfrac{\bar{M}_W}{\bar{M}_N} - 1\right)$ Ref. [3]	5 $\Phi = \left(\dfrac{\bar{M}_W}{\bar{M}_N} - 1\right)^{1/2}$ Ref. [4]	6 $\theta = \left(\dfrac{\bar{M}_Z}{\bar{M}_W} - 1\right)^{1/2}$ Ref. [5]	7 Remarks
1. $P(M) = A\exp\left[-\dfrac{1}{c^2}\ln^2\dfrac{M}{M_o}\right]$	$M_o e^{-\beta^2/4}$	$M_o e^{+\beta^2/4}$	$M_o e^{3\beta^2/4}$	$e^{\beta^2/2} - 1$	$(e^{\beta^2/2} - 1)^{1/2}$	$(e^{\beta^2/2} - 1)^{1/2}$	Lansing-Kramer
2. $P(M) = Ac^n M^{n-2} e^{(cM)^n}$ Adjustable parameters are c, n	$\dfrac{1}{c}\left[\dfrac{1}{n}\Gamma\left\{\dfrac{1-1}{n}\right\}\right]^{-1}$	$\dfrac{1}{c}\dfrac{1}{n}\Gamma\left\{\dfrac{1+1}{n}\right\}$	$\dfrac{1}{c}\dfrac{1}{n}\Gamma\left\{\dfrac{1+2}{n}\right\}$	$\dfrac{\pi/n}{\sin \pi/n} - 1$	$\left[\dfrac{\pi/n}{\sin \pi/n} - 1\right]^{1/2}$	$\dfrac{[\Gamma(1+2/n) - \Gamma(1+1/n)]^{1/2}}{[\Gamma(1+1/n)]^{1/2}}$	Tung Dist. Breadth Parameter $= n$
3. $P(M) = Ac^{n+1} M^n e^{-cM}$ Adjustable parameters are c, n	$\dfrac{(n+1)}{c}$	$\dfrac{(n+2)}{c}$	$\dfrac{(n+3)}{c}$	$\dfrac{1}{(n+1)}$	$\left[\dfrac{1}{(n+1)}\right]^{1/2}$	$\left[\dfrac{1}{(n+2)}\right]^{1/2}$	Breadth Parameter $= n$ Schultz Dist.

where M represents an "average" and p_i the probability that the value M_i is the *average* value of the sample i. Alternatively, measure several different molecular weight averages on the bulk sample.

We note finally that different chain lengths of a chemically homogeneous polymer *are separable* from one another, thereby leading to a data base indicated in Figure 1. Continous chromatographic separations can provide an experimental curve similar to Figure 1(B) and discrete, batch-type separations provide information as shown in Figure 1(A). It is noted, however, that the experimental data are never provided in "probability" form and to utilize the simple expressions (1) and (2), the dimensional experimental data must be converted and normalized.

Mathematical Distribution Functions

A complex distribution curve as indicated in Figure 1(B) can be approximated by various mathematical forms. Some useful mathematical distribution functions, each with adjustable parameters, are shown in Table I. If a continuous probability density function $p(M)$ on M is known, then by Equation (3)

$$\langle M^k \rangle = \int_0^\infty M^k p(M)\,dM \tag{3}$$

the moments of the distribution $\langle M^k \rangle$ can be calculated. The first four moments $k = 0,1,2,3$ are related to the three experimentally measurable averages as follows: the number average $\overline{M}_n = \langle M^1 \rangle / \langle M^0 \rangle$, the weight average $\overline{M}_w = \langle M^2 \rangle / \langle M^1 \rangle$ and the z average $M_z = \langle M^3 \rangle / \langle M^2 \rangle$ or, more simply, various average molecular weights are given by $\overline{M} = \langle M^k \rangle / \langle M^{k-1} \rangle$ where $\overline{M}$ preserves the units of molecular weight as grams/mole. Each of these three averages is accessible experimentally. In theory, knowledge of only these three averages is not as precise as knowing the distribution through a "well-fit" mathematical function since an infinite number of moments are required to define precisely a distribution curve, whereas a "well-fit" mathematical function utilizes many pieces of experimental data. However, the term "well-fit" covers a range of experimental problems associated with choosing: (1) the best mathematical function with (2) the minimum number of adjustable parameters, and (3) minimization of errors in the experimental data. Some authors feel, based on current precision in the experimental quantities necessary to provide either curve 1(A) or 1(B), that it becomes imprudent to attempt fitting any function with more than two adjustable parameters. Table I is a set of prudent, highly tested functions with simple utility. Related average molecular weights are also given and other quantitative measures of the distribution which are preferred by various authors are identified. A measure of the breadth in a distribution is given by the parameters in columns 5, 6, and 7 in Table I. They are obtained through the molecular weight averages defined in the next section. These functions find utility for single-peak distributions only, and multiple-peak systems are best handled by Path I or mathematical-decomposition procedures.

Molecular Weight Averages

As stated previously, the first three moments of a molecular weight distribution can be experimentally determined by independent experimental methods to be discussed later. A simple generating function definition for the various measurable molecular weight averages is given in Equation (4), where

$$M_K = \sum_i P_K M_i \qquad (4\text{-}1)$$

$$P_K = \frac{N_i M_i^{K-1}}{\Sigma (N_i M_i)^{k-1}} \qquad (4\text{-}2)$$

and $K = 1, 2, 3$ (n, w, z) as denoted previously. The symbol N_i refers to the number of molecules of molecular weight M_i. The factor M^{K-1} is a weighting factor for the particular average. Reduction of data similar to that shown in Figure 1(A) into Equation (4), however, is not accomplished simply experimentally because each fraction [bar in Figure 1(A)] has a molecular weight distribution of its own about M_i and each probability is subject to an error Δp_i. Thus, Equation (4) is strictly an idealization of points (p_K, M_i). Methods for the proper reduction of data and experimental design have been suggested by many authors[7]. The sum should be made so that the discrete point idealization is approached.

We have, therefore, the notion of a distribution of molecular weights for a linear homogeneous system, some mathematical functions, which have been used to describe real systems, a definition of the molecular weight averages encountered either through the distribution function or other experimental methods, and some parameters descriptive of the breadth or (MWD). Our attention now turns to experimental measurement of these quantities.

Molecular Weight Determination Methods Based on Colligative Properties: The Number Average

End-group Determination

Conceptually, the simplest technique for determining the number average molecular weight of molecules in a sample of weight x is to count the number of ends in the sample. Experimentally, one tags each chain on one or both ends with a unique functional group found nowhere else on the molecule. Then analytically determine the number of equivalents y of end-groups in the sample by the best and most precise analytical technique available. The number average molecular weight M_n is then easily calculated from

$$\overline{M}_n = \frac{x f}{y} \tag{5}$$

where f is the number of ends per chain which has the unique functional group. The method is very useful for low-molecular-weight samples $< 10,000$, and the experimenter has the full chemical range of functional groups from which to choose and the full range of analytical techniques available to determine their numbers. It is imperative, however, that the chemistry be exceptionally clean, in that (1) the number of tags per chain must be known, (2) the number of ends per chain must be known, and (3) there must be no functional groups on small molecules in the sample as an impurity. Finally, to obtain reliable data at the limit of $\overline{M}_n = 10,000$, analytical techniques of high accuracy for measurement of small quantities of functional groups or atoms must be used. It is not possible to catalog all combinations of these two criteria (functional groups and analytical procedure) that have been reported in the literature. Where accuracy is the prime consideration, the uncertainties in the chemistry and the analytical procedure are usually checked by another independent molecular weight method. This method does not necessarily require the dissolution of the polymer.

Solutions, Ideal and Nonideal

In performing molecular weight measurements and in separating macro-molecules according to their chain length, it is imperative that the polymers be soluble in a small-molecule solvent. Further, dilute solutions are the usual experimental medium for measurement so that the macromolecules are uniformly dispersed (dissolved) in the solvent and intermacromolecular interactions are minimized. However, except under very special circumstances, dilute polymer solutions are nonideal in their behavior. Experimentally, the nonideal nature is handled by the following two principles:

1. Suppose one measures a property Q_i: for a series of binary solutions of solute concentration C_i with the property of the pure solvent given as Q_0. Then one ordinarily calculates a ratio of the measured difference $Q_i - Q_0$ and the concentration C_i, that is;

$$\frac{Q_i - Q_0}{C_i} = \frac{\Delta Q_i}{C_i} \tag{6}$$

This is the principle of reduced values.

2. Second, unless the solution is ideal, the data calculated by Equation (6) will not be constant with C_i.

Therefore, it is common practice to relate the reduced values to an empirical geometrical progression formula, termed a virial expansion;

$$\frac{\Delta Q_i}{C_2} = A_1 + A_2 C_2 + A_3 C_2^2 + \cdots \qquad (7)$$

Toward the limit of infinite dilution as $(C_2 \to 0)$ the first two terms represent the data in a linear fashion when $(\Delta Q_i/C_2)$ is graphed against C_2. A linear extrapolation to infinite dilution is performed and A_1 is determined from the intercept. The first virial coefficient A_1 usually contains the value of the property to be determined; it is a value independent of the nonideal behavior because of the extrapolation procedure, and A_1 for a specific measure is given by theory. The other coefficients A_2, A_3, etc. reflect the nonideal behavior of the solutions. Their interpretation follows from the theory developed to describe the method. This is the principle of extrapolation to infinite dilution. In our subsequent discussion, these two principles are routinely used to perform the measurements noted.

Ebulliometry

In a binary mixture, the temperature at which first-order phase transitions occur in solvents is related to the concentration of the solute and the thermodynamic properties of the two species. In the limit of infinite dilution, the relationship between the boiling point elevation and the number average molecular weight of the polymer is given in Equation (8):

$$\left(\frac{\Delta T_B}{c_2}\right)_{c_2 \to 0} = \frac{RT_B^2}{\rho_1 \Delta \overline{H}_1^v} \cdot \frac{1}{\overline{M}_n} = \frac{K_b}{\overline{M}_n} \qquad (8)$$

where T_B is the equilibrium boiling temperature of the solvent, $\Delta \overline{H}_1^v$ the latent molar heat of vaporization of the solvent at T_B, ρ_1 the density of the solvent at T_B. These individual quantities are usually merged into the boiling point elevation constant K_b. To simplify further, T_B is rarely measured directly but rather chart units ΔD are recorded, then Equation (8) is transformed to Equation (9), where

$$\frac{\Delta D}{c_2}_{c_2 \to 0} = \frac{K'}{\overline{M}_n} \qquad (9)$$

K' is a combined physical parameter and instrument constant determined on a daily basis (because of changes in atmospheric pressure) using a standard of known molecular weight in the same solvent as used for the polymer. In the $\overline{M}_n$ region up to 20,000, the method has a precision from

0.5–5 percent recorded in the literature[8,9]. Extrapolation of $c_2 \to 0$ is required when the mixtures are nonideal. A simple expansion such as Equation (7) is used in the extrapolation, and, if the concentration range is not too large, the first two terms are sufficient for extrapolation where A_1 is given in Equation (9). The virial coefficients are related to properties of the components and are contained in the thermodynamic theory of polymer solutions.

The measurement device consists of an ebulliometer with a Cottrell pump to promote rapid equilibrium between the boiling solution and the vapor phase. A differential temperature is measured directly by utilizing two temperature sensors in the ebulliometer, one at the point where pure solvent vapor condenses to liquid and one where a good equilibrium is established between boiling solution and its vapor. The differential signal is generally determined by null-type instrumentation; for example, two matched thermistor sensors attached to the two arms of a sensitive wheatstone bridge. The ebulliometer must be isolated from pressure fluctuations and designed to permit easy addition of samples to the solvent for successive measurements. No commercial instruments are marketed; however, excellent designs of the glass apparatus appear in the literature or are readily obtained from experts in the field[10].

The measurement is often complicated by foaming of the boiling macromolecular solutions. Various methods for breaking the foams have been suggested, and in some cases alternate data-analysis techniques are useful to extrapolate the data to zero concentration[11]. The upper range of molecular weights measurable with a precision of $5 \to 10$ percent is in the region of 30,000 $\overline{M}_n$; however, there have been reports in the literature of considerably higher weights. Table II clearly indicates the very small temperature differential that must be determined accurately to reach $\overline{M}_n$ of 50,000. Excellent reviews on ebulliometry appear in books and journals in which details of the method are given along with the best methods for data analysis[8,9].

Freezing-point depression measurements have also been utilized to determine molecular weights $\overline{M}_n$[12]. The method is based on the same thermodynamic principles as ebulliometry but has never evolved as a popular technique, probably because of the limited number of solvents avail-

TABLE II[14]
Colligative Properties of 1% solution,
$\overline{M}_n$ 50,000

Property	Value
Lowering of vapour pressure, mm Hg	3×10^{-3}
Depression of freezing point, °C	1×10^{-3}
Elevation of boiling point, °C	5×10^{-4}
Osmotic pressure, cm solvent	10

able in a reasonable working temperature range. The freezing point depression constants K_f are normally larger than boiling point elevation constants but the problems associated with supercooling of the freezing mixture can be serious. The most difficult experimental problems are related to determining the true equilibrium values of the crystallization temperature T_c.

Osmotic Pressure

When a solvent and a dilute solution of a macromolecule prepared from the same solvent are separated from one another by a semipermeable membrane, the chemical potentials of the solvent on both sides of the membrane will not be equal. Equilibrium will be developed, however, by a hydrostatic pressure differential being established across the membrane. If the chambers separated by the membrane are topped by capillaries, this hydrostatic head will be manifest in a difference in height Δh to which the fluids are driven up the capillaries. This difference is proportional to the osmotic pressure π developed in the system. It is essential that π be measured under equilibrium conditions and at a constant temperature.

A number average molecular weight for a nonideal mixture is determined by extrapolating the reduced osmotic pressure (π/c) to infinite dilution with the help of a virial expression such as Equation (7).

$$\frac{\pi}{c_2} = A_1 + A_2 c_2 + A_3 c_2^2 + \ldots \tag{10}$$

where the intercept at $c_2 = 0$; $A_1 = RT/\overline{M}_n$ is independent of the solvent. The other virial coefficients are obtained from the thermodynamic theory of dilute polymer solutions. In simple terms, the osmotic pressure is proportional to the number of solute molecules, hence a number average molecular weight is obtained.

The method has been used extensively despite some serious experimental problems which have been only partially solved. The major experimental problem is that a true semipermeable membrane which passes only solvent molecules does not exist. Therefore, macromolecular samples having low $\overline{M}_n$ ($< 10,000$) will have a sizable number fraction of polymer molecules diffuse through the membrane. This action depletes the number of solute molecules on the solution side, true equilibrium is never attained, and no precise and adequate extrapolation method exists to obtain the equilibrium osmotic pressure at $t = 0$. Unfortunately, these small solute molecules diffuse rapidly, the shorter chains nearly as rapidly as the solvent. The preparation of membranes with very small holes only slows the solvent diffusion rate so that the length of time needed to reach equilibrium becomes impractical for measurement.

In recent years, commercial osmometers have been marketed which help alleviate these problems by minimizing the time needed to establish equilibrium. The instruments are designed with a feedback signal coupled to a servomechanism which changes the hydrostatic pressure so that equilibrium is achieved rapidly and is maintained across the membrane. A pseudoequilibrium value can be obtained in a few minutes following introduction of the sample. However, these instruments are also subject to solute diffusion giving signal drift with time, especially for low-molecular-weight samples. Nevertheless, they represent a significant advance in the measurement and provide a precision previously obtainable in only a few laboratories. This instrumentation is especially attractive for measuring the $\bar{M}_n$ of samples in the 10^5 region since, as seen in Table II, the hydrostatic head obtainable for solutions of macromolecules in this region is of sufficient dimension to be measured accurately by a linear dimension sensing device. Electronic and optical devices now are more accurate in recording this dimension (Δh) than the equivalent measurement made by eye and with the aid of a high-quality traveling microscope. Excellent and reasonably current reviews of osmotic pressure measurements made by commercial rapid-osmometers are available in the literature[13].

A summary of colligative property measurements is shown in Table II. The data there indicate some of the major limiting factors in the various measurements, associated with the upper molecular weight limit.

Light Scattering: The Weight Average Molecular Weight

In the classical concept of light scattering, a light train of essentially monochromatic waves, upon transversing a sample, interacts with the medium providing it is not isotropic in structure. The EM field acts through the polarizability α of the medium, which causes a small fraction of the radiation—after interaction—to be scattered. This simple interaction transforms the medium into a weak secondary radiation source. Slightly over a century ago Rayleigh derived the fundamental laws for scattering providing the medium was composed of independent scatterers[15]. The intensity of scattered light by a single particle was given as

$$I(\theta_Z, r) = \frac{k^4 \alpha^2}{r^2} I_0 \sin^2 \phi_Z \qquad (11)$$

where r is the distance from the scattering particle, I_0 is the incident light intensity, k is the wave number of the incident light ($k = 2\pi/\lambda$), ϕ_Z is the angle between the electric field vector of the incident light and the scattered wave vector with the incident radiation polarized vertically and propagating in the x-direction. If the incident light is unpolarized and a volume of N scatters is illuminated, then the scattered light now viewed in the (x, y) plane at an angle θ from the direction of the incident beam is given by:

$$I(\theta,r) = \frac{Nk^4\alpha^2}{2r^2} I_0(1 + \cos^2\theta) \tag{12}$$

Since α is an experimental quantity, which is not easily measured, the approximate relationship $4\pi N\alpha = n^2 - 1$ is substituted into Equation (12) to give

$$I(\theta,r) = \frac{k^4(n^2 - 1)^2}{32\pi^2 r^2 N} I_0(1 + \cos^2\theta) \tag{13}$$

where n is the refractive index of the media. This equation was verified experimentally many years ago and Avogadro's number N_A was determined through scattering experiments. The experimentally determined quantity has become known as Rayleigh's ratio and is defined as

$$R_\theta = \frac{I(\theta,r)r^2}{I_0 V} \tag{14}$$

Much later Debye introduced another concept in total light scattered by defining a quantity τ based on the form of the Beer–Lambert law[16]. Debye called τ the turbidity, a parameter analogous to an extinction coefficient that is,

$$I_t = I_0 e^{-\tau x} \tag{15}$$

where I_t is the intensity of the light transmitted in the forward direction, and x is the length of the sample. It is easily shown that τ and R are related as

$$\tau = \frac{16\pi}{3} R_{(90°)} \tag{16}$$

The Rayleigh formulation with some modification can be utilized for liquids and for very dilute solutions where the refractive index of the solute differs from the solvent and where the solute particles behave ideally and act as independent scatterers. In this case, the Rayleigh ratio refers to the intensity difference between solution and pure solvent, i.e., ($\Delta I_{(\theta,r)} = I_{(\theta,r)\text{solution}} - I_{(\theta,r)\text{solvent}}$). In addition, α now corresponds to an average polarizability related to the binary system. Finally, the k vector is modified to reflect the value of λ in the medium where $\lambda = \lambda_0/n$, with n being the refractive index of the solvent and λ_0 the wavelength of the incident radiation in vacuuo.

When these factors are taken into consideration, Rayleigh's ratio ΔR_θ is

$$\Delta R_\theta = \frac{k^4 \eta_1^2}{32^2 N\pi^2} \left(\frac{\partial \eta}{\partial c_2}\right) M_2 c_2 (1 + \cos^2\theta) \tag{17}$$

where $(\partial n/\partial c_2)$ is denoted as the specific refractive index increment, c_2 is the concentration of solute, and M_2 the molecular weight of the solute. This formula can be simplified at $\theta = 90°$ by combining all constant quantities into K_{12}

$$\frac{K_{12}c_2}{\Delta R_{(90°)}} = \frac{1}{M_2} \tag{18}$$

Equation (18) holds only for ideal solutions of independent scatterers. Early in the century, Einstein provided a separate concept for scattered light by suggesting that the inhomogeneities result from density fluctuations in a liquid and density plus concentration fluctuations in a binary mixture. He viewed the inhomogeneities which give scattered light as fluctuations in the dielectric constant about the average[17]. For pure fluids the fluctuations in the dielectric constant within a small scattering volume δV were treated originally as functions of temperature and density. For dilute solutions, he recognized an additional fluctuating component arising from changes in concentration of solute in δV. The Einstein equation for binary mixtures is

$$\Delta R_{(\theta)} = \frac{k^4 n_1^2 \delta V}{32\pi^2 r^2}\left(\frac{\partial n}{\partial c_2}\right)^2 \overline{\Delta c_2^2}(1 + \cos^2\theta) \tag{19}$$

where new terms are $\overline{\Delta c_2^2}$, the mean square fluctuation in concentration, and δV, a small volume element in which the fluctuations occur. From thermodynamic considerations it can be shown that

$$\overline{\Delta c_2^2} = -\frac{kT\overline{V}_1^0 c_2}{\delta V}\frac{1}{(2\Delta\mu_1)/2c_2} \tag{20}$$

and further, that the osmotic pressure is given as $\pi = -\Delta\mu_1/\overline{V}_1^0$. These two equations provide an entrance into the theory of nonideal mixtures through the chemical potential. Thus a virial expansion is useful here also, where the coefficients are provided by the thermodynamic theory of dilute solutions:

$$\frac{K_{12}c_2(1 + \cos^2\theta)}{\Delta R_{(\theta)}} = \frac{1}{M_2} + 2A_2c_2 + 3A_3c_2^2 + \ldots \tag{21}$$

In this equation, K_{12} is the same as in Equation (18). It contains quantities which are either constants or are especially measurable that apply to both the solvent[1] and solute[2], $\Delta R_{(\theta)}$ is measured by scattering experiments, and the $(1 + \cos^2\theta)$ term is a correction term to account for a change in volume viewed as a function of angle θ. In dilute solution $K_{12}c_2/\Delta R_{(\theta)}$ plotted against c_2 should give as the intercept as $c_2 \to 0, 1/M_2$.

This equation is satisfactory for molecules whose largest characteristic dimension in solution is $<(1/20)\lambda$; however, near and above this size a new phenomenon arises in scattering which nullifies the effectiveness of the simple extrapolation above. The new interference phenomenon—called dissymmetry—arises when two waves are scattered from the same molecule, giving constructive interference at the forward scattering angles and destructive interference at the back scattering angles. This angular scattering phenomenon is seen also with X-rays, where the problem is treated through mathematical functions called "form factors." Debye calculated the form-factor equations for light scattering, but use of these equations requires the analyst to prejudge the proper form factor to use; that is, to preselect the solution geometry of the macromolecule. Zimm, however, noted that this choice can be eliminated by performing a double extrapolation of the data: (1) extrapolation to a condition $c_2 \rightarrow 0$ to eliminate the thermodynamic nonideality, and (2) extrapolation to zero angle $\theta \rightarrow 0$ to eliminate nonideality due to dissymmetry. This double extrapolation technique provides three experimental parameters: the molecular weight from the intercept; the slope of the zero angle line provides the second virial coefficient of the system; and the slope of the zero concentration line provides the mean-square end-to-end dimension $\overline{R}_G^2$, a characteristic dimension of the solute molecules. The Zimm equation takes the following form:

$$\frac{K_\theta c_2}{\Delta R_\theta} = \frac{1}{M_2} + 2A_2 c_2 + \frac{K' \overline{R}_G^2}{M_2} \frac{\sin^2 \theta}{2} + \ldots \qquad (22)$$

We mention finally that if the solute is polydisperse in molecular weight, then from Equation (22)

$$\lim_{\substack{c_2 \to 0 \\ \theta \to 0}} \frac{K c_i}{\Delta R_\theta^i} = \frac{1}{M_i} \qquad (23)$$

where the i refers to each species in the polydispersity. Therefore we see that:

$$\Delta R_\theta = \Sigma \Delta R_\theta^i = K_\theta \Sigma c_i M_i \qquad (24)$$

with the understanding that K_θ represents all species. Finally, by combining Equations (23) and (24) we find that,

$$\lim_{\substack{c_2 \to 0 \\ \theta \to 0}} \frac{K_\theta c_2}{\Delta R_\theta} = \frac{\Sigma c_i}{\Sigma c_i M_i} \qquad (25)$$

It is clear, therefore, that the molecular weight average obtained is

$$\overline{M}_2 = \frac{\Sigma c_i M_i}{\Sigma c_i} = \frac{\Sigma N_i M_i^2}{\Sigma N_i M_i} = \overline{M}_w$$

With the classical theory of light scattering and an analysis according to the Zimm technique, the experimental measurements become useful for providing physical measurements on macromolecules. A higher average molecular weight, that is, $\overline{M}_w$, and a parameter descriptive of the size of the solute in solution is obtained. Thus, light-scattering measurements coupled with colligative property measurements provide a measure of molecular weight polydispersity through the ratio of $\overline{M}_w/\overline{M}_n$, plus a dimensional parameter $\overline{R}_G^2$, which is very useful in correlating measurements of macromolecular sizes in solutions as determined with other size-dependent measurements such as viscosity.

Three commercial instruments have been developed for performing light-scattering molecular weight measurements, and these have wide usage. Each instrument has four major components: the optical source, the scattering cell, the collection optics, and the photomultiplier-electronic recording equipment. The experimental technique can be systemized but requires special care in execution, especially in the elimination of dust or foreign particulate matter which, because of its size, scatters an extraordinary amount of light. Instruments are calibrated to provide absolute measurement and considerable data has been published on the calibration procedure and its related error[18,19]. In spite of the care needed in making measurements, the detailed corrections which must be applied for reflection plus other geometrical factors and the extensive calculation, the reproducibility on standard samples sent to various laboratories around the world has been to within 5 percent[2]; however, large error was noted on industrial samples[1].

In recent years a number of advances have taken place in classical light scattering as related to molecular weight determinations. The Zimm extrapolation can provide more accurate molecular weights if data is available at very low angles; however, most commercial instruments do not perform well below $\theta = 25°$. Zimm has modified an instrument and reports low-angle measurements to $10°$ for a biological macromolecule (DNA)[20,21]. These low-angle extrapolations led to a molecular weight over twice that reported previously from the $30°$ lower limit angular studies. Billmeyer has designed and constructed an instrument which provides data below $10°$[22].

In other studies the Zimm plot technique has been performed also on dilute polymer solutions with X-rays[23]. The data obtained overlapped well with the previously reported light-scattering studies. In classical light-scattering photometry, considerable attention has been focused on scattering from polymers dissolved in mixed solvents[24,25], and on polymer solutions in the critical region[26], plus scattering in concentrated solutions[27]. Other advances have appeared in light-scattering experiments on ordered systems[28], simple amorphous systems[29] and multicomponent amorphous blends and mixtures[25,30,31].

Nonequilibrium Methods

Dilute Solution Viscosity

Viscosity measurements do not provide a direct measurement of the molecular weight of a polymer but rather are proportional to the volume.

Thus, the viscosity of a polymer solution not only will be sensitive to the thermodynamic properties of a binary system but is also dependent on polymer structure features such as branching or, in the case of co-polymers, their chemical heterogeneity and sequential composition.

This method of characterization is the most common of all since it requires a minimal investment in equipment, can be performed by laboratory technicians and has a higher precision than all other methods[1]. It is probably the most commonly used relative "measure" of molecular weight, and because of this interest it is probably subject to more misinterpretation on limited data than any other characterization measurement. In the scientific realm the number of careful viscosity studies on macromolecular solvents are now legion, with the result that the Fox–Flory basic hydrodynamic theory[32] with some modifications made by Fixman, Stockmayer and Kurata[33], and others rests on a solid basis of experimental verification.

No attempt will be made to develop a complete discourse on the theoretical interpretations, since may reviews exist in the literature[34]. We will establish only these few salient points, then move to some recent interesting data.

1. The important measurable entity is the intrinsic viscosity $[\eta]$ or the limiting viscosity index. This extrapolated quantity is ordinarily obtained by use of two different empirical equations:

$$\frac{\eta_{sp}}{C} = [\eta] + k_1[\eta]C \qquad (26)$$

$$\frac{1n\eta_{rel}}{C} - [\eta] - k_2[\eta]C \qquad (27)$$

Both η_{sp} and η_{rel} are determined by measurement in a viscometer and both are calculated from the same data sets on efflux time ($n_{rel} \cong t$ solution/t_0 solvent.

2. Empirically $[\eta] = KM^a$ for a given solvent–polymer pair at one temperature and for samples with similar MWD. The constants K and a are determined by calibrating $[\eta]$ against $\overline{M}_i$ as measured by another absolute method. This empirical equation has become known as the Mark–Houwink–Sakurada (MHS) equation.

3. The Fox-Flory theory shows that $[\eta] = KM^{1/2}\alpha^3$ where α is a ratio of the mean square end-to-end dimension of the polymer in the dilute mixture ($\langle R_G^2 \rangle^{1/2}$) to that under the conditions of the Flory temperature (θ), that is ($\langle R_G^2 \rangle_{(\theta)}^{1/2}$). (Here α is not to be confused with the same symbol used earlier to define polarizibility.) The concept of the Flory temperature has not been treated here, but it is the temperature at which dilute solutions behave ideally (θ is analogous to the Boyle temperature of a gas). The parameter α depends on thermodynamic interactions and also on chain-structure parameters. The constant K is a measure of internal polymer structure and chain torsional motion properties. Modifi-

cation of the Fox–Flory theory has been related to the factor α and its functional dependence on thermodynamic interaction parameters and molecular weight[33,34].

4. The intrinsic viscosity is related to a viscosity average molecular weight, but, since viscosity reflects volume, this molecular weight will be variable and dependent on all factors which influence the volume. It has been shown that

$$\overline{M}_v = \left[\frac{\Sigma M_i^a C_i}{\Sigma C_i}\right]^{1/a} \tag{28}$$

Thus, the viscosity average is dependent on all factors which influence the parameter a which is related in the Fox–Flory theory through α. The viscosity average molecular weight has little practical utility. We will simply note that the empirical parameter a normally has values between 0.5 and 0.8 in magnitude for most dilute polymer–solvent systems. Under these circumstances, the viscosity average lies closer to the weight average $\overline{M}_w$, which shows the intrinsic viscosity to be more dependent on the larger molecules in a distribution than on the smaller ones[35].

In the last decade, automated viscometric systems have been marketed and descriptions of automated devices have appeared in the literature[36]. The Ubbelholde viscometer has become a standard in measurement and it has seen some evolution, making it more suitable for automated and repetitive measurement with limited attention by the operator[37].

Early in the 1960s, Zimm and Crothers further developed the Couette rotational viscometer[38]. Their interest arose from the fact that the shear rate in an ordinary capillary (Ubbelholde type) viscometer was large enough to shear degrade many of the molecules whose molecular weight they wished to investigate. In addition for very large molecules on $MW > 10^6$, the intrinsic viscosity $[\eta]$ must be measured at several shear rates and extrapolated to zero shear ($\gamma \to 0$). The Zimm–Crothers design is based on the concept that a rotor can be suspended in a liquid and held in place by the surface forces supporting the meniscus. The rotor floating in the liquid is placed in a torque field and its period of revolution is noted. The torque applied can be made exceedingly small and periods of rotation can be in the order of hours if desired. Several design modifications of the Zimm–Crothers instrument have appeared in the literature, making it adaptable to more automated measurements with reduced technician input[39–42]. This development represents a significant contribution to viscosity measurement, since the range of measurable shear rates has been lowered by three orders of magnitude. This viscometer gives $[\eta]$ values equal to those obtained in capillary instruments when the shear rates are equal.

Zimm and his collaborators[43] have shown the utility of this viscometer in measuring the molecular weights of biological macromolecules which are exceedingly fragile to shear and whose molecular weights are in the region $\overline{M}_w = 10^7 \rightarrow 10^9$. The Couette viscometer has been used also to examine extraordinarily high molecular weight synthetic polymers ($10^6 \rightarrow 10^7$) in the region of very low shear rate γ, where there is considerable theoretical interest in the shape of the $(\eta_{sp})_\gamma/(\eta_{sp})_{\gamma=0}$ versus shear rate curve[44–46]. It should also be mentioned that Zimm has used the Couette principle to design an instrument to measure relaxation times of very large macromolecules in solution[47]. This important advance opens another realm of investigation.

Unlike some other measurements of molecular weight made in mixed solvent systems, the intrinsic viscosity can show a rather strong dependence on the composition of the two solvent components[48,49]. The change in the intrinsic viscosity with solvent composition at constant temperature is in qualitative agreement with the predictions based on the Fox–Flory theory, but the quantitative theory was found to be inadequate to predict the magnitude of the polymer's mean-square end-to-end dimension as measured by this technique. Other recent advances in viscosity measurement are noted in the section on molecular architecture.

Vapor Pressure Osmometry

Another nonequilibrium method for determining the molecular weight of macromolecules has been developed from a principle established many years ago. In the literature the method is given various names, gas–liquid osmometry, vapor-pressure osmometry, inverse ebulliometry, or the Hill–Baldes method. In this technique, a drop of solvent and a drop of solution are each placed on a separate temperature sensor (thermocouples or thermistors). After saturation of the closed cell space with solvent vapor, the solvent will begin to condense on the solution droplet since the chemical potential is lower there. A temperature difference develops between the two sensors because of the lower solvent vapor pressure of the solution, which induces a slight difference in the heat of condensation of the solvent.

The method has found considerable utility in analysis of molecular weights, and commercial instruments are marketed. But in spite of wide usage and the rather simple principle involved, a number of variables can affect results.

It is necessary to conduct the measurement under steady-state conditions where heat losses are equal to heat gains from continuous condensation. Under the assumption of steady-state heat flow, a simple expression appears valid in the limit of infinite dilution ($c_2 \rightarrow 0$);

$$\Delta T_{ST} = Km \qquad (29)$$

where m is the molality of the solution and K is a complex function dependent on the thermodynamic parameters of the solvent, some transport coefficients, a diffusion constant of the solvent vapor, a cell constant and the drop size. By rearranging Equation (29) one obtains

$$\overline{M}_N = \frac{KM_1}{1,000\Delta T_{ST}} \tag{30}$$

In practice, extrapolation of ΔT_{ST} to $(\Delta T_{ST})_{c_2 = 0}$ is required to utilize Equation (30).

Measurement problems related to the instrument include temperature fluctuations, drop-size dependence, drift in T with time as a result of non-steady-state conditions, and irreversible changes in the temperature sensors. In addition, solutions can influence the results if there is association or dissociation and/or inclusion of volatile components. The method produces results to a precision of 5 percent up to about 40,000 $\overline{M}_N$. The method has been reviewed in detail in the literature[50–52].

Ultracentrifugation

An instrumental technique widely used in the realm of characterization of biological macromolecules is the analytical ultracentrifugation. This sophisticated and rather expensive device can be used in either an equilibrium mode or a nonequilibrium mode. Depending on the mode in which the instrument is operated and the method of analysis, $\overline{M}_n$, $\overline{M}_w$, $\overline{M}_z$ can all be obtained[53]. The field has seen significant progress in recent years, and this is the primary molecular weight measurement method used in biological macromolecular research. However, the analysis of data is complicated considerably when the MWD of a polymer sample is broad in relation to the narrow MWD ordinarily found in biological systems. The ultracentrifugation method has not become a widely used characterization method in industrial laboratories or for use with broad MWD commercial polymer systems. However, in a recent careful study, Yamamoto showed that the ultracentrifuge resolved a bimodal distribution that was not resolvable by GPC or elution fractionation[54].

Fractionation

The original Flory–Huggins thermodynamic theory of macromolecular solutions established the fundamental relationships between the chemical potentials in a binary mixed system and their dependence on the chain length of the polymer. It is on the basis of this chain-length-dependent thermodynamic property that polymers of different molecular weights can be separated under the conditions of thermodynamic equilibrium[55]. Various experimental processes have evolved over the past few decades which are based on these fundamental principles. Table (III) has been prepared to collect these for review. In general, various purposes are

TABLE III
Fractionation Based on Equilibrium Thermodynamics

I. Fractional Precipitation
    (a) Addition of a precipitant (nonsolvent) to a dilute solution at constant temperature
         (3-component system)
    (b) Evaporation of a solvent (2 components)
    (c) Variation of temperature (2 components or 3 components)
II. Fractionation by Extraction
    (a) Extraction of swollen polymer by a solvent–nonsolvent mixtures (3 components)
    (b) Extraction of films (3 components)
    (c) Extraction by temperature variation in a thermodynamic poor solvent (2 components)
III. Fractionation by Column Elution (3 components and support)
     Experimental Variables:
     1. Deposition of polymer
     2. Choice of support
     3. Elution mixture and solvent gradient
     4. Elution dynamics
IV. Precipitation Chromatography: Fractionation by Column Elution with a Temperature Gradient:
    Experimental Variables: the same as III with a superimposed temperature gradient on the
    column (3 components)

served by fractionation; for example: (1) preparation of polymer fractions
with narrow MWD for further experimentation on molecular weight
effects; (2) separation of isomeric species; (3) to obtain information on the
MWD.

Fractionation by the schemes shown in Table III ordinarily produces
discrete samples each of which has a MWD of its own, although narrower
than that of the original polymer. If the separation is based exclusively on
the principles of thermodynamics as applied to two phases at equilibrium,
then the principles of this theory must be adhered to experimentally.
Critical aspects are: maintenance of thermal equilibrium, use of very
dilute solutions, the absence of diffusional or rate-controlled processes,
and noninterference from surface effects such as adsorption. The fundamentals of fractionation and separation of high molecular weight polymers is the subject of an excellent book edited by Cantow[7]. In recent
years Koningsveld, using the fundamentals of the Flory–Huggins original
theory, has developed the thermodynamics of phase separation considering a dilute solvent–polymer mixture as a multicomponent system rather
than a pseudo-binary one. His series of articles clearly outlines the fundamental principles associated with equilibrium phase separation and brings
to light some important differences between the binary and multicomponent treatment which have very practical consequences[56,57].

Fractional Precipitation

This procedure is the oldest of the methods and has many disadvantages:
it is slow; and for preparing fractions, large volumes of solvent must be

handled. The efficiency of fractionation is related to the fractionation scheme adopted, and many schemes have been proposed. Huggins and Okamota[58] have developed a simplified calculation which shows that the efficiency of fractionation is related to the percentage of the sample precipitated at each separation of the phases and highly dependent on the initial volume fraction of the polymer before the first phase separation. The experimental and theoretical work of Kamide also established some important concepts and principles of procedure in this subject[59,60].

Fractionation by Extraction

Matsumoto calculated and compared the two techniques of fractional precipitation and fractional extraction[61]. The extraction procedure has been shown to give narrower MWD for individual fractions. In particular, it was found that lower molecular weight species did not spread across the separate fractions as broadly as they do in the precipitation method. Many straightforward fractionations by extraction have been reported in the literature. A major advantage of extraction fractionation is that relatively small volumes are required in the extraction zone as compared to those required by fractional precipitation. Extraction fractionation was pioneered by Desreux and his colleagues, and the process was used regularly to help separate isomeric species in samples of mixed structure[62,63]. More recently, Harrington and Zimm reported an automatic procedure for separation by extraction of 1.4 grams of polystyrene. Even with automation, the process took one week[64].

Column-Elution Fractionation

Many practical problems associated with the MWD in polyethylene, the discovery and preparation of stereoregular polyolefins, and the need for more rapid fractionation procedures supplied ample motivation for the development of fractionation by column elution. It is also noteworthy that the origins of the practical aspects of this type of fractionation derived from many industrial laboratories[65,66]. Small columns are used for MWD measurement, and designs for large columns in the literature show that as much as 0.5 Kg can be fractionated by this procedure. The stepwise process and the related principles are noted below.

Deposition of the Polymer

The objective of the procedure is to deposit the polymer evenly in a highly swollen condition onto an inert support. In principle, the support provides increased surface area so that equilibrium is achieved rapidly during the elution process. There is evidence that deposition of polymer onto the support should be achieved slowly by precipitation, and that a deposition of molecular weights inverse to the elution procedure is helpful in achieving good column resolution[67].

The Support

The support is finely divided to offer a large surface area. It should not adsorb the polymer and should offer some integrity to the swollen sample to prevent channeling. Supports such as glass, sand, aluminum, and celite have been used successfully but pretreatment is required, depending on the material.

The Eluant Mixture

Each polymer sample has a particular MWD, whose differential distribution profiles we noted earlier. Likewise there is an eluant gradient system of two liquids, the solvent and the nonsolvent which corresponds best to a particular MWD. An infinite variety of continuous elution schemes can be effected, but, in principle, one would prefer the ratio of the two solvents in the mixture to change most slowly in that region of the MWD where the largest number of species lie. Finally, the amount of polymer collected in each fraction can be controlled by the eluant composition. Control of the eluant mixture (or solvent–nonsolvent ratio) can be accomplished in part by properly designing the vessels from which the solvent is discharged into the nonsolvent mixing vessel, by varying the entrance position in the mixing vessel, and by stirring. In general, a logarithmic variation of composition of the solvent–nonsolvent is applied to the column[68].

The speed of elution is of prime importance, for the polymer must distribute continuously along the column and equilibrium conditions must be maintained during the elution process. If the speed is too slow, fractionation inversion can result. Optimum speeds are usually determined through experimentation. Columns are operated at constant temperature, and fractions are collected largely as discrete samples. Column fractionation has proven to be a useful method for studying the MWD. Its largest handicap is the slow rate at which data can be gathered. In a critical comparative study, elution fractionation was found to give a resolution of molecular weights equal to gel permeation chromatography (GPC)[54].

Precipitation Chromatography

Precipitation chromatography is the name given by some to a column elution technique where the column is under a temperature gradient. This technique was developed by Baker and Williams[69] over a decade ago and investigated extensively by Schneider[67]. It combines two gradients: (1) solvent–nonsolvent, and (2) temperature. In a rather recent study, Dawkins and Peaker have demonstrated that if the principles associated with the conditions expected for thermodynamic equilibrium are maintained and a small temperature gradient of 5° C is maintained over a 61 cm column which was 2.6 cm in diameter, good fractionation is achieved[70]. Gel permeation chromatograms of the resulting fractions showed uniformly distributed fractions with a small overlap of species between successive fractions. The process can also be automated in the same manner as the straight elution procedure.

Gel Permeation Chromatography (GPC)

Perhaps no single polymer characterization method has received as much attention in such a short time as GPC. The rapid development of this separation technique, which provides so much information for the characterization of a polymer sample, derives from the very early development of a commercial instrument by Waters and Associates (who also fostered considerable research on the fundamental and practical aspects of GPC), the rapidity with which data can be produced, the relative ease of automation even for an instrument made indigenously, and the commercial availability of prepacked gel columns of different porosity[71].

Although the intimate details of the separation process are still under scientific investigation, it is generally concluded by experts that the separation proceeds in the columns by the differential motion of macromolecules through a gel packing which has a distribution of pore sizes. Ordinarily columns are prepared with a broad distribution of pore sizes to effect separation of the whole spectrum of the polymer sample. Large molecules move through rapidly since few of the pores accommodate the polymer, whereas with decreasing size, more and more pores are available to the smaller chains. The process is termed steric exclusion, and implies a condition of diffusion equilibrium. The process is shown not to depend directly on the molecular weight of the chains but on their hydrodynamic volumes[72] and, as a result, also on such factors as the solvent–polymer thermodynamic interaction, the temperature, polymer architecture, etc. Because of the nature of the process, it has been given names such as gel filtration, exclusion chromatography, diffusion chromatography, and others. Each name describes the physical process in part.

Experimentally, a very dilute solution of a very small amount of polymer (1–20 mg) is injected into one end of a column packed with material of varying pore size and saturated with solvent. The sample is moved through the columns by a pump pushing pure solvent. As the polymer moves, it spreads because of the differential rate of movement of the chains according to their hydrodynamic volume and their inclusion into the pores. As the polymer is discharged from the opposite end of the column in very dilute solution, a detector continuously records the concentration of the polymer in the eluant. A differential refractometer is the standard instrument for detection of polymer concentration, but spectroscopic detectors such as IR and UV are equally applicable, as are other analytical techniques[71]. An experimental curve similar to Figure 1(B) is the type of signal provided by the Waters instrument, with the vertical axis proportional to the concentration of polymer in the eluant and the horizontal axis recorded in volume output of the column. If the system is fed several times with the same charge and with the same instrumental conditions, the chromatograms are highly reproducible.

The reduction of the data to the various molecular weight averages and parameters related to the distribution requires knowledge of instrumental parameters which influence the shape of the distribution curve. Several instrumental factors influence curve spreading (instrumental convolution): (a) size of injected sample, (b) pump pressure or flow rate, (c) pore-size distribution of the columns, (d) temperature variation, (e) evaporation of solvent, (f) resolution of the detecting instrument. Each of the factors has been extensively investigated and descriptions are available to determine the best possible methods of eliminating each as an influence on spreading. Many reviews, chapters, and articles present details of each of these procedures[73–76].

Data are ordinarily reduced to digital form rather than fitting the curve to a continuous mathematical function, and a computer program developed by Pickett, Cantow, and Johnson[77] is widely used to calculate various molecular weight averages and distribution parameters. One of the equation sets in the program is shown below:

$$\overline{M}_N = \left[\int_{M_L}^{M_H} \frac{1}{M} \left(\frac{db}{dM} \right) dM \right]^{-1} \tag{31}$$

where M_L and M_H are the lowest-to-highest molecular weight species in the sample, b is the weight fraction of polymer between M_L and M. Obviously, since M is not known, this integral must be related to experimental parameters. For example,

$$\overline{M}_N = \left[\int_{Z_L}^{Z_H} \left(\frac{Z}{K} \right)^{-\frac{1}{\epsilon + 1}} \left(\frac{db}{dZ} \right) dZ \right]^{-1} \tag{32}$$

where Z is a hydrodynamics volume parameter given by

$$Z = [\eta] M = K M^{\epsilon + 1} \tag{33}$$

and (db/dZ) given by Equation (3)

$$\frac{db}{dZ} = \frac{F_{(v_Z)}}{Z} \frac{1}{(df/dv)_{v_Z}} \log_{10} e \tag{34}$$

where the factor $F_{(v_Z)}$ is related to the normalized baseline adjusted chromatogram height at volume v_Z, and the second factor in the denominator (df/dv) is the slope of the curve at v_Z. It is also noted that b in Equation (34) is the weight of polymer between the new limits Z_L and Z, that is, related to the hydrodynamic volume rather than the molecular weight directly. Equation (33) shows the hydrodynamic volume to be related to the MHS expression for intrinsic viscosity.

The calculation therefore depends on the instrumental curve, through $(F_{(v_Z)}, v_Z)$ data pairs, the normalization of the curve (baseline, very important) and the factors K and ϵ. Equation (33) shows that to transform to Z, K and ϵ have to be determined by a set of independent measurements. Thus, the GPC calibration procedure is related directly to that used for intrinsic viscosity $[\eta]$ and molecular weight (M). Data calculated include $\overline{M}_n$, $\overline{M}_w$, $\overline{M}_z$ plus dispersion parameters similar to those in columns 5, 6, and 7 of Table I. Benoit[72] first showed that $[\eta]M = Z$, or the hydrodynamic volume is constant for a particular polymer solvent pair, thus leading to the possibility of constructing a calibration curve for a given polymer. Once the calibration is completed, the averages can be determined. Benoit showed also that since Z is a hydrodynamic volume (HDV), the calibrated curve applies equally to branched chains of the same chemical species since the effect of branching is to change the (HDV). Although much has been written about a single calibration curve for all polymers, it is now generally agreed that calibration curves for each chemical species is required to give accurate values.

Calibrating the instrument for a given system requires considerable effort since many viscosity and absolute molecular weight determinations must be made. However, the literature abounds with excellent papers on this subject for many different polymer systems[76-79]. Much of the recent progress in the field is related to better control of the instrumental parameters[78], investigation of various new gel types, more rapid analysis of the data and its attendant errors, attempts to improve resolution of the device, and combination of GPC directly with a direct MW determining device[80-83].

It has been shown that a chain branching index can be obtained by study of branched systems with GPC[1]. Rodridgez and workers, by using IR detectors, have been able to determine the relative fraction of a given monomer in the chromatograms of a copolymer, thus giving information on the chemical heterogeneity[84]. Finally, considerable attention has been shown to preparative GPC devices which can resolve samples as large as 20 grams of polymer per day[85].

The Effects of Molecular Architecture

In the introduction we posed some very elementary questions about polymer characterization. We turn now to the question of the dissimilarity in molecular architecture within a sample and what can be learned from the measurement resources we have reviewed. For the chemically homogeneous species we can distinguish the architectural varieties associated with stereoisomerism, geometrical isomerism, and chain branches where the latter can be subdivided into structures like cruciform, random, comb, etc. If we assume that all these varieties are constructed from the same monomer unit, then we would expect no aberration in the direct measurement of molecular weight through colligative properties, light scattering, or by vapor pressure osmometry. But these various structural entities have different thermodynamic properties from their linear homogeneous analogues, giving rise to a variation of their hydrodynamic volume in solution and in the temperature variations in HDV. Therefore, GPC, viscosity, column fraction, and ultracentrifugation will show an effect associated with architecturally different chemical homogeneous molecules. In addition, dimensional properties measured by light scattering—that is, the end-to-end dimension—will differ according to the structural type.

When two or more monomers are combined to form copolymers, then additional indices related to the heterogeneity of chemical composition are needed to characterize the system. In addition, copolymers can be formed in all the architectural varieties noted above plus some unique combinations which relate specifically to two or more units combined; for example, block copolymers. In the case of copolymers, only colligative-type properties which measure the number average are unaffected by chemical composition. However, all other methods outlined here will be dependent on chemical heterogeneity since dilute solutions now become three component mixtures. The question arises, "what indices exist to measure these more complex systems and how are these indices recorded?"

Stereo and Geometrical Isomerism

These structural variations, many of which have been prepared in the last two decades, are best indexed through measurements associated with the atomic and functional group level. Thus the classical methods of infrared, Raman, and NMR spectroscopy, are the best means to index; for instance, the fraction of various isomers, etc. This mode of characterization is important and is the subject of another portion of the symposium. Problems associated with separation of mixtures of isomeric species have received considerable attention in the literature. The intrinsic viscosity changes with these indices and the MHS parameters developed for linear random homogeneous chains may not apply.

Homogeneous Branched Polymers

Branching in macromolecules can alter the hydrodynamic properties considerably. This was recognized very early in the era of the development of polymer characterization[86]. The developments of high pressure poly-ethylene brought intense interest to the characterization of branched poly-mers. The standard branching index is:

$$g = \frac{[\eta]_{BR}}{[\eta]_L} \tag{35}$$

where $[\eta]_{BR} < [\eta]_L$ and g is related further to indices which are descriptive of the branched species; for example, in a comblike structure, a parameter is needed to describe the average length of chains between the teeth and another to describe the average length of the teeth. The factor g would be related to these two parameters[86-88].

Zimm, Kilb, Stockmayer, and Orofino have developed theories to show how g changes as a function of a branching index for systems of equal mass[86-88]. Benoit proved that the GPC calibration curve for a homologous series which is based on the hydrodynamic volume was ap-plicable to pure branched systems also[72]. However, the fundamental progress in this field of characterization awaited the preparation of well-characterized branched polymer systems where structural content was known and could be controlled. Several groups have prepared a series of well-characterized branched molecules (comb-type, star-type) and the hydrodynamic and thermodynamic properties have been investigated[89]. Also, further theory in heterogeneity in branched structures was de-veloped by Orofino[87].

For a branched system, GPC gives an apparent molecular weight M^*, based on a linear calibration curve. Benoit shows that the true M is given by

$$M = M^* g^{-(1/1+a)} \tag{36}$$

where a is the molecular weight exponent in the MHS expression[1]. If one considers only long chain branching, which provides a framework for the use of Gaussian statistics, then from corresponding measurement of M and M^* plus $[\eta]_{BR}$, $[\eta]_L$ then one obtains Equations (37) and (38)

$$g_M = \left(\frac{M^*}{M}\right)^{1+a} \tag{37}$$

$$g_{[\eta]} = \left(\frac{[\eta]_{BR}}{[\eta]_L}\right)^{1+a} \tag{38}$$

Benoit and Prechner have determined this type of data on the IUPAC high-density polyethylene test series[1]. For polydisperse branched sys-tems, Benoit further defines these polydisperse indices as:

$$g_M = \left(\frac{M_w^*}{M_w}\right)^{1+a} = \frac{\Sigma w_i M_i g_i^{(1/1+a)}}{\Sigma w_i M_i^{(1+a)}} \tag{39}$$

$$g_{[\eta]} = \left(\frac{[\eta]_{BR}}{[\eta]_L}\right) = \frac{\Sigma w_i [\eta_i] g_i^{(1/1+a)}}{\Sigma w_i [\eta_i]_{BR}} \qquad (40)$$

These two indices are not equal, but are measurable and are purely experimental without resting on any hypotheses. It is suggested that these become the standard for comparison in future years[1].

Thus, properties of pure branched systems are now understood better and the principles relating hydrodynamic behavior to some forms of branching are known, but practical methods to separate and easily characterize the branch type and content in a mixed linear–branched system or a heterogeneous branched system have not been developed.

Copolymers

Light scattering offers considerable information about the molecular masses of copolymers and, in addition, some exceptionally useful indices of chemical heterogeneity[90]. In the Rayleigh equation (18), the Rayleigh ratio depends on the square of the specific refractive index increment (dn/dc). In the three component copolymer–solvent system the $(dn/dc)_{cop}$ is an average value taken over the (dn/dc) of each of the homopolymers and the composition of the copolymer. The $(dn/dc)_{cop}$, which we symbolize by ν_C, is given by the well-established additive relationship

$$\nu_c = \bar{x}_A \nu_A + (1 - \bar{x}_A) \nu_B \qquad (41)$$

where the subscript on ν refers to the (dn/dc) value from the homopolymers in the same solvent and $\bar{x}_A$ is the average weight fraction of component A in the copolymer. Since Rayleigh ratios depend on the square of this quantity, the molecular weight obtained by a Zimm plot extrapolation will yeild an apparent molecular weight. The apparent molecular mass M_{ap} is given by:

$$\bar{M}_{ap} = \frac{1}{\nu_C^2} \frac{\Sigma \nu_i^2 c_i M_i}{\Sigma c_i} \qquad (42)$$

if we use Equation (17) and subscript each species in the system as i, then by squaring ν_i and combining with Equation (42) we have one definition of M_{ap} in terms of the weight averages of components.

$$\bar{M}_{ap} = \frac{1}{2\nu_C} [\bar{x}_A \bar{M}_n^A \nu_A^2 + (1 - \bar{x}_A) \bar{M}_w^B \nu_B^2 + 2\nu_A \nu_B M_w^{AB}] \qquad (43)$$

The quantity M_w^{AB} is related to the true weight average molecular weight of the copolymer (M_w) sample by

$$\bar{M}_w^{AB} = 1/2[\bar{M}_w - \bar{x}_A \bar{M}_w^A + (1 - \bar{x}_A) \bar{M}_w^B] \qquad (44)$$

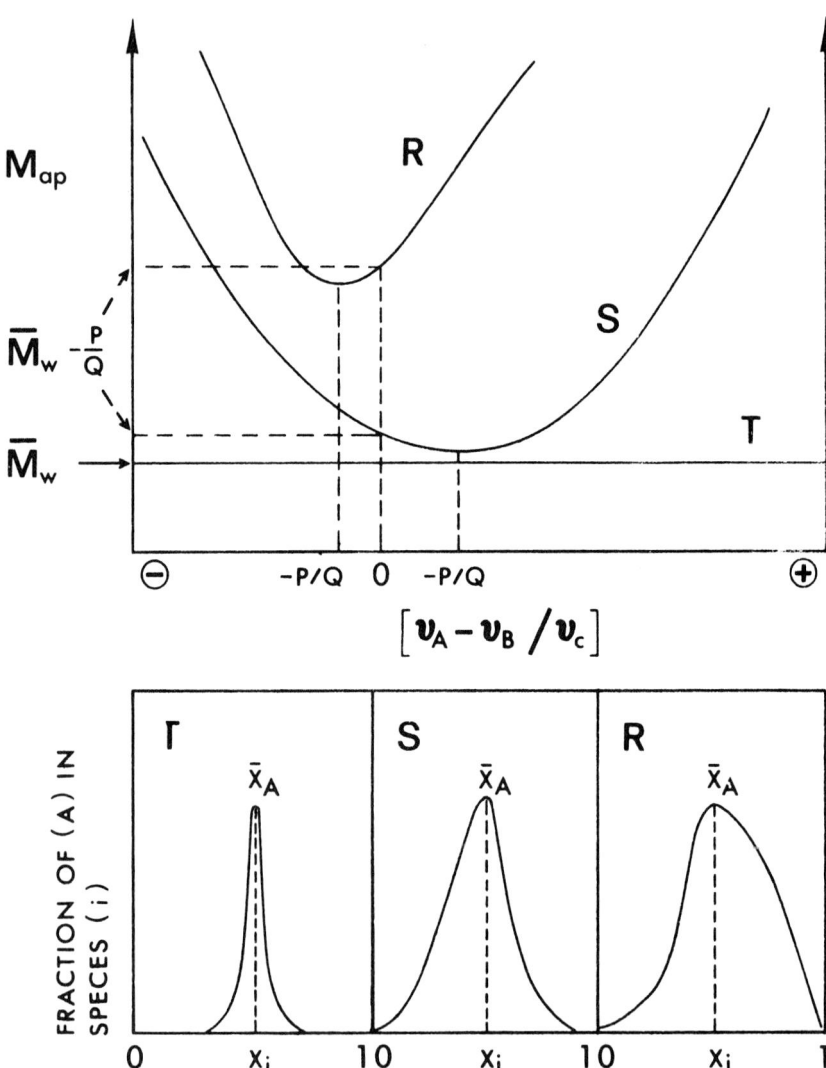

Figure 2. Copolymer apparent molecular weight as a function of $(\nu_A - \nu_B/\nu_c)$. Lower curves schematically represent related heterogeneity.

However, in the practical sense, the averages $\overline{M}_w^A$ and $\overline{M}_w^B$ are not particularly useful. Therefore, Benoit and Bushuk introduced two new but related parameters (P, Q) associated with the chemical heterogeneity of the copolymer species[91]. These definitions are given below, where $\Delta x_i = (\overline{x}_A - x_i)$:

$$P = \frac{\Sigma\, c_i M_i \Delta x_i}{\Sigma\, c_i} \tag{45}$$

$$Q = \frac{\Sigma\, c_i M_i (\Delta x_i)^2}{\Sigma\, c_i} \tag{46}$$

P represents a parameter descriptive of the skewness of the chemical heterogeneity about the average $\bar{x}_A$, and Q is a parameter descriptive of the breadth of the distribution. Figure 2 shows some possible variations on these parameters. By combining these definitions with Equations (43) and (44), Benoit provides the following quadratic form for the apparent molecular weight:

$$\overline{M}_{ap} = \overline{M}_w + 2(\nu_A - \nu_B/\nu_C)\,P + (\nu_A - \nu_B/\nu_C)^2\,Q \tag{47}$$

Measurable quantities are $\overline{M}_{ap}$, ν_A, ν_B, and ν_C. If these measurements are performed in three different solvents, then the three parameters $\overline{M}_w$, P, and Q are determined. Although this requires considerable effort, the method is very useful for indexing the chemical composition and molecular weight of a copolymer. In Figure 2 are displayed some theoretical curves based on the parabolic shape of Equation (47).

If the chemical heterogeneity is uniform and narrow about the average $\bar{x}_A$, then the parabola is essentially flat. Azeotropic copolymers have been shown to fit this special case[90]. The value of the true $\overline{M}_w$ is found

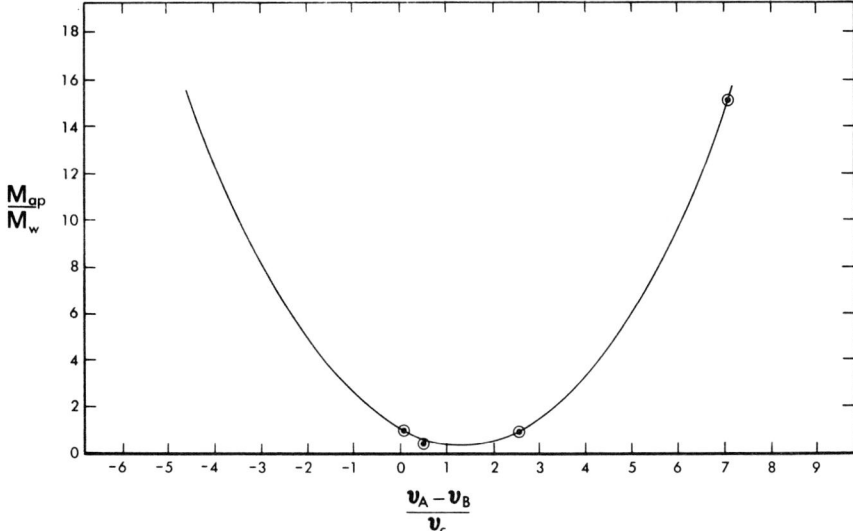

Figure 3. Copolymer of isobutylene and vinylidene chloride. M_{ap} measured in four solvents[131].

TABLE IV

Molecular Parameters for Copolymers I and II[93]

Copolymer	$M_w \times 10^{-4}$	$P \times 10^{-4}$	$Q \times 10^{-4}$	P/M_w	Q/M_w	$M_A^\alpha \times 10^{-4}$	$M_B^b \times 10^{-4}$	Q/Q_{max}
I	5.61	−3.46	2.44	−0.62	0.44	0.74	21.3	0.54
II	5.27	−2.81	1.97	−0.53	0.37	1.17	17.4	0.44

α = A-vinylidene chloride
b = B-isobutylene

where the variable $[(\nu_A - \nu_B)/\nu_C] = 0$, and if $P = 0$, the parabola is symmetrical about this point. It was shown many years ago that for steady-state copolymerization addition kinetics that $P = 0$[92]. The parameter P can take on minus or plus values, indicating a preference for monomer A in the high- or low-molecular-weight species, respectively. Figure 3 shows some data taken on a copolymer of isobutylene and vinylidine chlorides in our laboratories. Table IV lists the values of the calculated parameters.

Corresponding equations for the mean square end-to-end dimension have also been reported. It is not possible to repeat the same determination of an equivalent set of related parameters, since the radius of gyration of the copolymer chain varies with each solvent whereas the molecular weight averages and parameters P and Q are not solvent-dependent. However, a plot of $M_{ap}[R_G^{ap}]^2$ against a combined composition and solvent parameter $y = x_A \nu_A/\nu_C$ gives a revealing curve of a size–mass relationship as a function of the composition[90].

Viscosity of Copolymers

The viscosity of copolymers in a solvent depends on the composition of the copolymer, the composition of the solvent, the temperature, and, as shown in recent years, the polymer structure also[90]. For example, given two copolymers of equal average composition or equal molecular weight, their intrinsic viscosity may differ considerably depending on the structure or the distribution of sequences. Viscosity temperature relationships show a similar dependence also. These differences in viscosity are magnified in certain kinds of solvents and minimized in others. For instance if the two units A and B are highly soluble (low ΔH of mixing) in a solvent, structural differences will be minimized. If one of the units is very poorly soluble or even insoluble in a solvent, structural differences may be quite large and solubility inversion with temperature or solvent composition can occur[94–96].

Finally, we note that if the MHS expression is known (that is, K and a) for the two pure homopolymers in the same solvent and temperature, there is no current method to predict K and a accurately as a function of composition or structure for a copolymer. Likewise, no practical index of copolymer viscometric behavior has evolved. Only in the special case of azeotropic copolymers and regular alternating copolymers is viscometric behavior partially predictable from physical quantities measured on the two homopolymers[92]. It has been shown that copolymers of equal $\bar{x}_A$ and equal masses but different in structure (one a block copolymer and one a statistical copolymer) have the same end-to-end dimensions at their respective Flory (θ) temperatures[90]. The consequence of this finding is that solvent effects are very short range in nature.

New Methods

Measurements Based on Scattering of Coherent Light

The invention and technological development of the laser as an optical source has had an immense impact on the physics of measurement. The laser has many unique properties which make it ideal as a probe for measurement, and it has been responsible for the birth of a new interest in optical physics. Table V represents some of the special properties of laser light as compared to thermal light and relates its use in various kinds of scattering measurements.

We will emphasize here some new measurements based on the coherence properties of the laser. It is an intrinsic property of electromagnetic radiation that its mathematical description in time–space can be represented by wave equations in these variables. The concept of coherence properties has been oversimplified by many authors, and the term coherence often is used inaccurately in the literature. It is not possible to provide a dissertation on the theory of partially coherent light and its consequences here; however, anyone seriously interested in coherence and its physical significance should refer to the many excellent accounts available[97–99]. Fortunately, it is a consequence of measurements made with laser radiation that coherence effects associated with the time variable can be separated in many applications from those associated with the spatial coherence aspect.

In the present context, therefore, we will begin with measurements made possible through the excellent time coherence of laser radiation. In a simplified sense, a train of light waves will be coherent over some time interval $\Delta\tau$. This interval is often referred to as the coherence time. It will be associated with a spectral width $\Delta\nu$ so that $\Delta\tau\Delta\nu = K$ where K is a constant. If the emitting source produces a Gaussian wave train, $K = \pi/4$, and $\Delta\nu$ refers to the half-width at half-height of the wave-train spectrum. Most often the value of $\Delta\nu$ for laser radiation from a continuous wave gas laser (cw laser) is cited as being a few Hertz in bandwidth, producing a coherence time in the order of 10^{-1} sec. Using the further simple notion that $\Delta\ell = c/\Delta\tau$, we can estimate that the coherence

TABLE V
Properties of Optical Sources

	Thermal Optical Source	CW Laser Optical Source	Uses for Laser
1. Bandwidth	10^{12} Hz (1 part in 10^3). Must be narrowed by filter. Concommitant intensity loss	10 Hz in single mode (1 part in 10^{14})	(a) Brillouin Spectroscopy (b) Raman Spectroscopy (c) Light Beating Spectroscopy
2. Spatial Coherence	Very low $\gamma_{12}(0) \to 1$ as diffraction slight $< 1\,\mu$	$\gamma_{12}(0) \to 1$ over full beam width in single mode. Very high spatial coherence in single mode. Property of cavity, not plasma.	(a) Holography (b) Optical imaging (c) Correlation distance measurements (d) Small displacements measurements
3. Divergence	High	Very low (related to 2)	
4. Polarization	Polarized with polarizers or by reflection, sizeable losses.	Highly polarized without loss	(a) Various polarization measurements
5. Energy Density	Low, must be focused.	Can be focused to a small region due to spatial coherence	
6. Photon Counts in a coherence area	Low	Very high	(a) Optical beating spectroscopy (b) Velocimetry

length $\Delta \ell > 10^{10}$ cm for cw laser radiation[100]. This parameter is often confused with spatial coherence effects but is not associated with that phenomenon since it arises strictly from the concept of time coherence. Coherence length $\Delta \ell$ is perhaps better understood in terms of the Michelson interferometer experiment set into a time coherence measurement mode[97]. Then $\Delta \ell$ becomes the minimum path difference needed between the split wave trains to take the recombined waves out of interference.

Because of the extremely narrow frequency bandwidth of cw laser radiation, it is an excellent source for probing motional processes which develop small-frequency Doppler shifts from the incident frequency of the source. With a cw laser, new methods for measuring small shifts have appeared and older methods which suffered from the limited properties associated with thermal optical sources have received a renewal of interest. In particular, we refer to the great increase in Raman spectroscopy and its application to macromolecules, the development of light-scattering spectroscopy through Fabry–Perot interferometry with subsequent measurement of Brillouin lines as well as the depolarized spectrum of scattered light, and, finally, the development of light-beating spectroscopy sometimes called quasielastic scattering. Since we know of no application of Raman spectroscopy applicable to the measurement of the weight or size of macromolecules, it will not be included further in this review. However, a recent chapter on coherent scattering includes broader information on use of laser Raman scattering on macromolecules[101].

Rayleigh–Brillouin Spectroscopy

If one irradiates a liquid with a vertically polarized frequency-stabilized single-mode beam from a cw laser, and if the light scattered at 90° is viewed in the same plane as the beam, with a high-resolution spectrum analyzer, one sees a three-peak spectrum similar to that shown in line C of Figure 4. The central peak is related to the isobaric fluctuations associated with Rayleigh scattering and, if the spectrum analyzer is a Fabry–Perot interferometer, this line represents the shape function of the total instrument. Actually the Rayleigh line is Doppler-broadened by thermal motion, but this effect is unmeasurable for pure liquids and macromolecular solutions with a Fabry–Perot interferometer since the shifts are an order of 10^4–10^9 Hz smaller than the Brillouin shifts. Rayleigh line shapes are measured through light-beating spectroscopy and will be covered in a later section.

The two side bands are the frequency-shifted lines due to inelastic scattering from pressure waves which course through the liquid at equilibrium, and which are pumped continuously by the thermal energy kT. The shift in these lines from the frequency of the incident radiation was

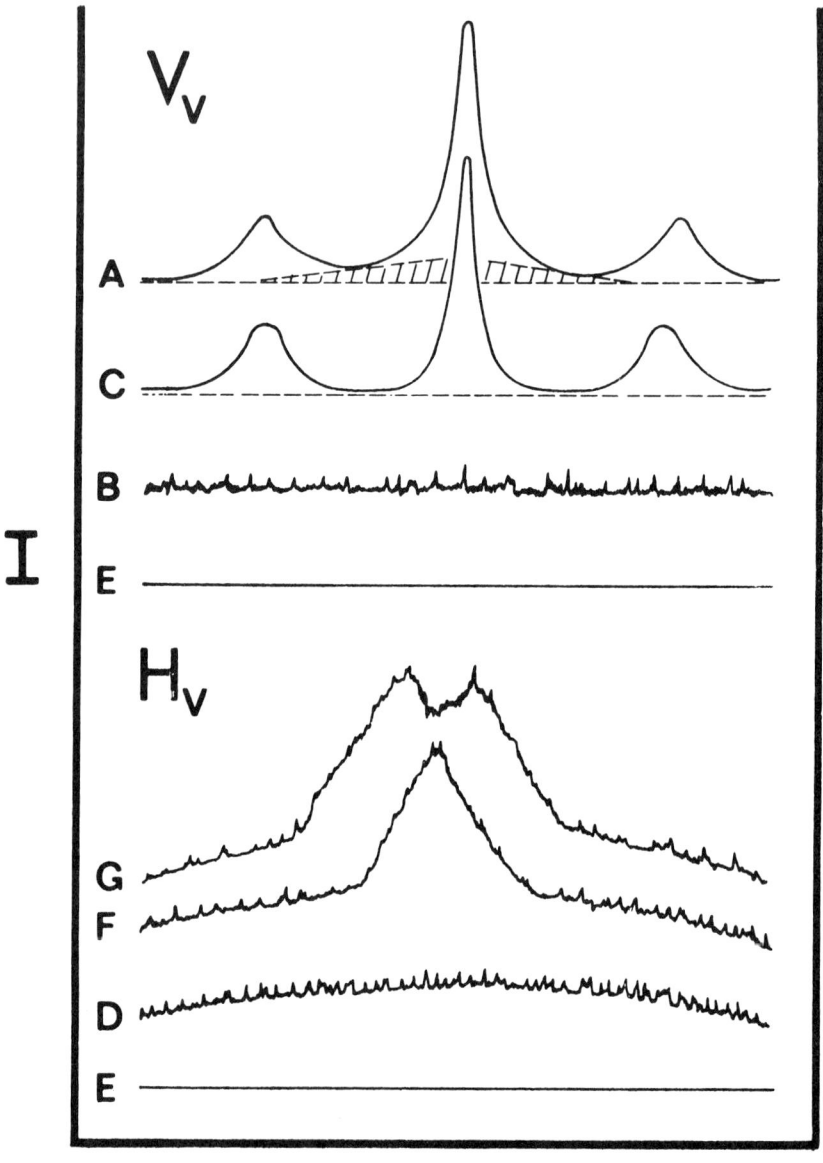

Figure 4. C. Rayleigh–Brillouin spectra, single relaxational mode for each: (A) Rayleigh–Brillouin spectra, with an additional relaxational mode; (B) vertically depolarized light due to anisotropy; (C) Rayleigh–Brillouin spectra, single relaxational mode for each; (D) horizontally depolarized light due to anisotropy; (E) base line; (F) horizontally depolarized light due to anisotropy plus rotational relaxation; (G) horizontally depolarized light due to rotational relaxation, plus shear wave.

proposed in 1922 by Brillouin[102] to be:

$$\Delta\nu = \pm\, 2\nu_o\eta\left(\frac{V_s}{c}\right)\sin\left(\frac{\theta}{2}\right) \tag{48}$$

where ν_o is the frequency of the incident light, V_s is the speed of sound in the medium at frequency $\Delta\nu$, η the refractive index of the medium, c the speed of light in vacuo, and θ the viewing angle measured in the plane of the beam from the propagating direction of the beam to the line of view. It is seen that the shift is angular-dependent, related to the wavelength of the incident radiation and the refractive index at incident wave length. This equation results from the proposition that the incident light vector k meets the Bragg-scattering condition since the pressure wave can be considered as a moving periodic boundary. The sonic frequency probed therefore depends on θ, and in some liquids V_S changes with $\Delta\nu$ and the fluid is said to have a dispersion region at $\Delta\nu$; that is, V_s (velocity of sound) will be frequency-dependent in this region. It is possible to measure V_s by Rayleigh–Brillouin spectrum analysis to within less than 1 percent precision. For pure liquids, $\Delta\nu$ are typically 10^9–10^{10} Hz. A typical mirror spacing in the Fabry–Perot interferometer gives a free spectral range of about 1–2 cm^{-1} for Brillouin spectra.

The second feature to note is that the Brillouin peaks are broader than the central Rayleigh peak (instrumental linewidth). This increase in bandwidth is related to the attenuation of the sound waves in the medium. The attenuation is related to the increase in the bandwidth (the deconvoluted bandwidth) at half height $(\Gamma_B)^{1/2}$ in units of frequency or, if divided by the velocity of sound, a parameter in cm^{-1}, usually designated α, is obtained. A similar parameter is measurable also by ultrasonic techniques and is especially useful for the classification of liquids[103]. It is important to note, however, that in light scattering α represents a time-decay process taking place in the fluid under equilibrium conditions, whereas an ultrasonic wave is energy of an unnatural frequency forced into a liquid by a transducer and its spatial decay or damping is associated with a frequency that may or may not be present in the natural molecular motions. For this reason, ultrasonic physicists prefer to compare frequency-reduced attenuation parameters, that is, $\alpha/\Delta\nu^2$. The parameters V_s, $(\Gamma_B)^{1/2}$, and α are highly dependent on the molecular structure of small molecules and the temperature dependence of α can be positive or negative depending on local liquid structure. Since these important properties are not widely appreciated by macromolecular chemists, a table of information is included for a partial familiarization of these properties for selected small-molecule liquids (Table VI).

In the case where only one relaxational process broadens the Rayleigh line and only one relaxational process is responsible for the Brillouin

TABLE VI
Sonic Properties of Fluids

Liquid	Temp	$10^{17}\,\alpha/\Delta\nu^2$	α/α_{class}	$d\alpha/dT$	Type Liquid	Ref.
Argon	85.2°K	10.0	1.0	–	Classical Fluid	[100]
CS_2	25°C	5800	1150	+	Large Polarizability (Knesser)	[100]
C_6H_6	25°C	800	100	+	Knesser	[100]
nC_6H_{14}	21°C	77	7.7	+	Knesser	[100]
H_2O	15°C	24	2.9	–	Associated	[100]
DMSO*	29.5°C	380	–	–	Associated	[103]

*Measured by Brillouin Scattering; all others are ultrasonic values.

lines, the parameter $(\Gamma_B)^{1/2}$ is given by:

$$(\Gamma_B)^{1/2} = V_S\alpha = \frac{\Delta\nu^2}{2\rho\,V_S}\left[\frac{4}{3}\,\eta_s + \eta_v + \kappa C_p(\gamma - 1)\right] \qquad (49)$$

where V_S is the speed of sound, η_s and η_v are the shear and bulk viscosities of the pure fluid, κ is the thermal diffusivity and c_p is the heat capacity, and $\gamma = c_p/c_V$ for the fluid. We will not enlarge on this parameter except as it relates to a measure of the area under the curve. Further details regarding these parameters are covered in excellent reviews[101,105,106]. In the model above, both relaxational processes lead to Lorentzian-shaped frequency-distribution curves and both curves are convoluted by an instrumental component. A third measurable parameter is the Landau–Placezk Ratio J which is a ratio of the total intensity under the Rayleigh peak to that under the two Brillouin peaks.

$$J = I_R/2I_B = \gamma - 1 \qquad (50)$$

This parameter for the simple relaxational model noted above is related only to the heat capacities of the fluid. Very few experimental studies on Rayleigh–Brillouin scattering for pure liquids have been reported. Most fluids which have been measured are related to that group for which values of many of the physical and hydrodynamic constants can be found in the literature[105]. The Landau–Placezk ratio for pure fluids is deceptively difficult to measure with high precision and thus wide discrepancies in their values exist. There are other factors, however, which interfere with or change this simple two-relaxation-mode picture, and these account in part for the irreproducibility of J for pure fluids.

An Additional Relaxation Mode

Any additional relaxation modes in the liquid that fall within the 0 → 5 cm^{-1} range will change the Rayleigh–Brillouin spectrum. An exaggerated example is shown in line A of Figure 4. The kind of mode shown has been explained by Mountain as deriving from a coupling process in which energy in the pressure waves (isentropic fluctuations) and the Rayleigh isobaric fluctuations are interconverted through internal molecular vibrational relaxations. Carbon tetrachloride has such a mode, and CCl_3CF_3 recently investigated in our laboratories also has a similar mode. Mountain shows that, with this kind of additional mode at very high frequencies[107,108],

$$J = \frac{V_\infty^2}{V_0^2}(\gamma - 1) = (\gamma - 1)\left(1 + \frac{c_I}{c_v - c_I}\right) \tag{51}$$

where c_I is the heat capacity for the internal modes, v_∞ is the static velocity of sound.

Anisotropic Scattering

If the molecule or the liquid has an anisotropic structure, a significant proportion of the scattered light can have its polarization vector rotated from the vertical. This component of depolarized light is discernible by viewing the output of the Fabry–Perot with a horizontally oriented analyzer. (H_v, where large letter refers to analyzer and small subscript to vector orientation of incidental light.) Such a spectrum is shown in line D of Figure 4. However, this anisotropically scattered light is not rotated fully to the 90° position, therefore a portion of this scattered light which gives a vertical component shows as a broad band under the Rayleigh–Brillouin spectrum, as indicated in line B of Figure 4. This vertically scattered light is not included in the theoretical Landau–Placezk ratio or for that matter, the intensity function developed by Mountain shown in Equation (51).

A second very interesting component of depolarized light often shows a narrower frequency spectrum when viewed with a horizontal analyzer (H_v) than that due to molecular anisotropy, and this is considered to result from a rotational relaxation of the dipole moment in the molecule. An example of this phenomenon is shown in line F of Figure 4. These lines have been the subject of intensive interest in recent years and have been measured for several benzene and toluene derivatives[109,110]. Finally, some fluids have a symmetrical splitting of this peak[111,112]. The latter phenomenon is said to result from shear waves in the liquid. The contribution to vertically polarized scattering of these motions have not been measured independently.

All of the foregoing relates to measurements on macromolecules only in this way: small molecules represent the media in which a macromolecule must by dissolved in order to perform high-resolution light-scattering spectral measurements. Therefore, their scattering constitutes a portion of the scattered light in any mixture and must be reconciled in spectral analysis of macromolecular mixtures and especially in the J value for the pure fluid. Second, although the polymer may have no effect at low concentration on the Brillouin peaks, its effect on others is unknown.

Finally, it should be noted that a theory for Rayleigh–Brillouin scattering of simple binary fluid mixtures has been formalized[113,114]. However, very few measurements have been reported on mixtures and the formal theory over the full compositional range has not been tested. A very large number of optical and hydrodynamic parameters on the pure components and the mixtures must be determined for a complete analysis of the theory.

Determination of Molecular Weight

G. Miller recognized very early that certain approximations could be made in very dilute solutions of small molecules, thereby leading to the measurement of the activity coefficients in a binary mixture by determining the Landau–Placzek ratio[115]. Once this was proved useful, it was a natural extension to consider a dilute macromolecular mixture with the thermodynamic theory of macromolecules easily entered through the chemical potential or the activity coefficient calculations of the Flory–Huggins and other more advanced statistical thermodynamic theories.

It was Miller who first suggested Rayleigh–Brillouin scattering as a useful technique for measuring the molecular weight of a macromolecule and who published the first absolute molecular weight measured in this fashion[116].

The method differs from classical light scattering in the following way: (1) no instrument calibration is needed; (2) the Rayleigh ratio ΔR_θ is not determined but rather the Landau–Placzek ratio ΔJ_θ; (3) the Brillouin peaks of the solvent provide an internal standard to give absolute values to the scattering; (4) a frequency spectrum is analyzed rather than direct photointegrated intensities.

In the approximation given by Miller, Equation (52) represents the final equation used to determine the molecular weight of macromolecules

$$\frac{BKc_2}{J_v^{\text{soln}} - J_v^o} = \frac{1}{\overline{M}_w} + 2A_2c_2 + 3A_3c_2^2 + \cdots \tag{52}$$

where J_v^o is the Landau–Placzek ratio of the pure solvent, for vertically polarized, vertically analyzed light, (J_v^{soln}) is the Landau–Placzek ratio for the solutions, B and K are constants related only to the solvent or to separate measurable quantities. This equation applies only to molecules without dissymmetry. It is based on the fundamental assumption that the Brillouin peaks are unaffected by the macromolecule and hence reflect properties of the solvent exclusively.

Nordhaus and Kinsinger have investigated this assumption for benzene by measuring the frequency shift $\Delta\nu$ and the attenuation parameters $(\Gamma_B)^{1/2}$ for benzene in solutions of polystyrenes of various molecular weights and various concentrations[117]. Within their range of variables, the Brillouin peaks are independent of the macromolecule. They conclude that the Miller approximate equation is an adequate representation of the phenomenon and is useful for molecular weight determinations. Figures 5 and 6 represent the Brillouin shift and line-width data taken on samples of different polystyrenes reported by Nordhaus and Kinsinger. Their data on standard samples of polystyenes is listed in Table VII and shown in Figure 7.

Nordhaus, in this thesis, shows that the molecular weights determined by the Miller equation are subject to an uncertainty associated with the Landau–Placzek ratio for the pure benzene J_v^o. As mentioned previously, literature values for J_v^o of pure fluids vary considerably due to the problems outlined at the beginning of this section. Miller recognized, for instance, the contribution due to the depolarized light and made a correction for it. Nordhaus, however, has taken data on an instrument with higher resolution (higher fineness) which, coupled with other data in the literature, leads him to believe that a coupling mode contributes to the benzene spectrum. This effect contributes significantly to J_v^o even if the percentage of light in the mode is as little as 3 percent. Thus, more investi-

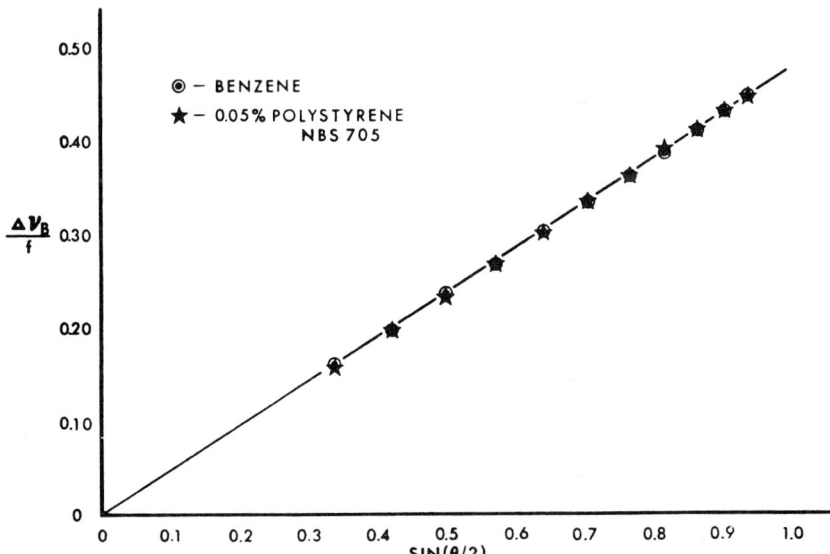

Figure 5. Relative Brillouin peak separation for benzene and a polystyrene solution.

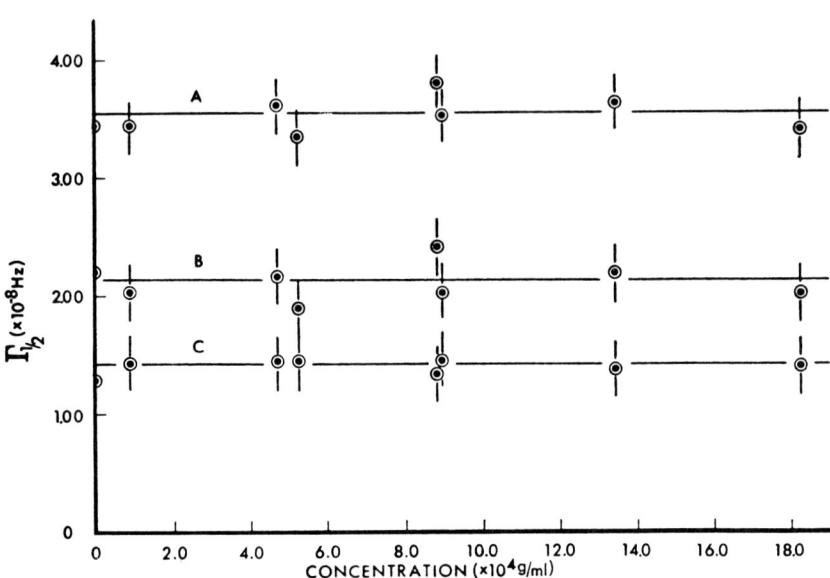

Figure 6.   Concentration dependence of Rayleigh and Brillouin peak half-widths for polystyrene in Benzene: (A) Brillouin, (B) Rayleigh, (C) true Brillouin half-width.

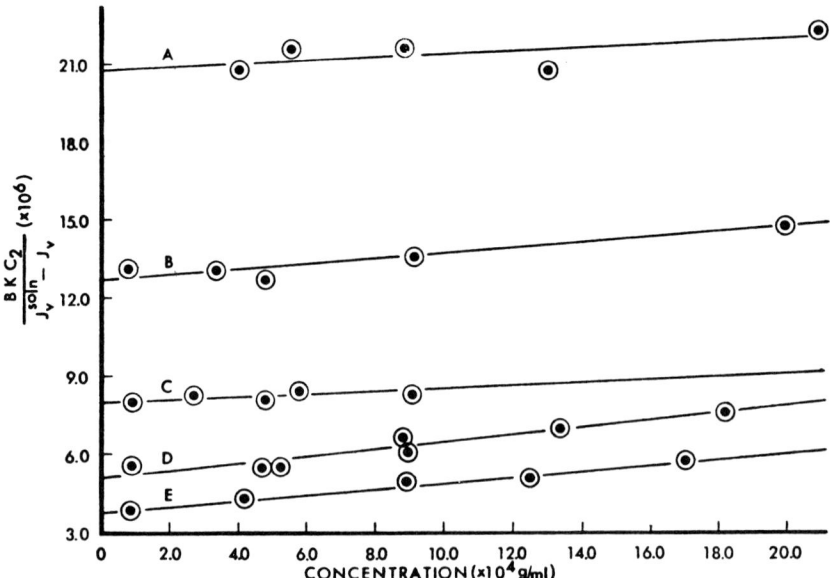

Figure 7.   Molecular weight measurements of polystyrene in benezene: (A) $\bar{M}_w$ = 48,000; (B) $\bar{M}_w$ = 78,400; (C) $\bar{M}_w$ − 124,000; (D) $\bar{M}_w$ = 194,000; (E) $\bar{M}_w$ = 260,000.

TABLE VII
Molecular Weight Measurement of Polystyrenes By Brillouin Spectroscopy

Type	$\bar{M}_w$ (Brillouin)	$\bar{M}_w$ (Viscosity)	$\bar{M}_w$ (Source)	$\bar{M}_w/\bar{M}_N$	$A_2 \times 10^4$ ml/g
Dow S-3	48,000	49,000	—	—	3.1
Dow S-102	78,400	89,400	82,500	1.05	5.3
Dow S-6	124,000	117,000	—	—	1.5
NBS-705	194,000	—	179,300	1.07	7.1
	(173,000)*				
NBS-706	260,000	—	257,800	2.1	5.5
Pressure					
Chemical	1,850,000	—	2,000,000	1.20	4.2

*Data of Miller[112]

gation of pure fluids needs to be performed at higher resolution before these questions can be settled.

Nordhaus has coupled the Miller equation with the Zimm technique to include effects due to dissymmetry[118]. This equation is given below, and the first Landau–Placzek–Zimm plot for a polystyrene $\bar{M}_w > 1.5 \times 10^6$ is shown in Figure 8.

$$\frac{BKc_2}{J_v^{\text{soln}} - J_v^o} = \frac{1}{\bar{M}_w} + 2A_2c_2 + \frac{K'\langle \bar{R}_G \rangle^2}{\bar{M}_w} \sin^2(\theta/2) \qquad (53)$$

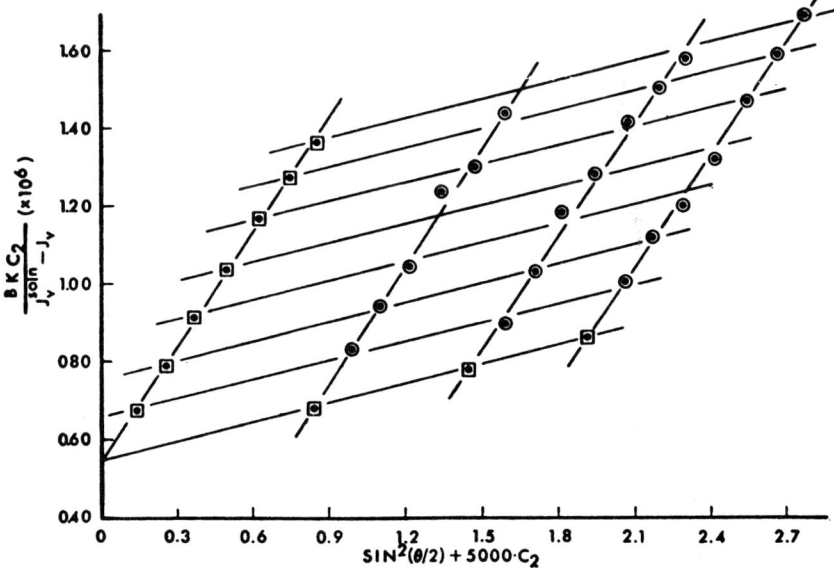

Figure 8.   Molecular weight measurement of polystyrene in benzene; $\bar{M}_W = 1.85 \times 10^6$.

Until a greater number of solvents and polymers are investigated, it is not apparent that Rayleigh–Brillouin spectrum measurements will supersede classical light scattering for several reasons, largely experimental. Fabry–Perot interferometers are notorious for their instability in alignment through causes such as vibration, mechanical and electrical relaxation, and small temperature fluctuations. Until second-generation, well-stabilized Fabry–Perot interferometers are available, it is doubtful that this technique will evolve rapidly for molecular weight measurements. Brillouin scattering from amorphous systems holds considerable promise for characterization of the inhomogeneities and motional properties of macromolecules[101,105,119].

Light-Beating Spectroscopy

In 1947 Forrester[120] suggested that the beat signal between light waves with similar but not identical frequency might be detectable from the output of a phototube. This concept of light frequencies was based on the measurement and use of the phenomenon of beats in electromagnetic radiation at longer wavelengths. It was Forrester's contention that if sufficient light of different frequencies could be focused onto a small area of a phototube (which acts as a nonlinear detector) and if the dc signal were blocked, the beat signal could be detected above the noise. The astounding nature of this prediction was that light signals in the range $\nu = 10^{15}$ Hz could be used to detect frequency shifts as small as $\nu = 10^3 \rightarrow 10^6$ Hz. Forrester had his skeptics, but in 1955 proved his contention in a now-classic experiment[121]. However, the method of measuring beats at optical frequencies did not evolve into a standard physical technique because of the very poor S/N ratio obtained with the optical sources and various electronic components available at the time. In 1961 Forrester[122] published another classic paper where he noted that laser radiation is the perfect optical source for light-beating experiments because of its very high time coherence and a high intensity that could provide much larger S/N ratio in a small coherence area. In this paper he outlined several different possible experimental arrangements which could be used to detect this effect and essentially presented the scientific community with a blueprint for success. Forrester noted two types of detection systems, the heterodyne spectrometer* in which the light spectrum to be measured is beat against a local oscillator and a homodyne* spectrometer in which the Fourier components of the spectrum to be measured are simply beat against each other. The first method raises the

*These words are used differently in various accounts. See reference [101] for a clarification.

signal level but provides a smaller half-width in the beat spectrum, and the second method suffers from a weaker signal but has a broader spectrum[123].

The first application of beat spectroscopy·to light scattering was reported by Cummins[123] and followed shortly by Benedek[124]. Both scientists measured the beat spectrum from solutions of biological macromolecules (TMV, bovine serum albumin).

The beat spectrum (now often called quasielastic scattering) is the Doppler-shifted spectrum of the Rayleigh peak. As noted previously, the Rayleigh peak in a pure small-molecule liquid is broadened by thermal motions which are entropic and random in nature, and propagate only over short distances compared with the wavelength of light. The beat spectrum of a pure liquid spreads from 10 Hz to approximately $10^6 \rightarrow 10^8$ Hz. If a large molecule or a phase-separated species such as a polystyrene latex sphere is suspended in a liquid phase, the light scattered from this large inhomogeneity will be orders of magnitude larger than the scattering from the liquid and the beat spectrum will lie largely in the range $0 \rightarrow 10^3$ Hz. Dissolved or homogeneously dispersed macromolecules gave an intermediate-range beat spectrum. The spectrum is a measure of the diffusional motions of the spheres or macromolecules in the media at equilibrium. Early in the 1960s, Percora[125] provided the theoretical base to describe this phenomenon for macromolecular systems.

The output of the phototube, whose frequency components are measured with a wave or spectrum analyzer, provides directly an intensity versus frequency spectrum. The half-width at half-height $\Gamma_{1/2}$ is proportional to the squared wave vector, $K(\lambda,\theta) = 2\pi/\lambda \sin(\theta/2)$, and the proportionality constant relating these quantities is a diffusion constant.

$$\Gamma_{1/2} = DK^2 \qquad (54)$$

In the case of independent spheres, D is the translational diffusion constant of the spheres in the fluid and is related to the radius by the Stokes–Einstein relationship.

$$D = \frac{kT}{6\pi\eta r} \qquad (55)$$

where η is the viscosity of the media, k is Boltzmann's constant, T is absolute temperature, and r is the radius of the spheres. Thus, the beat spectrum depends on a size parameter and the size parameters are related to the structure. In the decade since the first work in the area, the technique has been applied to particles which are spherical, oblong, cylindrical, and coiled[101]. In addition to the translational modes, low-frequency "bending or breathing" modes are subject to measurement by the beat spectrum[126,127]. This method opens the investigation of molecular motions of a diffusional or low-frequency cooperative type in the energy range covered by frequency range of 10^1–10^8 Hz. The diffusion constant will be proportional to dimensions related to the geometry of the particle or molecule. Some new motions in macromolecular systems have been measured by this technique and the accuracy for measuring the diffusion constant of large molecules has risen by an order of magnitude. The beat spectrum of the horizontally depolarized light is a measure of the rotational motion in macromolecules. This effect is also noted in the vertically polarized scattered light at larger angles. This phenomenon has been measured accurately on only a relatively few biological macromolecular systems.

It is also worth mentioning that Doppler-shift spectroscopy has found use in flow measurements and liquid suspensions under the influence of a force field[101,128]. Chen has determined the velocity profile of bacterial motion in aqueous solution, for example, and others have investigated motion in living organisms[129]. Recently, Flygare has combined electrophoresis and beat spectroscopy[130]. The electrophoretic effect is used to separate macromolecules of different charge density, and the velocity and molecular weight of the separate species is obtained from the beat spectrum. Where the beat spectrum has a half-width of 50 Hz or less, there is a definite advantage in moving the measurements into the time domain and performing a Fourier transform to the frequency domain thereafter. The time analysis is performed by an autocorrelator which can feed information directly to a computer, which performs the Fourier transform. A half-width of 10 Hz appears to be the experimental low-frequency limit.

We note two applications of beat spectroscopy to solutions of synthetic macromolecules. The first is an application to the solutions of polystyrene and cyclohexane in the region of an upper critical temperature at phase transition and to polyisobutylene in pentane at a lower critical temperature in the region of a phase transition[131-132]. In both studies, the approach of the macromolecular system to the critical state provides very high scattering from the large concentration fluctuations in the media. The measured diffusion constants D reflect the size of the inhomogeneities in the critical region, which provides a relationship between the diffusion constant and temperature. Both authors also note a molecular weight effect on D.

Ford, Karasz, and Owen[133] used a homodyne spectrometer to study the power spectrum of dilute solutions of polystyrene in 2-butanone and obtained a translational diffusion constant D as a function of concentration and molecular weight. The molecular weight-dependence of D is given by the simple form

$$D = K_D M^{-b} \tag{56}$$

where K and b are calibration constants obtained in the same manner as the MHS parameters. Also, it is noted that $3b = a + 1$, where a is the MHS constant from intrinsic viscosity measurements.

D was found to be independent of concentration above a level of

3×10^{-3} g/cm^3 but below this range a simple linear expression (57) was observed:

$$D_{(c)} = D_o(1 + k_D c) \qquad (57)$$

where k_D was a constant dependent on molecular weight but which changed signs minus to plus from low to high molecular weights. This work and other different instrumentally measured data on diffusion constants shows concentration regions where D is independent of c and other regions where linear relationships are found. Ford et al. estimated the average end-to-end dimensions from their work and found them in agreement with those determined by other methods (viscosity, sedimentation).

Frederick[134] was unable to fit his power spectra measured on high-molecular-weight polystyrene at the Flory temperature (θ) to a single Lorentzian curve. At molecular weights in the range of 1×10^6 and at angles 45° or higher, he was led to the conclusion that several effects such as intramolecular interaction, critical scattering effects, intramolecular motions, or molecular weight heterogeneity could be responsible for the deviations. In a second paper[125] he analyzed the effect of molecular weight dispersity using the theory of Pecora and noted that the calculations show a modified Lorenztian shape as predicted in the light-beating spectrum at very low frequencies.

Several authors have suggested that new low-frequency internal motions in large macromolecules in solution might be detected in beat-spectrum analysis. In particular, the region where $KR \gg 1$ (where K is the scattering vector and R the characteristic dimension of a molecule) has not been investigated experimentally, whereas at $KR \ll 1$ a number of measurements show $\Gamma_{1/2}$ is proportional to K^2. However, in the former

TABLE VIII
Coherent Light-Scattering Measurements

C = Classical; R = Rayleigh–Brillouin spectroscopy;
LB = Light-beating spectroscopy

1. Molecular weight determinations C, R, LB (II)
2. Size of molecules, C, R, LB (II)
3. Correlation distances low-angle scattering C
4. Concentration fluctuations LB
5. Diffusion constants, translational, rotational, self, and mutual LB
6. Critical phenomena C, R, LB
7. Hypersonic velocity and sonic attenuation R
8. Diffusional process, jump motion vs. segmental R
9. Rotational relaxation and diffusion R, LB
10. Collisional broadening R
11. Shear waves R
12. Velocimetry LB
13. Doppler shift LB

(II) = Implies secondary rather than primary measurement.

region $\Gamma_{1/2}$ is predicted by theory to be dependent on K to powers different from 2, depending on the models chosen. Huang and Frederick[136] investigated this region with samples of polystyrene 4.4 × $10^7 \overline{M}_w$, over a wide range of viewing angles (change in K) and concentration. The exponential factor varied with concentration from 2.2 to 2.9. As Frederick noted, this phenomenon makes it difficult to distinguish which theory and model correctly fits and explains any additional Lorentzian shape seen in the experimental data. Light-beating spectroscopy holds enormous potential for investigating new motions in macromolecules and for optical signal processing in general. Table VIII summarizes various areas of research investigated by laser scattering.

Other New Methods

There are many reports in the literature where various laboratories have developed a special technique for determining MW or MWD. Some are worthy of wider note. For instance, Benoit reports in his comprehensive study that several laboratories reported some determination of MW by spectrographic techniques such as IR or NMR[1]. In almost all these instances, the method is specific to a particular polymer due to a highly identifiable group, and, generally, spectroscopy is useful only for molecular weights to several thousand. For example, we note the paper of F. L. Ho, who was able to determine the $\overline{M}_N$ of poly(propylene-glycol) by NMR with the aid of the europium (EuDPM) chemical shift reagent[137].

Electronmicrographs of individual polyacrylamide polymers deposited from very dilute solutions and shadowed with platinum–carbon mixtures have been utilized to determine the MW and MWD of this polymer. Quayle reports, however, that ghosts are often found on the micrographs, but with proper technique they can be eliminated[138]. The micrography method has also received considerable attention at Mainz in the laboratory of G. V. Shulz[139–141] and by C. Ruscher[142]. There, they have developed methods for eliminating ghosts. They utilize freeze-dried polystyrene and control the shadow casting with carefully defined aperatures. This group has determined the MW and MWD of polystyrene by this technique. With the high cost of the equipment, the knowledge of the technology required, and the care needed in making the micrographs, it is unlikely that this technique will supersede other more established methods.

Some years ago Dole reported that he was able to place macrosized ions in the gas phase, rendering them capable of separation and measurement by mass spectroscopy[143]. Several groups are working on high-molecular-weight mass spectrographs, but nearly all reports in the literature on MW by this method relate to the analysis of decomposition products. There is very high interest in this method within the field of biopolymers.

Molecular weights of telemers of a homogeneous species can be determined through isothermal distillation[144]. The work of Inagaki, Belen'kii, Gankiva, Otocks, and Hellman may prove very useful. These investigators have developed thin-layer chromatography for (a) determining MW, (b) MWD, (c) composition of copolymers, and (d) for separating graft and block copolymers from mixtures. A rather recent review of the subject should stimulate others to try the very promising technique[145-147].

Two reports of a characterization of synthetic macromolecules conducted under the auspices of the IUPAC are recorded[1,2]. The first dealt largely with narrow-distribution polystyrenes and the second, in 1971, reported on a variety of measurements on commercial samples of several polymers (polystyrene polyvinylchloride, low- and high-density polyethylene). These two studies constitute the recorded results available for accurate comparison of methods. Results differed in the two studies; for example, precision for MW by light scattering on the commercial polymers was much worse than on the previous study, but, on the other hand, osmotic-pressure measurement improved in precision. These studies have very high value to the community of polymer scientists and certainly should be continued at least once a decade. It is hoped that the next sampling will include a branched polymer and a copolymer. Many of the generalizations in this article derive from these data.

We have tried in this chapter to present a generalized overview of experimental methods for characterizing MW and MWD of polymers. In addition, some emphasis has been placed on some recent progress in various fields. The review is not comprehensive in any area, but the bibliography should lead more interested investigators to in-depth reviews of specific topics. Our attempt has been to present the material within the framework and aspirations of the conference.

The author wishes to thank his students, Dr. H. Yuen, Dr. D. Nordhaus, Dr. Mary Tannahill, Dr. S. Gaumer, Dr. N. Stiso, and Dr. William Toth for providing some of the data contained herein.

REFERENCES

1. Strazielle, C. and Benoit, H., "Molecular Characterization of Commerical Polymers," *Pure Appl. Chem.*, 26 (1971), 451–79.

2. Atlas, S.M. and Mark, H.F., "Report on Molecular Weight Measurements of Samples, 1961," Commission on Macromolecules of IUPAC, Paris, 1963.

3. Schulz, G.V., "Molecular Weight Determinations on Macromolecular Compounds. VIII. Distribution Functions of Multimolecular Compounds and Their Determination by Fractionation," *Z. Phys. Chem.*, B47 (1940), 155–93.

4. Lowry, G.G., "Polydispersity Index Having Particular Applicability to Narrow Molecular Weight Distribution Polymers," *J. Polym. Sci., Part B*, 1 (1963), 489–90.

5. Hosemann, R.W. and Schramek, W., "Statistische Methoden zur Ermittlung der

Verteilung der Molekulgrossen in hoch polymeren polymolekularen Systemen aus Experimentalergebnissen der fraktionierten Fallung. I. Mitteilung: Mathematische Grundlagen," *J. Polym. Sci.,* 59 (1962), 13–27; "II. Mitteilung: Auswertungsmethoden," *J. Polym. Sci.,* 59 (1962), 29–50; "III. Mitteilung: Der Abbrucheffekt und seine Abschatzung," *J. Polym. Sci.,* 59 (1962), 51–70.

6. Pickett, H.E., Cantow, M.J.R. and Johnson, J.F., "Column Fractionation of Polymers. VII. Computer Program for Determination of Molecular Weight Distributions from Gel Permeation Chromatography," *J. Appl. Polym. Sci.,* 10 (1966), 917–24.

7. Cantow, M.J.R., *Polymer Fractionation,* New York: Academic Press, 1967.

8. Glover, C.A., "Determination of Molecular Weights by Ebulliometry," *Advan. Anal. Chem. Instrum.,* 5 (1966), 1–67.

9. Mikhel'son, V.Ya., "Microebulliometric Determination of Molecular Weight," *Tr. Tallin. Politekh. Inst., Ser. A,* 263 (1968), 59–70. (In Russian)

10. Daniels, T. and Lehrle, R.S., "Three Thermistor Ebulliometer," *J. Polym. Sci., Part C,* No. 16 (1965), 4533–35.

11. Ezrin, M., "Determination of Molecular Weight by Ebulliometry," in *Characterization of Macromolecular Structure,* Publication 1573, Washington, D.C.: National Academy of Sciences (1968), 3–9.

12. Newitt, E.J. and Kokle, V., "Molecular Structure of Polyethylene. XIII. An Improved Cryoscopic Method for Determining Number-Average Molecular Weight of Polyethylene," *J. Polym. Sci., Part A-2,* 4 (1966), 705–14.

13. Armstrong, J.L., "Modern Methods for Determining Number Average Molecular Weights," *Appl. Polym. Symp.,* No. 8 (1969), 17–36.

14. Moore, W.R. and Tidswell, B.M., "Instrumentation of Molecular Weight Measurements," *Chem. Ind.* (1967), 61–68.

15. Rayleigh, Lord, and J.W. Strutt, "On the Light from the Sky, Its Polarization and Colour," *Phil. Mag.,* 41 (1871), 107–20, 274–79 and "On the Scattering of Light by Small Particles," *Phil. Mag.,* 41 (1871), 447–54.

16. Debye, P., "Light Scattering in Solutions," *J. Appl. Phys.,* 15 (1944), 338–42.

17. Einstein, A., "Theorie der Opaleszenz von homogenen Flüssigkeiten und Flussigkeitsgemischen in der Nähe des kritischen Zustandes," *Ann. Phys. (Leipzig),* 33 (1910), 1275–98.

18. Kratohvil, J.P., "Calibrations of Light Scattering Instruments," *J. Colloid Interface Sci.,* 21 (1966), 498–512.

19. Kratohvil, J.P., "Absolute Calibration in the Light Scattering Measurements," in *Characterization of Macromolecular Structure,* Publication 1573, Washington, D.C.: National Academy of Sciences (1968), 59–67.

20. Harpst, J.A., Krasna, A.I. and Zimm, B.H., "Low Angle Light-Scattering Instrument for DNA Solutions," *Biopolymers,* 6 (1968), 585–94.

21. Harpst, J.A., Krasna, A.I., and Zimm, B.H., "Molecular Weight of T7 and Calf Thymus DNA by Low Angle Light Scattering," *Biopolymers,* 6 (1968), 595–603.

22. Livesey, P.J. and Billmeyer, F.W., "Particle Size Determination by Low Angle Light Scattering: New Instrumentation and a Rapid Method of Interpreting Data," *J. Colloid Interface Sci.,* 30 (1969), 447–72.

23. Wunderlich, W., "Comparison of Zimm Diagrams Measured by Light Scattering and X-Ray Small Angle Scattering in the Same System," *Makromol. Chem.,* 108 (1967), 315–17.

24. Yamakawa, H., "Light Scattering from Solutions of Polymers in Mixed Solvents," *J. Chem. Phys.,* 40 (1967), 973–78.

25. Casassa, E.F., "Interpretation of Rayleigh Scattering by Polymers in Mixed Solvents," *Makromol. Chem.,* 150 (1971), 251–54.

26. Borchard, W., "Critical Opalescence in Solutions of Polystyrene and Cyclohexane," *Ber. Bunsenges. Phys. Chem.*, 76 (1972), 224–27.

27. Scholte, T.G., "Light Scattering of Concentrated Polydisperse Polymer Solutions," *J. Polym. Sci., Part C*, No. 39 (1972), 281–91.

28. Baranov, V.G., "Small-Angle Light Scattering by Ordered Polymer Solutions," *Discuss. Faraday Soc.*, 49 (1970), 137–43.

29. Ross, G., "Light Scattering in Amorphous Media," *Opt. Acta*, 16 (1969), 95–109.

30. Moritani, M., Inoue, T., Motegi, M. and Kawai, H., "Light Scattering from a Two-Phase Polymer System. Scattering from a Spherical Domain Structure and Its Explanation in Terms of Heterogeneity Parameters," *Macromolecules*, 3 (1970), 433–41.

31. Yuen, H.K., "Low Angle and Brillouin Light Scattering from Inhomogeneous Amorphous Polymers," unpublished Ph.D. dissertation, Michigan State University, 1973.

32. Fox, T.G. and Flory, P.J., "Intrinsic Viscosity—Molecular Weight Relationships for Polyisobutylene," *J. Phys. Colloid Chem.*, 53 (1949), 197–212.

33. Stockmayer, W.H. and Fixman, M., "On the Estimation of Unperturbed Dimensions from Intrinsic Viscosities," *J. Polym. Sci., Part C*, No. 1 (1963), 137–41.

34. Kurata, M. and Stockmayer, W.H., "Intrinsic Viscosities and Unperturbed Dimensions of Long Chain Molecules," *Fortschr. Hochpolym. Forsch.*, 3 (1963), 196–312.

35. Kinsinger, J.B., "Viscometry," in *Encyclopedia of Polymer Science and Technology*, Vol. 14. *Thermogravimetric Analysis to Wire and Cable Coverings*, H.F. Mark, N.G. Gaylord, and N. Bikales, eds., New York: John Wiley & Sons (1971), 717–39.

36. Gramain, Ph. and Libeyre, R., "Automatic Recording Capillary Viscosimeter with Dilution," *J. Appl. Polym. Sci.*, 14 (1970), 383–91.

37. Hughes, J. and Rhoden, F., "A Modified Ubbelohde Viscometer for Measurement of the Viscosity of Polymer Solutions," *J. Phys. E: J. Sci. Instrum.*, 2 (1969), 1134–35.

38. Zimm, B.H. and Crothers, D.M., "Simplified Rotating Cylinder Viscometer for Deoxyribonucleic Acid," *Proc. Natl. Acad. Sci. U.S.*, 48 (1962), 905–11.

39. Berry, G.C., "Thermodynamic and Conformational Properties of Polystyrene. II. Intrinsic Viscosity Studies on Dilute Solutions of Linear Polystyrenes," *J. Chem. Phys.*, 46 (1967), 1338–52.

40. Stork, W.H.J. and DeVroome, H., "Automatic Zimm-Crothers Type Viscometer," *J. Phys. E: J. Sci. Instrum.*, 5 (1972), 314–16.

41. Lin, O.C.C., "Rotational Viscometer for Studying Non-Newtonian Solutions and Its Application to Pneumococcal Deoxyribonucleic Acid," *Macromolecules*, 3 (1970), 80–83.

42. Corey, H. and Creswick, N., "Versatile Recording Couette-Type Viscometer," *J. Texture Stud.*, 1 (1970), 155–66.

43. Hays, J.E. and Zimm, B.H., "Flexibility and Stiffness in Nicked DNA," *J. Mol. Biol.*, 48 (1970), 297–317.

44. Noda, I., Yamada, Y. and Nagasawa, M., "The Rate of Shear Dependence on the Intrinsic Viscosity of a Monodisperse Polymer," *J. Phys. Chem.*, 72 (1968), 2890–98.

45. Dunlevy, J.E. and Middleman, S., "Correlation of Shear Behavior of Solutions of Polyisobutane," *Trans. Soc. Rheol.*, 10 (1966), 157–68.

46. O'Donnell, R., "Shear Rate Effects in Determination of Viscosity Average Molecular Weight of Poly (1-butene sulfonate)," *Polymer*, 9 (1968), 567–73.

47. Chapman, R.E.Jr., Klotz, L.C., Thompson, D.S., and Zimm, B.H., "Instrument for Measuring Retardation Times of Deoxyribonucleic Acid Solutions," *Macromolecules*, 2 (1969), 637–43.

48. Dondos, A. and Benoit, H., "Effect of Temperature on the Nonperturbed Dimensions of Polymers Dissolved in Solvent Mixtures," *Eur. Polym. J.,* 6 (1970), 1439–50.

49. Dondos, A. and Benoit, H., "Unperturbed Dimensions of Polymers in Binary Solvent Mixtures," *Eur. Polym. J.,* 4 (1968), 561–70.

50. Bekhli, E.Yu., Novikov, D.D. and Entelis, S., "Thermoelectric Method of Determination of Molecular Weight from the Temperature Maximum," *Poly. Sci. USSR,* 9 (1967), 3117–28.

51. Dohner, R.E., Wachter, A.H. and Simon, W. "Apparatus for Determination of Molecular Weight of Very Dilute (10^{-4}M) Solutions by Vapor-Pressure Osmometry," *Helv. Chim. Acta,* 50 (1967), 2193–2200.

52. Van Dam, J., "Vapor-Phase Osmometry," in *Characterization of Macromolecular Structure,* Publication 1573, Washington, D.C.: National Academy of Sciences (1968), 336–42.

53. Creeth, J.M. and Palu, R.H., "The Determination of Molecular Weights of Biological Macromolecules by Ultracentrifuge Methods," *Progr. Biophys. Mol. Biol.,* 17 (1967), 217–87.

54. Yamamoto, A., Noda, I. and Nagawasa, M., "Comparison of Various Methods of Determining Molecular Weight Distribution," *Polym. J.,* 1 (1970), 304–11.

55. Flory, P.J., *Principles of Polymer Chemistry,* Ithaca, N.Y.: Cornell University Press, 1953.

56. Koningsveld, R. and Staverman, A.J., "Polymer Fractionation. I. The Preparative Problem," *J. Polym. Sci., Part A-2,* 6 (1968), 367–81.

57. Koningsveld, R., "Preparative and Analytical Aspects of Polymer Fractionation," *Advan. Polym. Sci.,* 7 (1970), 1–69.

58. Huggins, M.L. and Okamota, H., "Theoretical Considerations," in *Polymer Fractionation,* M.J.R. Cantow, ed., New York and London: Academic Press (1967), 1–42.

59. Kamide, K. and Sugamiya, K., "Role of Concentration Dependence of Polymer/Solvent Interaction Parameter in the Polymer Fractionation by Successive Precipitational Method," *Makromol. Chem.,* 139 (1970), 197–220.

60. Kamide, K. and Nakayama, C., "Efficiency of Precipitational Fractionation," *Makromol. Chem.,* 129 (1969), 289–93.

61. Matsumoto, M. and Oyanagi, Y., "The Successive Fractionation Method," *Kobunshi Kagaku,* 11 (1954), 7–13.

62. Desreux, V. and Spiegels, M.C., "Fractionation of Polyethylene by Extraction," *Bull. Soc. Chim. Belges,* 59 (1950), 476–89.

63. Desreux, V. and Oth, A., "Theory and Practice of Fractionation," *Chem. Weekblad.,* 48 (1952), 247–59.

64. Harrington, R.E. and Zimm, B.H., "Precise Molecular Weight Distributions of High Polymers By Semi-Automatic Solvent Extraction," *Amer. Chem. Soc., Div. Polym. Chem., Polym. Prepr.,* 6 (1965), 346–48.

65. Francis, P.S., Cooke, R.C.Jr. and Elliott, J.H., "Fractionation of Polyethylene," *J. Polym. Sci.,* 31 (1958), 453–66.

66. Henry, P.M., "Fractionation of Polyethylene," *J. Polym. Sci.,* 36 (1959), 3–19.

67. Schneider, N.S., "Review of Solution Methods and Certain Other Methods of Polymer Fractionation," *J. Polym. Sci., Part C,* No. 8 (1965), 179–204.

68. Kokle, V. and Billmeyer, F.W., "The Molecular Structure of Polyethylene," *J. Polym. Sci., Part C,* No. 8 (1965), 217–32.

69. Baker, C.A. and Williams, R.J.P., "A New Chromatographic Procedure and Its Application to High Polymers," *J. Chem. Soc.,* (1956), 2352–62.

70. Dawkins, J.V. and Peaker, F.W., "Fractionation of Poly(methyl methocrylate) by Precipitation Chromatography," *Eur. Polym. J.*, 6 (1970), 209–18.

71. Johnson, J.F. and Porter, R.S., "Gel Permeation Chromatography," *Prog. Polym. Sci.*, 2 (1970), 203–56.

72. Grubisic, Z., Rempp, P., and Benoit, H., "A Universal Calibration for Gel Permeation Chromatography," *J. Polym. Sci., Part B*, 5(1967), 753–59.

73. Tung, L.H., "Molecular Characterization of Polymers," *Tech. Pap., PAG Tech. Conf., Soc. Plast. Eng., Polym. Struct. Prop., Prof. Activ. Group* (19 Sept. 1968), 1–6.

74. Stuchbury, J.E., "Gel Permeation Chromatography," *Proc. Roy. Aust. Chem. Inst.*, 38 (1971), 293–97.

75. Tung, L.H., "Recent Advances in Polymer Fractionation," *J. Macromol. Sci., Rev. Macromol. Chem.*, C6 (1971), 51–84.

76. Dawkins, J.V., Denyer, R., and Maddock, J.W., "Molecular Weights by Gel Permeation Chromatography: Unperturbed Dimensions Calibration for Polyisoprene," *Polymer*, 10 (1969), 154–58.

77. Provder, T., Woodbrey, J.C. and Clark, J.H., "Gel Permeation Chromatography Calibration. I. Use of Calibration Curves Based on Polystyrene in THF and Integral Distribution Curves of Elution Volume to Generate Calibration Curves for Polymers in 2, 2, 2-Trifluoroethanol," *Separ. Sci.*, 6 (1971), 101–36.

78. Weiss, A. R. and Cohn-Ginsberg, E., "Universal Calibration Curve for Gel Permeation Chromatography," *J. Polym. Sci., Part B*, 7 (1969), 379–81.

79. Whitehouse, B.A., "Gel Permeation Chromatography Calibration. Intrinsic Viscosity-Polydispersity Effect," *Macromolecules*, 4 (1971), 463–66.

80. Meyerhoff, G., "Extension of GPC (gel permeation chromatography) Techniques," *Separ. Sci.*, 6 (1971), 239–48.

81. Scholtan, W. and Kwoll, F.J., "Application of a Flame Ionization Detector to Solvent Gradient Chromatography," *Makromol. Chem.*, 151 (1972), 33–48.

82. Lambert, A., "Extrapolation Method for Eliminating Overload Effects from Gel Permeation Chromatograms," *Polymer*, 10 (1969), 213–32.

83. Berger, K.C. and Schulz, G.V., "Absolute Error in Determination of Molecular Weight Distributions by GPC (gel permeation chromatography), Experiments with Poly(methyl methacrylates)," *Makromol. Chem.*, 136 (1970), 221–39.

84. Rodriguez, F. and Clark, O.K., "A Model for Molecular Weight Distributions," *Ind. Eng. Chem., Prod. Res. Develop.*, 5 (1966), 118–21.

85. Peyrouset, A. and Panaria, R., "New Instrument for Preparative Gel Permeation Chromatography," *J. Appl. Polym. Sci.*, 16 (1972), 315–28.

86. Zimm, B.H. and Stockmayer, W.H., "The Dimensions of Chain Molecules Containing Branches and Rings," *J. Chem. Phys.*, 17 (1949), 1301–14.

87. Orofino, T.A., "Branched Polymers. II. Dimensions in Non-Interacting Media," *Polymer*, 2 (1961), 305–14.

88. Orofino, T.A. and Wenger, F., "Dilute Solution Properties of Branched Polymers. Polystyrene Trifunctional Star Molecules," *J. Phys. Chem.*, 67 (1963), 566–75.

89. Altares, T.Jr., Wyman, D.P., Allen, V.R. and Meyerson, K., "Preparation and Characterization of Some Star- and Comb-Type Branched Polystyrenes," *J. Polym. Sci., Part A*, 3 (1965), 4131–51.

90. Benoit, H., "Investigation of Copolymer Solutions by Light Scattering and Viscosity," *Ber. Bunsenges. Phys. Chem.*, 70 (1966), 286–96. (In French)

91. Buskuk, W. and Benoit, H., "Light-Scattering Studies of Copolymers. I. Ef-

fect of Heterogeneity of Chain Composition on the Molecular Weight," *Can. J. Chem.*, 36 (1958), 1616–26.

92. Stockmayer, W.H., Moore, L.D.Jr., Fixmann, M. and Epstein, B.N., "Copolymers in Dilute Solution. I. Preliminary Results for Styrene-Methyl Methacrylate," *J. Polym. Sci.*, 16 (1955), 517–30.

93. Klimsch, H.M., "Light Scattering of Co-polymers. I. Effect of Dilution During Copolymerization on Chain Composition and Molecular Weight," unpublished M.S. thesis, Michigan State University, 1970.

94. Dondos, A., Rempp, P. and Benoit, H., "Solution Behavior of Graft Copolymers," *J. Polym. Sci., Part B*, 4 (1966), 293–300.

95. Dondos, A., "An Anomaly in the Variation of the Intrinsic Viscosity of Grafted Copolymers as a Function of Temperature," *Makromol. Chem.*, 99 (1966), 275–78. (In French)

96. Dondos, A., Rempp, P. and Benoit, H., "The Morphology of Graft Copolymers in Solution," *J. Chim. Phys.*, 62 (1965), 821–26. (In French)

97. Born, M. and Wolf, E., "Interference and Diffraction with Partially Coherent Light," in *Principles of Optics*, 4th ed., Oxford and London: Pergamon Press (1970), 491–555.

98. Beran, M.J. and Parrent, G.B.Jr., *Theory of Partial Coherence*, Englewood Cliffs, N.J.: Prentice-Hall, Inc., 1964.

99. Mandel, L. and Wolfe, E., eds., *Selected Papers on Coherence and Fluctuations of Light*, 2 vols., New York: Dover, 1970.

100. Birnbaum, G., *Optical Masers*, New York: Academic Press, 1964.

101. Peticolas, W.L., "Inelastic Laser Light Scattering from Biological and Synthetic Polymers," *Advan. Polym. Sci.*, 9 (1972), 285–333.

102. Brillouin, L., "Diffusion de la lumière et des rayons X par un corps transparent homogène influence de l'agitation thermique," *Ann. Phys. (Paris)*, 17 (1922), 88–122.

103. Herzfeld, K.F. and Litovitz, T.A., *Absorption and Dispersion of Ultrasonic Waves*, New York: Academic Press, 1959.

104. Tannahill, M.M., "Investigation of the Liquid Structure of Dimethyl Sulfoxide-Pyridine Mixtures with Brillouin Spectroscopy," unpublished Ph.D. dissertation, Michigan State University, 1973.

105. Fleury, P.A. and Boon, J.P., "Laser Light Scattering in Fluid Systems," in *Advances in Chemical Physics, Vol. 24*, S. Rice, ed., New York: John Wiley & Sons (1973).

106. McIntyre, D. and Sengers, J.V., "Light Scattering in Simple Liquids," in *Physics of Simple Liquids*, H.N.V. Temperley, J.S. Rowlinson, and G.S. Rushbrooks, eds., Amsterdam: North Holland (1968), 449.

107. Mountain, R.D., "Thermal Relaxation and Brillouin Scattering in Liquids," *J. Res. Nat. Bur. Stand.*, 70A (1966), 207–20.

108. Toth, W.J.Jr., "Investigation of Brillouin Light Scattering and Thermal Relaxation in CF_3CCl_3 Using a Computer Interfaced Spectrometer," unpublished Ph.D. dissertation, Michigan State University, 1973.

109. Bucaro, J.A. and Litovitz, T.A., "Rayleigh Scattering: Collisional Motions in Liquids," *J. Chem. Phys.*, 54 (1971), 3846–53.

110. Jackson, D.A. and Simic-Glavaski, B., "Depolarized Rayleigh Scattering in Liquid Benzene Derivatives," *Mol. Phys.*, 18 (1970), 393–400.

111. Andersen, H.C. and Pecora, R., "Kinetic Equations for Orientational and Shear Relaxation and Depolarized Light Scattering in Liquids," *J. Chem. Phys.*, 54 (1971), 2584–96.

112. Stegeman, G.I.A. and Stoicheff, B.P., "Spectrum of Light Scattering from Thermal Shear Waves in Liquids," *Phys. Rev. Lett.,* 21 (1968), 202–206.

113. Mountain, R.D. and Deutch, J.M., "Light Scattering from Binary Solutions," *J. Chem. Phys.,* 50 (1969), 1103–08.

114. Lekkerkerker, H.N.W. and Laidlaw, W.G., "The Landau-Placzek Ratio for Multi-Component Fluids," *Phys. Chem. Liquids,* 3 (1972), 175–80.

115. Miller, G.A., "Equations for the Calculation of Activity Coefficients of Solutions from the Intensity of Brillouin Scattering," *J. Phys. Chem.,* 71 (1967), 2305–08.

116. Miller, G.A., San Fillippo, F.I., and Carpenter, D.K., "Brillouin Spectra of Solutions. I. Molecular Weight Determination of Standard Polystyrene 705," *Macromolecules,* 3 (1970), 125–27.

117. Nordhaus, D.E. and Kinsinger, J.B., "Brillouin Spectroscopy of Macromolecular Solutions," *J. Polym. Sci., Part C,* No. 43 (1973), 251–65.

118. Nordhaus, D.E., "Brillouin Spectroscopy of Macromolecular Solutions," unpublished Ph.D. dissertation, Michigan State University, 1973.

119. Yuen, H.K., "Low-Angle and Brillouin Light Scattering from Inhomogeneous Amorphous Polymers," unpublished Ph.D. dissertation, Michigan State University, 1973.

120. Forrester, A.T., Parkins, W.E. and Gerjuoy, E., "On the Possibility of Observing Beat Frequencies between Lines in the Visible Spectrum," *Phys. Rev.,* 72 (1947), 728.

121. Forrester, A.T., Gudmundsen, R.A., and Johnson, P.O., "Photoelectric Mixing of Incoherent Light," *Phys. Rev.,* 99 (1955), 1697–1700.

122. Forrester, A.T., "Photoelectric Mixing As a Spectroscopic Tool," *J. Opt. Soc. Amer.,* 51 (1961), 253–59.

123. Cummins, H.Z., Knable, N. and Yeh, Y., "Observation of Diffusion Broadening of Rayleigh Scattered Light," *Phys. Rev. Lett.,* 12 (1964), 150–53.

124. Ford, N.C.Jr. and Benedek, G.B., "Observation of the Spectrum of Light Scattered from a Pure Fluid Near Its Critical Point," *Phys. Rev. Lett.,* 15 (1965), 649–53.

125. Pecora, R., "Quasi-Elastic Light Scattering from Macromolecules," *Ann. Rev. Biophys. Bioeng.,* 1 (1972), 257–76.

126. Fujime, S., "Quasi-Elastic Light Scattering from Solutions of Macromolecules. II. Doppler Broadening of Light Scattered from Solutions of Semi-Flexible Polymers, F-Actin," *J. Phys. Soc. Jap.,* 29

127. Prins, W., Rimai, L. and Chompff, A.J., "An Audiofrequency Resonance in the Quasielectric Light Scattering of Polymer Gels," *Macromolecules,* 5 (1971), 104–06.

128. Angus, J.C., Morrow, D.L., Dunning, J.W.Jr. and French, M.J., "Motion Measurement by Laser Doppler Techniques," *Ind. Eng. Chem.,* 61 (1969), 8–20.

129. Nossal, R. and Chen, S.H., "Light Scattering from Motile Bacteria," *J. Phys. (Paris),* 33, Suppl. C1 (1972), 171–76.

130. Ware, B.R. and Flygare, W.H., "Light Scattering in Mixtures of BSA, BSA Dimers, and Fibrinogen Under the Influence of Electric Fields." *J. Colloid Interface Sci.,* 39 (1972), 670–75.

131. Kuwahara, N., Fenby, D.V., Tamsky, M. and Chu, B., "Intensity and Linewidth Studies of the System Polystyrene-Cyclohexane in the Critical Region," *J. Chem. Phys.,* 55 (1971), 1140–48.

132. Stiso, S.N., "An Analysis of the Spectral Distribution of Light Elastically Scattered from High Molecular Weight Polymer Solutions in the Low Critical Temperature Region," unpublished Ph.D. dissertation, Michigan State University, 1973.

133. Ford, N.C.Jr., Karasz, F.E. and Owen, J.E.M., "Rayleigh Scattering from Polystyrene Solutions," *Discuss. Faraday Soc.,* 49 (1970), 228–37.

134. Frederick, J.E. and Reed, T.F., "Rayleigh Line-Broadening Studies of the Motions of Polystyrene in Dilute Solution," *Macromolecules*, 4 (1971), 72–79.

135. Frederick, J.E., Reed, T.F. and Kramer, O., "The Effect of Polydispersity on Rayleigh Line-Broadening Measurements of Diffusion Constants of Random-Coil Macromolecules," *Macromolecules*, 4(1971), 242–46.

136. Huang, W-N. and Frederick, J.E., "Rayleigh Line Spectrometry of Very Large Macromolecules in Dilute Solution," *J. Chem. Phys.*, 58 (1973), 4022–23.

137. Ho, F.F.L., "Application of the Eu(DPM)$_3$ Chemical Shift Reagent to the Determination of the Molecular Weight of Polypropylene Glycol by NMR," *J. Polym. Sci., Part B*, 9 (1971), 491–95.

138. Quayle, D.V., "Molecular Weight Determinations on Polymers by Electron Microscopy," *Brit. Polym. J.*, 1 (1969), 15–23.

139. Barnikol, I., Barnikol, W.K.R., Beck, A., Campagnari-Terbojevic, M., Janovic, N. and Schulz, G.V., "Determination of the Molecular Weight Distribution of Noncrystalline Polymers by Electron Microscopy. I. Methods of Preparations and Controls," *Makromol. Chem.*, 137 (1970), 111–21. (In German)

140. Barnikol, I., Barnikol, W.K.R., Janovic, N. and Schulz, G.V., "Determination of Molecular Weight Distribution of Noncrystalline Polymers by Electron Microscopy. II. Quantitative Characterization of Shadow-Casting," *Makromol. Chem.*, 137 (1970), 123–31. (In German)

141. Barnikol, W.K.R. and Schulz, G.V., "Determination of the Molecular Weight Distribution of Noncrystalline Polymers by Electron Microscopy. III. Definitions and Statistical Considerations on the Evaluation of Exposures Obtained by Electron Microscopy," *Makromol. Chem.* 145 (1971), 299–308. (In German)

142. Ruscher, C., "Electron Microscopic Studies on High Polymer Solutions," *J. Polym. Sci., Part C*, No. 16 (1967), 2923–30.

143. Dole, M., Hines, R.L., Mack, L.L., Mobley, R.C., Ferguson, L.D. and Alice, M.G., "Gas Phase Macroins," *Macromolecules*, 1 (1968), 96–97.

144. Callot, P. and Bandered, A., "Isothermal Distillation as a Method of Studying Macromolecular Solutions. Definition of the Possibilities," *J. Chim. Phys.*, 64 (1967), 1260–70.

145. Otooks, E.P. and Hellman, M.Y., "Fractionation of Polymers by Thin Layer Chromatography. I. Separation," *Macromolecules*, 3 (1970), 362–65.

146. Belen'kii, B.G. and Gankina, E.S., "Thin-Layer Chromatography of Polymers. Introductory Lecture," *J. Chromatogr.*, 53 (1970), 3–25.

147. Inagaki, H., Kamiyama, F. and Yagi, T., "A Note on Fractionation of Polymers by Thin Layer Chromatography," *Macromolecules*, 4 (1971), 133–34.

H. JAMES HARWOOD
University of Akron
Akron, Ohio

Chapter 11

Characterization of Sequence Distribution and Tacticity

ABSTRACT

Knowledge of the microstructures of polymers is a basic requirement for fundamental considerations of the chemical and physical properties of polymers as well as of polymerization mechanisms. This chapter surveys methods used to characterize the arrangements of monomer units in copolymers and the configurational structures of homopolymers and copolymers.

The chemical methods useful for such determinations include: selective degradation followed by analysis of the degradation fragments, pyrolysis–gas chromatography, pyrolysis–mass spectroscopy, intra- and intersequence cyclization, and studies of polymer reactivity. Physical methods useful for such determinations include: infrared and nmr spectroscopy, studies on heats of formation of copolymers, heats of polymerization, dipole moment measurements, melting points, and degrees of crystallization. The glass transition temperatures of copolymers, and their solution properties have also been shown to be influenced by microstructures. Selected examples of these various approaches are provided.

Introduction

Copolymerization of monomers $A + B$ can yield a variety of polymers having the same composition. Thus the monomer units in a copolymer may be distributed randomly along the polymer chain, they may tend to alternate, or they may cluster in blocks of like units.

$$\sim ABABABABABAB \sim$$
$$A + B \rightarrow \ \sim ABBAAABABBBAAB \sim$$
$$\sim AAAAAABBBBBBB \sim$$

315

The properties of polymers having such different structures can be quite different, even when they have the same composition. It is therefore important to be able to characterize the way monomer units are distributed in copolymers.

Most properties of copolymers are dependent on the relative amounts of various dyads, triads, tetrads, pentads, or higher sequences present. These various possibilities are listed in Table I. Measurement of such properties thus provides information about the structure of the copolymers.

In most polymers, several configurational structures are possible for each of the possible dyads, triads, etc., and it is necessary also to consider such structures in studies of polymer properties. Thus, meso and racemic

TABLE I
Recognizable Features of Copolymer Structure

	A		B
Monomers	A		B
Dyads	AA	AB BA	BB
Triads	AAA		BBB
	BAA		ABB
	AAB		BBA
	BAB		ABA
Tetrads	AAAA	AABA ABAA	ABBA
	BAAA	AABB BBAA	BBBA
	AAAB	BABA ABAB	ABBB
	BAAB	BABB BBAB	BBBB
Pentads	AAAAA	AAABA ABAAA	ABABA
	BAAAA	BBAAA BBAAA	BBABA
	AAAAB	BAABA ABAAB	ABABB
	BAAAB	BAABB BBAAB	BBABB
	ABBBA	AABBA ABBAA	AABAA
	BBBBA	AABBB BBBAA	BABAA
	ABBBB	BABBA ABBAB	AABAB
	ABBBB	BABBB BBBAB	BABAB

dyads as well as isotactic, heterotactic, and syndiotactic triads are possible in many homopolymers.

$$\begin{array}{cccc} & H & & X \\ & | & & | \\ \text{---}CH_2\text{---} & C & \text{---}CH_2\text{---} & C\text{---} \\ & | & & | \\ & X & & H \end{array} \qquad \begin{array}{cccc} & H & & H \\ & | & & | \\ \text{---}CH_2\text{---} & C & \text{---}CH_2\text{---} & C\text{---} \\ & | & & | \\ & X & & X \end{array}$$

racemic dyad   meso dyad

$$\begin{array}{cccccc} & H & & H & & H \\ & | & & | & & | \\ \text{---}CH_2\text{---} & C & \text{---}CH_2\text{---} & C & \text{---}CH_2\text{---} & C\text{---} \\ & | & & | & & | \\ & X & & X & & X \end{array}$$

isotactic triad

$$\begin{array}{cccccc} & H & & X & & X \\ & | & & | & & | \\ \text{---}CH_2\text{---} & C & \text{---}CH_2\text{---} & C & \text{---}CH_2\text{---} & C\text{---} \\ & | & & | & & | \\ & X & & H & & H \end{array}$$

heterotactic triad

$$\begin{array}{cccccc} & H & & X & & H \\ & | & & | & & | \\ \text{---}CH_2\text{---} & C & \text{---}CH_2\text{---} & C & \text{---}CH_2\text{---} & C\text{---} \\ & | & & | & & | \\ & X & & H & & X \end{array}$$

syndiotactic triad

In the case of copolymers of A and B units, meso- and racemic- $A-A$, $A-B$, $B-A$, and $B-B$ dyads are possible. The characterization of copolymer structure is thus a very complex problem, since many structural entities can be considered. Fortunately, many of the structural entities are inter-related, and the various aspects of copolymer structure can often be de-fined by a few statistical parameters. The use and evaluation of such pa-rameters will be discussed in this chapter.

Methods of characterizing the distributions of monomer sequences in copolymers have been surveyed previously[1,2]. This chapter will provide recent applications of these methods, but no attempt will be made to pre-sent an exhaustive survey.

Statistical Considerations

Characterization of the configurational structure of a homopolymer or the distribution of monomer units in a copolymer, including configurational aspects where necessary, involves specifying the minimum number of statistical quantities necessary to calculate any structural aspect of the polymer that would be of interest. The number of parameters that need to be specified depends on the number of different fundamental structural entities present in the polymer and on the complexity of the statistics required to describe relationships among these entities.

The simplest situation is encountered when the fundamental structural entities are distributed randomly along a polymer chain. In such a case, only the composition of the polymer is needed for calculating structural aspects. Thus, if σ is defined as the fraction of dyads in a homopolymer that are of the meso type, the relative amounts of isotactic, heterotactic and syndiotactic triads are given by σ^2, $2\sigma(1 - \sigma)$ and $(1 - \sigma)^2$, respectively. Similarly, if P_A, the mole fraction of A units in an A-B copolymer is known, the probabilities of various dyads, triads, etc., may be calculated, as is illustrated by the following examples.

$$P_{A-A} = P_A^2 \tag{1}$$

$$P_{A-B} + P_{B-A} = 2 P_A P_B \tag{2}$$

$$P_{ABA} = P_A^2 P_B \tag{3}$$

$$P_{BABAB} = P_A^2 P_B^3 \tag{4}$$

If σ_{AA}, $\sigma_{AB} = \sigma_{BA}$ and σ_{BB} represent the probabilities that AA, $AB + BA$, and BB dyads have meso configurations, then the probabilities of configurational structures may also be calculated. An example of this is provided below

$$P\left(\underset{A \quad B \quad A}{\underline{|\quad\quad|\quad\quad|}}\right) = \sigma_{AB}^2 \cdot P_{ABA} = \sigma_{AB}^2 \cdot P_A^2 P_B \tag{5}$$

In considerations of the properties of monomer units in copolymers, the fractions of monomer units of a given type that are in particular environments are of interest more than the distributions of particular sequences containing the monomers. Thus, it is common to discuss many properties of copolymers in terms of monomer centered triad or pentad fractions, rather than in terms of particular triad or pentad distributions. Monomer-centered triad or pentad fractions (e.g., f_{ABA}, f_{BBABB}) are calculated by dividing the appropriate triad or pentad distributions by the mole fraction of the unit central to such structures, viz.:

$$f_{ABA}^* = P_{ABA}^* / P_B = P_A^2 \tag{6}$$

$$f_{BBABB}^* = P_{BBABB}^* / P_A = P_B^4 \tag{7}$$

The above considerations apply to copolymers having random distributions of monomer units. Such copolymers are said to obey Bernoullian or zero-order Markoffian statistics. The structures of most copolymers must be characterized by first-order Markoffian statistics, however. In such cases, two parameters need to be specified to characterize their structures (σ_{AA}, etc., must also be specified if dyad configurations are considered). These may include any of the following: conditional probabilities such as $P(A/B)$, the probability that an A unit follows a B unit in the chain; the mole fraction of A units, P_A; the reactivity ratio product, $r_A r_B$, which describes the tendency of monomers $A + B$ to copolymerize; and the "run number," R.

We have found it convenient to characterize the structures of copolymers in terms of their compositions (P_A) and run numbers, R. The run number is defined[3] as the average number of monomer sequences (runs) occurring per 100 monomer units in a copolymer chain. It can easily be calculated for copolymers prepared in low conversion by Equation (8), where A_f/B_f is the ratio of monomers A and B in the polymerization mixture and where r_A and r_B are reactivity ratios for monomers A and B, respectively. Run numbers for copolymers having random structures can be calculated with the aid of Equation (9).

$$R = \frac{200}{2 + r_A A_f/B_f + r_B B_f/A_f} \tag{8}$$

$$R_{random} = 200\, P_A(1 - P_A) \tag{9}$$

Various features of copolymer structure are easily calculated from R and P_A, as the following examples demonstrate.

$$P_{AA} = P_A - R/200 \tag{10}$$

$$P_{BB} = P_B - R/200 \tag{11}$$

$$P_{(AB+BA)} = R/100 \tag{12}$$

$$f^*_{AAA} = (1 - R/200P_A)^2 \tag{13}$$

$$f^*_{BAB} = (R/200P_A)^2 \tag{14}$$

$$f^*_{BABAB} = (R/200P_A)^2(R/200P_B)^2 \tag{15}$$

Thus, if P_A is known and any aspect of the copolymer structure can be measured, R can be determined and used to calculate any other feature of the copolymer structure.

In copolymers whose structures must be characterized by second-order Markoffian statistics, four parameters are required to specify their structures. These are generally conditional probabilities such as $P_{(A/AA)}$, $P_{(A/BA)}$, $P_{(B/AB)}$ and $P_{(B/BB)}$. The fundamental entities of interest for structure calculations are then dyads instead of monomer units. If the conditional probabilities are known, the dyad probabilities (P_{AA}, etc.) can be calculated by solving Equations (16)–(19) simultaneously.

$$P_{AA} = P_{AA}P_{(A/AA)} + P_{BA}P_{(A/BA)} \tag{16}$$

$$P_{AB} = P_{AA}(1 - P_{(A/AA)}) + P_{BA}(1 - P_{(A/BA)}) \tag{17}$$

$$P_{BB} = P_{AB}P_{(B/AB)} + P_{BB}P_{(B/BB)} \tag{18}$$

$$P_{AA} + P_{AB} + P_{BA} + P_{BB} = 1 \tag{19}$$

Dyad probabilities so calculated can then be combined to obtain mole fractions of monomer units (P_A, P_B) or combined with conditional probabilities to calculate triad, tetrad, and pentad distributions or triad and pentad fractions.

$$P_A = P_{AA} + P_{AB} = P_{AA} + P_{BA} \tag{20}$$

$$P_B = P_{AB} + P_{BB} = P_{AB} + P_{BB} \tag{21}$$

$$P_{ABA} = P_{AB}P_{(A/AB)} \tag{22}$$

$$P_{ABAA} = P_{AB}P_{(A/AB)}P_{(A/BA)} \tag{23}$$

$$f^*_{ABA} = P_{AB}P_{(A/AB)}/P_B \tag{24}$$

$$f^*_{BABAB} = P_{BA}P_{(B/BA)}P_{(A/AB)}P_{(B/BA)}/P_B \tag{25}$$

The conditional probabilities required for these calculations can be calculated from penultimate reactivity ratios or by comparing dyad and triad concentrations when they can be measured. Thus, if r_{AAA} is the ratio of rate constants k_{AAA} and k_{AAB} in the reactions shown below, $P_{(A/AA)}$ can be calculated by Equation (26).

$$\sim\sim AA \cdot + A \xrightarrow{k_{AAA}} \sim\sim AA \cdot$$

$$\sim\sim AA \cdot + B \xrightarrow{k_{AAB}} \sim\sim AB \cdot$$

$$P_{(A/AA)} = \frac{k_{AAA}(\sim\sim AA \cdot)(A)}{k_{AAA}(\sim\sim AA \cdot)(A) + k_{AAB}(\sim\sim AA \cdot)(B)} = \frac{r_{AAA}A_f/B_f}{(r_{AAA}A_f/B_f) + 1} \tag{26}$$

If P_{AA} and P_{AAA} are known, $P_{(A/AA)}$ can be calculated by Equation (27).

$$P_{(A/AA)} = P_{AAA}/P_{AA} \tag{27}$$

In subsequent discussion, only copolymers whose structures are consistent with zero- and first-order Markoffian statistics will be considered. Before proceeding it should be noted that computer programs[4,5] are available for calculating aspects of copolymer and terpolymer structure, provided that zero-, first- or second-order Markoffian statistics can be used to characterize their structures. This should be true for most copolymers, although Markoffian statistics have been claimed to be unsuitable for some copolymer systems.[6]

Chemical Methods

A variety of chemical methods are available for studying the structures of copolymers. These methods include selective degradation reactions, pyrolysis, cyclization reactions, and studies of polymer reactivity. Examples of these approaches have been discussed in an earlier review[1], so this chapter will cover some of the work done since 1965.

Selective Degradation Reactions

The classical approach for investigating any organic structure is to degrade the structure into small fragments, to elucidate the structures of the fragments, and then to deduce the structure of the original material. This approach works quite well in copolymer structure analysis, provided that methods are available for selectively cleaving certain types of bonds in the copolymer.

Ozonolysis was used in some of the earliest studies on copolymer structure, but it is only recently that interest in this approach has been resumed. Brock and Hackathorn[7] have developed a very convenient microoozonolysis–gas chromatographic procedure for analyzing the structures of polymers derived from dienes, and it would seem that application of their method to copolymer structure studies would be very fruitful.

Olefin–polymer methathesis[8,9,10], followed by hydrogenation and gas-chromatographic analysis of the products obtained provides an interesting alternative to ozonolysis for investigating the microstructure of polydienes and diene-containing copolymers. Thus, when styrene-butadiene copolymers are allowed to metathesize with 2-butene, the yield of 2,6-octadiene obtained provides a measure of the 1,4-butadiene-1,4-butadiene present. The yields of phenylcyclohexene and 5-phenyldodecadiene-2,8 provide a measure of the butadiene–styrene–butadiene triads present in the copolymers

$$CH_2—CH{=}CH—CH_2—CH_2—CH{=}CH—CH_2 \sim \;\; +$$

$$2\,CH_3CH{=}CH—CH_3 \;\; \xrightarrow[Et_2AlCl]{WCl_6} \;\; \sim CH_2—CH{=}CH—CH_3 \;\; +$$

$$CH_3CH{=}CH—CH_2—CH_2—CH{=}CH—CH_3 \;\; +$$

$$CH_3CH{=}CH—CH_2\sim$$

Copolymers derived from cyclic ethers and SO_2, cyclic anhydrides or lactones contain ether and ester linkages. Hydrolysis of the latter yields polyether oligomers that can be analyzed by gas chromatography. The following equations describe recent applications[11–14] of this approach.

$$\text{⁓OSO—(CH}_2\text{—CH—O)}_n\text{—H} \;\longrightarrow\; \text{H}_2\text{SO}_3 \;+\; \text{HO—(CH}_2\text{—CH—O)}_n\text{—H}$$

$$\underset{\text{CH}_3}{\big|} \qquad\qquad\qquad\qquad \underset{\text{CH}_3}{\big|}$$

(sulfonate: $\text{O}=\text{S}=\text{O}$)

$$\text{⁓O—C—CH=CH—C—O—(CH}_2\text{—CH—O)}_n\text{—O—C—CH=CH—C—O⁓} \;\longrightarrow$$

$$\overset{\text{O}}{\|}\qquad\overset{\text{O}}{\|}\qquad\qquad\underset{\text{CH}_3}{\big|}\qquad\overset{\text{O}}{\|}\qquad\overset{\text{O}}{\|}$$

$$2\ \text{HOC—CH=CH—C—OH} \;+\; \text{HO(CH}_2\text{—CH—O)}_n\text{—H}$$

$$\overset{\text{O}}{\|}\qquad\overset{\text{O}}{\|}\qquad\qquad\qquad\underset{\text{CH}_3}{\big|}$$

$$\text{⁓O—CH}_2\text{—CH}_2\text{—C—O—(CH}_2\text{—CH—O)}_n\text{—C—O—CH}_2\text{—CH}_2\text{—C—O⁓} \;\longrightarrow$$

$$\overset{\text{O}}{\|}\qquad\qquad\underset{\text{CH}_3}{\big|}\qquad\overset{\text{O}}{\|}\qquad\qquad\qquad\overset{\text{O}}{\|}$$

$$2\ \text{HOCH}_2\text{CH}_2\text{C—OH} \;+\; \text{HO(CH}_2\text{CH—O)}_n\text{—CH}_2\text{CH}_2\text{COOH}$$

$$\overset{\text{O}}{\|}\qquad\qquad\qquad\underset{\text{CH}_3}{\big|}$$

Sulfonate–carbonate copolymers derived from bisphenols can also be characterized by selective degradation, since the carbonate linkages react readily with aqueous ammonia whereas the sulfonate linkages are resistant to this reagent [14]. Thus the amount of bisphenol A liberated when sulfonate–carbonate copolymers derived from this material are ammonolyzed provides a convenient measure of their carbonate–carbonate dyad contents. The isotope dilution technique can be employed very effectively in such studies.

Finally, it should be noted that synthetic polypeptide and polysaccharide copolymers can sometimes be selectively hydrolyzed with the aid of enzymes. A limited amount of work has been done in this area [15,16].

Pyrolysis

A large variety of products is usually obtained when polymers are pyrolyzed, and it is difficult to obtain quantitative information about copolymer structure from analysis of pyrolysis products. Some copolymers decompose to monomers almost exclusively, however, and the relative yields of the products obtained can provide reliable information about structure. Thus, methyl methacrylate(I)/α,β,β-trifluorostyrene(II) copolymers decompose to yield I, II, methyl β,β-difluoromethacrylate(III) and α-fluorostyrene(IV). Products III and IV can arise only from I–II and II–I junctions in the copolymer and the yields of these materials are dependent on run number of the copolymer and on θ, the fraction of bonds originally present in I and II that are broken during pyrolysis (Equation 28). For a random pyrolysis process, $\theta = 1$. Studies on the pyrolysis of methyl methacrylate–trifluorostyrene copolymers [17] indicate that θ is slightly greater than one, for reasons not presently understood. This approach has also worked well in studies [18] on the structures of methyl methacrylate/α-trideuteriomethyl-β,β-dideuteriostyrene copolymers.

$$CH_2\!\!=\!\!C\!\!\begin{smallmatrix} CH_3 \\[2pt] COOCH_3 \end{smallmatrix} \qquad + \qquad CF_2\!\!=\!\!CF\phi$$

(I)   (II)

$\downarrow\ \uparrow$

$$\overset{\displaystyle CH_3}{\underset{\displaystyle \underset{\displaystyle OCH_3}{C=O}}{\sim\!\!CH_2\!-\!\overset{|}{\underset{|}{C}}\!-\!CF_2\!-\!\overset{\displaystyle F}{\underset{\displaystyle \phi}{\overset{|}{\underset{|}{C}}}}\!-\!CH_2\!-\!\overset{\displaystyle CH_3}{\underset{\displaystyle \underset{\displaystyle OCH_3}{C=O}}{\overset{|}{\underset{|}{C}}}\!-\!CF_2\!-\!\overset{\displaystyle F}{\underset{\displaystyle \phi}{\overset{|}{\underset{|}{C}}}}\!\sim}$$

$\downarrow$

(III) $CF_2\!\!=\!\!C\!\!\begin{smallmatrix} CH_3 \\[2pt] COOCH_3 \end{smallmatrix}$ $+$ $CH_2\!\!=\!\!CF\phi$ (IV)

YIELD III   =   YIELD IV   = $R\theta/4$

In those cases where certain pyrolysis products arise only or predominantly from particular monomer sequences, the relative product yields provide semiquantitative information about copolymer structure. Examples of such approaches are the use of butane and pentane yields to characterize the structures of ethylene-propylene copolymers[19] and the characterization of the structures of chlorinated polyethylene and chlorinated poly(vinyl chloride) by determining the relative amounts of benzene and halogenated benzenes formed when the polymers are pyrolyzed[20, 21]. Similarly, the amount of methanol formed when copolymers containing methyl acrylate are pyrolyzed is related to the proportion of methyl acrylate units centered in acrylate–acrylate–acrylate triads [22].

Pyrolysis–gas chromatography also has been used to characterize fragments containing up to 24 carbon atoms that are formed during the pyrolysis of ethylene–propylene copolymers[23]. The relative yields of these fragments provide an approximate indication of the distributions of ethylene sequences in the copolymers.

Cyclization Reactions

When neighboring substituents on copolymer chains can react to form cyclic structures containing 5- or 6-membered rings, the reactions often occur smoothly and the extents of cyclization obtained are dependent on the way the monomer units are distributed along the chains. When the reactions involve like units (intrasequence cyclization), the fraction of A units uncyclized f_A(uncy.) after the maximum amount of cyclization has been obtained is given by Equation (29), where $\%A$ is the mole percentage of the cyclizable units present.

$$f_A(\text{uncyc.}) = e^{-2P(A/A)} = e^{-2(1-R/2\%A)} \tag{29}$$

Similarly, f_A(uncyc.) is related to R, $\%A$ and $\%B$ through Equation (30) when the cyclization reaction involves pairs of A and B units (intersequence cyclization).

$$f_A(\text{uncyc.}) = \left[\cosh\frac{R}{2}\left(\frac{\%A}{\%B}\right)^{-1/2} - \left(\frac{\%B}{\%A}\right)^{1/2}\sinh\frac{R}{2}\left(\frac{\%A}{\%B}\right)^{-1/2}\right]^2 \tag{30}$$

The extent of cyclization obtained in a copolymer cyclization reaction is often very sensitive to R, and cyclization reactions can be used very effectively in studies of copolymer and terpolymer structure[24–29]. Listed below are cyclization reactions that have been used for this purpose.

$$
\begin{array}{ccc}
& \overset{H}{\underset{|}{}} & \overset{H}{\underset{|}{}} \\
\sim\!\!CH_2\!-\!\!C\!-\!\!CH_2\!-\!\!C\!\sim & \longrightarrow & \sim\!\!CH_2\!-\!\!CH\!-\!\!CH_2\!-\!\!CH\!\sim \quad +\ ROH \\
\underset{|}{}NH_2 \quad \underset{|}{}NHCOOR & & \underset{|}{}NH\!-\!\!CO\!-\!\!NH
\end{array}
$$

$$
\sim\!\!CH_2\!-\!\!CH\!-\!\!CH_2\!-\!\!CH\!\sim \longrightarrow \sim\!\!CH_2\!-\!\!CH\!-\!\!CH_2\!-\!\!C\!\sim
$$

with OH and COOH below, producing O—C=O ring.

Chemical Reactivity

Steric effects, neighboring group participation, and ionic interactions cause the reactivity of functional groups on polymers to depend on the nature of neighboring units. Thus the reactivity of A units in an AB copolymer may differ depending on whether the units are centered in AAA, AAB (BAA) or BAB triads. Configurational effects may also cause A units in the various types of triads to have different reactivity. For example, the methyl methacrylate (M) units centered in $M\overset{*}{M}M$ triads in copolymers are highly hindered sterically and are much less reactive than M units in $X\overset{*}{M}X$ and $X\overset{*}{M}M$ ($M\overset{*}{M}X$) triads, where X = methyl acrylate, styrene, vinylpyridine, butadiene, isoprene, ethylene, etc. Studies [30–32] on the hydrolysis of methyl methacrylate units in copolymers with styrene, methyl acrylate or isoprene indicate that some of the M units are much less reactive than others. The proportion of M units with low reactivity correlates rather well with $f_{M\overset{*}{M}M}$ values calculated for the copolymers, provided that M units in isotactic MMM triads are considered to be reactive. It has been known for some time [33] that M units in isotactic poly(methyl methacrylate) are considerably more reactive than those in the atactic polymer, which contains mostly heterotactic and syndiotactic $M\overset{*}{M}M$ triads. In studies on the hydrolysis of M units in methyl methacrylate/vinylpyridine copolymers, the M units in XMX triads were found to be considerably more reactive than those in $XMM(MMX)$ and MMM triads[34]. Other examples of the influence of neighboring units on polymer reactivity are provided in an earlier review [1]. Attention might also be called to studies on the ultraviolet degradation of vinyl acetate (A)/ethyl acrylate copolymers in which differences in the rates of scission of AA, AB, and BB bonds were noted [35].

It is thus possible that measurements of the relative amounts of "fast" and "slow" reacting units in a polymer can be used to characterize the structure of the polymer. However, the reader should be cautioned that there are potential pitfalls associated with the use of conventional kinetic methods to measure the relative amounts and relative reactivities of units in different environments. Changes in polymer-solvent interaction, polyelectrolyte effects, copolymer solubility, etc., that occur during the course of a reaction can influence the reactivity of the groups present on the chain, so that their reactivity may alter during the course of the reaction. It is possible that autoinhibition might result from such effects and that it would be misinterpreted to indicate the presence of groups of different reactivity in a polymer when all groups have the same reactivity, for example. A radiotracer technique has been developed to circumvent this difficulty in studies of polymer hydrolysis reactions[36].

Physical Properties

The physical properties of copolymers are often dependent on the distribution of monomer sequences. Sometimes nmr or infrared spectroscopy can be used to measure particular structural aspects, but most physical properties of copolymers are observed as averages of the contributions of the various types of structural entities present. For example, a property of the A units in an A–B copolymer may be observed as an average (X_{AV}) of the properties of A units centered in $AAA(X_{AAA})$, $(AAB + BAA)$ $(X_{(BAA+AAB)})$ and $BAB(X_{BAB})$ triads. Equation (31) describes how X_{AV} should depend on the A-centered triad fractions if configurational effects can be neglected and if only nearest neighbors influence the properties of the A units.

$$X_{AV} = f_{AAA} \cdot X_{AAA} + f_{(AAB+BAA)} X_{(AAB+BAA)} + f_{BAB} X_{BAB} \qquad (31)$$

Since f_{AAA}, etc., are functions of copolymer composition and run number (or of other parameters, *vide supra*), X_{AV} can be used to characterize the structure of the copolymer, provided that X_{AAA}, etc., are known. X_{AAA} can be obtained from studies on the homopolymer of A and X_{BAB} can be evaluated from studies on copolymers containing only small amounts of A units ($f_{BAB} \sim 1$). $X_{(BAA+AAB)}$ is obtained from the properties of copolymers having low B contents. When the content of B units is sufficiently low that they may be assumed to be present only as isolated units, only $A\overset{*}{A}A$ and $(B\overset{*}{A}A + A\overset{*}{A}B)$ triads need to be considered and the relative amounts of these can be estimated from the composition of the copolymer. Thus, $f_{AAA} \sim (1 - \%B/\%A)^2$ and $f_{(BAA+AAB)} \sim 2(\%B/\%A)$ $(1 - \%B/\%A)$ when the B units are isolated ($R = 2\%B$). Figure 1 shows how $X_{(AAB+BAA)}$ can be estimated from X_{AV} values obtained from a series of copolymers with low B contents.

Similarly, when the properties of AA, $(AB+BA)$ and BB dyads differ, the average value (X_{AV}) of a property dependent on dyad distributions will

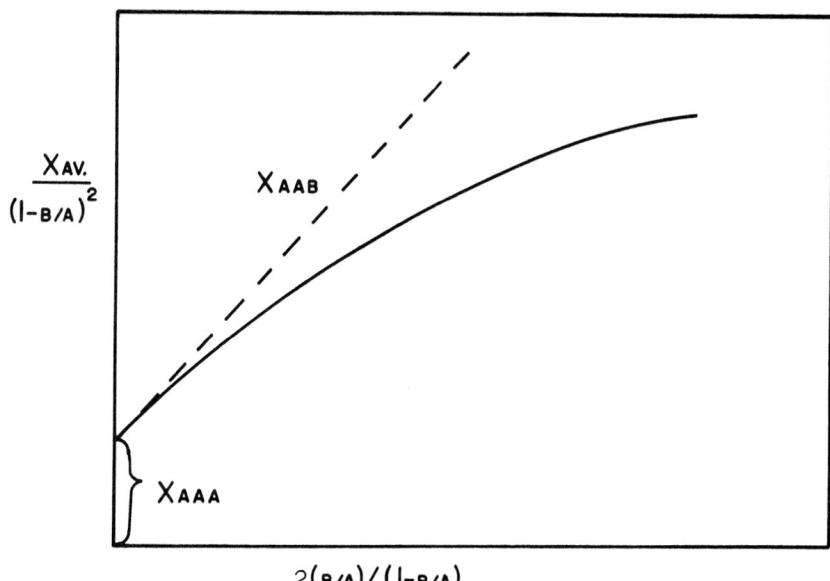

Figure 1. Evaluation of $X_{AAB} = X_{BAA}$ from the average properties (X_{AV}) of copolymers having low contents of monomer B. B/A = % B/% A.

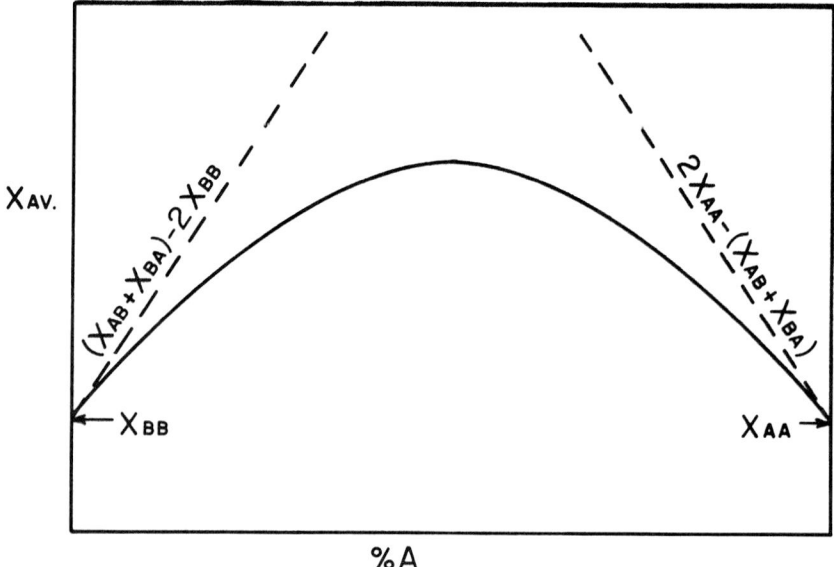

Figure 2. Evaluation of $X_{(AB+BA)}$ from the average properties (X_{AV}) of copolymers having low contents of monomer A or monomer B.

be given by Equation (32), where X_{AA}, etc., represent values of the property due to particular dyads.

$$X_{AV} = P_{AA}X_{AA} + P_{(AB+BA)}X_{(AB+BA)} + P_{BB}X_{BB} \qquad (32)$$

X_{AA} and X_{BB} can be evaluated from studies on the appropriate homopolymers and $X_{(AB+BA)}$ can be estimated by studies on the properties of copolymers with low A and B contents. Figure 2 shows a graphical procedure for estimating $X_{(AB+BA)}$ from X_{AV} values obtained from copolymers with low A or B contents.

Infrared Spectroscopy

The infrared spectra of copolymers can sometimes provide information about the arrangements of their monomer units. The analysis of such spectra is complicated by the fact that the absorptions of interest occur very close together and are often overlapped. It is also difficult sometimes to decide whether the differences observed in a given region in the spectra of a series of copolymers are due only to changes in copolymer composition or to sequence distribution effects. A graphical technique has been developed[37] to help in the recognition of sequence distribution sensitive bands, however.

In analyzing the infrared spectra of a series of copolymers, it is possible to use only the absorption observed at a single wavelength, in conjunction with Equations (30) or (31), and the methods described in the previous section, to evaluate the structures of the copolymers, but this would involve utilizing only a small part of the information available. A better approach is to measure the copolymer adsorbtivity (X_{AV}) at three wavelengths (these being those most sensitive to the various dyads or triads being determined), to evaluate absorbtivities (X_{AA} or X_{AAA}, etc.) for the various dyads or triads at each of these wavelengths, and then to solve simultaneously three equations having the form of Equations (30) or (31) and obtain as a solution the various dyad distributions or triad fractions. Enomoto[38] and Germar[39] have provided examples of this approach in studies on the structure of vinyl chloride/vinylidine chloride copolymers.

Although it remains to be done, this approach could be improved by collecting data at many wavelengths over a sequence-distribution-sensitive region and then analyzing such data by regression analysis to obtain dyad distributions or triad fractions. The availability of devices for obtaining spectra in digital form and the availability of computers and computer programs for conducting such analyses make this a desirable extension of previously employed methods.

The distributions of monomer sequences in ethylene/propylene copolymers have been evaluated from methylene sequence absorption (720–815 cm^{-1} and from methyl absorptions (937–960 cm^{-1}). The absorption of methylene sequences[40] could not be analyzed in terms of dyad distributions or triad fractions, and it was necessary to study model compounds to evaluate the absorbtivities of methylene groups in methylene sequence of various length. Druschel[41] used a computer programmed with a least squares routine to help overcome the problem of peak overlap.

Nuclear Magnetic Resonance Spectroscopy

Nuclear magnetic spectroscopy is the most powerful of the methods available for determining the microstructures of polymers. It can be used routinely to study the environments of hydrogen (H^1), fluorine (F^{19}) and carbon (C^{13}) atoms in polymers. Little work has been done on the resonance of (N^{15}) nuclei in polymers to date, but this situation can be expected to change rapidly. The method is very discriminating, and it can be used to obtain information about monomer sequence distributions and the relative configurations of monomer units. The method is easily applied quantitatively, since resonance intensities are directly proportional to the concentrations of the nuclei responsible for the resonance. (This may not be generally true for C^{13} spectra, due to the Nuclear Overhauser Effect, although NOE enhancements seem to be fairly uniform for carbon atoms of a given type—*e.g.*, methine, carbonyl—in polymers).

The resonances observed in the nmr spectra of polymers are much broader than those of low molecular compounds, partly because of conformational effects, but mostly because a large number of closely related resonances occur in each region. This is due to the large numbers of structural environments, (various configurations of pentad or heptad environments, associated with a given triad, for example) that are possible for a given nucleus. The nmr spectra of polymers with very regular structures, *e.g.,* isotactic poly(3,4,5-trideuteriostyrene), consist of very narrow resonances. The degree to which the various resonances are resolved in polymer spectra determines the amount of information about their structures that can be obtained. The best results are obtained when the spectra are recorded at high temperatures to minimize conformational effects and dipolar broadening, when high field spectrometers are used, and when the spectra are not complicated by spin–spin coupling effects. The individual transitions resulting from spin–spin coupling are seldom resolved in copolymer spectra; their net effect is broadening of the resonances of interest and poor resolution. Thus, well-resolved spectra that provide measures of monomer-centered tetrad and hexad fractions are obtained in studies on the H^1 spectra of vinylidine chloride/isobutylene copolymers[42,43] whereas the spectra of vinylidine chloride/vinyl chloride[44] and of vinyl chloride/ethylene[45] copolymers are much less well defined. The nmr spectra of many copolymers are influenced simultaneously by sequence distribution and configuration effects. This leads to very complicated spectra, but it has been possible to interpret such spectra in terms of their expected structures and reasonable σ values[46–49]. In certain cases it is possible to study copolymers in which all the units have the same configuration[50] or to study the configurations of highly alternating copolymers[51–52]. The spectra are then often easily interpreted.

Numerous examples of the use of nmr spectroscopy to characterize copolymers are present in the literature and no attempt will be made to refer to all of them here. As an example of the general approach taken in such studies, we may consider the interpretation of the nmr spectra of methyl acrylate (A)/methacrylonitrile (M) copolymers[53]. Figures 3 and 4 compare the 60-, 100-, and 220-MHz nmr spectra of several such copolymers. The advantage of studying spectra recorded with the highest field spectrometer (220 MHz) is easily appreciated if the spectra are compared in the 2.5–3.5 δ region. Figures 5 and 6 show the 220 MHz spectra observed for a series of such copolymers. The relative intensities of the three methine proton resonances observed in these spectra ($\delta = 2.8$–3.2) provide measures of the relative proportions of methyl acrylate units centered in

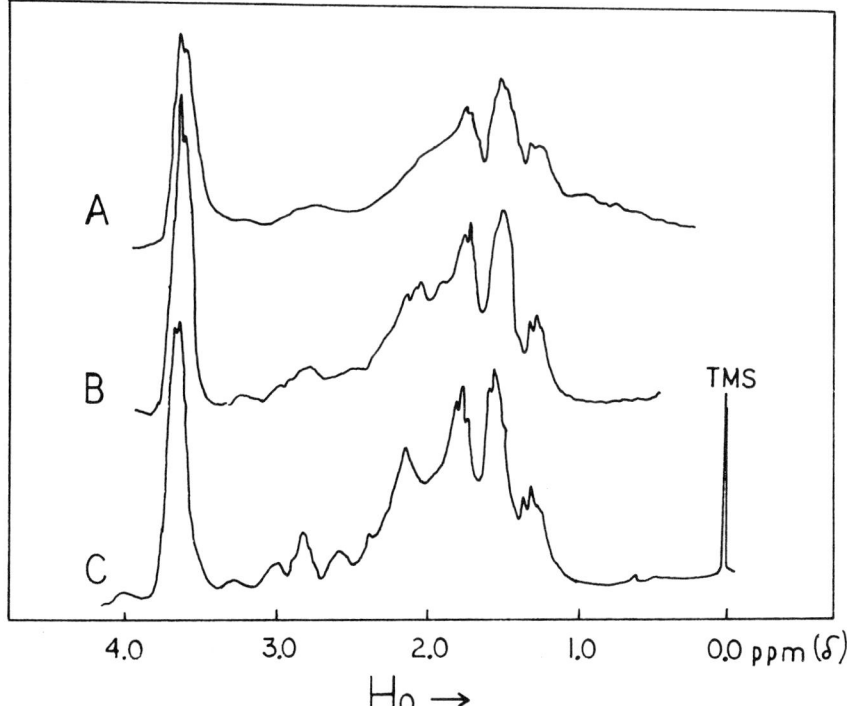

Figure 3. NMR spectra of a 49:51 methyl acrylate–methacrylonitrile copolymer in pyridine at 100° C as recorded by 60 MHz (A), 100 MHz (B), and 220 MHz (C) spectrometers.

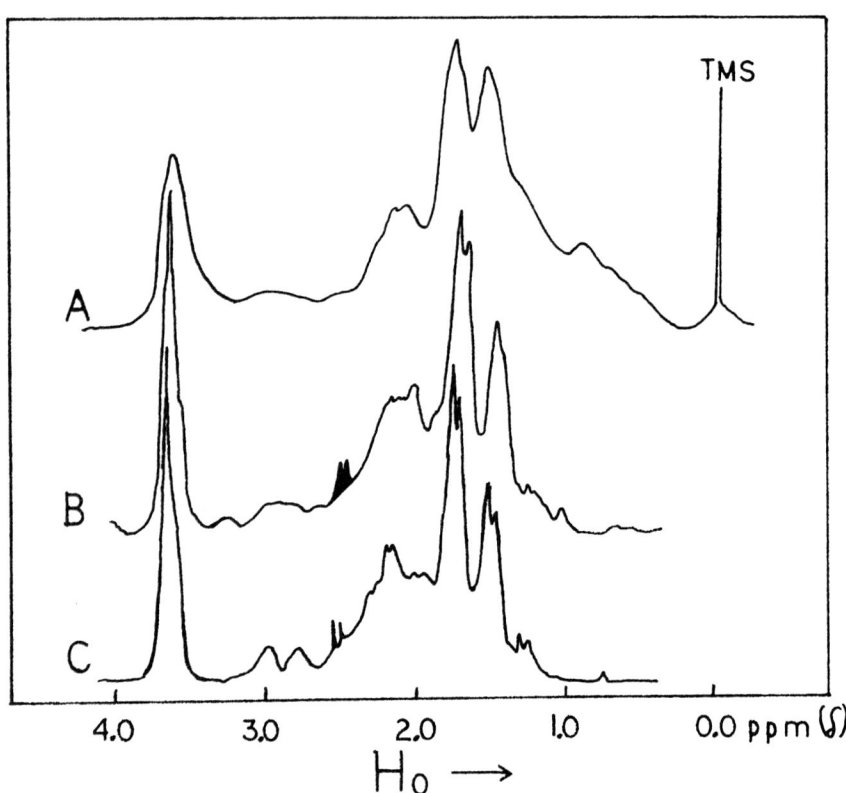

Figure 4.   NMR spectra of a 30:70 methyl acrylate–methacrylonitrile copolymer in pyridine at 100° C as recorded by 60 MHz (A), 100 MHz (B), and 220 MHz (C) spectrometers.

$AAA(\delta \sim 2.8)$, $(AAM + MAA)(\delta \sim 2.9)$ and MAM $(\delta \sim 3.1)$ triads. In addition, resonance of α-methyl protons of methacrylonitrile units is observed in three general areas, which are attributed to M units centered in $MMM(\sim 1.8\ \delta)$, $(MMA + AMM)$ $(\sim 1.6\ \delta)$ and $AMA(\sim 1.4\ \delta)$ triads. Each of the α-methyl proton resonance areas contains several resonances. These are probably due to different configurations of the various triads, but they may also reflect pentad effects. The assignments given were made by noting the chemical shifts of resonances observed in the spectra of homopolymers and of copolymers with high and low acrylate contents. The nmr spectra of methyl acrylate/methacrylonitrile copolymers thus provide information about all six possible triad distributions.

    The most recent developments in the use of nmr to characterize copolymer structure include studies of C^{13} resonance [54–60] and the use of "shift reagents"[61,62] to enhance the separation of resonances. One

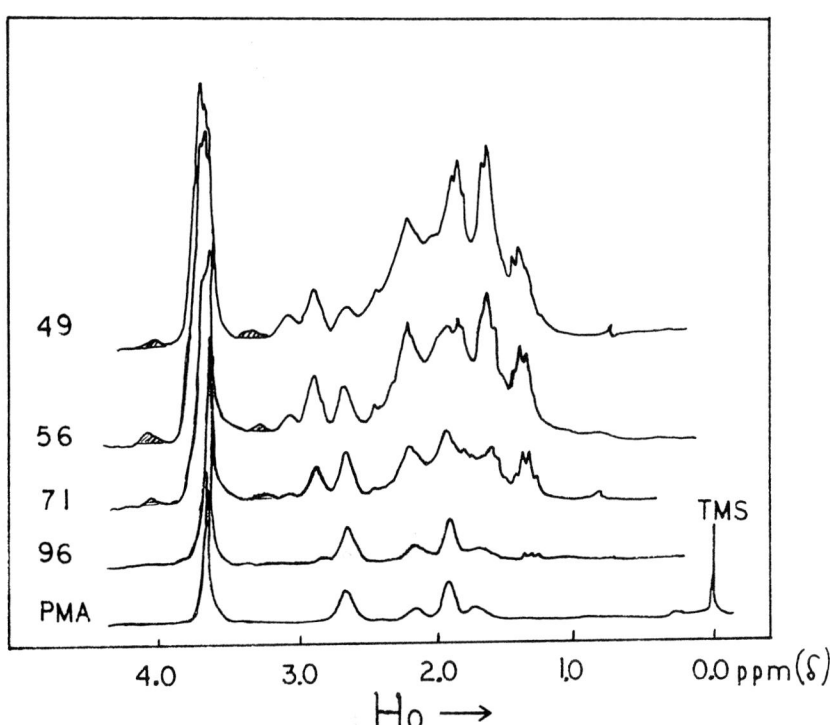

Figure 5. 220 MHz NMR spectra of poly(methyl acrylate), PMA, and of methyl acrylate–methacrylonitrile copolymers in pyridine at 100° C. The molar percentages of methyl acrylate in the copolymers are indicated by the numbers at the left of the spectra.

might also call attention to the "ESCA" technique, which has recently been applied to polymer characterization problems[63].

Other Physical Properties

Other physical properties of copolymers that are sensitive to sequence distribution effects include heats of formation (actually measured as heats of copolymerization), solution properties[64–67], glass transition temperature[68–72], dipole moments[73–74], melting points, crystalline contents, and properties dependent on crystalline content such as hardness. The reader is referred to an earlier review[2] for a discussion about the relationships of these properties to microstructure. References to some recent work in this area are provided above as an aid to the reader, but no attempt has been made to be exhaustive. It should be mentioned that, at the present time, only a few of these properties can be used for microstructure

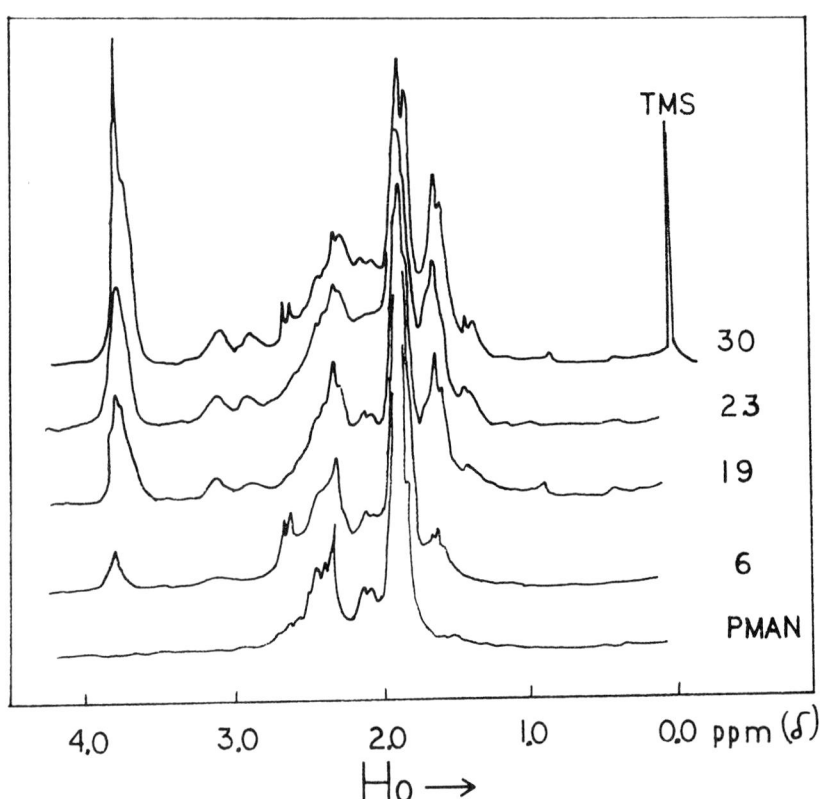

Figure 6.   220 MHz NMR spectra of polymethacrylonitrile, PMAN, and of methyl acrylate–methacrylonitrile copolymers in pyridine at 100° C. The molar percentages of methyl acrylate in the copolymers are indicated by the numbers at the right of the spectra.

characterization purposes. Most studies on the properties included in this section attempt only to understand the properties in terms of microstructure.

Concluding Comments

This chapter attempts to provide a general survey of the methods available for characterizing the microstructures of copolymers. To a large extent, the examples presented represent work done by the writer and his students. Were more time available, a large number of excellent other examples would have been presented. The writer apologizes to those whose work was not included herein. A complete review of this subject is being prepared.

References

1. Harwood, H.J., "Sequence Distribution in Copolymers. Chemical Studies," *Angew. Chem., Int. Ed. Engl.,* 4 (1965), 394–98.

2. Harwood, H.J., "Sequence Distribution in Copolymers. Physical Property Studies," *Angew. Chem., Int. Ed. Engl.,* 4 (1965), 1051–60.

3. Harwood, H.J., "The Characterization of Sequence Distribution in Copolymers," *J. Polym. Sci., Part B,* 2 (1964), 601–607.

4. Harwood, H.J., "A FORTRAN II Program for Conducting Sequence Distribution Calculations," *J. Polym. Sci., Part C,* No. 25 (1968), 37–45.

5. Harwood, H.J., Kodaira, Y., and Newman, D.L., "Stochastic Calculations of Polymer Structure," in *Computers in Chemistry and Instrumentation,* Vol. 6 (Polymer Science), edited by J.S. Mattison, H.C. MacDonald, Jr., and H.B. Marck, Jr. New York: Dekker, in press.

6. Schaefer, J., Bude, D.A. and Katnik, R.J., "Non-Markoffian Monomer Distributions in Copolymers. The Copolymerization of Ethylene Oxide and Maleic Anhydride," *Macromolecules,* 2 (1969), 289–95.

7. Hackathorn, M.J. and Brock, M.J., "Alternating Structures in Copolymers As Elucidated by Microozonolysis," *Amer. Chem. Soc., Div. Polym. Chem., Polym. Prepr.,* 14 (1973), 42–45.

8. Michajlov, L. and Harwood, H.J., "Characterization of Diene Polymer Microstructure By Olefin-Polymer Metathesis," *Polymer Characterization—Interdisciplinary Approaches,* edited by C.D. Craver. New York: Plenum, 1971, pp. 221–29.

9. Hummel, K. and Ast, W., "Reaktion an Ziegler-Nattla-Katalysatoren, 6. Metathese-Abbau von Polybutadien mit 4-Octen," *Makromol. Chem.,* 166 (1973), 39–44.

10. Hummel, K., Wewerka, D., Lorber, F. and Zeplichal, G., "Reaktion an Ziegler-Nattla-Katalysatoren, 7. Metathese-Abbau von Polybutadien mit 6-Dodecen," *Makromol. Chem.,* 166 (1973), 45–49.

11. Schaefer, J., Kern, R.J. and Katnik, R.J., "The Monomer Distribution of Propylene Oxide-Sulfur Dioxide Copolymers," *Macromolecules,* 1 (1968), 107–11.

12. Schaefer, J., Katnik, R.J. and Kern, R.J., "The Monomer Distributions of Propylene Oxide-Maleic Anhydride Copolymers," *J. Amer. Chem. Soc.,* 90 (1968), 2476–80.

13. Okada, M., Yamashita, Y. and Ishii, J., "Cationic Copolymerization of 1.3-Dioxolane with Styrene," *Makromol. Chem.,* 94 (1966), 181–93.

14. Kodaira, Y. and Harwood, H.J., "Distributions of Monomer Sequences in Sulfonate-Carbonate Copolymers," *Amer. Chem. Soc., Div. Polym. Chem., Polym. Prepr.,* 14 (1973), 323–28.

15. Kanakkanatt, A.T., "Measurement of Sequence Distribution in Synthetic Polypeptide Copolymers," unpublished Ph.D. dissertation, University of Akron, Ohio, 1963; and Kreiner, J.G., "Measurement of Sequence Distribution in Synthetic Polypeptide Copolymers," unpublished Ph.D. dissertation, University of Akron, Ohio, 1967.

16. Lin, J.W-P. and Schuerch, C., "Copolymerization of 1,6-Anhydro-2,3,4–tri-0-(p-methybenzyl)-β-D-glucopyranose and 1.6-Anhydro-2,3,4-tri-0-benzyl-β-D-galacto-pyranose. Possible Application to the Synthesis of Stereoregular Heteropolysaccharides and Oligosaccharides," *Macromolecules,* 6 (1973), 320–24.

17. Zymonas, J. and Harwood, H.J., "Pyrolysis Behavior of Poly(α,β-β Trifluorostyrene-Co-Methyl Methacrylate)," paper presented at the 16th Canadian High Polymer Forum, Waterloo, Ontario, 1971.

18. Lindvay, M.W. Jr., Quinn, M.H. and Harwood, H.J., "Pyrolysis of Partially Deuteriated Polymers Containing Methyl Methacrylate and α-Methylstyrene Units. Eval-

uation of Gaylord's Mechanism for Metal-Halide Modified Copolymerization Systems," *Amer. Chem. Soc., Div. Polym. Chem., Polym. Prepr.*, 13 (1972), 1035–40.

19. Nencini, G., Giuliano, G., and Salvatori, T., "Pyrolysis-Mass Spectrometry Method for Blockiness Determination of Ethylene-Propylene Copolymers," *J. Polym. Sci., Part B*, 3 (1965), 483–85.

20. Tsuge, S., Okumoto, T. and Takeuchi, T., "Pyrolysis-Gas Chromatographic Investigation on the Structure of Chlorinated Poly(vinyl chlorides)," *Macromolecules*, 2, (1969), 277–80.

21. Tsuge, S., Okumoto, T. and Takeuchi, T., "Pyrolysis-Gas Chromatographic Studies on Sequence Distribution of Vinylidene Chloride-Vinyl Chloride Copolymers," *Makromol. Chem.*, 123 (1969), 123–29.

22. Grassie, N. and Torrance, B.J.D., "Thermal Degradation of Copolymers of Methyl Methacrylate and Methyl Acrylate, II. Chain Scission and the Mechanism of the Reaction," *J. Polym. Sci., Part A-1*, 6 (1968), 3315–26.

23. Michajlov, L., Zugenmaier, P. and Cantow, H.J., "Structural Investigations on Polyethylenes and Ethylene-Propylene Copolymers by Reaction Gas Chromatography and X-Ray Diffraction," Polymer, 9 (1968), 325–43.

24. Johnston, N.W. and Harwood, H.J., "Intersequence Cyclization Reactions in Methyl Methacrylate-Vinyl Halide Copolymers and Terpolymers," *J. Polym. Sci., Part C*, No. 22 (1969), 591–610.

25. Johnston, N.W. and Harwood, H.J., "Intersequence Cyclization in Methyl Methacrylate-Vinyl Chloride-Styrene Terpolymers," *Macromolecules*, 2 (1969), 221–24.

26. Johnston, N.W. and Harwood, H.J., "Intersequence Cyclization in Brominated Methyl Methacrylate-Butadiene Copolymers," *Macromolecules*, 3 (1970), 20–23.

27. Shepherd, F. and Harwood, H.J., "Equimolar Alternating Vinyl Chloride-Methyl Methacrylate Copolymers: Synthesis and Proof of Structure," *J. Polym. Sci., Part B*, 9 (1971), 419–23.

28. Shepherd, F. and Harwood, H.J., "Crystallinity in Equimolar Highly Alternating Butadiene-Methyl Methacrylate Copolymers," *J. Polym. Sci., Part B*, 10 (1972), 799–804.

29. Guilbault, L.J. and Harwood, H.J., "Intersequence Cyclization in Copolymers of n-Vinyl Carbamates," *J. Polym. Sci., Polymer Chem. Ed.*, 12 (1974):1461.

30. Baines, F.C. and Bevington, J.C., "A Tracer Study of the Hydrolysis of Methyl Methacrylate and Methyl Acrylate Units in Homopolymers and Copolymers." *J. Polym. Sci., Part A-1*, 6 (1968), 2433–40.

31. Bevington, J.C., Brinson, R. and Hunt, B.J., "A Tracer Study of the Hydrolysis of the Pendant Ester Groups in Polymers and Co-Polymers of Methacrylates," *Makromol. Chem.*, 134 (1970), 327–30.

32. Bevington, J.C. and Ebdon, J.R., "The Alkaline Hydrolysis of Methyl Methacrylate/ Isoprene Copolymers. I. Influence of Copolymer Composition on the Hydrolysis Behavior," *Makromol. Chem.*, 153 (1972) 165–71.

33. Glavis, F.J., "Hydrolysis of Crystallizable Poly(Methyl Methacrylate)," *J. Polym. Sci.*, 36 (1959) 547–49.

34. Thall, E. and Harwood, H.J., "Chemical Reactivity of Copolymers," paper presented at the International Symposium on Macromolecules, Helsinki, Finland, July 1972.

35. Kuist, C.H. and Maxim, L.D., "The Ultra-violet Degradation of Scissioning Copolymers," *Polymer*, 6 (1965), 523–30.

36. Robertson, A.B. and Harwood, H.J., "A Radiotracer Study of the Acidic Hydrolysis of Poly(Methyl Methacrylate)," *Amer. Chem. Soc., Div. Polym. Chem., Polym. Prepr.*, 12 (1971), 620–27.

37. Harwood, H.J., Ang, T.L., Bauer, R.G., Johnston, N.W. and Shimizu, K., "The Selection of Infrared Bands for Use in Copolymer Composition and Sequence Distribution Determinations," *Amer. Chem. Soc., Div. Polym. Chem., Polym. Prepr.*, 7 (1966), 980–86.

38. Enomoto, S., "A Study of Vinylidene Chloride-Vinyl Chloride Copolymers," *J. Polym. Sci.*, 55 (1961), 95–112.

39. Germar, H., "Determination of the Molecular Chain Structure of Vinylidene Chloride-Vinyl Chloride Copolymers from the Infrared Spectrum," *Makromol. Chem.*, 84 (1965), 36–50.

40. Bucci, G. and Simonazzi, T., "Contribution to the Study of Ethylene-Propylene Copolymers by Infrared Spectroscopy. Distribution of the Monomeric Units," *J. Polym. Sci., Part C*, No. 7 (1964), 203–12.

41. Drushel, H.V., Ellerbe, J.S., Cox, R.C. and Lane, L.H., "Monomer Sequence Distribution in Ethylene-Propylene Copolymers by Computer Analysis of Infrared Spectra," *Anal. Chem.*, 40 (1968), 370–79.

42. Kinsinger, J.B., Fischer, T. and Wilson, C.W., "Microstructure in Copolymers. II. Analysis of Tetrad Sequences in Vinylidene Chloride-Isobutylene Copolymers by NMR Spectroscopy," *J. Polym. Sci., Part B*, 5 (1967), 285–94.

43. Hellwege, K.H., Johnsen, U. and Kolbe, K., "Nuclear Spin Resonance of Diads, Triads, Tetrads, Pentads, and Hexads in Vinylidene Chloride-Isobutylene Copolymers," *Kolloid Z-Z Polym.*, 214 (1966), 45–52.

44. Johnsen, U., "The Determination of the Microstructure of Vinyl Polymers by High Resolution Proton Magnetic Resonance," *Kolloid Z-Z Polym.*, 210 (1966), 1–15.

45. Wilkes, C.E., Westfahl, J.C. and Backderf, R.H., "Microstructure of Ethylene-Vinyl Chloride Copolymers. I. Direct Verification of Terminal Radical Effects on Copolymerization by a Simplified NMR Approach," *J. Polym. Sci., Part A-1*, 7 (1969), 23–33.

46. Ito, K., Iwase, S., Umehara, K. and Yamashita, Y., "Copolymer Microstructure by High-Resolution Nuclear Magnetic Resonance Studies," *J. Macromol. Sci., Chem.*, A1 (1967), 891–908.

47. Chang, R.C. and Harwood, H.J., "Methine Proton Resonance Patterns of Styrene-Methacrylonitrile Copolymers and Their Deuteriated Analogs," *Amer. Chem. Soc., Div. Polym. Chem., Polym. Prepr.*, 12 (1971) 338–45; and "Methine Proton Resonance of Styrene-Methacrylonitrile Copolymers Prepared in the Presence of $ZnCl_2$," *Amer. Chem. Soc., Div. Polym. Chem., Polym. Prepr.*, 14 (1973), 31–35.

48. Bockrath, R.E. and Harwood, H.J., "Proton Magnetic Resonance Spectra of Styrene-Methacrylic Acid Copolymers and of Styrene-MMA Copolymers Derived from Them," *Amer. Chem. Soc., Div. Polym. Chem., Polym. Prepr.*, 14 (1973), 1163–68.

49. Klesper, E., Gronski, W. and Johnsen, A., "Complete Triad Assignment of Methylmethacrylate-Methacrylic Acid Copolymers," in *NMR-Basic Principles and Progress*, Vol. 4. *Natural and Synthetic High Polymers*. P. Diehl, E. Fluck, and R. Kosfeld, eds., New York: Springer-Verlag (1971), 47–69.

50. Klesper, E., Gronski, W. and Barth, V., "Triad Statistics in Dependence of the Reaction Conditions During Partial Hydrolysis of Syndiotactic Poly(methyl methacrylate)," *Makromol. Chem.*, 139 (1970), 1–16.

51. Blouin, F.A., Chang, R.C., Quinn, M.H. and Harwood, H.J., "300 MHz NMR Spectra of Highly Alternating Copolymers Containing Styrene, or α-Methylstyrene, and Methyl Acrylate or Methyl Methacrylate Units," *Amer. Chem. Soc., Div. Polym. Chem., Polym. Prepr.*, 14 (1973), 25–30.

52. Lindsay, G.A., Santee, E.R. Jr. and Harwood, H.J., "300 MHz NMR Spectra of Highly Alternating 1,1,4,4-Tetradeuteriobutadiene-Acrylonitrile Copolymers," *Amer. Chem. Soc., Div. Polym. Chem., Polym. Prepr.*, 14 (1973), 646–51.

53. Udipi, K., Harwood, H.J., Friebolin, H. and Cantow, H.J., "Methine and α-Methyl Proton Resonance in Methyl Acrylate/Methacrylonitrile Copolymers," *Makromol. Chem.*, 164 (1973), 283–94.

54. Mochel, V.D., "Carbon-13 NMR of Polymers," *J. Macromol. Chem., Rev. Macromol. Sci.*, C8 (1972), 289–347.

55. Schaefer, J., "High-Resolution Pulsed Carbon-13 Nuclear Magnetic Resonance Analysis of the Monomer Distribution in Acrylonitrile-Styrene Copolymers," *Macromolecules*, 4 (1971), 107–10.

56. Schaefer, J., "Carbon-13 Nuclear Magnetic Resonance Analysis of Ethylene Oxide-Maleic Anhydride Copolymers," *Macromolecules*, 2 (1969), 210–14.

57. Wilkes, C.E., Carman, C.J. and Harrington, R.A., "Monomer Sequence Distribution in Ethylene-Propylene Terpolymers Measured by ^{13}C Nuclear Magnetic Resonance," *J. Polym. Sci., Part C*, No. 43 (1973), 237–50.

58. Delfini, M., Segre, A.L. and Conti, F., "Sequence Distributions in Ethylene-Vinyl Acetate Copolymers. I. ^{13}C Nuclear Magnetic Resonance Studies," *Macromolecules*, 6 (1973), 456–59.

59. Whipple, E.P. and Green, P.J., "Sequence Analysis of Ethylene Oxide-Propylene Oxide Copolymers by Carbon-13 Nuclear Magnetic Resonance," *Macromolecules*, 6 (1973), 38–42.

60. Randall, J.C., "Carbon-13 NMR of Ethylene-1-Olefin Copolymers: Extension to the Short-Chain Branch Distribution in a Low Density Polyethylene," *J. Polym. Sci., Polym. Phys. Ed.*, 11 (1973), 275–87.

61. Fleischer, D. and Schulz, R.C., "Sequenzanalyse bei Trioxan/Dioxolan-Copolymeren durch NMR-Spektroskopie unter Zusatz von Lanthaniden-Komplexen," *Makromol. Chem.*, 152 (1972), 311–15.

62. Fleischer, D. and Schulz, R.C., "Sequenzanalyse bei Trioxocan/Dioxolan-Copolymeren durch NMR-Spektroskopie unter Zusatz von Lanthaniden-Komplexen," *Makromol. Chem.*, 162 (1972), 103–11.

63. Clark, D.T., Feast, W.J., Kilcast, D. and Musgrave, W.K.R., "Applications of ESCA to Polymer Chemistry. III. Structures and Bonding in Homopolymers of Ethylene and the Fluoroethylenes and Determination of the Compositions of Fluoro Copolymers," *J. Polym. Sci., Polym. Chem. Ed.*, 11, (1973), 389–411.

64. Biskup, U. and Cantow, H.J., "Zur Berechnung der ungestörten Dimensionen von Vinylcopolymeren," *Makromol. Chem.*, 148 (1971), 31–40.

65. Mark, J.E., "Configurational Characteristics of Ethylene-Vinyl Chloride Copolymers," *Amer. Chem. Soc., Div. Polym. Chem., Polym. Prepr.*, 14 (1973), 1101–07.

66. Tanaka, T. and Kotaka, T., "Molecular Dimensions of Block Copolymers in Solution," *Bull. Inst. Chem. Res., Kyoto Univ.*, 50 (1972), 107–16.

67. Kamiyama, F., Inagaki, H., and Kotaka, T., "Thermodynamic and Conformational Properties of Styrene-Methyl Methacrylate Block Copolymers in Dilute Solution. VI. Chain Conformations Disclosed by Thin-Layer Chromatography," *Polym. J.*, 3 (1972), 470–75.

68. Kraus, G., Childers, C.W. and Gruver, J.T., "Properties of Random and Block Copolymers of Butadiene and Styrene. I. Dynamic Properties and Glassy Transition Temperatures," *J. Appl. Polym. Sci.*, 11 (1967), 1581–91.

69. Johnston, N.W., "The Use of Alternating Copolymers in Sequence Distribution-Glass Transition Predictions," *Amer. Chem. Soc., Div. Polym. Chem., Polym. Prepr.*, 14 (1973), 46–51; and "Sequence Distribution-Glass Transition Effects. V. Acrylonitrile Co- and Ter-Polymers," *Amer. Chem. Soc., Div. Polym. Chem., Polym. Prepr.*, 14 (1973), 634–39.

70. Johnston, N.W., "Sequence Distribution-Glass Transition Effects. III. α-Methyl-styrene-Acrylonitrile Copolymers," *Macromolecules,* 6 (1973), 453–56.

71. Johnston, N.W., "Sequence Distribution-Glass Transition Effects. II. Alkyl Meth-acrylate/Vinyl Chloride Copolymers," *J. Macromol. Sci., Chem.,* A7 (1973), 531–45.

72. Kenney, J.F., "Effects of Monomer Unit Arrangement on the Properties of Copoly-mers," *Amer. Chem. Soc., Div. Polym. Chem., Polym. Prepr.,* 14 (1973), 964–69.

73. Mark, J.E., "Random-Coil Dimensions and Dipole Moments of Propylene-Vinyl Chloride Copolymers," *J. Polym. Sci., Polym. Phys. Ed.,* 11 (1973), 1375–83.

74. Mark, J.E., "The Use of Dipole Moments to Characterize Chemical Sequence Dis-tributions in Vinyl Copolymers," *J. Amer. Chem. Soc.,* 94 (1972), 6645–50.

J. E. MARK
The University of Michigan
Ann Arbor, Michigan

Chapter 12

Statistical Properties of
Chain Molecules

ABSTRACT

Theoretical methods recently developed for the calculation of configuration-dependent statistical properties of chain molecules are briefly outlined. A rotational isomeric state scheme is used to describe configurations of the chains, and the required averaging over all such configurations is carried out with proper account of the interdependence of bond rotational states. The statistical properties chosen for illustration of these methods are the mean-square dimensions and dipole moments of several polymer chains in the randomly coiled state. The calculation and interpretation of these qualities in the case of homopolymers are illustrated using poly(vinyl chloride) and poly(p-chlorostyrene) chains of various stereochemical compositions. The more elaborate analysis of copolymeric chains is demonstrated using binary copolymers containing the comonomeric pairs p-chlorostyrene–p-methylstyrene, p-chlorostyrene-styrene, ethylene–propylene, propylene–vinyl chloride, and ethylene–vinyl chloride. In configuration studies of either type, comparison of theoretical and experimental values of statistical properties and their temperature coefficients yield conformational energies of the chain molecule; such results are very useful in a variety of areas in polymer physical chemistry, including the thermodynamic and molecular interpretation of rubberlike elasticity. In the case of copolymeric chains, it is also possible to obtain information on chemical sequence distribution, a structural variability in copolymers which has an extremely important effect on their properties.

Introduction

The properties of polymeric materials both in solution and in the bulk, undiluted state are largely determined by the configurational characteristics of the long-chain molecules of which they are composed. Thus, the correlation of properties with structure in this important class of materials has increasingly focused on studies of chain configurations, *i.e.,* spatial arrangements of the molecules which differ one from another only through rotations about skeletal bonds. Such studies involve the measurement and theoretical interpretation of configuration-dependent properties such as the random-coil dimensions, dipole moments, optical rotation, strain birefringence and dichroism, and stereochemical and cyclization equilibria.

Theoretical methods[1–3] recently developed for the interpretation of such statistical properties are based on a rotational isomeric state model, in which each rotatable skeletal bond is restricted to one of a small number of discrete rotational states, generally of significantly different energy. Although the theoretical study of chain configurations is usually complicated by the interdependence of rotations about neighboring bonds, the application and elaboration of Ising lattice methods to this problem now permits the calculation of many configuration-dependent properties[2]. The first requirement in these calculations is, of course, the bond lengths, bond angles, dipole moments, etc., associated with the units making up the chain. In addition, since the desired statistical properties must be averaged over all configurations of the chain, with each configuration weighted with its relative probability of occurrence, the energies of various chain conformations or rotational states is obviously also necessary. In favorable cases, such energies can be estimated from semiempirical, interatomic potential functions. More reliably, however, this information is obtained by comparison between calculated and experimental values of several statistical properties, or their temperature coefficients[1–3]. Alternatively, the same information can be obtained frequently from the relative amounts of diastereoisomers present at stereochemical equilibrium in a suitably catalyzed mixture of chain oligomers; this is possible because the equilibrium composition in such mixtures also depends on chain intramolecular interactions[2]. Once established, these energies may then be used in the prediction of other properties or in the interpretation of any process in which chain configurations are altered (for example, the deformation of an elastomeric polymer network)[1–4].

Configurational studies of this type are of particular interest in the case of chemical copolymers, where statistical properties may be studied as a function of chemical sequence distribution. Such materials are of great current interest[5,6] since they frequently undergo microscopic phase separation if their chemical sequences are of sufficient length and of sufficiently different chemical structure. It is then often possible to control the phase separation so as to yield materials having highly desirable physical properties. For example, in the case of copolymers of styrene with butadiene, two-phase systems in which rubberlike polybutadiene regions are dispersed in a polystyrene matrix have unusually high impact resistance; when these roles are reversed, the glasslike polystyrene regions act as easily reformable cross links in the polybutadiene matrix and the resulting material is a "thermoplastic" elastomer with exceedingly attractive processing characteristics[6]. These comments merely serve to illustrate the point that chemical sequence distribution is a very important characteristic of chemical copolymers and its study through configurational analysis is thus of considerable importance.

In this review, the rotational isomeric state theory of the dimensions and dipole moments of chain molecules is briefly described. Calculated values of these properties for several vinyl homopolymers and chemical copolymers are then presented and compared with pertinent experimental results in the literature, and the information obtained in this way is discussed in some detail.

Theory

In the study of the statistical properties of vinyl homopolymers, which have variable stereochemical $(d, \mathcal{l})$ structure, and vinyl copolymers, which have variable chemical structure as well, it is necessary first to generate or simulate chain molecules which are closely similar in these structural features to the polymers actually prepared and experimentally investigated. The stereochemical structure is generally characterized by the replication probability p_r, which is the probability of "isotactic" placements dd or $\mathcal{l}\mathcal{l}$ rather than "syndiotactic" placements $d\mathcal{l}$ or $\mathcal{l}d$. The chemical structure, on the other hand, is characterized by the probability p_2 that a given unit is of comonomer type 2 rather than type 1, and also the average lengths n_1 and n_2 of sequences of the two types of units in the chain. As described in detail elsewhere[2,7–10] Monte Carlo techniques can readily be applied to p_r, and to p_2 and the reactivity ratio product[11] $r_1 r_2$ controlling the chemical sequence distribution, in order to generate chains of the desired stereochemical and chemical structure.

The chain molecules to be discussed are all composed of vinyl units CHR—CH$_2$—, or combinations of such units with those of ethylene CH$_2$—CH$_2$—, and thus all have backbones consisting simply of singly bonded C atoms. These molecules may therefore be described in terms of the three rotational states, *trans* (t), *gauche* positive (g^+), and *gauche* negative (g^-) assumed to occur at rotational angles φ of 0°, 120°, and −120°, respectively, except in those cases in which intramolecular interactions are sufficiently large to displace some of these states by an amount $\Delta\varphi$ the order of 10–20°[2,7]. In order to take into account the interdependence of rotational states, statistical weights must be assigned to pairs of consecutive skeletal bonds. Each such statistical weight u_{ij} is customarily expressed as a product of Boltzmann factors of the form[2]

$$\zeta_k = \exp\left[-\frac{E(\zeta_k)}{RT}\right] \tag{1}$$

where $E(\zeta_k)$ is the energy characterizing one of the intramolecular interactions occurring in the associated conformation (*i.e.*, pair of consecutive
rotational states) relative to some arbitrarily designated conformation of
zero energy. The statistical weight of any particular pair of consecutive
states will then be given by the appropriate product of such factors, one
for each intramolecular interaction requiring characterization. The entire
set of statistical weights required to describe all such conformations about
a particular bond i in the chain is then given by the statistical weight
matrix U_i defined by[2]

$$
U_i = \begin{bmatrix} u_{tt} & u_{tg^+} & u_{tg^-} \\ u_{g^+t} & u_{g^+g^+} & u_{g^+g^-} \\ u_{g^-t} & u_{g^-g^+} & u_{g^-g^-} \end{bmatrix} \tag{2}
$$

where rows are associated with states about bond i-1 and columns with
states about bond i. By way of illustration, the element $u_{g^+g^-}$ in U_i is the
statistical weight for the conformational sequence consisting of a g^+ rotational state followed by a g^- rotational state about bonds i-1 and i, respectively. (This conformation is one which illustrates particularly clearly
the interdependence of rotational states. These conformations, and frequently others as well, must be suppressed in most molecules by assignment of statistical weights near zero because of "pentane-type" interferences[1–3] between groups separated by four bonds in such conformations.
More detailed illustrations will be given below.)

The configurational partition function Z for a chain of n bonds is
then simply[2]

$$
Z = J^* \left[\prod_{i=1}^{n} U_i \right] J \tag{3}
$$

where J^* and J are row and column vectors, respectively, of the form
required to extract the desired terms from the matrix resulting from the
specified multiplications[2]. The statistical properties of interest here are
the mean-square dimensions $\langle r^2 \rangle_0$ and dipole moments $\langle \mu^2 \rangle$ obtained by
averaging over all configurations of the chains. The zero subscript in
$\langle r^2 \rangle_0$ indicates that this quantity will pertain only to chains unperturbed
by long-range interactions[12]; it will serve as a reminder that, in general,
only those experimental values of the chain dimensions obtained under
Θ-conditions[12], where the effects of such interactions are nullified, will
be suitable for comparison with theory. On the other hand, the chain
molecules to be considered are of such structural symmetry that long-
range interactions have no effect at all on the dipole moments[13,14];
thus experimental values of this quantity need not pertain to measurements obtained under Θ-conditions and no restrictive subscript is required on $\langle \mu^2 \rangle$.

The mean-square unperturbed dimensions are given by the equation[2,10]

$$\langle r^2 \rangle_0 = 2Z^{-1} \mathbf{I}^* \left[\prod_{i=1}^{n} \mathbf{G}_i \right] \mathbf{I} \qquad (4)$$

where the generator matrix $\mathbf{G}_i$ for a particular skeletal bond i contains the appropriate statistical weight matrix $\mathbf{U}_i$, a coordinate transformation matrix $\mathbf{T}_i$ which depends only on the skeletal bond angle and allowed rotational angles φ_i, and the skeletal bond vector $\mathscr{l}_i$[2,10]. The row and column vectors $\mathbf{I}^*$ and $\mathbf{I}$ serve the same purpose as $\mathbf{J}^*$ and $\mathbf{J}$ in Equation (3). Equation (4) also serves for the calculation of $\langle \mu^2 \rangle$, when the bond vectors $_i$ are replaced with group dipole vectors $\mathbf{m}_i$[2,8,9]. The matrices $\mathbf{U}_i$ and $\mathbf{G}_i$ depend, in general, on the chemical and stereochemical nature of both the monomer unit in which bond i is located and its predecessor along the chain. The precise sequence of $\mathbf{U}$'s and $\mathbf{G}$'s to be used in Equations (3) and (4) are specified by Monte Carlo methods, as has been mentioned earlier. For convenience, the unperturbed dimensions will be expressed as the "characteristic ratio" $\langle r^2 \rangle_0 / n \mathscr{l}^2$ where n is the number of skeletal bonds each of the same length $\mathscr{l} = |\mathscr{l}|$. The dipole moments of the homopolymers will be expressed by the analogously defined "dipole moment ratio" $\langle \mu^2 \rangle / x m^2$ where $x = n/2$ is the degree of polymerization and, in the cases to be considered, also the number of bond dipoles, each of magnitude $m = |\mathbf{m}|$. In the vinyl copolymers, $\mathbf{m}$ is different for the two types of units in the chain and the ratio $\langle \mu^2 \rangle / x$, the mean-square dipole moment per unit, is a more convenient quantity for interpretation in these cases.

Values of the temperature coefficients of $\langle r^2 \rangle_0$ and $\langle \mu^2 \rangle$ can be obtained from values of these properties calculated for different values of the Boltzmann factors ζ_k. For example, in the case of $\langle r^2 \rangle_0$,

$$d \ln \langle r^2 \rangle_0 / dT = \sum_k \left(\frac{\partial \ln \langle r^2 \rangle_0}{\partial \ln \zeta_k} \right) (\partial \ln \zeta_k / \partial T)$$

$$= -\left(\frac{1}{T} \right) \sum_k \ln \zeta_k \left(\frac{\partial \ln \langle r^2 \rangle_0}{\partial \ln \zeta_k} \right) \qquad (5)$$

The temperature dependence of $\langle r^2 \rangle_0$ is of particular importance and interest because of its intimate relationship to the thermodynamic quantities characterizing the elastic deformation of an amorphous polymer network. Specifically, the energetic contribution f_e to the total retractive force f exhibited by a deformed network is related to the coefficient $d \ln \langle r^2 \rangle_0 / dT$ of the network chains according to the equation[4,15]

$$\frac{f_e}{f} = T \frac{d \ln \langle r^2 \rangle_0}{dT} \qquad (6)$$

Knowledge of $d \ln \langle r^2 \rangle_0 / dT$ thus directly indicates the relative importance of energetic and entropic changes occurring in the deformation of a network consisting of the polymer chains thus characterized.

Illustrative Results and Discussion

Homopolymers

The two polymer chains chosen to illustrate the application of this type of configurational analysis to homopolymers are poly(vinyl chloride) CH_3—$[CHCl$—CH_2—$]_x H$ and poly(p-chlorostyrene) CH_3—$[CH(p$-$C_6H_4Cl)$—CH_2—$]_x H$. This pair of chain molecules is of particular interest since the two chains are virtually identical in geometric and structural features and, thus, differences in characteristic ratio or dipole moment ratio would be entirely due to differences in conformational energy resulting from differences in nonbonded interactions along the chains[8,16].

Figure 1 shows a sequence of units in a vinyl chloride chain, and also serves to illustrate the nature of the stereochemistry of vinyl chains in general. (It further serves to illustrate the nature of pentane-type interferences[2]; the tt conformation shown about bonds i–1 and i places the neighboring pair of Cl atoms on the same side of the chain, in close proximity.) Representative chains characterized by values of the stereochemical replication probability p_r of 0.05 (highly syndiotactic), 0.50 ("atactic" or random), and 0.95 (highly isotactic) were generated using the Monte Carlo technique, and these results were used for the vinyl sequences in all the chains investigated. At the appropriate[8,16] value of the degree of polymerization, $x = 100$, used throughout the calculations, these chains have average sequence lengths of units of identical stereochemistry (isotacticity) of 1.07, 1.97, and 16.4, respectively[10]. For poly-(vinyl chloride), and the other chains as well, bond lengths and bond

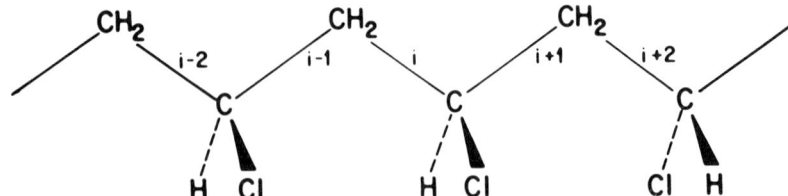

Figure 1.   The planar (all-*trans*) conformation of a portion of a vinyl chloride chain consisting of an isotactic (dd) dyad followed by a syndiotactic (dℓ) dyad. Bonds extending toward the reader are represented by lines of increasing thickness, and those extending away, by dotted lines. (From[8].)

angles were assigned values of 1.53 Å and 112°, respectively. In the specific case of poly(vinyl chloride), the pendant group R is the relatively small Cl atom and the displacement $\Delta\varphi$ of rotational states was therefore considered negligible. Group dipoles of magnitude m = 2.00 D were located on the C—Cl bonds, which were assumed to be tetrahedrally bonded to the chain backbone[8]. Conformational energies of vinyl chloride chains are available from the stereochemical equilibration results obtained for vinyl chloride oligomers by Flory and coworkers[17,18]. The most important of these results are qualitatively summarized here. In syndiotactic sequences *trans* states are very strongly favored because interactions of the small Cl atom with the chain backbone are of relatively low energy in this conformation. Pentane-type interferences between Cl atoms separated by four bonds are sufficiently large to diminish significantly this preference in the case of isotactic sequences; they are not large enough, however, to restrict such sequences to conformations (*e.g.*, *tg*)[2] giving rise helices (3_1)[2] to the extend shown by isotactic sequences of units having pendant groups the size of CH_3 or larger[2,7].

The results of these circumstances are shown by the values of the characteristic ratio shown in Figure 2[8]; shown also, for purposes of comparison, are some similarly calculated results for polypropylene (R = CH_3)[2,7,8]. The characteristic ratio of poly(vinyl chloride) is much

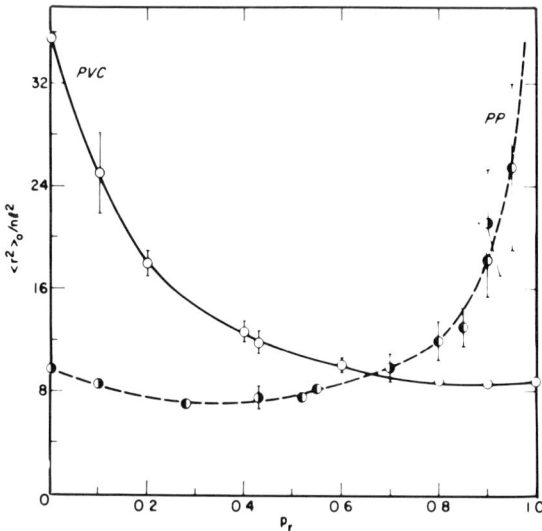

Figure 2. Calculated values of the characteristic ratio $\langle r^2 \rangle_0 / n \ell^2$ shown as a function of the probability p_r of replication (isotactic placement) for poly(vinyl chloride) and polypropylene. In this and the following figures, all results pertain to a degree of polymerization of x = n/2 = 100 and a temperature of 25°C, unless specified otherwise. (From[8].)

larger than that for polypropylene in the region of high syndiotacticity but the reverse is true for high isotacticity. These differences are directly due to the relatively small size of the Cl atom in poly(vinyl chloride). This is the origin both of the preference for spatially extended *trans* conformations in syndiotactic sequences and the alleviation of the strong preference for extended helical conformations which occur in isotactic poly(propylene) sequences and are caused by repulsions between the relatively large CH_3 groups[2,8,17,18]. Values of the dipole moment ratio for poly(vinyl chloride) are shown in Figure 3[8]. The behavior of this statistical property is very similar to that shown by the characteristic ratio since high chain extension generally corresponds to large dipole moment in this chain molecule. As is illustrated by comparison of Figures 2 and 3, however, the dipole moment ratio is generally much smaller than the characteristic ratio of the same chain because of the considerable attenuation occurring among group dipole vectors. The experimental value of the characteristic ratio of atactic poly(vinyl chloride) at 155° is reported to be 9.8[8];* the calculated result for a chain of these characteristics, at this temperature, is 9.5. Experimental and calculated results for the temperature coefficient of $\langle r^2 \rangle_0$ are also in good agreement, with both giving the value of $d \ln \langle r^2 \rangle_0 / dT = -1.5 \times 10^{-3} \deg^{-1}$[8]. This coefficient is

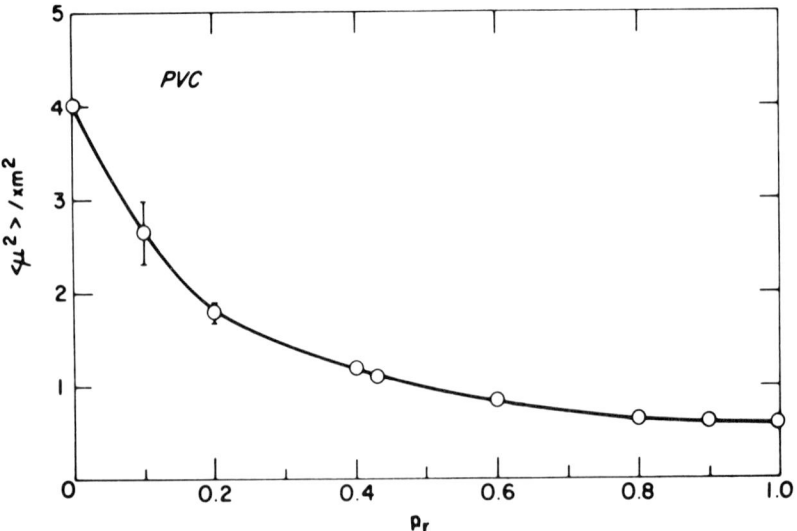

Figure 3.   The dependence of the dipole moment ratio $\langle \mu^2 \rangle / xm^2$ on the replication probability for poly(vinyl chloride) chains. (From[8].)

*In general, pertinent experimental studies are numerous and will not be cited directly; they are cited, discussed in detail, and evaluated in the theoretical studies of each chain molecule.

negative in the case of poly(vinyl chloride) chains because, as already described, conformations of lowest energy are of relatively high chain extension in these molecules[8]. Comparison between experimental and theoretical values of $\langle \mu^2 \rangle$ and its temperature coefficient are not feasible at the present time because of the very wide range of experimental values obtained for these quantities in the case of poly(vinyl chloride).

As is obvious from the above discussion, poly(vinyl chloride) is not a typical vinyl polymer, because of the unusually small size of its pendant group. As can be seen from Figure 4, the *para*(*p*)-substituted styrene chains are closely related polymers structurally, but have significantly larger side chains. Of interest here is the particular case of poly(*p*-chlorostyrene), where the *para* substituent X = Cl. Differences in characteristic ratio between poly(*p*-chlorostyrene) and poly(vinyl chloride) should be due entirely to differences in conformational energy since the skeletal bond vectors and bond angles are the same in both chains, and the rotational state displacement $\Delta \varphi$ may again be set equal to zero in first approximation. The same is true for the dipole moment ratio because the nature of the bonding and the symmetry of the *para* isomer results in tetrahedral orientation of the group dipole relative to the chain backbone, as is the case in poly(vinyl chloride). (Styrene units in which substitution is at the *para* position also have the advantage of having dipole moments which are independent of side chain rotation, as can be seen from Figure 4.) The fact that the value m = 1.68 D of the group dipole moment appropriate for this chain is somewhat smaller than that for vinyl chloride

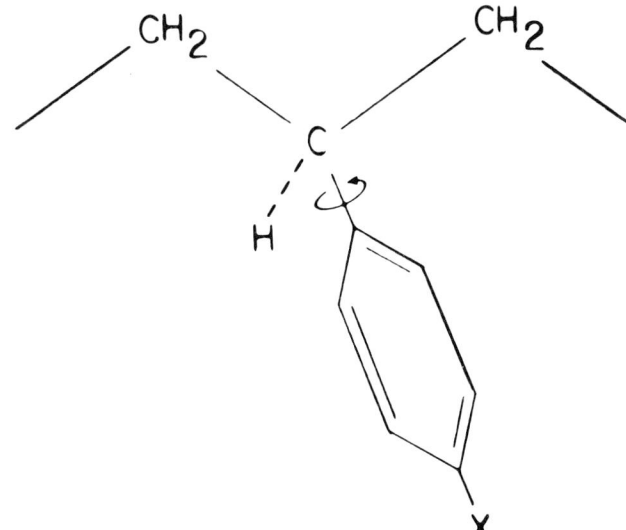

Figure 4. A schematic diagram of the repeat unit in *para*-substituted styrene chains.

chains is irrelevant since calculated dipole moments are expressed relative to the corresponding group dipole moment in ratios such as $\langle \mu^2 \rangle / xm^2$. The conformational energies appropriate for poly(p-chlorostyrene) chains are the same, to very good approximation, as the corresponding energies for unsubstituted styrene chains and these are available from stereochemical equilibration experiments carried out on styrene oligomers by Williams and Flory[19]. The validity of this approximation can be seen from detailed analysis of the intramolecular interactions occurring in such chains, and is confirmed by examination of values of the characteristic ratio for a variety of atactic polymers having the structure CH_3—$[CH(p\text{-}C_6H_4X)\text{—}CH_2\text{—}]_x H$. Such results for pertinent *para*-substituents X are given in Table I and indicate that such substituents must be much larger than a Cl atom in order to give a value of the characteristic ratio significantly different from that of polystyrene itself. It is thus permissible to use the conformational energies determined for styrene oligomers in the calculation of the configuration-dependent properties of p-chlorostyrene chains. In brief, these energies indicate only a slight preference for *trans* states in syndiotactic sequences and the essentially complete suppression of conformations giving rise to pentane-type interferences. Calculations based on these conformational energies and the structural information already cited give the values of the characteristic ratio and dipole moment ratio for p-chlorostyrene chains shown in Figures 5 and 6, respectively[16]. Included for purposes of comparison are the corresponding results calculated for vinyl chloride chains[8]. Poly(p-chlorostyrene) has relatively small values of the characteristic ratio and dipole moment ratio in the region of high syndiotacticity but very large values at high isotacticity, the reverse of the situation pertaining to poly(vinyl chloride). The relatively large size of the C_6H_4Cl pendant group in poly(p-chlorostyrene) greatly diminishes the preference for extended *trans* states in syndiotactic sequences but also strongly suppresses departures from extended helical conformations in isotactic sequences. The same arguments apply to the dipole moment ratio. The low energy conformations for syndiotactic sequences in these chains are pairs of *trans* states separated by pairs of

TABLE I

Effect of the Substituent X on the Characteristic Ratios[a] of Atactic Polystyrenes

CH_3—$[CH(p\text{-}C_6H_4X)\text{—}CH_2\text{—}]_x H$

Polymer	X	$R(X)^b$	$\langle r^2 \rangle_0 / n\ell^2$
Polystyrene	H	1.20	10.0($\pm$0.2)
Poly(p-chlorostyrene)	Cl	1.75	10.6($\pm$0.4)
Poly(p-bromostyrene)	Br	1.85	12.3($\pm$1.1)
Poly(p-methylstyrene)	CH_3	~2.0	10.7($\pm$0.1)
Poly(p-cyclohexylstyrene)	C_6H_{11}	$\geq$2.0	13.7

[a] The appropriate experimental studies are cited and discussed in[16].
[b] Van der Waals' radius, in Å.

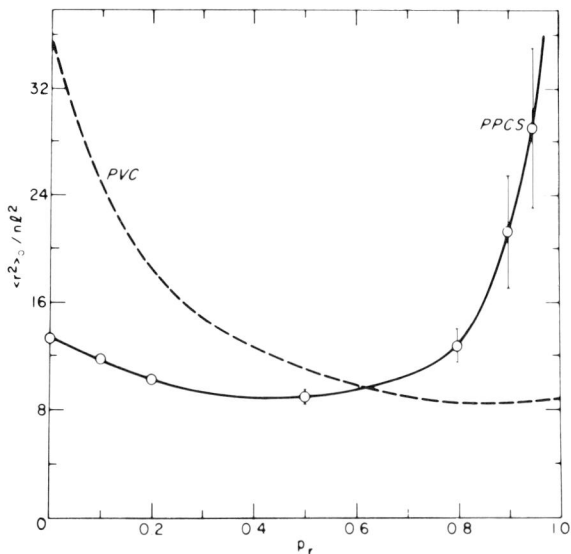

Figure 5. The characteristic ratio shown as a function of stereochemical composition for poly(p-chlorostyrene) chains; included for purposes of comparison are the similarly calculated results for poly(vinyl chloride) chains. (From[16].)

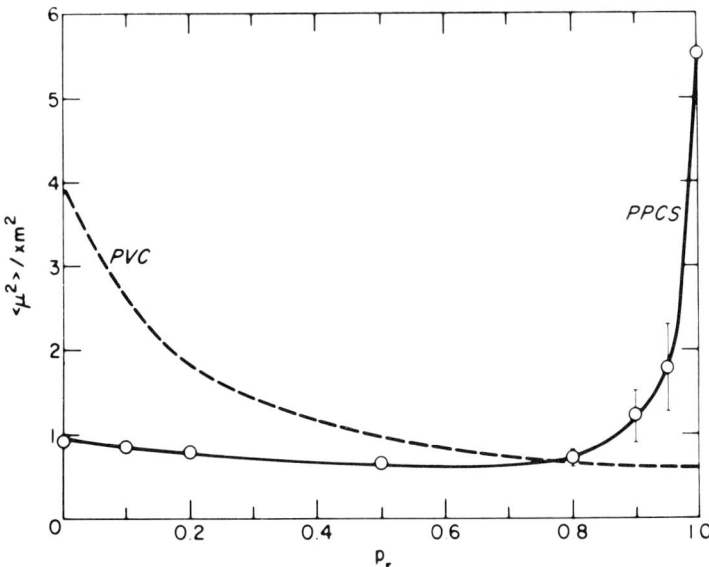

Figure 6. The dipole moment ratio for poly(p-chlorostyrene) and poly(vinyl chloride) chains. (From[16].)

gauche states of the same sign, as well as the all-*trans* conformation[2,19]. The former conformation is present to a very significant extent in *p*-chlorostyrene chains, and since it has essentially a zero dipole moment, as illustrated in Figure 7, $\langle\mu^2\rangle/xm^2$ is very small in this region of stereochemical structure. This ratio is, however, very large in the case of highly isotactic *p*-chlorostyrene chains since each group dipole in a helical sequence has a component pointing in the same direction along the helix axis[16]. Relocation[2] of some rotational states by an amount $\Delta\varphi = 10°$ because of the relatively large steric interactions[19] has a pronounced effect on these statistical properties only for chains which are significantly isotactic in structure; in this case, both the characteristic ratio and the dipole moment ratio decrease in value, primarily because of changes in the geometry of the helical conformations[16].

These calculations suggest a value of the characteristic ratio of approximately 10 for atactic poly(*p*-chlorostyrene) (and polystyrene) in the vicinity of room temperature. As can be seen from Table I, this calculated result is in good agreement with the experimental values reported for these two polymers[16]. The calculated value of the dipole moment ratio for atactic poly(*p*-chlorostyrene) is 0.73, a result in good agreement with the average value 0.68 ($\pm$0.10) obtained from a number of experimental stud-

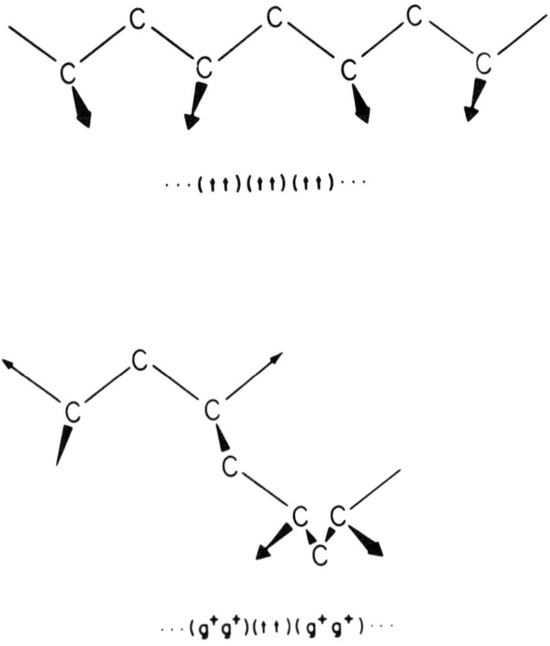

Figure 7. Relative orientations of group dipoles in *p*-chlorostyrene chains in the two low-energy conformations for syndiotactic sequences. (From[16].)

ies of this polymer[16]. There is also, in general, satisfactory agreement between theoretical and experimental values of the temperature coefficient of the dipole moment[16].

Chemical Copolymers

The polymers discussed in this section were chosen to illustrate a number of important features of copolymeric chains. The copolymers[9] poly(p-chlorostyrene-p-methylstyrene) ($R_1 = C_6H_4Cl$, $R_2 = C_6H_4CH_3$ in Figure 8) and poly(p-chlorostyrene-styrene) ($R_1 = C_6H_4Cl$, $R_2 = C_6H_5$) are of interest because each can serve to illustrate a situation which is the reverse of that characterizing the pair of homopolymers poly(vinyl chloride) and poly(p-chlorostyrene). The two homopolymers were interpreted as having identical structural features but very different conformational energies; in the case of the two copolymers, however, the conformational energy is independent of chemical composition, for reasons described above, and thus it is possible to study changes in the dipole moment ratio due to variation of the magnitude of some of the group dipoles at constant energy[9]. Ethylene-propylene copolymers[10] ($R_1 = H$, $R_2 = CH_3$) are commercially very important elastomers and are discussed here in order to illustrate particularly the use of configurational calculations to elucidate the nature of rubberlike elasticity. The last two copolymers discussed in detail are poly(propylene–vinyl chloride)[20] ($R_1 = CH_3$, $R_2 = Cl$) and poly(ethylene–vinyl chloride)[21] ($R_1 = H$, $R_2 = Cl$). They were chosen in order to remedy an unavoidable limitation in the studies of the other copolymers mentioned. In none of these preceding cases is it possible to use both $\langle r^2 \rangle_0 / n\ell^2$ and $\langle \mu^2 \rangle / x$ to illustrate the dependence of statistical properties on chemical composition and sequence distribution. In the case of the styrene-substituted styrene chains[9], changes in these variables have no effect on the skeletal bond lengths, conformational energies, or values of the characteristic ratio. On the other hand, the conformational energy and thus the characteristic ratio would be expected to depend on chemical composition and chemical sequence distribution in the case of ethylene–propylene copolymers[10], but the essentially nonpolar nature

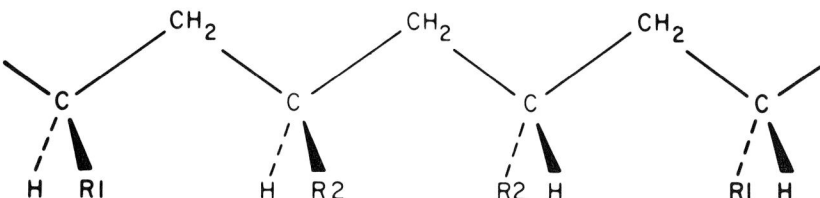

Figure 8. The planar conformation of a portion of a vinyl copolymer chain consisting of the chemical sequence M1, M2, M2, M1. (From[9].)

of both repeat units precludes calculation of useful values of $\langle\mu^2\rangle/x$ for this copolymer. In the case of propylene–vinyl chloride[20] and ethylene–vinyl chloride[21] chains, however, both statistical properties are useful for characterizing chain configurations, and it is thus possible in these cases to get a better understanding of the relative sensitivity of $\langle r^2\rangle_0/n\ell^2$ and $\langle\mu^2\rangle/x$ to chemical sequence distribution by calculations carried out on the *same* copolymeric chain. Such information is, of course, essential to the possible use of statistical properties to characterize chemical sequence distributions in chemical copolymers.

As already mentioned, the chemical sequence distribution in a copolymer may be characterized in part by the average sequence lengths n_1 and n_2 of chemically identical units of type 1 and 2, respectively. These characteristics are controlled by the chemical composition variable p_2 (fraction of units of type 2) and the reactivity ratio product r_1r_2[9,10]. Values of n_1 and n_2 for selected values of p_2 and r_1r_2 are tabulated in detail elsewhere, but the following results for equimolar copolymers ($p_2 = 0.50$) are presented here for purposes of illustration. For values of $r_1r_2 = 0.0, 1.0, 1000.0$, and ∞, the representative values of $n_1 = n_2$ were, respectively, 1.0 (perfectly alternating copolymers), 2.0 ("ideal" copolymers), 24.0 ("block" copolymers), and ∞ (mixtures of homopolymers)[9,10,21].

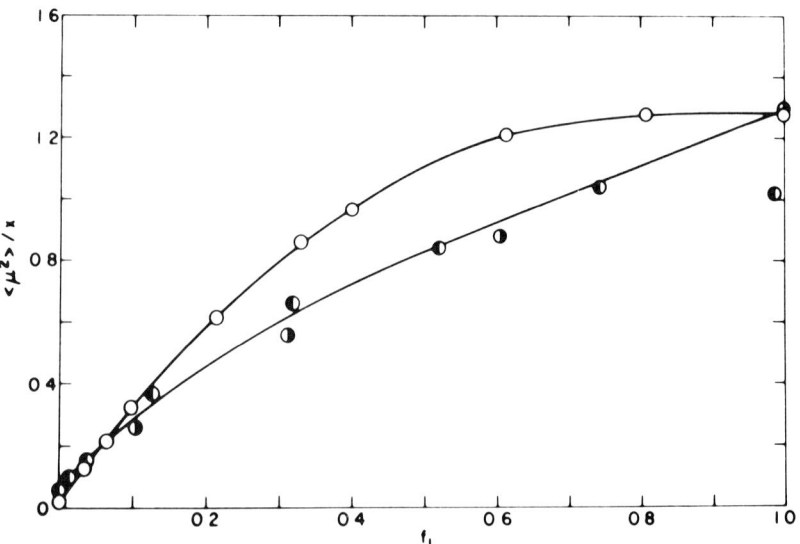

Figure 9. Experimental values of the mean-square dipole moment per monomer unit for atactic ($p_r = 0.50$) poly(p-chlorostyrene–p-methylstyrene) at 155°C (open circles) and atactic (p-chlorostyrene–styrene) at 130°C (half-filled circles) shown as a function of the mole fraction $f_1 = p_1$ of p-chlorostyrene in the copolymer. (From[9].)

Mean-square dipole moments per unit reported for atactic p-chloro-styrene–p-methylstyrene and p-chlorostyrene–styrene chains at 155° and 130°, respectively, are presented in Figure 9[9]. There are two very note-worthy features in these experimental results, which, as will be seen, can be understood readily in terms of configurational calculations. Specifi-cally, both curves of $\langle \mu^2 \rangle / x$ against the fraction f_1 of p-chlorostyrene units in the chain show marked positive deviations from simple linearity be-tween the value of $\langle \mu^2 \rangle / x$ characterizing the corresponding homopoly-mers, and the replacement of a p-methylstyrene unit of dipole moment zero by a styrene unit of dipole moment 0.36 D *decreases* $\langle \mu^2 \rangle / x$. Calcula-tions[9] for the correspondingly atactic p-chlorostyrene–p-methylstyrene co-polymer were carried out using the information give above, $\Delta \varphi = 10°$, and the appropriate value 0.75 for $r_1 r_2$[9]. The values of $\langle \mu^2 \rangle / x$ obtained in these calculations are shown by the points in Figure 10 and are seen to be in good agreement with the experimental relationship, which is shown by the solid line. The dashed lines show results calculated for other, illustra-tive values of $r_1 r_2$ and demonstrate that $\langle \mu^2 \rangle / x$ is quite sensitive to chemi-cal sequence distribution at constant chemical composition. Therefore, in this case, measurements of dipole moments could have been used to ob-tain reliable values of $r_1 r_2$, and thus values of the average chemical se-

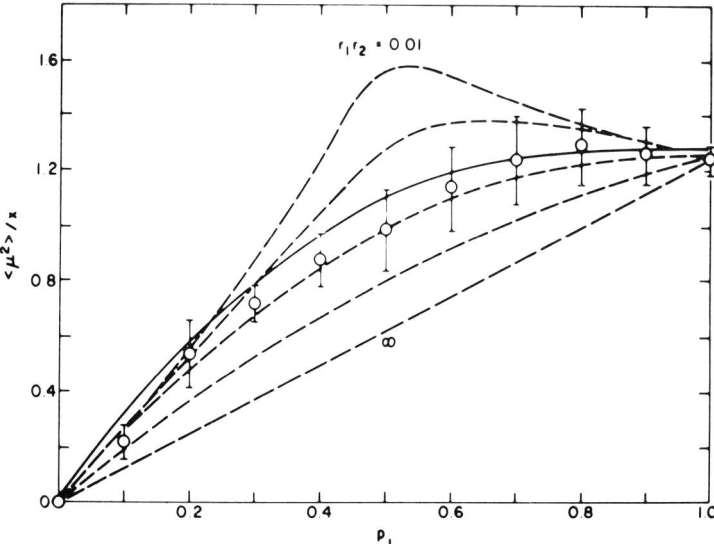

Figure 10. Values of $\langle \mu^2 \rangle / x$ for atactic poly(p-chlorostyrene–p-methylstyrene) copolymers at 155°C shown as a function of chemical composition. The solid curve shows the experimentally observed dependence, the points show the calculated results for the appropriate value of $r_1 r_2$, and the dashed curves show additional, illustrative results for values of $r_1 r_2$ of 0.01, 0.10, 1.00, 10.0, and ∞, respectively. (From[9].)

quence lengths in the copolymer. Calculations[9] for the p-chlorostyrene–styrene copolymers were carried out using the value of the reactivity ratio product, 0.85, reported for this comonomer pair[9]. Theoretical and experimental results for these chains, shown in Figure 11, are also in good agreement. Detailed analysis of the calculated results indicate that the bulkiness of the C_6H_4 groups causes neighboring group dipoles to be largely in orientations in which there is a great deal of attenuation. Such attenuation would be expected to be less in p-chlorostyrene–p-methylstyrene copolymers, in which one type of unit has a zero dipole moment, thus explaining the observation that replacement of a p-methylstyrene unit by one of larger dipole moment decreases $\langle \mu^2 \rangle / x$. This circumstance is also the origin of the convex nature of the curves shown in Figures 9–11; the cited attentuation would obviously be less in a copolymer than in a mixture of homopolymers having the same overall composition[9]. Additional results obtained for p-chlorostyrene–p-methylstyrene chains, shown in Fig. 12, indicate that the dependence of $\langle \mu^2 \rangle / x$ on chemical sequence distribution is particularly strong in the case of syndiotactic chains; this can be understood readily from the fact that *trans* sequences are low-energy conformations in chains of this stereochemical structure. In these conformations, as shown in Figure 13, decrease in $r_1 r_2$ toward zero causes alternation in the placement of the group dipoles with the result that all of the nonzero $[m(R_1)]$ dipoles assume parallel, perfectly additive orientations. The ratio $\langle \mu^2 \rangle / x$ thus becomes very much larger than it is for the

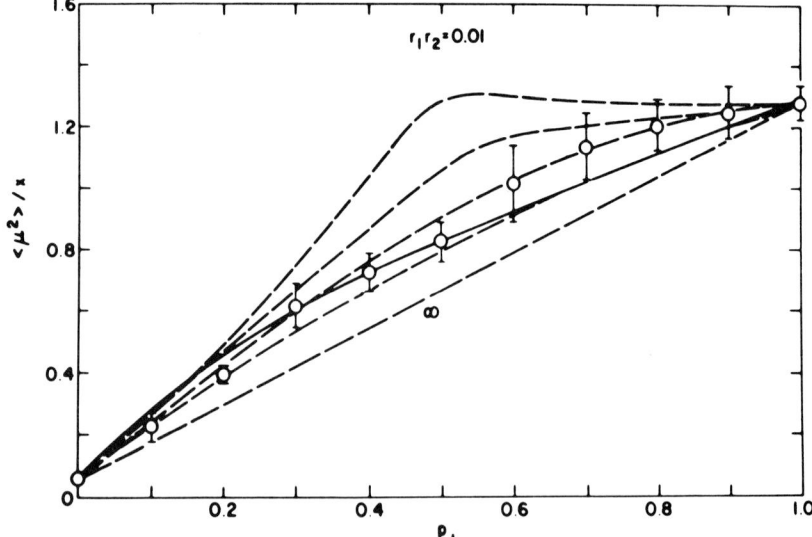

Figure 11. Values of $\langle \mu^2 \rangle / x$ for atactic poly(p-chlorostyrene–styrene) copolymers at 130°C; see legend to preceding figure. (From[9].)

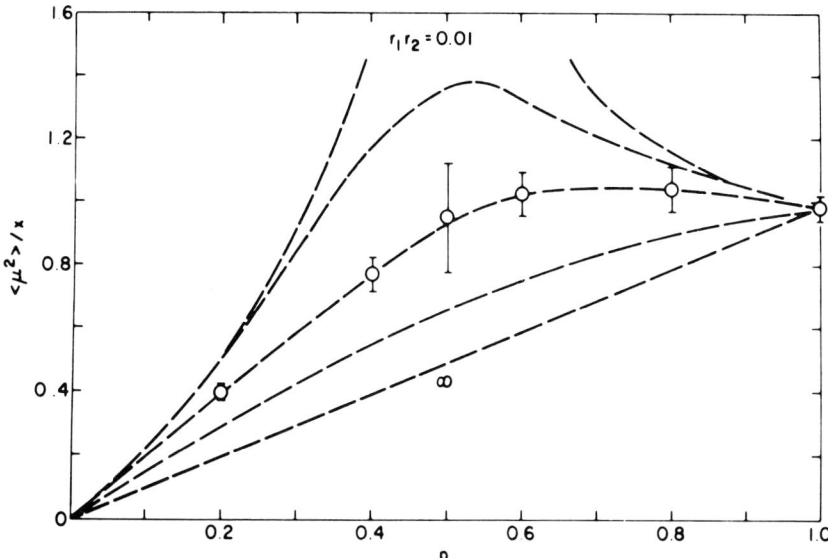

Figure 12. Calculated values of $\langle \mu^2 \rangle / x$ for poly(p-chlorostyrene–p-methylstyrene) copolymers of high syndiotacticity (p_r = 0.05) at 155°C. The results are shown as a function of chemical composition for values of $r_1 r_2$ of 0.01, 0.10, 1.00, 10.0, and ∞, respectively; for purposes of clarity, only those points used to locate the curve corresponding to $r_1 r_2$ = 1.00 are shown. (From[9].)

SYNDIOTACTIC VINYL CHAINS
$m(R1) > 0$, $m(R2) = 0$

A. $r_1 r_2 = \infty$

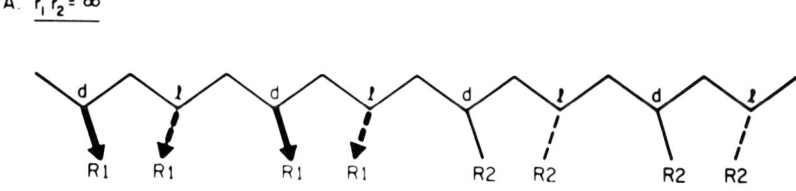

B $r_1 r_2 = 0$

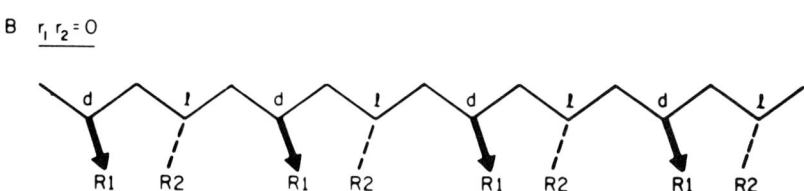

Figure 13. The relative orientations of the group dipoles $m(R_1) > 0$ and $m(R_2) = 0$ in the all-*trans* conformation of a syndiotactic vinyl copolymer. Case A ($r_1 r_2 \rightarrow \infty$) represents a block copolymer, and case B ($r_1 r_2 = 0.0$) represents an alternating copolymer.

357

case of the corresponding block copolymer $(r_1r_2 \rightarrow \infty)$ where there is considerable attenuation among group dipole vectors[9].

In the case of ethylene–propylene copolymers[10], *trans* states in ethylene sequences are of lower energy than *gauche* states $(g^{\pm})$ of either sign, but in propylene sequences they are of the same energy as one of the *gauche* states. Also, all pentane-type interactions are strongly suppressed in these chains, except in those cases where one of the interacting species is the relatively small H atom. Calculations[10] of values of the characteristic ratio of these copolymers were carried out on this basis, and the results for chains in which the propylene sequences are highly isotactic are shown in Figure 14. It is interesting to note that the characteristic ratio of either homopolymer is expected to decrease, in general, upon the addition of units of the other type. In the case of copolymer chains that contain predominantly ethylene, the propylene units cause departures from the extended all-*trans* conformations, which are the low-energy forms for ethylene sequences; this results, of course, from the fact that there are two conformations (t and either g^+ or g^-) which are of essentially equal probability in the case of propylene units[2,7]. For chains consisting mostly of (highly isotactic) propylene, the ethylene units disrupt the extended 3_1 helical sequences, thus again causing a decrease in the characteristic ratio. Experimental values of $\langle r^2 \rangle_0/n\ell^2$ are, unfortunately,

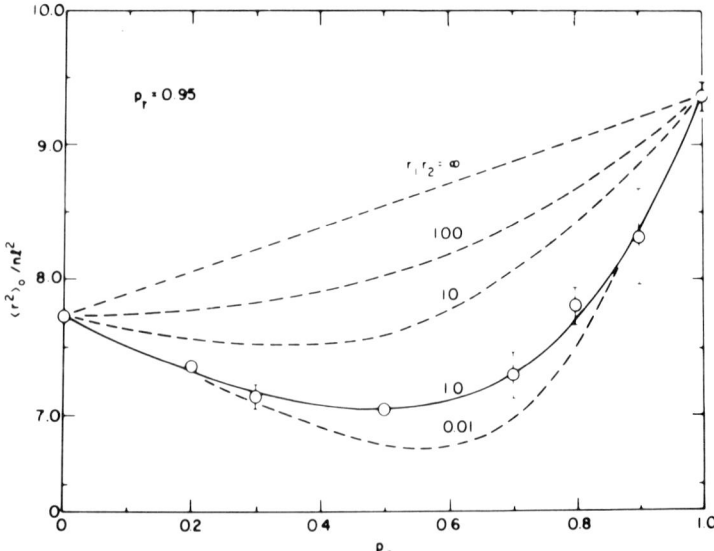

Figure 14. Calculated values of the characteristic ratio for ethylene–propylene copolymers having highly isotactic propylene sequences, shown as a function of the mole fraction p_2 of propylene units in the chain. (From[10].)

not available at the present time, but there are a number of reported values of the coefficient $d \ln \langle r^2 \rangle_0 / dT$, which have been obtained from force–temperature measurements on deformed ethylene–propylene networks[4,10]. The most reliable of these experimental results are summarized in the first two columns of Table II. Appropriately calculated values[10] of this coefficient, given in the next column of the table, are in good agreement with these experimental results. The calculations indicate that $d \ln \langle r^2 \rangle_0 / dT$ is strongly negative largely because of two types of transitions, to higher energy conformations, which accompany an increase in temperature. In ethylene sequences these conformational transitions correspond to an increase in the population of relatively compact *gauche* states; in isotactic propylene sequences the result is an increase in the number of similarly compact nonhelical conformations[10]. Values of the fraction f_e/f of the elastic force due to changes in intramolecular energy upon network deformation were calculated from the experimental results by means of Equation 6, and are presented in the last column of the table. The origin of the negative values of this quantity are also easily understood from the configurational calculations. Deformation of any elastomer requires an increase in the end-to-end distance of its constituent chains. Since, as already mentioned, high chain extension corresponds to low conformational energy in ethylene–propylene chains, a negative contribution to the elastic force arises from the energy changes accompanying the extension of these network chains[4,10].

Some selected results on both the dimensions and dipole moments of propylene–vinyl chloride and ethylene–vinyl chloride chains are presented here, in part to illustrate the relative sensitivity of these quantities to chemical sequence distribution. Calculated values[20] of $\langle r^2 \rangle_0 / n \ell^2$ and $\langle \mu^2 \rangle / x$ for propylene–vinyl chloride chains of high syndiotacticity are shown in Figure 15. These results, and others calculated for other stereochemical compositions, indicate that $\langle \mu^2 \rangle / x$ is much more sensitive to both chemical composition and chemical sequence distribution than is

TABLE II
Temperature Coefficients of $\langle r^2 \rangle_0$ for Ethylene-Propylene Copolymers[a]

p_2^b	$-10^3 d \ln \langle r^2 \rangle_0 / dT$		$-f_e/f$ (Obs'd.)
	Obs'd.	Calc'd.	
0.32	1.4($\pm$0.6)	1.3	0.45($\pm$0.19)
0.49	1.6($\pm$0.4)	1.5	0.52($\pm$0.13)
0.50	1.4($\pm$0.6)	1.5	0.47($\pm$0.20)
0.50	0.9($\pm$0.2)	1.5	0.30($\pm$0.05)
0.55^c	~2.0	1.7	~0.5

[a] The experimental results are cited and discussed in reference 4.
[b] Mole fraction of propylene units in the copolymer.
[c] Partially chlorinated to prevent crystallization.

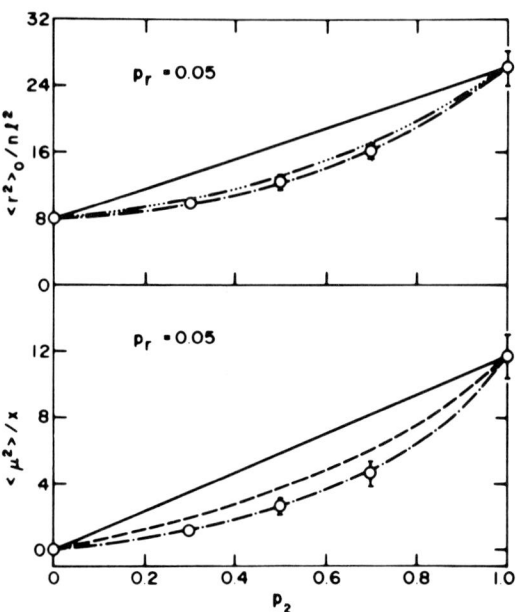

Figure 15. The characteristic ratio and mean-square dipole moment per monomer unit for propylene–vinyl chloride chains of highly syndiotactic structure shown as a function of the mole fraction p_2 of vinyl chloride units in the chain. The illustrative values chosen for $r_1 r_2$ are 0.01 (——— ———), 1.0 (——— · ———), 100.0 (——— · · · · ———), and ∞ (———). (From[20].)

$\langle r^2 \rangle_0 / n \ell^2$, a result of some importance with regard to the possible characterization of such distributions by measurements of configuration-dependent properties. This enhanced sensitivity in the case of $\langle \mu^2 \rangle / x$ is obviously due to the fact that in the calculation of $\langle r^2 \rangle_0 / n \ell^2$, chemically different units differ in conformational energy but have essentially identical values of the quantity being averaged, the skeletal bond vector, as expressed in the skeletal bond coordinate system. On the other hand, in the calculation of $\langle \mu^2 \rangle / x$, *both* the conformational energy and the group dipole moment depend on the chemical nature of the comonomeric unit.

Figures 16 and 17 show values of $\langle r^2 \rangle_0 / n \ell^2$ and $\langle \mu^2 \rangle / x$ plotted as a function of log $r_1 r_2$ at $p_2 = 0.50$ for propylene–vinyl chloride chains[20] of high syndiotacticity and high isotacticity, respectively. The first of these figures is of interest because it shows that $\langle r^2 \rangle_0 / n \ell^2$ and $\langle \mu^2 \rangle / x$ can go through a minimum with increase in $r_1 r_2$, thus complicating the use of experimental values of the properties to characterize chemical sequence distributions in these copolymers. (The dashed lines in this and the following figure pertain to results obtained for a value of the rotational state displacement of 10° instead of 0°. These results are included

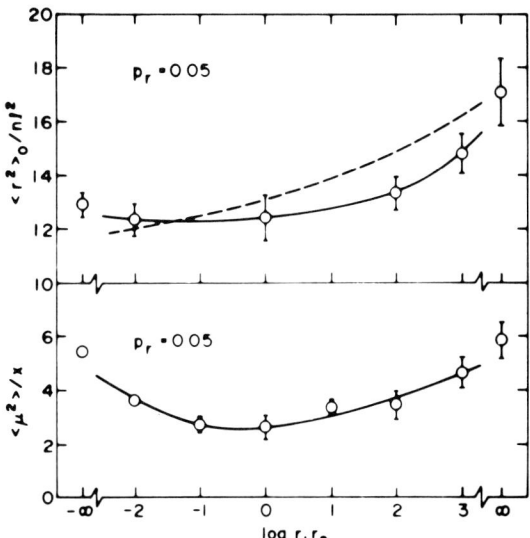

Figure 16. The dependence of $\langle r^2 \rangle_0 / n \ell^2$ and $\langle \mu^2 \rangle / x$ on the reactivity ratio product for highly syndiotactic propylene–vinyl chloride chains of equimolar composition ($p_2 = 0.50$). The dotted lines show results obtained on altering[2] the locations of some of the bond rotational states by an amount $\Delta\phi = 10°$, when these results differ significantly from those calculated for $\Delta\phi = 0°$. (From[20].)

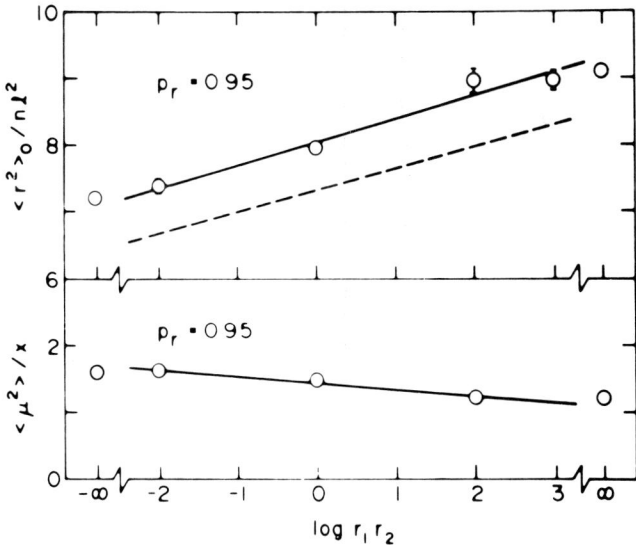

Figure 17. The dependence of $\langle r^2 \rangle_0 / n \ell^2$ and $\langle \mu^2 \rangle / x$ on $r_1 r_2$ for highly isotactic propylene–vinyl chloride chains of equimolar composition; see legend to preceding figure. (From[20].)

361

merely to show that the general conclusions described here are not critically dependent on the precise locations chosen for the bond rotational states.) Figure 17 has been included to show the unexpected result that increase in chemical sequence length in chemical sequence length in highly isotactic propylene–vinyl chloride copolymers should increase $\langle r^2 \rangle_0/n\ell^2$ but *decrease* $\langle \mu^2 \rangle/x$. As has already been stated, vinyl chloride isotactic sequences have much less tendency to form extended 3_1 helical conformations than do the propylene isotactic sequences. Therefore, segregation of the vinyl chloride units by increase in $r_1 r_2$ increases $\langle r^2 \rangle_0/n\ell^2$ because of the corresponding increase in the average length of the propylene sequences, which are in more nearly perfect helical conformations. This increase in the average length of the (nonpolar) propylene sequences does not, of course, increase $\langle \mu^2 \rangle/x$. In fact, as seems reasonable, the corresponding segregation of the vinyl chloride units decreases $\langle \mu^2 \rangle/x$ because of the mutual attenuation occurring among group dipole vectors when they are closely spaced along the chain.

Similarly calculated results[21] for ethylene–vinyl chloride chains support the conclusion that $\langle \mu^2 \rangle/x$ is generally much more sensitive to chemical sequence distribution than is $\langle r^2 \rangle_0/n\ell^2$. Values of both statistical properties for the particular case of ethylene–vinyl chloride chains in which the vinyl chloride units are highly isotactic are shown in Figure 18. These results are included here because of the unusually complex behavior shown in the dependence of $\langle r^2 \rangle_0/n\ell^2$ and $\langle \mu^2 \rangle/x$ on $r_1 r_2$. The origin of the maxima and minima is readily understood by considering in detail the effect of chemical composition on, for example, the characteristic ratio. The addition of vinyl chloride units to a polyethylene chain increases its value of $\langle r^2 \rangle_0/n\ell^2$ because of the stronger preference for extended *trans* states in vinyl chloride units. The addition of ethylene units to a highly isotactic vinyl chloride chain, however, decreases its value of $\langle r^2 \rangle_0/n\ell^2$ because of disruptions in the extended helical conformations. Since the values of the characteristic ratio for the two homopolymers are not very different, the dependence of $\langle r^2 \rangle_0/n\ell^2$ exhibits the pronounced maximum and minimum shown in Figure 18. Similar arguments apply to $\langle \mu^2 \rangle/x$[21].

One general conclusion from these results on chemical copolymers is that block copolymers should have configuration-dependent properties which vary approximately linearly with (molar) chemical composition [9,10,20,21] This simple dependence of such properties on composition obviously results from the fact that, in this context, block copolymers differ from mixtures of homopolymers of the same overall composition only in the relatively infrequent junction points between long sequences of chemically identical units. The fact that such linearity should be observed for this type of copolymer was first pointed out by Stockmayer and coworkers[22] many years ago, and now has been verified in a number of experimental studies. For example, as shown in Figure 19, values of the characteristic ratio of styrene–isoprene block copolymers[23] do show a simple linear dependence on composition. Similarly, the values of the

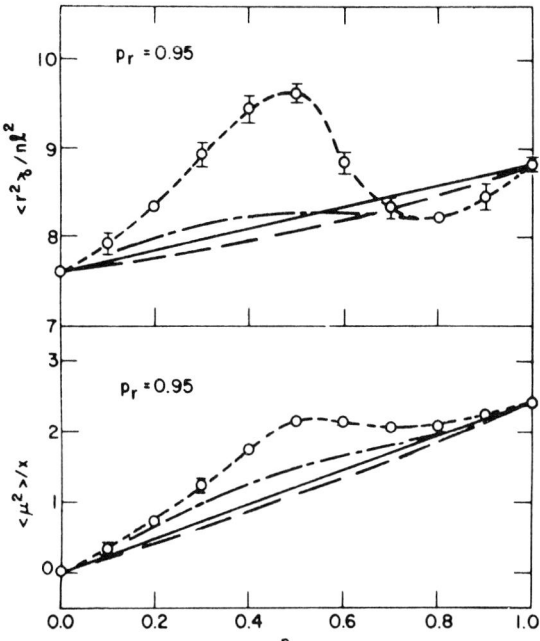

Figure 18. Values of $\langle r^2 \rangle_0 / n\ell^2$ and $\langle \mu^2 \rangle / x$ for ethylene–vinyl chloride copoly-mers having vinyl chloride sequences of highly isotactic structure shown as a function of the mole fraction p_2 of vinyl chloride units in the chain. The illustrative values chosen for $r_1 r_2$ are 0.01 (— — — —), 1.0 (—— — ——), 100.0 (———), and ∞ (——); for purposes of clarity, only the points used to locate the curves corresponding to $r_1 r_2 = 0.01$ are shown. (From [21].)

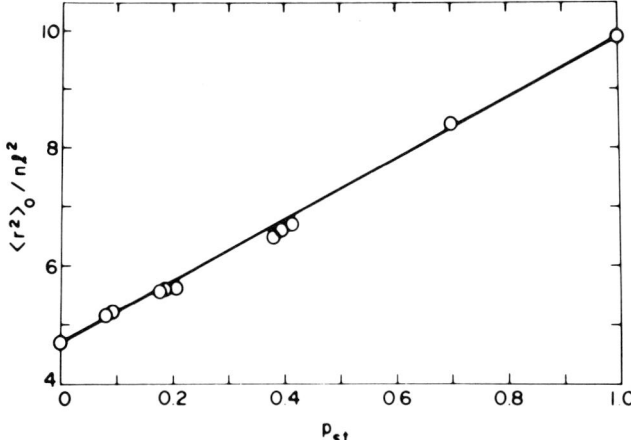

Figure 19. The characteristic ratio of styrene–isoprene block copolymers shown as a function of the mole fraction p_{st} of styrene units in the chain. (Results obtained from [23].)

363

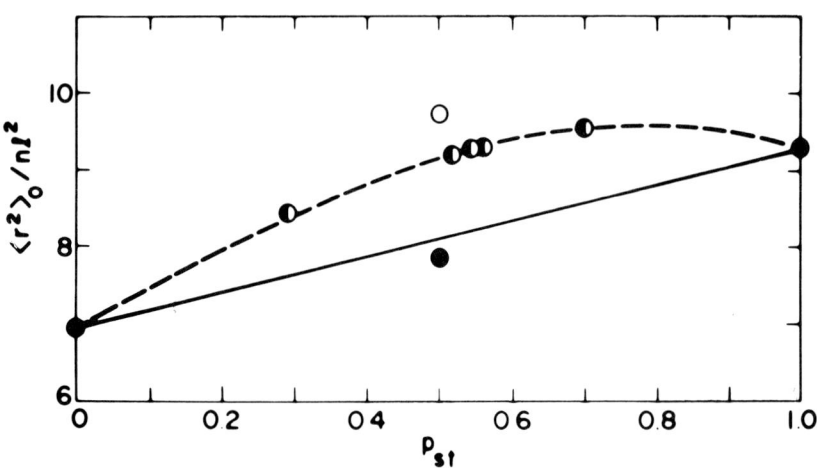

Figure 20. The characteristic ratio for styrene–methyl methacrylate copolymers shown as a function of the mole fraction of styrene units in the chain. The open, half-filled, and filled circles locate results obtained for alternating, ideal, and block copolymers, respectively. (Results obtained from[24].)

characteristic ratio of styrene–methyl methacrylate copolymers[24] show very significant departures from a linear dependence on composition in the case of alternating or ideal copolymers, but show approximate linearity in the case of the block copolymer, as expected.

Summary

The configurational analysis of a variety of representative polymers has been used to demonstrate the fact that the study of the configurations of chain molecules is of the greatest importance with regard to the understanding of their properties. Such interpretation of properties in terms of chain configuration is of particular interest in the case of chemical copolymers since many commercial materials now fall into this category and most polymers of biological interest are, of course, chemically copolymeric.

Acknowledgments

It is a pleasure to acknowledge that much of the work carried out by the author in this area of research has been supported by the National Science Foundation. The author also wishes to thank the members of the Stanford University Department of Chemistry for their hospitality during his 1973-74 sabbatical year there, when this manuscript was written.

References

1. Flory, P.J., "Configurational Statistics of Chain Molecules," in *Proceedings of the Robert A. Welch Foundation Conferences on Chemical Research, X. Polymers.* W.O. Milligan, ed., Houston: Robert A. Welch Foundation (1967), 133-67.

2. Flory, P.J., *Statistical Mechanics of Chain Molecules,* New York: Interscience Publishers, 1969.

3. Flory, P.J., "Configuration-Dependent Properties of Polymer Chains," *Pure Appl. Chem.,* 26 (1971), 309-26.

4. Mark, J.E., "Thermoelastic Properties of Rubberlike Networks and Their Thermodynamic and Molecular Interpretation," *Rubber Chem. Technol.,* 46 (1973), 593-618.

5. Aggarwal, S.L., ed., *Block Polymers,* New York: Plenum Press, 1970.

6. Molau, G.E., ed., *Colloidal and Morphological Behavior of Block and Graft Copolymers,* New York: Plenum Press, 1971.

7. Flory, P.J., Mark, J.E. and Abe, A., "Random-Coil Configurations of Vinyl Polymer Chains. The Influence of Stereoregularity on the Average Dimensions," *J. Amer. Chem. Soc.,* 88 (1966), 639-50.

8. Mark J.E., "Random-Coil Dimensions and Dipole Moments of Vinyl Chloride Chains," *J. Chem. Phys.,* 56 (1972), 451-58.

9. Mark, J.E., "The Use of Dipole Moments to Characterize Chemical Sequence Distributions in Vinyl Copolymers," *J. Amer. Chem. Soc.,* 94 (1972), 6645-50.

10. Mark, J.E., "On the Configurational Statistics of Ethylene-Propylene Copolymers," *J. Chem. Phys.,* 57 (1972), 2541-48.

11. Ham, G. E., *Copolymerization,* New York: Interscience Publishers, 1964.

12. Flory, P.J., *Principles of Polymer Chemistry,* Ithaca, N.Y.: Cornell University Press, 1953.

13. Nagai, K. and Ishikawa, T., "Excluded-Volume Effect on Dipole Moments of Polymer Chains," *Polym. J.,* 2 (1971), 416-21.

14. Liao, S.C. and Mark, J.E., "The Effect of Excluded Volume on the Dipole Moments of Polymer Chains," *J. Chem. Phys.,* 59 (1973), 3825-30.

15. Flory, P.J., Cifferri, A. and Hoeve, C.A.J., "The Thermodynamic Analysis of Thermoelastic Measurements on High Elastic Materials," *J. Polym. Sci.,* 45 (1960), 235-36.

16. Mark, J.E., "Random-Coil Dimensions and Dipole Moments of p-Chlorostyrene Chains," *J. Chem. Phys.,* 56 (1972), 458-64.

17. Flory, P.J. and Williams, A.D., "Stereochemical Equilibrium and Configurational Statistics in Oligomers of Poly(vinyl chloride)," *J. Amer. Chem. Soc.,* 91 (1969), 3118-21.

18. Flory, P.J. and Pickles, C.J., "Stereochemical Equilibrium in 2,4,6-Trichloro-n-heptane with Applications to Poly(vinyl chloride)," *J. Chem. Soc., Faraday Trans. II,* 69 (1973), 632-43.

19. Williams, A.D. and Flory, P.J., "Stereochemical Equilibrium and Configurational Statistics in Polystyrene and Its Oligomers," *J. Amer. Chem. Soc.,* 91 (1969), 3111-18.

20. Mark, J.E., "Random-Coil Dimensions and Dipole Moments of Propylene-Vinyl Chloride Chains," *J. Polym. Sci., Polym. Phys. Ed.,* 11 (1973), 1375-83.

21. Mark, J.E., "Configurational Characteristics of Ethylene-Vinyl Chloride Copolymers," *Polymer*, 14 (1973), 553–57.

22. Stockmayer, W.H., Moore, L.D.Jr., Fixman, M. and Epstein, B.N., "Copolymers in Dilute Solution. I. Preliminary Results for Styrene-Methyl Methacrylate," *J. Polym. Sci.*, 16 (1955), 517–30.

23. Urwin, J.R. and Girolamo, M., "The Configuration of Block Copolymers of Polyisoprene and Polystyrene in Dilute Solution," *Makromol. Chem.*, 160 (1972), 183–94.

24. Kotaka, T., Tanaka, T., Ohnuma, H., Murakami, Y. and Inagaki, H., "Dilute Solution Properties of Styrene-Methyl Methacrylate Copolymers with Different Architecture," *Polym. J.*, 1 (1970), 245–59.

CHARACTERIZATION OF POLYMERS IN BULK

Moderator: Edward A. Collins

B.F. Goodrich Chemical Company

L. MANDELKERN
Florida State University
Tallahassee, Florida

Chapter 13

Morphology of
Semicrystalline Polymers

ABSTRACT

The wide range in properties that can be developed in polymers crystallized from the pure melt and from dilute solution are analyzed in terms of the principal morphological structures that have been revealed by the electron microscope. From this type of coordinated and complementary analysis the molecular nature of the major structural features of semicrystalline polymers can be established. Associated with the lamellalike crystallite formed from dilute solution is a disordered interfacial layer comprising from 10 to 20 percent of the chain units. The concept of a regularly structured, regularly chain-folded interface, or variations thereof, must be rejected in light of the properties of such crystals. A lamellalike crystallite is also the predominant morphological form of bulk crystallized polymers. In this case the crystallites are characterized by a very distorted interfacial region which is many units thick. The crystallites are connected by chain units in nonordered conformations which form interzonal or amorphous regions. All properties of bulk crystallized polymers can be shown to be a reflection of the relative proportion of the three major regions present. The relative amounts of these regions—and, consequently, the resulting properties—can be controlled in a demonstrable way by the molecular weight and the crystallization conditions.

Introduction

Theoretical and experimental investigations have demonstrated rather conclusively that the crystallization of polymers can be formally treated in a classical manner[1,2,3]. The melting–crystallization process has been shown to be a first-order phase transition[2,4] and the crystallization process adheres to the general mathematical formulation for the kinetics

369

of phase changes[4,5]. Although the establishment of these principles marked a major advancement in the understanding of this class of substances, it left many important problems unresolved. As might be intuitively apparent, any realizable crystallization procedure must be conducted under conditions well removed from equilibrium. Consequently, the complete transformation to the crystalline state is rarely if ever attained. A polycrystalline state, with an associated complex morphology, usually develops. Virtually all properties, from mechanical and physical to thermodynamic ones, depend on the morphological structures formed. Therefore, a molecular description of the morphological structures and crystalline forms is of crucial importance in comprehending the underlying basis for the observed properties. There has been a great deal of activity directed toward the observation, by electron microscopy and related techniques, of the structures in the crystalline state. These studies have lead to certain valid general observations. They have, however, also fostered attempts at molecular interpretations. In this chapter we review these microscopic observations and, in addition, concomitantly examine appropriate physical chemical properties. The primary purpose is to see if any general relations exist between morphological forms, molecular structure, and the properties of semicrystalline polymers. For convenience, we divide the subsequent discussion into crystallization from dilute solution and crystallization from the pure melt.

Crystallization from Dilute Solution

General Morphology

When crystallized from dilute solution, long-chain molecules of sufficiently regular structure form the well-known platelets or lamellalike crystallites[6,7,8]. A typical example of such a crystallite is shown in Figure 1. The lateral dimensions of such crystallite are usually of the order of several microns while they are about a hundred angstroms thick. The crystallites are very uniform with respect to thickness, but the magnitude of this dimension depends on the crystallization temperature. Although linear polyethylene has been the most extensively studied polymer, the same crystal habit has been observed in such a wide variety of chains of regular structure as to attest to the universality of this morphological form[9,10]. Selected area electron diffraction studies[6,7,11] have unequivocally established that the chain axes are preferentially oriented perpendicular to the wide faces of the lamella and that the direction of the *a* and *b* crystallographic axes are preserved throughout the platelet structure. Thus these structures have been commonly called "single crystals." Since the platelets are usually no more than the order of a few hundred

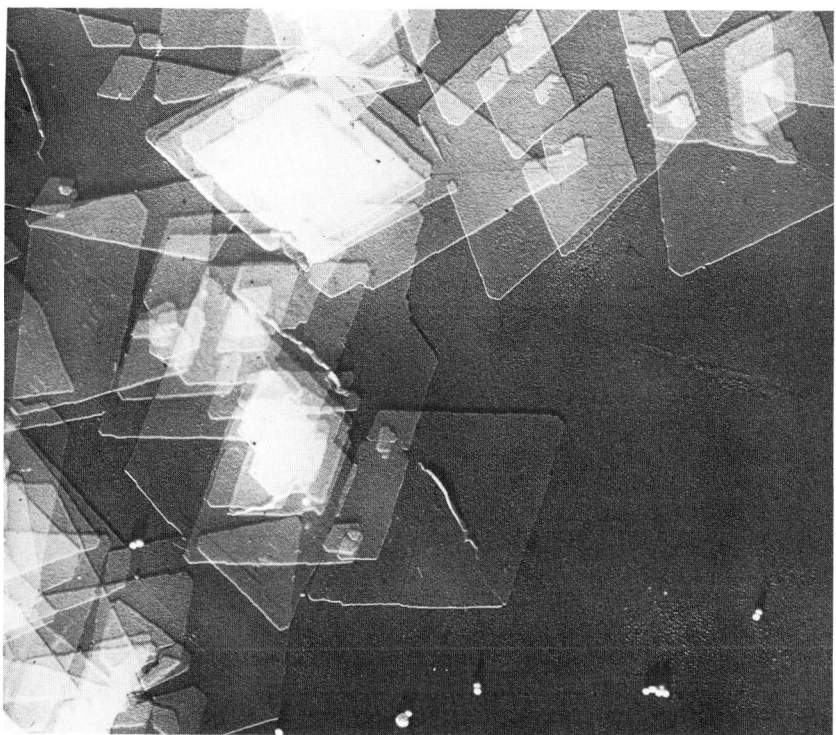

Figure 1.   Typical solution grown crystals of polyethylene.

angstroms thick and the lamella crystal habit is observed for very high molecular weight chains, it is necessary that a given molecule traverse a crystallite many times in order to satisfy the requirements of chain length, crystallite thickness, and chain orientation.

With the above facts well established, the detailed nature of the interfacial structure associated with the basal plane of the lamella crystallites becomes a matter of importance. The nature of the interface is of interest as a structural problem in itself, as well as having strong implications with regard to the morphology and properties of bulk crystallized polymers. The reentrant requirement demands some type of "folded chain," the detailed nature of which is not at all obvious from the examination of electron micrographs. (As the subsequent discussion will make clear, the use of the phraseology "folded-chain" or "fold-period" conveys a certain connotation which is not in accord with the molecular structure. Much confusion and misinterpretation of intent would be alleviated if the use of such phraseology were discontinued.) However, based on the external shape of the platelet crystallites and the apparent smoothness of the basal planes, it was postulated that the chains were regularly folded. Under this assumption, crystallization is complete except for the number of chain elements which are required to make the hairpin connections between adjacent crystalline sequences. In this manner it is proposed that a regularly folded, pleated, smooth 001 interface is formed[6,8,12,13,14]. From this thesis, which was based solely on electron-microscope observations, theories evolved as to how regularly folded crystallites could form. One such theory was based on equilibrium considerations[15–18] while others found their origin in kinetic concepts[19,20,21]. In essence, the latter assumed that regularly folded critical-size nuclei were favored and classical nucleation theory appropriate to monomeric systems was invoked. Despite these diverse attempts at a theoretical rationale, it should be recognized that the present state of electron-microscopic techniques is such that a distinction cannot be made between a regularly structured interface and one that is disordered[10,12]. Thus, despite, a very strong aesthetic appeal, the molecular nature of the interfacial structure, *i.e.,* the disposition of the chain units, cannot be established solely by electron-microscope observations. Other kinds of measurements and deductive reasoning, which take into account the general morphological features, need to be brought to bear on the problem.

A detailed discussion of a variety of properties of solution-formed crystals has been given elsewhere[9,23,24]. Hence, they will be only briefly reviewed here and brought up to date. Regularly folded crystals require a density close to that of the unit cell. For linear polyethylene, this density is 1.00 g/cm^3. Although some confusion had developed, because of experimental inadequacies[25,26,27a] and theoretical misconceptions[26,27b], as to the density of solution-formed crystals, the matter has been resolved in a consistent and satisfying manner. The evidence is now rather overwhelming that the density of polyethylene crystals formed from dilute solution is in the range of 0.96 to slightly greater than 0.97[28,29,30]. These values are significantly less than would be expected from a regularly folded, regularly structured smooth interface. Except for very low molecular weights, there is no dependence on chain length. Even for the very low molecular weights and, as shall be discussed subsequently, even under the very special conditions where dislocation networks are formed, the density values do not indicate the approach to complete crystallinity[31].

The low density values, relative to that of the unit cell, indicates that a substantial amount of disorder is associated with the lamella crystallites. As is most natural, and will in fact be substantiated by a large number of other physical–chemical measurements, the density deficiency must be attributed to a disordered interfacial structure, *i.e.,* the lamellalike crystallites contain an amorphous overlayer[32,33]. By assuming the additivities of the crystalline and disordered portions, the density measurements yield a level of crystallinity of about 80–85 percent. This means that about 15–20 percent of the chain units exist in disordered or nonordered conformations as compared to the only 5 percent that are required for a regular folded structure.

The concept that the density deficiency can be attributed to an amorphous overlayer is supported by the relationship between the density and crystallite thickness shown in Figure 2[33,34,35,36]. The density is inversely proportional to the crystallite thickness and extrapolates to a value of 0.998 for an infinite-size crystal. This is the expected value for the perfectly crystalline material. The functional form observed is easily derived on the basis of an amorphous overlayer coexisting with the ordered crystalline regions.

Measurements of the absolute intensity of the low-angle X-ray scattering of such crystals further support the model described[37,38]. It is found that the fluctuations in electron density are directly related to the macroscopic difference between the crystalline and amorphous states. From these measurements the difference in density between the crystalline and amorphous states is found to be 0.16 g/cm^3 as compared to the accepted value of 0.14. Thus these measurements give strong support to the concept that the disorder, as reflected in the macroscopic density measurements, represents the structure of the surface layer. The intensity of

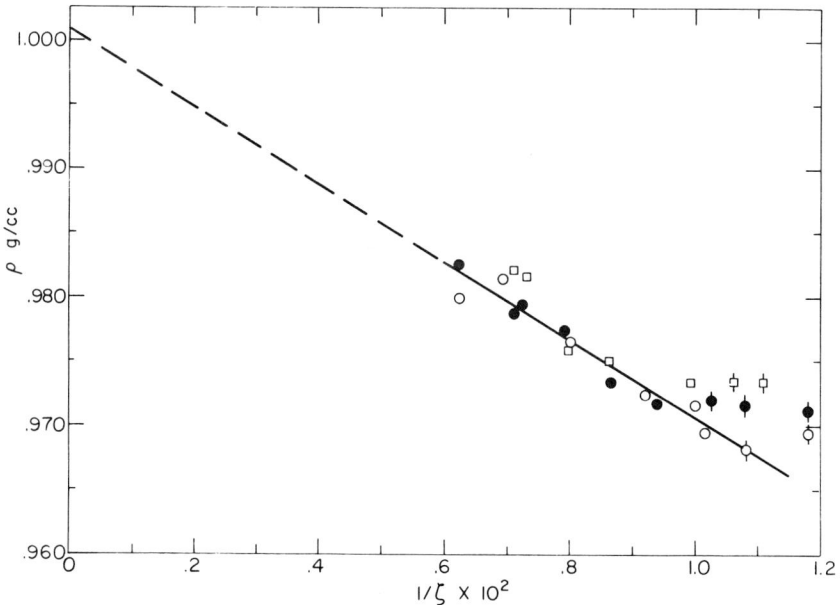

Figure 2. Plot of density as a function of the reciprocal crystallite thickness ζ, in CH_2 units for samples crystallized from different solvents: toluene, $\circ$; n-octane, $\bullet$; n-hexadecene, $\square$. Short vertical lines represent nonisothermal crystallization[36].

scattering cannot be attributed to voids or similar types of defects within the interior of the crystalline structure. (A measured density of 0.96 could, in the absence of any other information, be attributed to the presence of 4 percent voids within the interior of the crystallite[13,39]. This conclusion, however, is merely an exercise in arithmetic. Such a concentration of defects is unprecedented for any condensed system, and this concept is in fact refuted by a large number of physical and direct chemical observations.)

The enthalpy of fusion of solution-formed polyethylene crystals depends only on the crystallite thickness[36,40]. Formally, the data can be explained by either of the two extremes in interfacial structure. The same extrapolated value for the enthalpy of fusion of a macroscopic crystal is found for either model. This extrapolation indicates that internal defects or voids, if they exist at all, make no contribution to the measured enthalpy of fusion. A discrimination between the two models can only be made on the basis of the deduced values of the interfacial enthalpy and entropy[36,40]. These values are much too high to be reconciled with a regularly folded interface, but are consistent with a disordered interfacial structure[46].

Conventional determination of the degree of crystallinity by infrared analysis yields levels of crystallinity in the range of 77–85 percent[41]. These results are thus in very good agreement with the deductions from the density measurements. The extinction coefficient for the 1894 cm^{-1} crystalline band is the same for solution formed crystals and the normal paraffins.

A study of the infrared spectrum of the cyclic hydrocarbon $C_{34}H_{68}$ has shed additional light on the interfacial structure of solution formed crystals[42]. The conformation of this hydrocarbon in the crystalline state is that of a fully collapsed hoop with tight regular folds, characteristic of a regularly folded, adjacent reentry, interfacial structure. Because of the sequence of bond orientations necessary to make this tight fold, an intense absorption band is observed at 1340 cm^{-1} in the crystalline hydrocarbon. This band is barely detectable in the polyethylene-solution crystals. In fact, it is found that the infrared spectra of the *gauche* bands of polyethylene solution crystals are virtually identical to that of the melt of normal hydrocarbons. We can, therefore, conclude that similar liquid or disordered structures exist in both cases.

Crystals formed from mixtures of hydrogenated and deuterated polyethylene display a characteristic splitting in the 720, 730, and 1300 cm^{-1} infrared modes. These results have been interpreted as being caused by a regularly folded interfacial structure; or at least one where the chain reentry is always adjacent[43]. Theoretically, splitting occurs when the mass distribution of the two species within the crystal lattice is nonuniform. This requirement will of course be met by a regularly folded structure. However, this is not a unique explanation. The melting temperatures of the hydrogenated polymer and monomeric analogue are 4–6° greater than the corresponding deuterated samples[44]. Consequently, from simple considerations of phase equilibria and of crystallization kinetics, it follows that the composition of the crystals and the solution from which they are formed cannot be the same. Thus, there must automatically be a nonuniform mass distribution within the crystals. The splitting of these infrared bands cannot therefore given an unequivocal interpretation to the interfacial structure.

It was proposed that because of the influence of successive bonds in *gauche* orientation the relative intensities of the 1303 and 1352 cm^{-1} bands in linear polyethylene would be influenced by the sequential distribution of such units[45]. In particular, major differences in band intensities were expected from a regularly folded interface, which requires five successive bonds in *gauche* orientations, as compared with an irregularly structured interfacial region. However, when detailed studies were conducted[46], it was found that the conclusion of a regularly structured interface is not tenable. In fact, the intensity ratio of the *gauche* bands of solution-formed crystals is identical with bulk-crystallized samples having crystallite thicknesses comparable to the extended chain length[4]. Despite the inadequacy of the basic premise and the original conclusion with regard to polyethylene, the same idea has been carried over to other polymers[47, 48,49]. It can be anticipated that ultimately similar conclusions will be reached with these systems also.

The analysis of the lowest mode Raman vibration of polyethylene chains in oriented mats of solution crystals indicates that there is no coherent crystal lattice extending from one surface to another. The experimental data are in good support of the concept of a less dense amorphous surface layer in the basal planes of the crystals[50].

Wide-angle X-ray diffraction patterns obtained from dilute-solution crystals are essentially the same as those obtained from bulk-crystallized polyethylene with a density corresponding to an amorphous content of 15–20 percent. Originally, the noncrystalline content was attributed to chain folds, branch points, and other gross imperfections within the crystal lattice[51]. They clearly must now be assigned to the amorphous overlayer. This result is consistent with other properties of solution-formed crystals and similar measurements with bulk-crystallized samples.

Compelling evidence for a disordered interfacial structure on the surface of solution-formed crystals comes from the direct observations of glass formation in such systems. Irrespective of theories and arguments that have been put forth with respect to the mechanisms of glass formation, it is universally agreed that it is a property of the disordered or amorphous regions. Extensive dynamic mechanical measurements on mats of solution-formed crystals for ten different polymers[52,53] have demonstrated that loss peaks are found at temperatures which correspond to the glass temperatures of each polymer[54]. The glass temperature being established, of course, by independent methods. In a different kind of experiment, Fischer and collaborators[55,56] have studied the temperature-dependence of the absolute intensity of the low-angle-scattering maximum for solution crystals of polybutene-1 and linear polyethylene. This experiment is an indirect measure of the expansion coefficient of the sample. Abrupt discontinuities are found at temperatures corresponding to the glass temperature. The very definitive display of glass formation for solution crystals cannot be reconciled with any kind of regularly structured interface. If this were so, then regularly folded chains would also represent the structure of completely amorphous polymers.

A refined analysis of broad-line nmr spectra require the contribution of three different types of structures and thus three different types of motion to the observed spectra. These can be considered to be contributions from the crystalline regions, interfacial regions, and interzonal or amorphous regions. For a particular sample of crystals formed in dilute solution, Bergmann and Nawotki[57] found it to be only 89 percent crystalline. The remaining contribution to the spectra came primarily from the motions of the interfacial region.

When solution crystals are placed in contact with liquids at room temperature, the magnitude of the low-angle-diffraction maxima increases. This process is reversible, for upon removal of the solvent the original X-ray pattern is regained. The increase in spacing depends on the thermodynamic interaction parameter between the monomeric liquid and the disordered chain units, as determined by classical thermodynamic methods. Since the crystal structure and lattice parameters remain intact, the spacing change must be attributed to the reversible deformation of the interfacial layer. The magnitude of the deformation requires the chain units to be in nonordered conformations. Similar conclusions are reached from studies of solvent imbibition on the broad-line nmr[60,61] and from surface-decorating techniques[62]. Although the chain length and crystallization temperature influence the magnitude of the swelling[58], it becomes independent of molecular weight in the range 10^5-10^6[59]. Before this asymptotic limit is reached, the variation in the amount of swelling must reflect subtle differences in interfacial structure, which are not as yet understood and which have not manifested themselves in the other properties studied.

Chemical methods have also proven to be useful in elucidating the interfacial structure. Palmer and Cobbold[63] have shown that fuming nitric acid selectively oxidizes portions of bulk-crystallized polyethylene. The oxidative process preferentially degrades the noncrystalline amorphous material. The end-product, or debris, resulting from this degradative process is a collection of lamellalike crystals. The same technique has been applied to polyethylene grown from dilute solution[64–68]. For these crystals, the scission takes place in two stages. During the first stage, the interfacial layers are selectively removed and the internal crystalline core remains as a residue. This process is reflected in weight loss, molecular weight decrease, increases in the density and enthalpy of fusion, and the decrease of the characteristic low-angle X-ray diffraction maxima to a limiting value. All of these properties change very precipitously with initial exposure to the fuming nitric acid. About 15–25 percent of the material is removed during this stage. This is about the same amount of noncrystalline material in the interfacial layer that has been deduced from other measurements. However, after a well-defined time interval of exposure, the changes in properties occur at a much more retarded rate. The same conclusion is reached from a comparison of the broad-line spectra of oxidized and nonoxidized solution crystals. The onset of segmental motion that is detected at −100°C in the nonoxidized solution crystals, which correspond to the motion of disordered chain units, disappear for the selectively oxidized crystals[65]. It should be obvious that because of the drastic chemical reactions that occur with reagents such as fuming nitric acid, conclusions reached with respect to the molecular structure should be severely tempered and restricted to very general considerations. Attempts to deduce more quantitative information, as, for example, the molecular weight distribution within the debris, are fraught with great difficulties[67,68].

The reaction of bromine[69] and of ozone[70] with the vinyl end-groups of unfractionated linear polyethylene indicates that at least 90 percent of the terminal units are located within the interfacial region. This distribution of terminal units precludes a regularly folded interfacial structure. It is consistent with the type of interfacial structure that has been deduced from other measurements. The absence of end-groups within the crystal lattice rules out the presence of the particular type of lattice vacancy, or void, that has been very popular and has been invoked to explain properties[39].

Woodward and collaborators have studied the properties of solution-formed crystals of poly(trans 1-4 butadiene) crystals[71,72,73]. Advantage was taken of the epoxidation reaction of the double bonds. The fraction of double bonds reacting reaches an asymptotic limit relatively rapidly. Electron microscopy and selected-area diffraction indicate that both the basic morphology and crystal structure are not changed because of the reaction. These results strongly suggest that only the double bonds located on the crystal surface participate in the chemical reaction[73]. Depending on the crystallizing solvent and temperature, the fraction of noncrystalline material determined by this method ranges from 15 to 25 percent. For high temperatures of crystallization, agreement is found between this chemical method and other physical methods. For low crystallization temperatures, the measurements indicate the possibility of the presence of internal amorphous regions. More quantitative analysis is complicated in this situation by the two different polymorphic forms.

Summary

The analysis of this wide array of properties leads to the very strong conclusion that crystals formed from dilute solution do not possess a regularly structured interface. All of the properties studied demand that the interfacial region be formed by chain units in nonordered conformations. For molecular weights greater than about 15,000, the crystallite thickness is independent of molecular weight and depends only on the crystalliza- tion temperature[31,59,74,75]. Since all properties are independent of molecular weight, even for very high molecular weights, there can be no meaningful contributions to the interfacial structure of "cilia," "loose- loops," or other imaginary structures that have been postulated[58,76]. Although a very refined and detailed structure of the interfacial region cannot be given at present, its general characteristics are consistent with well-established principles and properties of chain molecules.

In retrospect, the major and perhaps the only reason that chains were thought to be regularly folded arose from an aesthetically pleasing and imaginative effort to explain the external shapes of the platelets. As has been pointed out, there is a very large scientific gap in attempting to relate the external habit of crystals to the detailed molecular chain structure. Under particular and specified crystallization conditions truncated lamella displaying distinct sectors are observed[12,76]. In these cases, it has been supposed that the crystals grow as hollow pyramidal structures, rather than the flat platelets that finally evolve. These latter structures are pre- sumed to result from drying. This type of crystal habit has been inter- preted in terms of regularly folded chains with well-defined crystalline domains[13]. Although this may be the simplest interpretation, it is not necessarily a unique interpretation. There is no reason that such struc- tures need be incompatible with a disordered interface.

When the density deficiency from that of the unit cell was accepted, it was realized that a regularly folded chain of uniform periodicity was no longer a tenable structure. It was then postulated that while the chains were still sharply folded the lengths of the crystalline sequences were not uniform[76,77]. Although it is conceivable that this model could be made to explain the measured densities, the distribution of crystalline sequence lengths must be fairly sharp in order to account for the many orders of low-angle X-ray diffraction that are observed[6]. Any model that re- quires as a major feature a sharp, hairpinlike fold is an untenable one, since it cannot account for the high concentration of *gauche* units, glass formation, segmental mobility, and interfacial deformation (among other properties) that need to be accounted for.

Interfacial Dislocation Networks

With the discussion of the properties and the structures deduced there- from we can now examine some detailed aspects of the morphology of solution-formed crystals. When two lamella crystallites are superposed upon one another forming bilayer crystals which are slightly rotated with respect to one another, the usual Moire fringe pattern is observed[76,78]. This is a common phenomenon which is found over a wide range of

Figure 3. Electron micrograph of polyethylene solution crystals illustrating interfacial dislocation networks. $M_w = 1142; M_n = 1055[82]$.

molecular weights and crystallite sizes. Unfortunately, these observations do not yield any unique or detailed information with respect to the interfacial structure. However, when there is an interaction between the two layers, with sufficient penetration so that there is a strong tendency to maintain the lattice continuity over a good portion of the overlapping area, a set of dislocation networks are developed, as is shown in Figure 3. Such dislocations require specific crystallographic interaction between the layers. This requirement has lead to the supposition that a crystallographically regular interface is also necessary[76,79,80]. Therefore, it is not surprising that it was thought at one time that interfacial dislocation phenomena represent the most direct evidence in existence for fold regularity. However, careful delineation of the conditions under which interfacial dislocation networks are observed indicate that this far-sweeping statement is incorrect. As is shown in the following discussion, the conditions under which such structures can be observed are extremely restricted. It is very doubtful, in fact, whether they involve any type of chain-folded interface.

Interfacial dislocation loops between a set of bilayer crystallites was first observed by Holland and Lindenmeyer[79], utilizing a polyethylene sample whose molecular weight was about 10,000. In a later study by Sadler and Keller[78] it was concluded that only a sample having a molecular weight lower than about 3,000 is capable of displaying interfacial dislocation networks. When such dislocations are observed in samples of apparently higher molecular weight, it is caused by the lower molecular-weight component which fractionates during crystallization. The molecular-weight fractionation of the lower molecular-weight species during dilute-solution crystallization is now well established[31,81]. This process is, in fact, a very sensitive method of fractionation so that very sharp fractions must be used to establish the conditions under which dislocation networks will be observed. In a recent study utilizing very sharp fractions, it was confirmed that in order to observe this type of dislocation the molecular weight must be less than 3,000[82]. Higher molecular-weight fractions, which also yield lamellalike crystallites, do not display dislocation networks. However, it is found that the stipulation of molecular weight is only a necessary condition; it is not a sufficient one. Within the molecular-weight requirement, interfacial dislocation networks are only observed when the crystallization conditions are such as to yield particular relations between the crystallite thickness and the extended chain length. When the crystallite thickness is comparable to the extended chain length, interfacial dislocation networks are observed. (This result would appear to be the main reason for the molecular weight restriction. For molecular weights of 3,000 or greater, kinetic difficulties preclude the development of such "extended-chain" crystals from dilute solution[82].) In this situation, the interface is composed of chain ends and disordered units which obviously must still allow for the appropriate crystallographic contact and continuity. Except for one special case, all other crystallization conditions which yield crystallite sizes less than the extended chain length do not yield interfacial dislocation networks despite the fact that the molecular weights are less than 3,000. The special case is for a sample $M_w = 1,756$; $M_n = 1,586$ when crystallized at room temperature. The thickness in this instance is approximately half of the extended molecular length and dislocation networks are observed. However, a sample $M_w = 3,140$; $M_n = 2,859$ crystallized so that a similar relation exists between crystallite thickness and extended chain length does not display the dislocations. Thus, there is only one case, besides those where "extended-chain" crystals are formed, where dislocation networks are observed. Irrespective of the detailed mechanism for the formation of interfacial dislocation networks, they clearly represent a very special and unique situation restricted to very low molecular-weight chains under particular crystallization conditions. Their relation to any type of folded chain structure is quite nebulous. It has also been reported that sectorization exists in crystals showing these dislocations[76]. If this observation is sustained, then sectorization does not require a folded chain structure.

Annealing

When crystals prepared from dilute solution are separated from the mother liquor, major changes are observed in morphological and thermodynamic properties when the samples are heated above a certain temperature. Although crystals formed from linear polyethylene have been most extensively studied[33,51,83–85] similar behavior has been observed in several other polymers[10]. The most dramatic change in properties that takes place is illustrated in Figure 4. In this figure, the low-angle-diffraction maximum is plotted against the heating or annealing temperature. The long spacing remains invariant with increasing temperature until a critical temperature range is reached. At temperatures above this range, the diffraction maxima increases very rapidly and as much as a three- to fourfold increase in thickness is observed. An electron-micrograph observation of a sample which has undergone this kind of treatment is shown in Figure 5. The orientation of the chain axis is presumably still maintained. To quote the original observers, "many holes develop in single crystal regions which were originally one lamella thick and the lamella increase in thickness. The polymer required for the increase in thickness is drawn from localized regions of the same lamella"[51].

Two quite different explanations have been offered for these observations. Starting with the assumption that a regularly folded interfacial structure is initially present, it is further argued that this structure is maintained during the heating (annealing) process. The polymer molecule would therefore have to possess an extraordinary high and coordinated mobility. Undaunted by this extraordinary characteristic that has been endowed to chains situated within the crystal lattice, various theories were advanced as to how this thickening process occurs[14,86–88]. These range from the propagation of point defects through the lattice to collective longitudinal motion involving the whole sequence of crystalline units. A corollary of this proposition soon developed in that it was argued that thickening of solution-formed crystals takes place during isothermal crystallization. (This argument was invoked, without any experimental evidence in hand, because the crystallite thickness was assumed to be controlled by a two-dimensional or Gibbs-type nucleation process[20]. Since mature crystallites of these dimensions are unstable at temperatures infinitesmally above the crystallization temperature, either the interfacial energy must decrease or the crystallite size get larger. Since it is highly unlikely that the interfacial free energy will decrease as mature crystallites develop from nuclei, this assumption with regard to size control requires growth or thickening at a constant crystallization temperature while an interface of regularly folded chains is maintained in order to allow for thermodynamically stable crystallites to be formed.) Alternatively, the very simple concept has been put forth that the observations are a con-

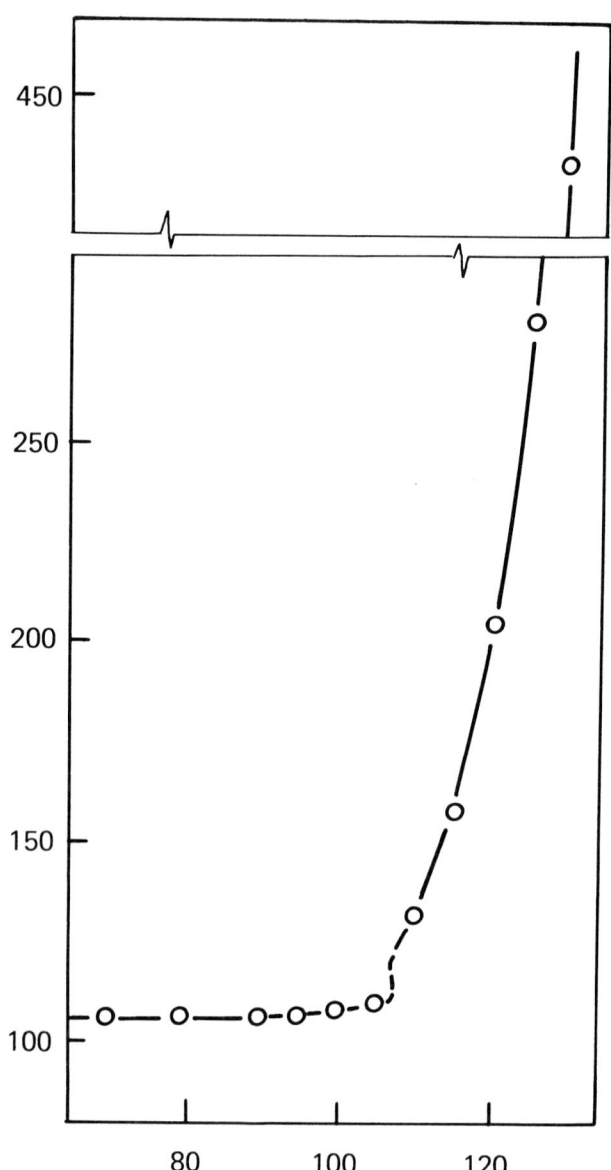

Figure 4.   Plot of long-period of polyethylene solution crystals as a function of annealing temperature. Annealing time 24 hrs[33].

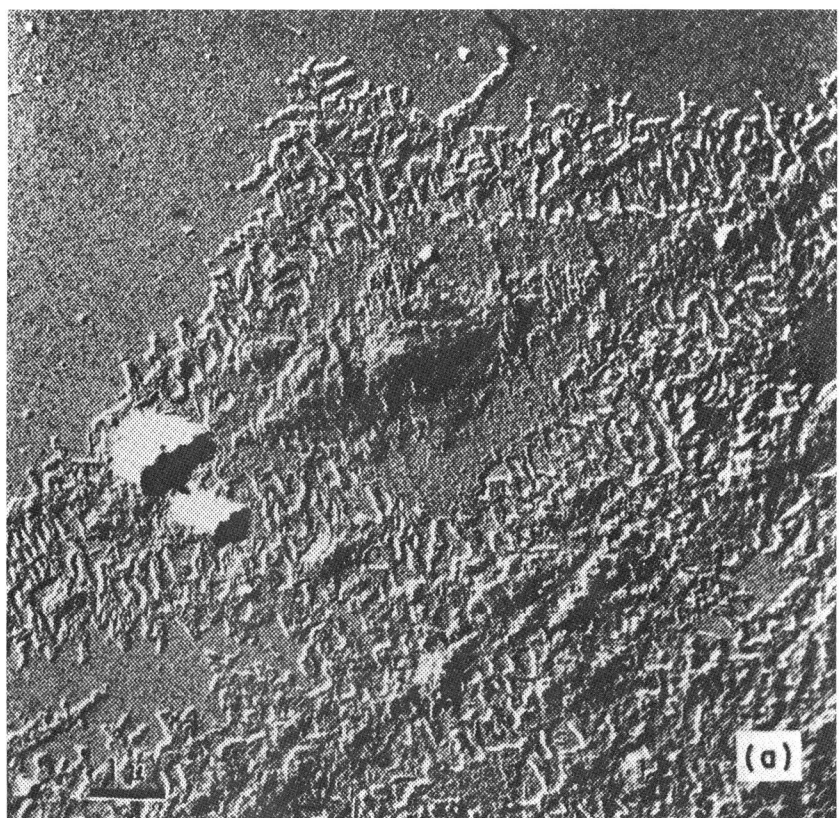

Figure 5. Electron micrograph of polyethylene solution crystals annealed at 125°C for 30 minutes[51].

sequence of melting, or partial melting, followed by recrystallization from the molten state[28,33,89,90].

We first examine the postulate that during isothermal crystallization, the low-angle spacing and, thus, the crystallite thickness increase with time. Experiments demonstrate[91,92] that for a wide range of linear polyethylene fractions, crystallized over a wide range of temperatures, the low-angle maximum and the dissolution temperature remain invariant with time. These results show conclusively that there is no change in the crystallite thickness or thermodynamic stability from the first observation of precipitate to times well beyond the completion of the process. Therefore, a Gibbs type of monomolecular nucleus (two-dimensional), or minor variations thereof cannot be size-controlling. Theories[13,14,20] that predict thickening or growth in the chain direction during isothermal crystallization from dilute solution are not in accord with experimental facts.

A critical examination of heating or annealing subsequent to crystallization indicates that the surprise and wonderment expressed at the exceptionally high mobility of ordered chains[51] was warranted, since this mechanism does not in fact represent the actual process. The most rudimentary considerations of this phenomenon, taking cognizance of what is known about the melting and crystallization kinetics of linear polyethylene indicates that the temperatures at which changes take place is close to the estimated melting temperature of the solution-crystallized sample. At these temperatures the recrystallization rates from the molten state are extremely rapid[89]. Therefore, the concept that the changes in properties on annealing are due to melting, or partial melting, followed by recrystallization should be given serious consideration. This concept is strongly supported by direct experimental observations. Measurements by Fischer and Schmidt[33] showed that during annealing, at appropriate temperatures, the changes in density, birefringence, and wide-angle X-ray diffraction clearly identified partial melting and recrystallization as being the process. Calorimetric studies of the fusion of such crystals also support this conclusion[90].

Detailed experiments of the type illustrated in Figure 4 show that the temperature above which major changes in the low-angle spacing take place can be defined to about $\pm 1°$[85]. Up to two degrees below this temperature, the X-ray patterns remain very sharp. At this point the patterns become more diffuse, although the original spacing is still maintained. Upon heating above this critical and defined temperature, the patterns become very diffuse and the spacings increase in magnitude. For sufficiently high annealing temperature, the set of maxima in the diffraction pattern are completely lost. The changes in the nature of the X-ray pattern is what is expected for partial melting and recrystallization. A straightforward thermodynamic analysis of the variation of this critical temperature with the inital crystallite thickness shows that this temperature can be identified with the melting temperature of the solution-formed crystals. The direct observation of the melting temperature by light-microscopy, utilizing rapid heating rates to the avoid structural reorganization, is in very good agreement with the critical temperature determined from the heating (annealing) experiments[93].

When annealed samples are allowed to imbibe decalin at room temperature, there is a small decrease in the amount of swelling for samples which were previously heated just below the critical temperature. After heating above this temperature, however, and concomitant with the increase in long spacing, the swelling becomes negligible[85]. The swelling behavior of the annealed samples is consistent with the crystals formed from dilute solution being transformed to a morphology more characteristic of crystallization from the melt. More detailed studies by Nagai and Kajikawa[94] have indicated that upon annealing the orientation of the crystallographic axis has become more random.

The premise that the annealing phenomenon, with the attendant changes in properties, is a consequence of fusion, or partial fusion, and subsequent recrystallization from the melt is borne out by the above analysis. This is a very natural explanation consistent with all other chain properties. It is not necessary that ordered chain units within the interior of crystals be endowed with any extraordinary mobility. The mobility and motion of the chain units are merely natural consequences of the fusion and recrystallization processes.

Crystallization from Melt

Introduction

When homopolymers are crystallized from the pure melt, lamellalike crystallites are also found to be the major morphological entity. Because of the common morphological forms, a direct connection was made between such crystallites and crystals formed in dilute solution [6,8,13,14]. Moreover, since it was also assumed that an interface of regularly folded chains was automatically associated with lamellae, such an interface was also assigned to crystallites formed in the melt. Consequently, all chain units in a crystalline homopolymer were assigned to either the interior of crystallites or to the interface. Chain units connecting crystallites were rare and, if they existed at all, were present in an ordered conformation. Deviations in properties from that expected from a macroscopic crystal were widely known and accepted. Contributions from the interface, as well as from defects within the interior of the crystallites, were postulated to account for the observed properties. A crystal-defect model was thus widely publicized to represent the crystalline state of polymers. It was proposed that a crystalline polymer is best viewed as consisting of disordered material or defects which are embedded within the crystalline matrix[39]. The alternate view that chain units in nonordered conformations connect crystallites, *i.e.*, so-called amorphous regions, was by necessity discarded since the assumed interfacial structure could not accommodate such units. However, the presence of disordered chain units is not incompatible with lamellalike crystallites. It is only the additional requirement of a regularly folded interfacial structure that ruled out the existence of disordered regions and consequently their contribution to properties. The question as to whether such regions are present, as well as other molecular features of the structure of crystalline polymers, is of prime importance for many reasons. Not only is a complex problem in morphology involved, but all of the properties of such systems are governed by the morphology and molecular disposition of the chain units. In an effort to help establish the molecular nature of the structure of crystalline homopolymers, and its relations to properties, we shall examine in some detail the major morphological features and their associated properties.

Crystallites

The presence of lamellalike crystallities in homopolymers can be inferred from a variety of experimental observations such as low-angle X-ray diffraction[38,89,95–98] and the examination of the internal structure of spherulites[99]. They can also be directly observed by electron-microscopic examination of thin films and replicas of fracture surfaces[6,8,100–103]. Microscopic examination of the residue or debris of preferentially oxidized crystalline samples of polyethylene[63,104] and polypropylene [105] also demonstrate lamella crystallites.

The early electron-microscope studies of thin films and fracture surfaces of bulk-crystallized polymers revealed very thin, nondescript lamella such as is illustrated in Figure 6 for linear polyethylene[101] The thickness of these types of lamella is only of the order of 100 Å. In retrospect, it was perhaps unfortunate that these kinds of micrographs were the first

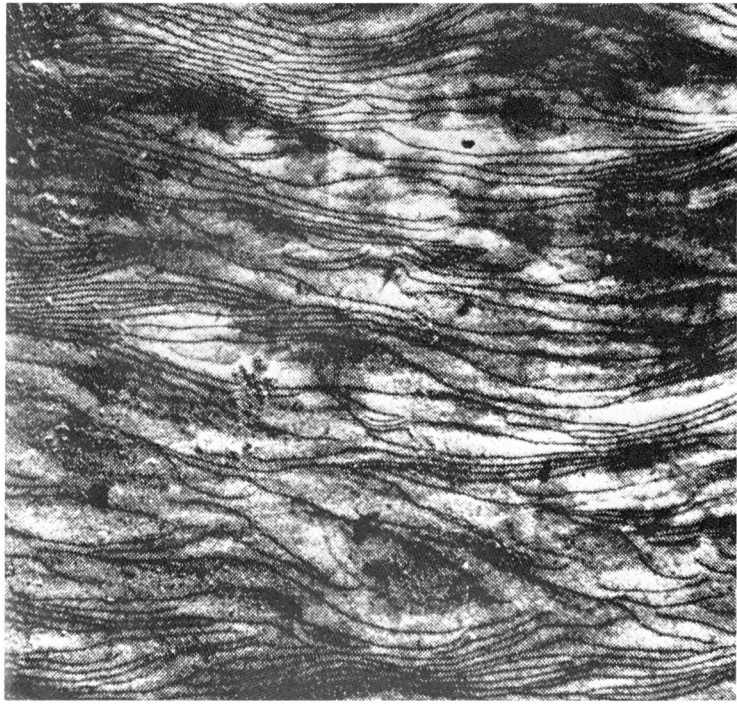

Figure 6.   Electron micrograph of surface film of linear polyethylene[101].

observed. Since the crystallite thicknesses were comparable to those observed in dilute-solution crystals, it is not difficult to see how the connection between the two was made. A systematic study of the character and thickness of the lamella as a function of molecular weight and crystallization conditions turns out to be quite revealing. It renders completely inadequate and incorrect this initial interpretation.

Figure 7. Electron micrograph of replica of fracture surface of linear polyethylene. M_n = 2890[102].

Electron micrographs of fracture surfaces of linear polyethylene fractions, crystallized under controlled conditions, and covering the molecular weight range from 2.9 × 10³ to 5.7 × 10⁵ have been reported[102,103]. For molecular weights of 56,000 or less, fracture can be easily accomplished at room temperature. Fractures of higher molecular-weight samples can only be accomplished at liquid-nitrogen temperature. For very high molecular weights, fracture studies are not available. However, in the debris, after selective oxidation, for samples in the several-million-range, lamella crystallites also predominate[106]. The drastic chemical treatment in this case obviates any detailed analysis of the lamella characteristics.

Figure 8. Electron micrograph of replica of fracture surface of linear polyethylene. M_n = 10,000[102].

Over the molecular weight range 2.9×10^3 to 5.7×10^5, a striated, banded, lamella-type crystallite is the predominant structural feature of linear polyethylene. Typical examples for this polymer are shown in Figures 7 to 10. Similar micrographs have been observed from fracture surfaces of slowly cooled polytetrafluoroethylene[107] as well as for a wide molecular-weight-range of polyethylene oxide samples[108]. As one examines the micrographs, without concern at the moment for the crystallite thickness, it is apparent that the predominant features are independent of molecular weight. Figure 11 represents a fracture surface for a very polydisperse unfractionated linear polyethylene sample (Marlex-50). The general character of this micrograph is very similar to the results obtained with the molecular weight fractions. For the high-molecular-weight samples ($>$ 200,000) much thinner lamella similar to those shown in Figure 6 are also found. These result from crystallization upon cooling subsequent to isothermal crystallization at an elevated temperature. The implications of this type of lamella will be discussed after we have analyzed the crystallite thickness as a function of molecular weight. The important conclusion that can be made at this point is that the character of the crystallites are independent of molecular weight and molecular weight distribution.

Figure 9. Electron micrograph of replica of fracture surface of linear polyethylene. M_n = 47,000[103].

Figure 10. Electron micrograph of replica of fracture surface of linear polyethylene. M_n = 200,000[103].

The average thickness of the lamella can be determined from the electron micrographs. The available data for high temperatures of isothermal crystallization are plotted in Figure 12[102,103]. For present purposes only the thicker, striated crystallites are counted. Here the crystallite thickness ζ, given in terms of the number of CH_2 units, is plotted against the degree of polymerization, x. Both the average size and the size range is given for each molecular weight. It is readily apparent that these quantities are very dependent on molecular weight. Initially, there is a very large increase in ζ with increasing chain length; at about x = 1,000 the increase in thickness becomes much more protracted and an asymptotic limit of about 1,000–1,200 CH_2 units is reached at a molecular weight of about 200,000. This size range, as determined by electron microscopy, is consistent with the low-angle X-ray maximum of about 1000 Å determined by Pollack *et al.*[109]. The concept that for bulk crystallized polmers the lamella thickness is only the order of 100 Å (as in the dilute solution case) is clearly a very misleading one which can lead to a great deal of misinterpretation.

The relation between ζ and x is of importance. For $x \leq 900$ the average value of ζ is very close to x. Such crystallites could be termed

Figure 11. Electron micrograph of replica of fracture surface of unfractionated linear polyethylene (Marlex-50)[102].

"extended chain crystals." However, it should be clearly understood that they are not to be thought of as molecular crystals. Molecular crystals of chain molecules requires that the terminal units be paired in the crystal lattice[110]. This condition requires that each chain be of precisely the same length; a condition which cannot be satisfied by any real polymer system. As the molecular weight increases from 12,000 to 56,000 the average lamella thickness increases from about 600 to 900 units. The ratio of the average crystallite thickness to extended chain length decreases from 0.7 to 0.25 in this molecular weight range. This ratio still represents a significant portion of the chain units being incorporated into the crystallite. For higher molecular weights, the crystallite thickness increases only slightly and represents an exceedingly small portion of the extended chain length. As will be discussed below, the quantity ζ/x is a very important determinant of the properties of crystalline polymers.

Despite the large variation in both the absolute and relative crystallite thicknesses that are observed with molecular weight, the general appear-

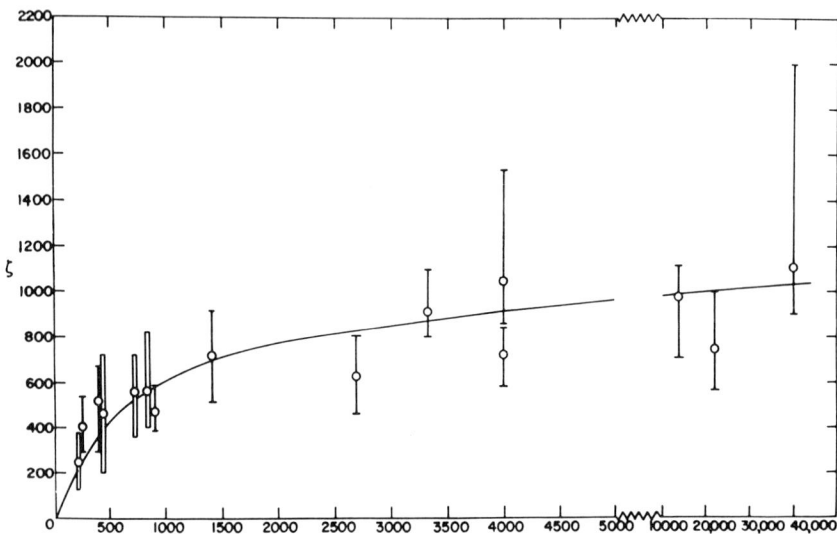

Figure 12. Plot of crystallite size ζ, in CH_2 units, against chain length x for molecular weight fractions of polyethylene crystallized at $130°C$ and cooled to room temperature [103].

ance of the lamella remain the same. Consequently, the observation of lamella structures cannot be interpreted by themselves as reflecting any kind of folded interfacial structure, or one of regularly folded chains in particular. Obviously, this cannot be the case when ζ/x is the order of unity. By the same token, the results for the higher molecular weight fractions show that the observation of banded or striated lamella cannot be taken as evidence for the formation of extended chains crystals. In unfractionated samples the fact that both striated lamella and thinner ones, where the striations are apparently absent, are found cannot, therefore, be interpreted as demonstrating molecular weight segregation, or fractionation, during crystallization from the melt[13,102,111]. As is seen in Figure 10, both types of lamella, if a distinction really needs to be made, are found in high-molecular-weight fractions. Striated crystallites are observed over the complete molecular weight range. Crystallization upon cooling, which is accentuated in the higher molecular weight range, naturally leads to the distribution of crystallite sizes that are observed. Therefore, other kinds of evidence have to be developed to support the contention that molecular weight fractionation takes place during crystallization in the bulk. (In very marked contrast, when crystallization is conducted from dilute solution extensive molecular weight fractionation has been demonstrated, particularly for molecular weight less than about 20,000 [78,81,82]. For bulk crystallization, as the discussion in the text indicates, there is no evidence for molecular weight fractionation. However, the possibility of fractionation of the extremely low molecular species still remains to be explored.)

It is pertinent to consider in this context some of the interpretations that have been given to multiple peaks commonly observed in differential calorimetry. These have on occasion been very casually attributed to "extended chain crystals" and "folded chain crystals" or to molecular weight segregation[102,111,112] Studies of homogeneous mixtures of molecular weight fractions show that the fusion process subsequent to isothermal crystallization, with no cooling, displays only one melting peak[113]. The fusion of mixtures displays multiple peaks only when the samples are first cooled after crystallization at an elevated temperature. This behavior is also found with molecular weight fractions[114] and can be shown to be the result of the fusion of the smaller size crystallites formed on cooling.

Interfacial Free Energy

From the measured crystallite thickness and melting temperature, it is a straightforward matter to calculate the interfacial free energy associated with the basal plane of the lamella crystallites. The quantity is obtained from the relation (1)[103,115].

$$\frac{1}{T_m^\zeta} - \frac{1}{T_m^0} = \frac{R}{\Delta H_u}\left[\frac{2\sigma_{ec}}{RT_m^\zeta} - \frac{1}{\zeta}\ln\left(\frac{x - \zeta + 1}{x}\right)\right] \tag{1}$$

Here T_m^ζ is the melting temperature of a crystallite of thickness ζ; T_m^0 is the equilibrium melting temperature for an infinite molecular weight chain and ΔH_u is the enthalpy of fusion per repeating unit. The interfacial free energy associated with the actual crystallites formed is σ_{ec}. This quantity should not be confused or identified with the corresponding equilibrium quantity or the one involved in nucleation. For large x, Equation (1) reduces to the classical result

$$\frac{1}{T_m^\zeta} - \frac{1}{T_m^0} = \frac{R}{\Delta H_u}\frac{2\sigma_{ec}}{RT_m^\zeta} \tag{2}$$

For $x \geq 2,000$, the second term on the right of Equation (1) becomes negligible and Equation (2) is more than adequate. Two independent sets of data for linear polyethylene are available from which σ_{ec} can be determined. In one of these, the crystallite thickness was determined by electron microscopy and the melting temperature by dilatometry[103]. In the other set of data, the crystallite thickness and melting temperatures were determined from the low-angle diffraction maximum and the temperature of its disappearance[96]. As is seen in Figure 13, both sets of data yield the same results. Here the interfacial free energy, in units of cal/mol of sequences emerging from the basal plane, is plotted against the molecular weight. For molecular weights greater than 10^5 the interfacial free energy

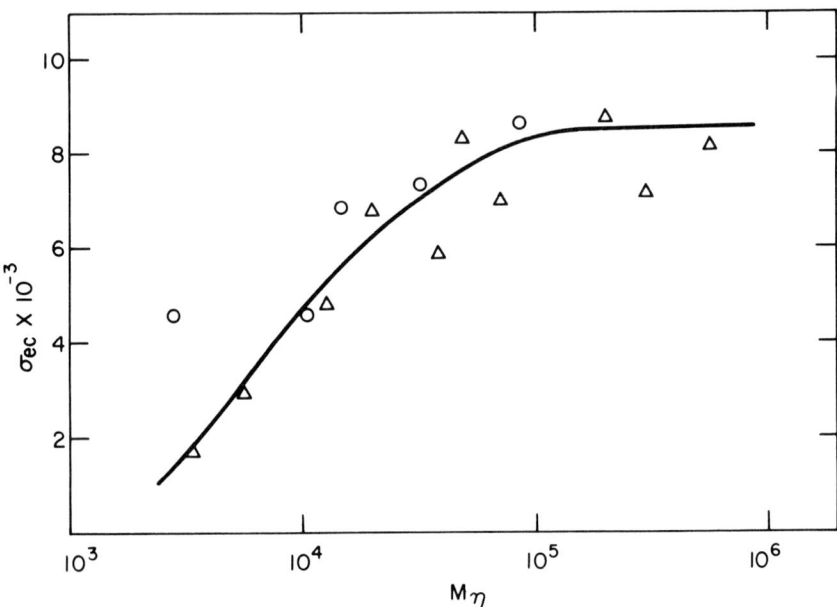

Figure 13. Plot of interfacial free energy σ_{ec} (cal/mole) against molecular weight for bulk crystallized linear polyethylene. Δ, electron microscopy from [103]; 0, low-angle X-ray diffraction from [96].

is about 8,000 cal/mol and is independent of molecular weight. The lowest interfacial free energies $\sim 2,000$ cal/mol is found for the lowest molecular weights. There is then a transition region between the low and high molecular weights. The interfacial free energy associated with the mature crystallites that are actually formed depends very markedly on the molecular weight. This result can be interpreted as being a reflection of the ratio of the crystallite thickness to the extended chain length. When this ratio approaches unity, σ_{ec} has its smallest value, although it is still relatively large when compared to monomeric systems. When the ratio of the crystallite size to the extended-chain length is very small, the corresponding interfacial free energy is large.

This kind of thermodynamic analysis cannot, of course, specify the molecular structure of the interfacial region. It points out rather forcefully, however, that the interfacial structures must be different in the different molecular weight ranges, despite the fact that lamella crystallites are always observed. The interfacial region associated with the lower molecular weights, ζ/x large, yields the most stable, higher melting system since it possesses the lowest interfacial free energy. Therefore, if the crystallites formed from the higher molecular weight chains possessed this interfacial structure, higher melting temperatures would be observed. In the molecular weight range where $\zeta/x \ll 1$ it is possible to postulate, solely on geometric grounds without regard to any other considerations, the existence of an interface of regularly folded chains. If such a structure did indeed exist, it would not do so because of any enhanced stability that is endowed to this type of interface.

Properties

In order to delve deeper into the morphological and molecular structure of crystalline polymers, it is necessary to analyze the properties that are associated with such systems. Although one should and must take cognizance of the microscopic observations, a limit is reached where more facts must be developed before rational progress can be made. We shall briefly discuss a variety of properties of such systems and develop a more detailed analysis of the structure which will be consistent with the morphological observations. In examining the properties of crystalline polymers, it is important that a wide range of molecular weights and crystallization conditions be considered. As we will find below, in contrast to crystals formed in dilute solution, all the properties of polymers crystallized from the pure melt are very dependent on molecular weight. It is, therefore, very important to recognize that there is no unique value to a given property. The range in properties that can be developed can be illustrated by studying molecular weight fractions of linear polyethylene which are crystallized under controlled conditions.

We examine first some simple thermodynamic properties such as the density and enthalpy of fusion. As can be seen in Figure 14, the density measured at room temperature after crystallization at 130° for long periods of time, or after rapid crystallization, are very dependent on molecular weight[24,115,116,117]. It is convenient to compare the observed densities with that of the unit cell, which is 1.00 gcm^{-3} at room temperature. After 130° crystallization, the observed densities range from 0.99 to somewhat less than 0.94 gcm^{-3}. After rapid crystallization, high-molecular-weight samples of linear polyethylene have densities which are less than 0.92. Thus, the observed densities are not restricted to any narrow limits, as had been implied. The densities change in a systematic manner with molecular weight and crystallization conditions. At low molecular weights, the values of the observed densities approach that expected for the unit cell. As the molecular weight is increased above 50,000–60,000, the density monotonically decreases up to a molecular weight of approximately 1.2×10^6. At still higher molecular weights, an apparently low value of the density is attained. The deviations in density from that of the unit cell cannot be attributed to end-groups acting as lattice imperfections [118,119] In fact, it is the lowest molecular weights, which possess the highest concentration of end-groups, where the densities are closest to that of the unit cell.

Enthalpy of fusion measurements yield essentially the same result

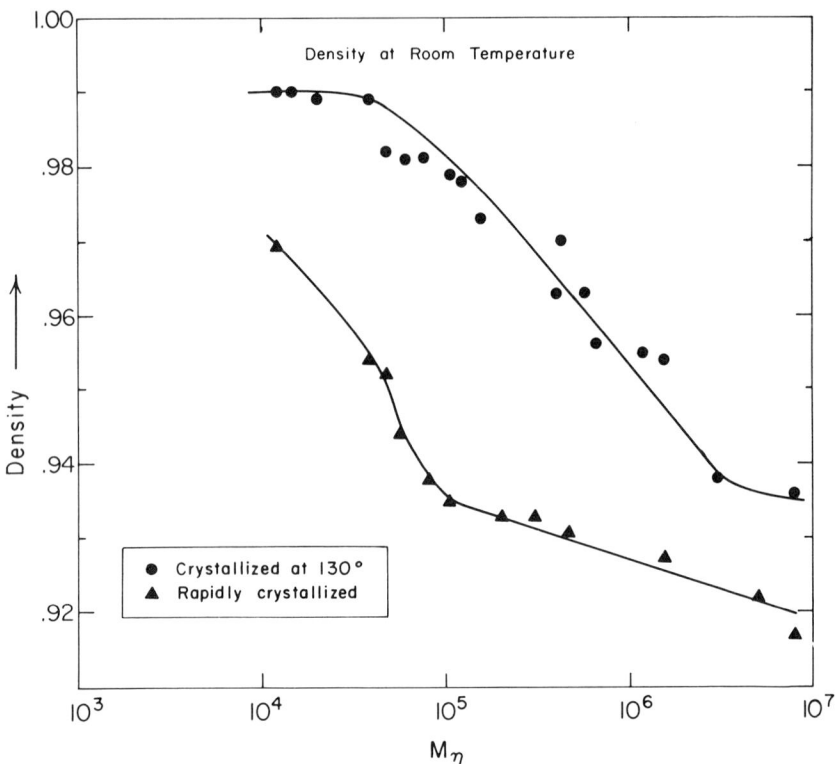

Figure 14. Plot of density, measured at room temperature, as a function of molecular weight for linear polyethylene fractions [24].

[40,117]. For the macroscopic perfect crystal, an enthalpy of fusion of 69 ± 1 cal/g is expected. After high-temperature crystallization, followed by cooling to room temperature, the measured enthalpies of fusion range from 69 cal/g for the lower molecular weights to 37.5 cal/g for the higher ones. The lower molecular weights thus yield values close to that expected for the macroscopic crystal. After rapid crystallization of high-molecular-weight samples, the enthalpy of fusion can be reduced to as low as 20–25 cal/g. There is therefore, again a large and systematic change in a property with molecular weight and mode of crystallization.

We can note in examining this data that there is a very strong correlation between the properties and the quantity $\langle \zeta \rangle / x$ discussed in the previous section. When $\langle \zeta \rangle / x$ is approximately unity, characteristic of the lower molecular weight range, the properties observed are comparable to those expected for the unit cell. For these samples fracture occurs rather easily at room temperature. The decrease of $\langle \zeta \rangle / x$ from unity coincides with the molecular weight range where the density and enthalpy of fusion begin to undergo their changes. As this ratio becomes progressively smaller, with an increase in molecular weight, the deviations of the thermodynamic quantities from that of the unit cell increase.

All other properties, many of which will be detailed in the subsequent discussion, show very similar behavior. The changes in mechanical properties, for example, can be correlated with the changes in density. In Figure 15 the variation in the logarithmic decrement, with samples of linear polyethylene of different densities, is shown[120]. The density changes are achieved by control of molecular weight and crystallization conditions. This figure shows quite clearly that the dynamic mechanical properties can also vary quite markedly. Particular attention is directed to the increased intensity of the loss peak with a decrease in the density.

With the systematic change in properties that are observed and the large deviations from those expected for the unit-cell, attention must be directed as to the molecular origin of this behavior. We have already seen that the deviations cannot be attributed to chain ends within the crystalline lattice, because the molecular weight effect is in the opposite direction. We next inquire as to whether this effect can be caused by other kinds of defects or imperfections within the lattice. (It should be recognized that the concentration of such defects that are required to explain the properties would be unprecedented for a condensed phase.) Such internal imperfections, if of any consequence to the problem at hand, should manifest themselves in changes in the lattice parameters. Consequently, in Figure 16 there is a plot of the unit cell density, as determined by wide-angle X-ray diffraction, as a function of the macroscopic density[121]. The macroscopic densities range from 0.92 to 0.99 gcm^{-3}. The results shown in this figure demonstrate rather conclusively that, over this very wide range in macroscopic densities for linear polyethylene, the unit cell density, and thus the lattice dimensions, remain invariant. We can, therefore, exclude the presence of any internal defects which influence the lattice parameters. It would hardly be expected that if they existed, and influenced the macroscopic properties, that they would not influence the lattice dimensions. Since it is shown that the deviations in properties from that of the unit-cell cannot be attributed to chain-ends or to major imperfections within the lattice, their origin will have to be sought outside of the crystalline portions.

The question immediately arises as to what kinds of structures can be involved. The previous analysis of the interfacial free energy leads to the expectation that there could be contributions from the interfacial structure. Although the interfacial free energy is relatively high, it reaches an asymptotic limit at a molecular weight of about 10^5[103]. Concomitantly, the crystallite thickness and melting temperature also become constant [103,116]. On the other hand, properties—such as density and enthalpy of fusion—continue to monotonically decrease toward values characteristic of liquidlike structures. Any new type of defect that might be postulated would have to be present in unprecedented concentration for a condensed

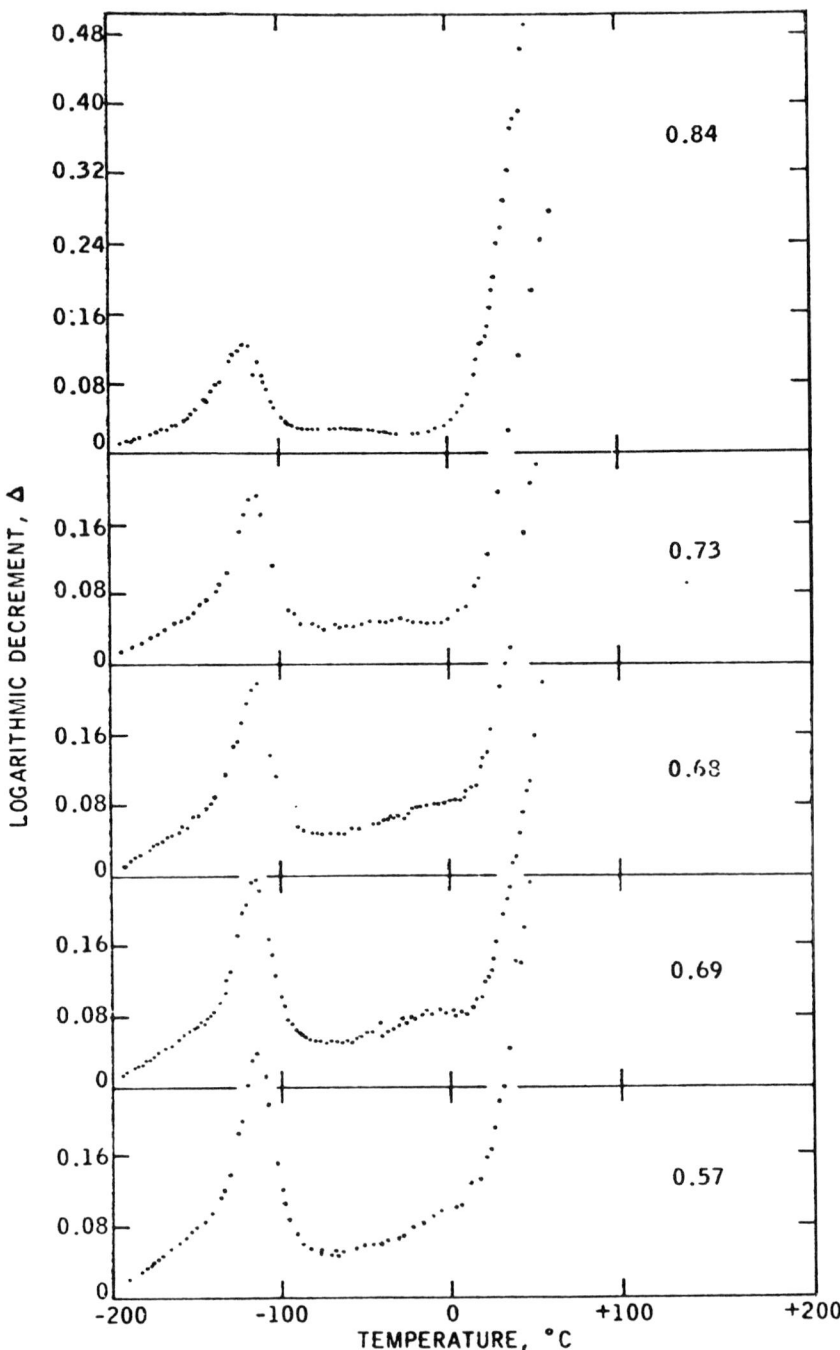

Figure 15. Plot of logarithmic decrement against temperature for linear polyethylene samples at indicated levels of crystallinity [120].

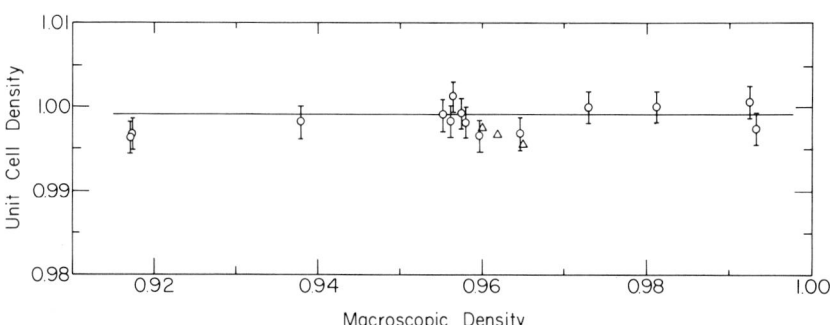

Figure 16. Plot of unit cell density against macroscopic density for linear polyethylene [121].

phase, while at the same time having no effect on the thermodynamic stability as manifested in the melting temperatures. It would not be contributing to the interfacial structure since the interfacial free energy remains constant. Until such time as defects having the required properties are found and quantitatively described, it seems more prudent to seek other structures which can be shown to explain the observed macroscopic properties. It is natural, therefore, to reexamine the possibility that amorphous regions exist. In these regions, chain units in nonordered conformation connect crystallites. At one time such regions were summarily dismissed[6,13,14,39] because of the observation of lamellalike crystallites and the assignment of a regularly folded interfacial structure. The wide range of properties that are now observed, while the lamella morphology is maintained, coupled with the restraints on the melting temperature and interfacial structure gives more than adequate reason to reexamine this old concept.

Degree of Crystallinity

Question as to whether amorphous regions actually exist depends to a large extent on the validity of the calculations of the degree of crystallinity. Classically, this type of calculation has been based on the proposition that the properties of the crystalline and amorphous regions are additive[4]. One must now also consider the possibility of contributions from the interfacial region, and perhaps defects within the crystalline interior which do not manifest themselves in alteration of the lattice parameters. The set of data generated by studying molecular weight fractions is ideally suited to test quantitatively the concept of the degree of crystallinity. For example, on a simple additivity basis, the degree of crystallinity, $1-\lambda$, from the density measurements range from 0.50 to 0.95 for linear polethylene. To be meaningful, the analyses of the degree of crystallinity must be made

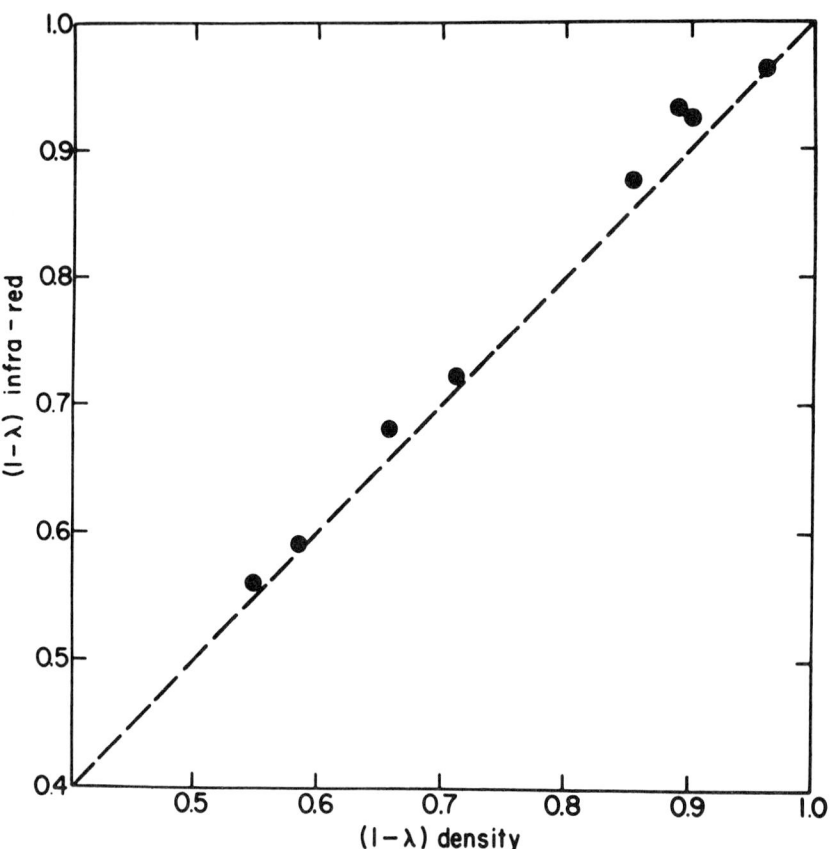

Figure 17. Plot of degree of crystallinity calculated from infrared absorption against degree of crystallinity calculated from specific volume for linear polyethylene [115].

utilizing a variety of techniques on similarly constituted samples since it is possible that the regions in the polymer have different sensitivities to various kinds of measurements.

To assess critically the validity of the degree of crystallinity calculations, it is advisable to examine representative sets of data. Figure 17 represents a comparison of the degree of crystallinity calculated from infrared absorption and density measurements. The infrared degree of crystallinity was calculated from a combination of the 1894 cm^{-1} crystalline and 1303 cm^{-1} noncrystalline bands[46]. Similar results are obtained if the bands are analyzed independently or in other combinations. Inherent in these results is the assumption that there are no interfacial contributions to the quantities in question. The agreement that is found between these two methods is extraordinarily good and gives strong quantitative support to the degree-of-crystallinity concept. The extinction co-efficient for the crystalline 1894 cm^{-1} band is the same as for the n-hydro-carbons which form molecular crystals. Thus, from infrared absorption measurements, the interior of the crystals is the same in both cases.

Comparison of the wide-angle X-ray diffraction and infrared absorp-tion for a variety of linear and branched polyethylenes, whose specific volumes ranged from 1.03 to 1.10 cm^3/g, showed very good concordance in the degree of crystallinity. These results were also in accord with the degree of crystallinity calculated from the density[122]. Other quantita-tive wide-angle X-ray studies show good agreement with the degree of crystallinity[97,123] calculated from the density. It appears, therefore, that there are very little, if any, detectable contributions from the inter-facial regions to the quantities of interest in these cases. The classical con-cept of the degree of crystallinity is thus valid by these techniques, for samples ranging from about 95 to 50 percent crystallinity.

This simple method of calculating the degree of crystallinity appears to be invalid when enthalpy of fusion measurements are analyzed. In Figure 18, the degree of crystallinity calculated from the enthalpy of fu-sion is plotted against the corresponding quantity calculated from the specific volume for polyethylene samples crystallized isothermally at 130°C and subsequently cooled to room temperature[40]. Samples crys-tallized in a less stringent way yield essentially the same results. Except for the very highest levels of crystallinity, 1-$\lambda \geq 0.90$, There is a large disparity between the two methods. As the density of the samples de-crease, the enthalpy measurements yield lower values for 1-λ than do the density values. Since the enthalpy of fusion is calculated on a mass basis, this discrepancy cannot be attributed to voids within the sample. This has been another popular, but unsupported postulate[14]. The reason for this discrepancy can be understood when samples are analyzed at the crystallization temperature, so that further crystallization upon cooling does not take place. As can be seen in Table I, under these conditions, very good agreement is obtained between the two methods over a very wide range in crystallinities[40]. The disparity illustrated in Figure 18 is thus a result of the samples being cooled to room temperature. Further crystallization occurs on cooling, with much smaller crystallites being formed. The increase in crystallinity on cooling is proportionately greater for the higher molecular weight samples[116]. Associated with the smaller crystallites is a substantial contribution from the interfacial enthalpy to the measured enthalpy of fusion. Analysis of dilute solution crystals[17] shows that the interfacial enthalpy is substantial and can thus be expected to make a significant contribution when small crystals are present. The interfacial enthalpy is of opposite sign to the bulk enthalpy of fusion. Consequently, a lower value for 1-λ will be deduced, from simple addi-tivity, when smaller crystallites are present. Although the degree of crys-tallinity is still a quantitatively meaningful concept in this case the simple additivity of the properties of two phases is inadequate.

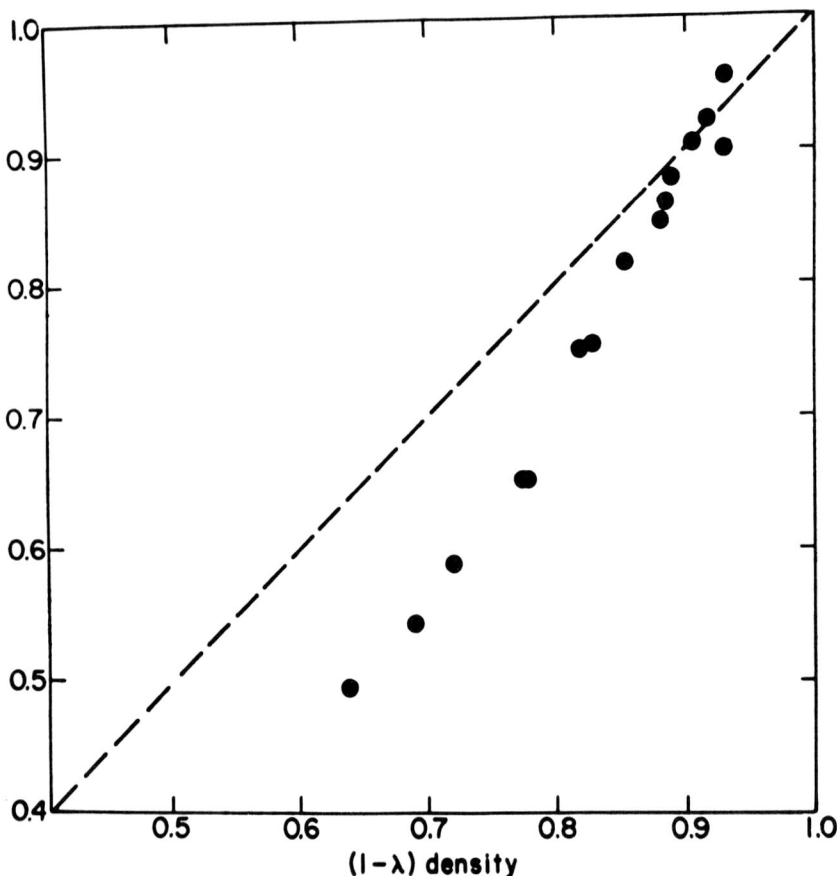

Figure 18. Plot of degree of crystallinity calculated from enthalpy of fusion against degree of crystallinity calculated from specific volume for polyethylene fractions crystallized at 130° C and cooled to room temperature [115].

The analyses of the broad-line nmr spectra of crystalline polymers is very helpful in structural elucidation. According to Bergman and Nawotki[57] and confirmed by more recent studies[124], three components are necessary to quantitatively account for the observed spectra. The spectra can be quantitatively decomposed into a broad component characteristic of the segmental motions within the crystalline regions; a narrow component resulting from the motions within the amorphous or liquid-like regions; and an intermediate component which can be identified with motions from the interfacial structure[124]. The broad and narrow components have also been termed the immobile and mobile fractions, respectively. The spectrum characteristic of the crystalline regions is the same as

TABLE I
Degree of Crystallinity from Enthalpy of Fusion for Non-cooled Samples[40]

Molecular Wt.	$T_c^0 C$	$(1 - \lambda)_{\Delta H}$	$(1 - \lambda)_d$	Ratio
7×20^6	125	0.27	0.28	0.96
4.7×10^5	125	0.50	0.53	0.94
2.0×10^5	128	0.65	0.67	0.97
5.6×10^4	128	0.74	0.76	0.97

that obtained with n-hydrocarbons. Recent results with linear polyethylene, covering the complete range of densities that were utilized in the studies of other properties, are summarized in Table II. The agreement between the nmr measurements and the density for the fraction crystalline is very good over the complete range. For the samples of high density, where $\langle \zeta \rangle / x$ is close to unity, virtually all of the noncrystalline material can be assigned to the amorphous regions. However, for the samples of low density, where $\langle \zeta \rangle / x$ is very small, about 10 percent of the chain units must be assigned to the interfacial region. This conclusion is consistent with the high value of the interfacial free energy that is associated with crystallites in this range. Thus, the broad-line nmr experiments are sensitive to contributions from the interfacial regions, as well as the crystalline and amorphous portions of the system.

Absolute intensity measurements of the low-angle X-ray scattering of linear polyethylene crystallized in the bulk have not as yet led to a consistent set of results. Shultz and Kavesh[97] studied a sample which was about 70 percent crystalline by other methods. However, when the intensities were analyzed on the basis of the electron-density difference between the crystalline and amorphous phases, a disparity is found in the assignment of the amorphous density. This disparity could be due to contributions to the scattering from the interfacial regions which were not explicitly taken into account. More recently, the opposite conclusion was reached. Roe and Gieniewski[98] concluded that the two-phase concept quantitatively described the scattering intensity of a system which was 80

TABLE II
Proportion of Major Regions as Determined by Broad-Line NMR[124]

Molecular Wt.	Fraction Crystalline		Fraction Interface	Fraction Interzonal
	Density	NMR		
1.3×10^4	0.94	0.96 ± 0.02	0.01 ± 0.005	0.01 ± 0.005
2.5×10^4	0.91	0.94 ± 0.02	0.02 ± 0.01	0.02 ± 0.02
2.15×10^5	0.83	0.83 ± 0.02	0.04 ± 0.02	0.13 ± 0.03
4.25×10^5	0.77	0.78 ± 0.03	0.07 ± 0.02	0.15 ± 0.03
1.1×10^6	0.63	0.63 ± 0.03	0.09 ± 0.02	0.28 ± 0.05
7×10^6	0.53	0.54 ± 0.04	0.10 ± 0.03	0.36 ± 0.07

percent crystalline. Moreover, from their analyses they concluded that the interface was not smooth. In a similar vein, from their studies of a bulk-crystallized branched polyethylene sample, which was 64 percent crystalline, Stroble and Müller[38] concluded that a rectangular electron-density profile satisfies the absolute scattering curve. This would imply that their data does not require a contribution from the interfacial regions.

From the above survey it can be concluded that the degree of crystallinity is a quantitative concept. The properties of disordered chains are used in these calculations so that there is a major contribution from amorphous or liquidlike regions. For some of the properties studied, small but quantitatively significant contributions from interfacial structures can be detected. A semicrystalline polymer cannot, therefore, be considered in any meaningful way a macroscopic crystal, replete with a large concentration of undefined defects.

Molecular Morphology

Based on a consideration of the observed properties and the lamellalike morphology that is revealed by the electron microscope, a schematic representation of a primitive crystallite can be developed. Such a representation is illustrated in Figure 19. Although highly schematic, and undoubtedly oversimplified, this fundamental morphological entity can explain the observed properties and their dependence on molecular weight and crystallization conditions. It is also consistent with the direct microscopic observations. Three major and distinct regions are of importance. These are the crystalline regions, the interfacial region or zone, and the amorphous or interzonal region.

The crystalline regions, as indicated by the vertical straight lines in the diagram, are composed of chain sequences in ordered or preferred conformation. There are undoubtedly imperfections and defected structures within the interior of the crystallites as there are in monomeric substances. The infrared[46] and broad-line nmr studies[124] indicate a strong similarity in the interior of polyethylene and n-hydrocarbon crystals. The interfacial region is a very diffuse and a not too well-defined zone. It is many units thick and the segmental packing is very distorted and crowded. In contrast to monomeric crystals or molecular crystals of chain molecules, it cannot be represented by a smooth plane. Some chains will pass through this zone and form the interzonal region. Other chains will return to the crystallite of origin but not necessarily in juxtaposition. Thus, a disordered, highly irregularly structured interfacial region results. Such an interfacial structure is consistent with the high value of the interfacial free energy associated with the basal plane when $\langle \zeta \rangle / x \ll 1$. On the other hand, when $\langle \zeta \rangle$ is comparable to x, the interzonal region will be essentially absent. There is no need in this situation for a chain to return to the crystallite of origin. The distortion and crowding of chain units

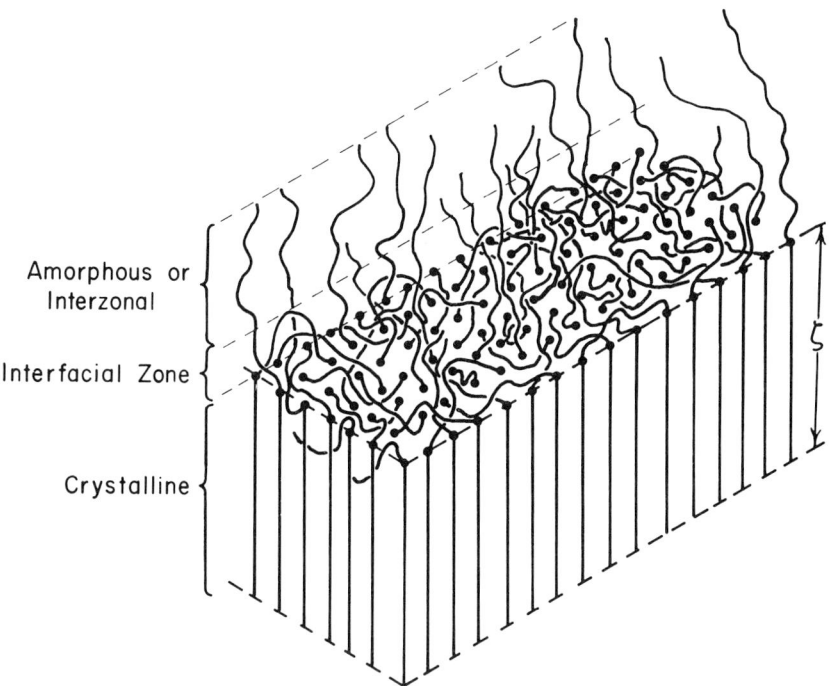

Amorphous or
Interzonal

Interfacial Zone

Crystalline

Figure 19. Schematic representation of crystallite [24].

within the interfacial region will thus be alleviated. As consequence, lower values of the interfacial free energy result.

The interzonal or amorphous regions are characterized by the chain units being in disordered or nonordered conformations. They have thermodynamic, spectral, and mechanical properties which are very similar to those in the pure melt. The chain units in this disordered array connect one crystallite with another. In addition to the evidence cited in the discussion of the degree of crystallinity, further substantiation for interzonal regions is found in other observations. For example, semicrystalline polymers display well-defined glass temperatures. For polymers which can be compared in the semicrystalline state as well as in the completely amorphous one, and for polymers whose degree of crystallinity can be controlled, the value of the glass temperature is essentially independent of the degree of crystallinity[120,125–127]. Although there is some controversy as to the details of the molecular mechanism involved in glass formation, there is complete agreement among the protagonists that it is a property of the liquid and disordered state. Glass formation thus involves chain elements in random conformation. Semicrystalline polymers must, therefore, contain such units. The fracture properties previously mentioned are consistent with the existence of interzonal regions and glass

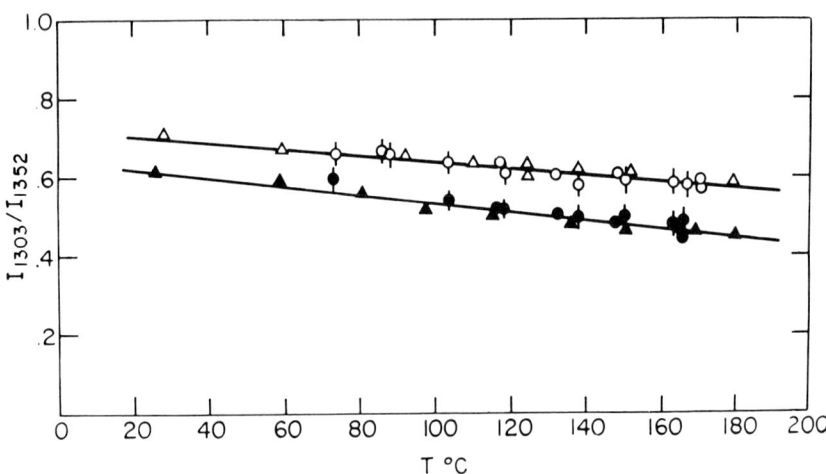

Figure 20. Plot of intensity ratio I_{1303}/I_{1352} as a function of temperature for $C_{32}H_{66}$ ◊◆; for $C_{94}H_{190}$ ◊◆; and for polyethylene $M_n = 250,000$; density $= 0.933$ △ [46].

formation. Low-molecular-weight samples of linear polyethylene, $M \leq 5 \times 10^4$, which are very highly crystalline are easily fractured at room temperature. However, for higher molecular weight samples, and other samples crystallized so that $\langle \zeta \rangle / x \ll 1$ fracture can only be accomplished at liquid-nitrogen temperature. This temperature is below the glass temperature of linear polyethylene.

Further understanding of the chain structure in the interzonal region is found in the analysis of the intensity ratio of the infrared absorption of the two noncrystalline *gauche* bands froun at 1303 cm^{-1} and 1352 cm^{-1}. In Figure 20 this intensity ratio, for a linear polyethylene of low density, is compared with that for the melt of *n*-hydrocarbons[46]. (The two straight lines in Figure 20 represent two different ways of analyzing the infrared data. The same functional dependence with temperature is found for all the samples by either method.) The intensity ratios for the two *n*-paraffins, $C_{94}H_{190}$ and $C_{32}H_{66}$, and for the linear polyethylene delineate the same straight line from room temperature to above the polymer melting temperature. Thus, as far as *gauche* bonds are concerned they are the same in the melts of *n*-hydrocarbons as in the noncrystalline regions of the polymer.

The phrase "tie-molecules" has been commonly used to describe the connections between crystallites. To be meaningful, this expression has to be clearly and carefully defined. Obviously, the connections are not molecules but portions of molecules. The properties that are observed require that the chain units be in nonordered conformations. They cannot be straightened out or completely extended portions of chains. Electron micrographs have been produced which purport to demonstrate that in linear polyethylene the connecting chain units are in the extended *trans* planar zigzag conformation[128,129]. Such structures are incompatible with all the known properties of semicrystalline polymers. They were observed as a result of the particular method of crystallization that was employed. In essence, the samples were prepared by crystallizing from a relatively concentrated mixture with the *n*-hydrocarbon $C_{32}H_{66}$. The diluent was removed at room temperature subsequent to the crystallization. As the most rudimentary considerations of phase equilibria show, this process is quite different than crystallizing from the pure melt. Consequently, major differences are found between thermodynamic properties of samples crystallized from the pure melt and those crystallized from the diluent mixtures in the manner indicated[130]. Significant differences in morphology and structure are to be expected. The identification of the interfacial and interzonal regions in the two cases is extremely tenuous.

Conclusion

The elementary crystallite illustrated in Figure 20 can explain the macroscopic properties in a systematic way depending upon the proportion of the chain units that are located in each of the regions. The critical quantity is then the relation between the crystallite thickness and the extended chain length. The latter is determined by the molecular weight; the former by the crystallization process. One must, therefore, seek from the actual crystallization process the answers to certain fundamental questions. Two of the more important ones are: (a) why are lamella crystallites formed, and (b) as the molecular weight increases why does the level of crystallinity decrease while the crystallite thickness and interfacial free energy is maintained? Although the answer to the second of these questions can be seen in crystallization kinetic studies[106], a detailed molecular explanation has not as yet been discerned.

With regard to the first question, for this particular phase transition nucleation processes have been shown to occupy a very dominant and influential role[4]. Therefore, the ratio of the critical-size nucleus in the chain direction ζ^* to the extended chain length becomes a crucial quantity. In addition, one must consider the consequences of a calculation made by Flory[32] where it was shown that, for polyethylene and other polymers of close-packed crystal structure, not all the chains that emerge from an infinite or very large crystal face can be accommodated in random conformation in the space above the basal plane. A portion of this emanating flux of chains must be dissipated. This dissipation can be accomplished by a variety of methods, including the formation of new nearby crystallites or by the return of the chain to the crystallite of origin. There is no necessity, however, for the returning position to be in juxtaposition to its emergence point. These spatial restrictions apply only to a large basal plane and would not restrain the formation of a critical-size nucleus which would require the bringing together of about 50–100 sequences in the usual range of undercooling[115].

Growth following nucleation will be expected to occur most rapidly by the lateral accretion of ordered sequences. However, when ζ^* is small compared to x, a concentration of chain units emerging from the 001 face will be reached which, for the reasons cited above, can no longer be accommodated outside the crystal face. In order for the crystallization to proceed, a portion of this flux must be dissipated. Some of the chains will penetrate the interfacial region and form new crystallites. The free-energy decrease per repeating unit that is achieved on further crystallization will allow other chains to return to the crystallite of origin despite the free-energy increase per sequence that will accompany the penetration of the interfacial region. Thus, a lamellalike crystallite will develop with an associated distorted and crowded interfacial region.

At the other end of the molecular weight scale, where ζ^* is comparable to the extended chain length, a relatively small number of units per chain will be excluded from the nucleus and the resulting crystallite. The stringent spatial problem above the basal plane no longer exists so that there is no impediment to continuous lateral growth. Lamella crystallites will again be formed but the associated interfacial region will be less distorted and will be characterized by a smaller value for the interfacial free energy. There will be minimal connections between crystallites. The absence of interzonal regions leads to very brittle, easily fractured material.

The crystallization process that has been outlined utilizes the same chain disposition within the nucleus irrespective of the molecular weight and the character and size of the crystallites that evolve. Lamella crystallites will be formed for all molecular weights without the need of changing any of the fundamental processes to coincide with the morphological observations. The change in the interfacial free energy with molecular weight receives a natural explanation as does the formation of interzonal regions without having to postulate alterations in the mode of nucleation. Utilizing nucleation theory pertinent to chains of finite length[131,132] a natural demarcation is found in ζ^* with chain length[115]. An asymptotic limit is reached in ζ^* so that for the higher molecular weights it becomes much smaller than the extended chain length.

Acknowledgment

This work was supported by the National Science Foundation under grant No. GH-33794.

References

1. Flory, P.J., "Thermodynamics of Crystallization in High Polymers. A Theory of Crystalline States and Fusion in Polymers, Copolymers, and Their Mixtures with Diluents," *J. Chem. Phys.,* 17 (1949), 223–40.

2. Flory, P.J., "Role of Crystallization in Polymers and Proteins," *Science,* 124 (1956), 53–60.

3. Mandelkern, L., "The Melting of Crystalline Polymers," *Rubber Chem. Tech.,* 32 (1959), 1392–1451.

4. Mandelkern, L., *Crystallization of Polymers,* New York: McGraw-Hill Book Publishing Company, 1964.

5. Avrami, M., "Kinetics of Phase Change. I. General Theory," *J. Chem. Phys.,* 7 (1939), 1103–12; and Avrani, M., "Kinetics of Phase Change. II. Transformation-Time Relations for Random Distribution of Nuclei," *J. Chem. Phys.,* 8 (1940), 212–24.

6. Keller, A., "A Note on Single Crystals in Polymers: Evidence for a Folded Chain Configuration," *Phil. Mag.,* 2 (1957), 1171–75.

7. Fischer, E.W., "Stufen-und spiral förmiges Kristallwachstum bei Hochpolymeren," *Z. Naturforsch. A.,* 12 (1957), 753–54.

8. Keller, A., "The Morphology of Crystalline Polymers," *Makromol. Chem.,* 34 (1959), 1–28.

9. Mandelkern, L., "Thermodynamic and Physical Properties of Polymer Crystals Formed From Dilute Solution," in *Progress in Polymer Science, Vol. 2,* A.D. Jenkins, ed., Oxford: Pergamon Press (1970), 165–200.

10. Fava, R.A., "Polyethylene Crystals," *J. Polym. Sci., Part D,* 5 (1971), 1–108.

11. Keller, A., "Morphology of Crystalline Polymers," in *Growth and Perfection of Crystals,* R.H. Doremus, B.W. Roberts, and D. Turnbull, eds., New York: John Wiley & Sons (1958), 499–532.

12. Keller, A., "Polymer Single Crystals," *Polymer,* 3 (1962), 393–421.

13. Lindenmeyer, P.H., "Crystallization and Molecular Folding," *Science,* 147 (1956), 1256–62.

14. Hoffman, J.D., "Theoretical Aspects of Polymer Crystallization with Chain Folds: Bulk Polymers," *SPE (Soc. Plast. Eng.) Trans.,* 4 (1964), 315–62.

15. Peterlin, A., "Chain Folding and Free Energy Density in Polymer Crystals," *J. Appl. Phys.,* 31 (1960), 1934–38.

16. Peterlin, A. and Fischer, E.W., "Thermodynamische Stabilität makromolekularer Kristalle. I. Der Einfluss der Longitudinalschwingungen der Kettenmoleküle," *Z. Phys.,* 159 (1960), 272–87.

17. Peterlin, A., Fischer, E.W. and Reinhold, C., "Thermodynamic Stability of Polymer Crystals. II. Torsional Vibrations of Chain Molecules," *J. Chem. Phys.,* 37 (1962), 1403–08.

18. Peterlin, A. and Reinhold, C., "Thermodynamic Stability of Polymer Crystals. III. Torsional and Longitudinal Chain Vibrations," *J. Polym. Sci., Part A,* 3 (1965), 2801–10.

19. Price, F.P., "Markoff Chain Model for Growth of Polymer Single Crystals," *J. Chem. Phys.,* 35 (1961), 1884–92.

20. Lauritzen, J.I. Jr. and Hoffman, J.D., "Theory of Formation of Polymer Crystals

with Folded Chains in Dilute Solution," *J. Res. Nat. Bur. Stand., Sect. A,* 64A (1960), 73–102.

21. Frank, F.C. and Tosi, M., "On the Theory of Polymer Crystallization," *Proc. Roy. Soc., Ser. A,* A263 (1961), 323–39.

22. Blackadder, D.A., "Ten Years of Polymer Single Crystals," *J. Macromol. Sci., Rev. Macromol. Chem.,* Cl (1967), 297–326.

23. Mandlekern, L., "Thermodynamic and Physical Properties of Crystals Formed from Dilute Solutions," *Polym. Eng. Sci.,* 9 (1969), 255–67.

24. Mandelkern, L., "Thermodynamic and Morphological Properties of Crystalline Polymers," *J. Phys. Chem.,* 75 (1971), 3920–28.

25. Kawai, T. and Keller, A., "On the Density of Polyethylene Single Crystals," *Phil. Mag.,* 8 (1963), 1203–10.

26. Kawai, T. and Keller, A., "On the Gradient Column Method for Measuring Densities; with Particular Reference to Its Application to Polymer Single Crystals," *Phil, Mag.,* 8 1963), 1973–76.

27a. Hamada, F., Wunderlich, B. Sumida, T., Hayashi, S. and Nakajima, A., "Density and Heat of Fusion of Folded-Chain Polyethylene Crystals," *J. Phys. Chem.,* 72 (1968), 178–85.

27b. Wasiak, A., "Surface Tension Corrections in Density Measurements Using Density Gradient Columns," *Kolloid Z-Z Polym.,* 215 (1966), 158–61.

28. Jackson, J.B., Flory, P.J. and Chiang, R., "Thermodynamic Stability of Solution-Crystallized Polyethylene," *Trans. Faraday Soc.,* 59 (1963), 1906–17.

29. Blackadder, D.A. and Lewell, P.A., "The Density of Polyethylene Single Crystals," *Polymer,* 9 (1968), 249–63.

30. Sharma, R.K. and Mandelkern, L., "The Density of Polyethylene Crystallized in the Bulk from Dilute Solution," *Macromolecules,* 2 (1969), 266–71.

31. Go, S., Kloos, F. and Mandelkern, L., to be published.

32. Flory, P.J., "On the Morphology of the Crystalline State in Polymers," *J. Amer. Chem. Soc.,* 84 (1962), 2857–67.

33. Fischer, E.W. and Schmidt, G., "Über Langperioden bei verstrecktem Polyäthylen, *Angew. Chem.,* 74 (1962), 551–62.

34. Fischer, E.W. and Hinrichsen, G., "Schmelz-und Rekristallisations-vorgänge bei Polyäthylen-Einkrystallen. Teil II: Untersuchun gen mit Hilfe der Differential thermoanalyse," *Kolloid Z-Z Polym.,* 213 (1966), 93–108.

35. Blackadder, D.A. and Roberts, T.L., "A New Model for the Fold Surfaces of Polyethylene Single Crystals," *Makromol. Chem.,* 126 (1969), 116–29.

36. Sharma, R.K. and Mandelkern, L., "Thermodynamic Properties of Linear Polyethylene Crystals Formed from Dilute Solution," *Macromolecules,* 3 (1970), 758–63.

37. Fischer, E.W., Goddar, H. and Schmidt, G.F., "A Remark on the Surface Structure of Polyethylene Single Crystals," *J. Polym. Sci., Part B,* 5 (1967), 619–24.

38. Strobl, G.R. and Muller, N., "Small-Angle X-Ray Scattering Experiments for Investigating the Validity of the Two-Phase Model," *J. Polym. Sci., Polym. Phys. Ed.,* 11 (1973), 1219–33.

39. Lindenmeyer, P.H., "Crystallization in Polymers," *J. Polym. Sci., Part C,* No. 1 (1963), 5–39.

40. Mandelkern, L., Allou, A.L. Jr. and Gopolan, N., "The Enthalpy of Fusion of Linear Polyethylene," *J. Phys. Chem.,* 72 (1968), 309–18.

41. Okada, T. and Mandelkern, L., "The Infrared Determination of the Degree of Crys-

tallinity of Polyethylene Crystallized from Dilute Solution," *J. Polym. Sci., Part B,* 4 (1966), 1043–48.

42. Schonhorn, H. and Luongo, J.P., "Fold Structure of Polyethylene Single Crystals," *Macromolecules,* 2 (1969), 366–69.

43. Bank, M.I. and Krimm, S., "Mixed Crystal Infrared Study of Chain Folding in Crystalline Polyethylene," *J. Polym. Sci., Part A-2,* 7 (1969), 1785–1809.

44. Stehling, F.C., Ergos, E. and Mandelkern, L., "Phase Separation in n-Hexatriacontane- n-Hexatriacontane-d_{74} and Polyethylene-Poly(ethylene-d_4) Systems," *Macromolecules,* 4 (1971), 672–77.

45. Koenig, J.L. and Whitenhafer, D.E., "Infrared Studies of Polymer Chain Folding. I. Linear Polyethylene," *Makromol. Chem.,* 99 (1966), 193–201.

46. Okada, T. and Mandelkern, L., "Effect of Morphology and Degree of Crystallinity on the Infrared Absorption Spectra of Linear Polyethylene," *J. Polym. Sci., Part A-2,* 5 (1967), 239–62.

47. Koenig, J.L. and Agboatwala, M.C., "Infrared Studies of Chain Folding in Polymers. V. Polyhexamethylene Adipamide," *J. Macromol. Sci., Phys.,* B2 (1968), 391–420.

48. Frayer, P.D., Koenig, J.L. and Lando, J.B., "Infrared Studies of Chain Folding in Polymers. VI. Polyhexamethylene Adipamide and Its Cyclic Oligomers," *J. Macromol. Sci., Phys.,* B3 (1969), 329–36.

49. Statton, W.O., Koenig, J.L. and Hannon, M., "Characterization of Chain Folding in Poly (Ethylene Terephthalate) Fibers," *J. Appl. Phys.,* 41 (1970), 4290–95.

50. Peterlin, A., Olf, H.G., Peticolas, W.L., Hibler, G.W. and Lippert, J.L., "Laser-Raman and X-Ray Study of the Two-Phase Structure of Polyethylene Single Crystals," *J. Polym. Sci., Part B,* 9 (1971), 583–89.

51. Statton, W.O. and Geil, P.H., "Recrystallization of Polyethylene During Annealing," *J. Appl. Polym. Sci.,* 3 (1960), 357–61.

52. Takayanagi, M., Imada, K., Nagai, A., Tatsumi, T. and Matsuo, T., "Viscoelastic Properties of Single Crystal Mats of Some Polymers and Polymer Crystal Obtained by Radiation-Induced Solid-State Polymerization," *J. Polym. Sci., Part C,* No. 16 (1967), 867–76.

53. Takayanagi, M., "Viscoelastic Behavior of Crystalline Polymers," in *Proceedings of the Fourth International Congress on Rheology, Pt. 1,* E.H. Lee and A.L. Copley, eds., New York: Interscience Publishers (1965), 161–87.

54. Stehling, F.C. and Mandelkern, L., "Glass Transitions in Polymer Crystals Prepared from Dilute Solution," *J. Polym. Sci., Part B,* 7 (1969), 255–65.

55. Fischer, E.W., Kloos, F. and Lieser, G., "Direct Evidence for Glass Transition Taking Place in the Surface Layers of Solution Grown Crystals of Polybutadiene-1," *J. Polym. Sci., Part B,* 7 (1969), 845–50.

56. Fischer, E.W. and Kloos, F., "Proof of the Existence of a Glass Transition in the Surface Layers of Polyethylene Single Crystals," *J. Polym. Sci., Part B,* 8 (1970), 685–93.

57. Bergmann, K., and Nawotki, K., "Eine neue Interpretion der Breitlinien-Kernresonanzspektren von linearem Polyäthylen," *Kolloid Z-Z Polym.,* 219 (1967), 132–44.

58. Udagawa, Y. and Keller, A., "Liquid-Induced Reversible Long-Spacing Changes in Polyethylene Single Crystals and Their Implications for the Fold-Surface Problem," *J. Polym. Sci., Part A-2,* 9 (1971), 437–54.

59. Ergoz, E. and Mandelkern, L. "Swelling of Solution Formed Polyethylene Crystals," *J. Polym. Sci., Part B,* 10 (1972), 631–35.

60. McCall, D.W. and Anderson, E.W., "Molecular Motion in Polyethylene. III," *J. Polym. Sci., Part A,* 1 (1963), 1175–84.

61. Fischer, E.W. and Peterlin, A., "Kernresonanzmessungen zur Untersuchung der Kettenbeweglichkeit in Polyäthylen-Einkristallen," *Makromol. Chem.*, 74 (1964), 1–28.

62. Bassett, G.A., Blundell, D.J. and Keller, A., "Surface Structure of Polyethylene Crystals as Revealed by Surface Decoration. I. Preliminary Survey," *J. Macromol. Sci., Phys.*, B1 (1967), 161–84.

63. Palmer, R.P. and Cobbold, A.J., "The Texture of Melt Crystallised Polyethylene as Revealed by Selective Oxidation," *Makromol. Chem.*, 74 (1964), 174–89.

64. Peterlin, A. and Meinel, G., "Fuming Nitric Acid Treatment of Polyethylene. I. Morphology of Single Crystals," *J. Polym. Sci., Part B*, 3 (1965), 1059–64.

65. Peterlin, A., Meinel, G. and Olf, H.G., "Fuming Nitric Acid Treatment of Polyethylene. II. The Fold Surfaces of Single Crystals," *J. Polym. Sci., Part B*, 4 (1966), 399–405.

66. Winslow, F.H., Hellman, M.Y., Matreyek, W. and Salovey, R., "Etching of Solution-Crystallized Polyethylene with Fuming Nitric Acid," *J. Polym. Sci., Part B*, 5 (1967), 89–93.

67. Blundell, D.J. and Keller, A., "Investigation of the Fold-Surface Problem in Polyethylene Single Crystals with the Nitric Acid Oxidation Technique," *J. Polym. Sci., Part A-2*, 5 (1967), 991–1012.

68. Williams, T., Blundell, D.J., Keller, A. and Ward, I.M., "Gel Permeation Chromatographic Studies of the Degradation of Polyethylene with Fuming Nitric Acid. I. Single Crystals," *J. Polym. Sci., Part A-2*, 6 (1968), 1613–19.

69. Witenhafer, D.E. and Koenig, J.L., "Effect of Bromine on Polyethylene Crystals," *J. Polym. Sci., Part A-2*, 7 (1969), 1279.

70. Keller, A. and Priest, D.J., "Experiments on Location of Chain Ends in Monolayer Single Crystals of Polyethylene," *J. Macromol. Sci., Phys.*, B2 (1968), 479–95.

71. Stellman, J.M. and Woodward, A.E., "Chain Folding in Poly(*trans*-1, 4-Butadiene) Crystals," *J. Polym. Sci., Part B*, 7 (1969), 755–59.

72. Stellman, J.M. and Woodward, A.E., "Chain Folding in Poly-*trans*-1, 4-Butadiene Crystals Grown from Various Solvents," *J. Polym. Sci., Part A-2*, 9 (1971), 59–66.

73. Ng, S.B., Stellman, J.M. and Woodward, A.E., "The Location of the Amorphous Component in Poly(*trans*-1, 4-butadine) Crystals," *J. Macromol. Sci., Phys.*, B7 (1973), 539–47.

74. Jackson, J.F. and Mandelkern, L., "Solubility of Crystalline Polymers. II. Polyethylene Fractions Crystallized from Dilute Solutions," *Macromolecules*, 1 (1968), 546–54.

75. Bair, H.E. and Salovey, R., "The Effect of Molecular Weight on the Structure and Thermal Properties of Polyethylene," *J. Macromol. Sci., Phys.*, B3 (1969), 3–18.

76. Keller, A., "Polymer Crystals," *Rep. Progr. Phys.*, 31 (1968), 623–704.

77. Hoffman, J.D., Lauritzen, J.I. Jr., Passaglia, E., Ross G.S., Frolen, L.J. and Weeks, J.J., "Kinetics of Polymer Crystallization from Solution and the Melt," *Kolloid Z-Z Polym.*, 231 (1969), 564–92.

78. Sadler, D.M. and Keller, A., "Polyethylene Crystals with Dislocation Networks; Their Origin Structure and Relevance to Polymer Crystallization. Part 1," *Kolloid Z-Z Polym.*, 239 (1970), 641–54.

79. Holland, V.F. and Lindenmyer, P.H., "Direct Observation of Dislocation Networks in Folded-Chain Crystals of Polyethylene," *J. Appl. Phys.*, 36 (1965), 3049–56.

80. Bassett, D.C., "On Fold Surfaces of Polymer Crystals," *Phil. Mag.*, 17 (1968), 37–50.

81. Sadler, D.M., "Fractionation during Crystallization," *J. Polym. Sci., Part A-2*, 9 (1971), 779–99.

82. Kloos, F., Go, S. and Mandelkern, L., to be published.

83. Statton, W.O., "Rate of Recrystallization of Polyethylene Single Crystals," *J. Appl. Phys.*, 32 (1961), 2332–34.

84. Takayanagi, M. and Nagatoshi, F., "Effect of Heat-Treatment upon Final Thickness of Single Crystals of Polyethylene," *Mem. Fac. Eng., Kyushu Univ.*, 24 (1965), 33–44.

85. Mandlekern, L., Sharma, R.K. and Jackson, J.F., "On the Annealing of Polyethylene Crystals Formed from Dilute Solution," *Macromolecules*, 2 (1969), 644–47.

86. Reneker, D.H., "Point Dislocations in Crystals of High Polymer Molecules," *J. Polym. Sci.*, 59 (1962), 539–542.

87. Hirai, N., Yamashito, Y., Mitshata, T. and Tamura, Y., "Thickening of Lamellar Single Crystals of Polyethylene by Heat Treatment," *Rep. Res. Lab. Surf. Sci., Okayama Univ.*, 2 (1961), 1–15.

88. Peterlin, A., "Thickening of Polymer Single Crystals During Annealing," *J. Polym. Sci., Part B*, 1 (1963), 279–84; and "Molecular Weight Dependence of Isothermal Long Period Growth of Polyethylene Single Crystals," *Polymer*, 6 (1965), 25–34.

89. Mandelkern, L., Posner, A.S., Diorio, A.F. and Roberts, D.E., "Low-Angle X-Ray Diffraction of Crystalline Nonoriented Polyethylene and Its Relation to Crystallization Mechanisms," *J. Appl. Phys.*, 32 (1961), 1509–17.

90. Mandelkern, L. and Allou, A.L. Jr., "The Fusion of Polyethylene Single Crystals," *J. Polym. Sci., Part B*, 4 (1966), 447–52.

91. Jackson, J.F. and Mandelkern, L., "Dimensional Properties of Polyethylene Crystals Formed During Isothermal Crystallization from Dilute Solution," *J. Polym. Sci., Part B*, 5 (1967), 557–64.

92. Nakajima, A. and Hayashi, S., "Lamellar Thickness of Polyethylene Single Crystals Isothermally Crystallized in Dilute Solution," *Kolloid Z-Z Polym.*, 225 (1968), 116–21.

93. Wunderlich, B., Sullivan, P., Arawaka, T., Dicyan, A.B. and Flood, J.F., "Thermodynamics of Crystalline Linear High Polymers. III. Thermal Breakdown of the Crystalline Lattice of Polyethylene," *J. Polym. Sci., Part A*, 1 (1963), 3581–96.

94. Nagai, H. and Kajikawa, N., "Annealing of Polyethylene Single Crystals," *Polymer*, (1968), 177–99.

95. Sella, C. and Trillat, J.J., "Structural and Physiochemical Modifications of Polyethylene by the Action of Ions," *Compt. rend.*, 246 (1958), 3246–48; "Periodic Structures in Polyethylene," *Compt. rend.*, 248 (1959), 410–13; "Study of Polyethylenes by Small-Angle X-Ray Diffraction," *Compt. rend.*, 248 (1959), 1819–22; "Crystalline Structure and Molecular Structure of Polyethylenes," *Compt. rend.*, 248 (1959), 2348–50.

96. Schultz, J.M., Robinson, W.H. and Pound, G.M., "Temperature-Dependent X-Ray Small-Angle Scattering from Melt-Crystallized Polyethylene," *J. Polym. Sci., Part A-2*, 5 (1967), 511–33.

97. Kavesh, S. and Schultz, J.M., "Lamellar and Interlamellar Structure in Melt-Crystallized Polyethylene. II. Lamellar Spacing, Interlamellar Thickness, Interlamellar Density, and Stacking Disorder," *J. Polym. Sci., Part A-2*, 9 (1971), 85–114.

98. Roe, R. J. and Gieniewski, C., "Small-Angle X-Ray Diffraction Study of Chlorinated Polyethylene Recrystallized from the Melt," *Macromolecules*, 6 (1973), 212–17.

99. Fujiwara, Y., "The Superstructure of Melt-Crystallized Polyethylene. I. Screwlike Orientation of Unit Cell in Polyethylene Spherulites with Periodic Extinction Rings," *J. Appl. Polym. Sci.*, 4 (1960), 10–15.

100. Geil, P.H., "Nylon Single Crystals," *J. Polym. Sci.*, 44 (1960), 449–58; and "Morphology of an Acetal Resin," J. Polym. Sci., 47 (1960), 65–74.

101. Eppe, R., Fischer, E.W. and Stuart, H.A., "Morphologische Shukturen in Poly-

äthylenen, Polamiden, und anderen kristallisieren den Hochpolymeren," *J. Polym. Sci.*, 34 (1959), 721–40.

102. Anderson, F.R., "Morphology of Isothermally Bulk-Crystallized Linear Polyethylene," *J. Appl. Phys.*, 35 (1964), 64–70; and "Fracture Studies of Isothermally Bulk-Crystallized Linear Polyethylene," *J. Polym. Sci., Part C*, No. 3 (1963), 123–34.

103. Mandelkern, L., Price, J.M., Gopalan, M. and Fatou, J.G., "Sizes and Interfacial Free Energies of Crystallites Formed from Fractional Linear Polyethylene," *J. Polym. Sci., Part A-2*, 4 (1966), 385–400.

104. Keller, A. and Sawada, S., "On the Interior Morphology of Bulk Polyethylene," *Makromol. Chem.*, 74 (1964), 190–221.

105. Hock, C.W., "Selective Oxidation with Nitric Acid Reveals the Microstructure of Polypropylene," *J. Polym. Sci., Part B*, 3 (1965), 573–76; and "Morphology of Polypropylene Crystallized from the Melt," *J. Polym Sci., Part A-2*, 4 (1966), 227–42.

106. Ergoz, E., Fatou, J.G. and Mandelkern, L., "Molecular Weight Dependence of the Crystallization Kinetics of Linear Polyethylene. I. Experimental Results," *Macromolecules*, 5 (1972), 147–57.

107. Bunn, C.W., Cobbold, A.J. and Palmer, R.P., "The Fine Structure of Polytetrafluoroethylene," *J. Polym. Sci.*, 28 (1965), 365–76.

108. Gopalan, M.R. and Mandelkern, L., unpublished observation.

109. Pollack, S.S., Robinson, W.H., Chiang, R. and Flory, P.J., "X-Ray Diffraction of Linear Polyethylene Crystallized at 131°C," *J. Appl. Phys.*, 33 (1962), 237–38.

110. Flory, P.J. and Vrij, A., "Melting Points of Linear-Chain Homologs. The Normal Paraffin Hydrocarbons," *J. Amer. Chem. Soc.*, 85 (1963), 3548–53.

111. Geil, P.H., Anderson, F.R., Wunderlich, B. and Arawaka, T., "Morphology of Polyethylene Crystallized from the Melt under Pressure," *J. Polym. Sci., Part A*, 2 (1964), 3707–20.

112. Kardos, J.L., Baer, E., Geil, P.H. and Koenig, J.L., "Differential Thermal Analysis of Pressure Crystallized Polyethylene," *Kolloid Z-Z Polym.*, 204 (1965), 1–7.

113. Jackson, J.F. and Mandelkern, L., "Calorimetric Studies of the Fusion of Linear Polyethylene," in *Analytical Calorimetry*, R.S. Porter and J.F. Johnson, eds., New York: Plenum Press (1968), 1–7.

114. Mandelkern, L., Fatou, J.G., Denison, R. and Justin, J., "A Calorimetric Study of the Fusion of Molecular Weight Fractions of Linear Polyethylene," *J. Polym. Sci., Part B*, 3 (1965), 803–07.

115. Mandelkern, L., "Thermodynamic and Morphological Properties of Bulk Crystallized Polymers," *Polym. Sci. Eng.*, 7 (1967), 232–52.

116. Fatou, J. G. and Mandelkern, L., "The Effect of Molecular Weight on the Melting Temperature and Fusion of Polyethylene," *J. Phys. Chem.*, 69 (1965), 417–28.

117. Ergoz, E.H., "Crystallization Kinetics and Thermodynamic Properties of Linear Polyethylene Over an Extended Molecular Weight Range," unpublished Ph.D. dissertation, Florida State University, Tallahassee, 1970.

118. Eby, R.K., "Thermal Generation of Vacancies and Substitutional Sites in Crystalline Polymers," *J. Appl. Phys.*, 33 (1962), 2253–56.

119. Predecki, P. and Statton, W.O., "Dislocations Caused by Chain Ends in Crystalline Polymers," *J. Appl. Phys.*, 37 (1966), 4053–59.

120. Stehling, F.C. and Mandelkern, L., "The Glass Temperature of Linear Polyethylene," *Macromolecules*, 3 (1970), 242–52.

121. Kitamaru, R. and Mandelkern, L., "Unit-Cell Dimensions of Bulk- and Solution-Crystallized Linear Polyethylene,: *J. Polym. Sci., Part A-2*, 8 (1970), 2079–87.

122. Hendus, H. and Schneel, G., "Röntgenographische und IR-spektroskopische kristallinitätsbestimmung an Polyäthylen," *Kunstoffe*, 51 (1961), 69–74.

123. Gopalan, M.R. and Mandelkern, L., "Degree of Crystallinity of Linear Polyethylene from Wide-Angle X-Ray Diffraction," *J. Polym. Sci., Part B*, 5 (1967), 925–29.

124. Chow, J., Glick, R.E., and Mandelkern, L., to be published.

125. Bekkedahl, N. and Matheson, H., "Heat Capacity, Entropy, and Free Energy of Rubber Hydrocarbon," *J. Res. Nat. Bur. Stand.*, 15 (1935), 503–15.

126. McCrum, N.G., "An Internal Friction Study of Polytetrafluoroethylene," *J. Polym. Sci.*, 34 (1959), 355–69.

127. Takayanagi, M., Yoshino, M. and Minami, S., "State of Crystallization of Polyethylene Terephthalate Determined by the Viscoelastic Absorptions," *J. Polym. Sci.*, 61 (1962), S7–S10.

128. Keith, H.D., Padden, F.J. Jr. and Vadimsky, R.G., "Intercrystalline Links in Polyethylene Crystallized from the Melt," *J. Polym. Sci., Part A-2*, 4 (1966), 267–81.

129. Keith, H.D., Padden, F.J. Jr. and Vadimsky, R.G., "Further Studies of Intercrystalline Links in Polyethylene," *J. Appl. Phys.*, 37 (1966), 4027–34.

130. Devoy, C. and Mandelkern, L., "Thermodynamic Properties of Polyethylene Crystallized from Diluent Mixtures," *J. Polym. Sci., Part B*, 6 (1968), 141–50.

131. Mandelkern, L., Fatou, J.G. and Howard, C., "The Nucleation of Long-Chain Molecules," *J. Phys. Chem.*, 68 (1964), 3386–91.

132. Mandelkern, L., Fatou, J.G. and Howard, C., "The Nucleation of Long-Chain Molecules in Monomolecular Layers," *J. Phys. Chem.*, 69 (1965), 956–59.

STUART B. CLOUGH
Lowell Technological Institute
Lowell, Massachusetts

Chapter 14

Characterization of Bulk Polymer Morphology by Optical Methods

ABSTRACT

The application of X-ray diffraction, electron microscopy, and light scattering to the study of polymer morphology is briefly presented. Results of these techniques used for the study of crystals of poly(n-hexyl isocyanate) are discussed.

Introduction

The designation "optical methods" is taken here to include X-ray diffraction and electron microscopy, in addition to techniques based on visible light. The use of wide- and small-angle X-ray scattering, electron microscopy and diffraction, and light scattering for the determination of structural organization of polymers will be presented. In addition to determination of crystal morphology, these methods are useful for the study of noncrystalline heterogeneous materials, *e.g.,* block copolymers containing microphases. Since these techniques are standard in material science, emphasis will be placed on special applications to polymer science, and on recent developments and applications in this field. Though the birefringence technique and the orientation of structure are important in studying and determining the material's physical properties, only occasional mention will be given to such measurements, as these topics were discussed at the previous Sagamore Conference[1].

Polymer crystals generally have a lamellar shape, with the chains oriented normal to the large dimensions. The thickness is on the order of 100 angstroms. If grown from dilute solution, the crystals can be rather distinct (single crystals) as seen in Figure 1, and there is evidence that under carefully controlled growth conditions the chains fold regularly as shown in Figure 2[2]. When the polymer is crystallized from the melt, the morphology is more complex, but the basic lamellar structure is pre-

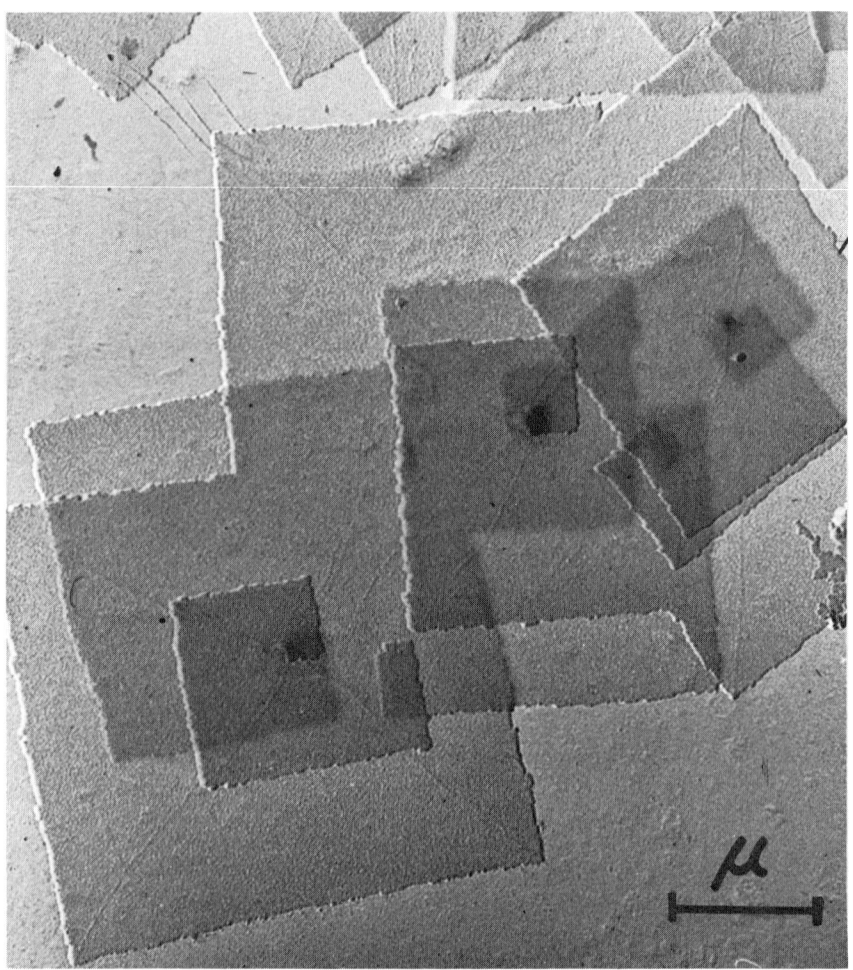

Figure 1.   Electron micrograph of poly(4-methyl-pentene-1) crystals grown from dilute solution.

served. Many molecules now run from one lamella to an adjacent one. The crystals frequently grow in spherulitic aggregates, as in Figure 3. Special preparation conditions, *e.g.*, crystallization under high pressure, can lead to thicker crystals with the chains predominantly extended rather that folded[3].

The application of some of the methods to the study of the crystal structure of poly(n-hexyl isocyanate) will be presented. The poly-isocyanate class of polymers behaves in solution as rather stiff chains, in contrast to the usual flexibly coiled macromolecule[4–6]. Han and Yu[7]

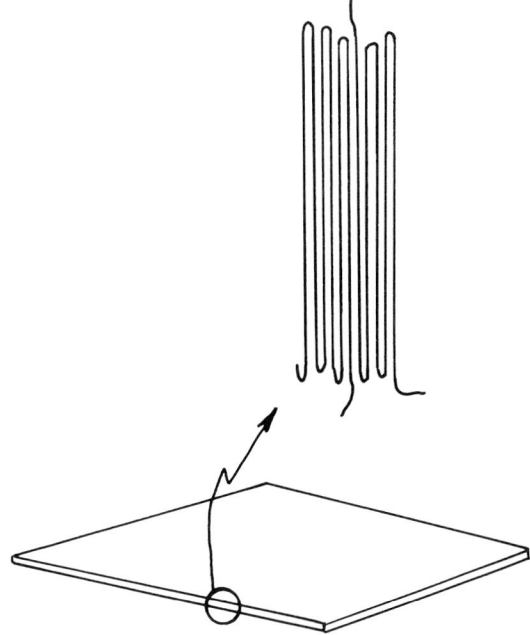

Figure 2.   A model of chain folding in polymer crystals.

Figure 3.   Optical micrograph of polyethylene spherulites.

concluded that the stiffness results from steric interactions. The question of interest is: do morphological differences (from the normal chain-folded polymer crystals) result when the crystals are prepared from solution, due to the chain stiffness?

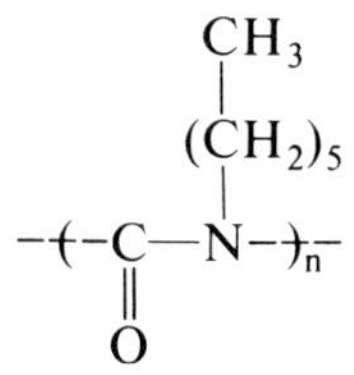

$$CH_3$$
$$(CH_2)_5$$
$$-(-C—N-)_n-$$
$$\parallel$$
$$O$$

Poly(n-hexyl isocyanate)

X-Ray Diffraction

Wide Angle

The principal applications of wide-angle X-ray diffraction to bulk poly-mers are for determination of
  1. degree of crystallinity,
  2. crystal and molecular structure, and
  3. orientation of crystallites.
In addition, recent studies have been performed in the area of dynamic X-ray diffraction in order to determine the response of crystals to imposed deformations.

Degree of Crystallinity

As compared, for example, to diffraction from ceramic materials, crystalline polymers give broad peaks, and in addition, a broad, weak halo (Figure 4). These are manifestations of small imperfect crystals, and the lack of complete crystallinity of the sample. In order to crystallize at all, the polymer must have some degree of both chemical and geometrical regularity along the chain. The degree of crystallinity (weight fraction crystallinity, x_c) depends on such order as well as the method of sample preparation. The term "degree of crystallinity" usually implies the existence of two phases, crystalline and amorphous. In some cases, the cause of the broadening and the halo is better represented by a paracrystalline model (a single phase with a large number of defects)[8]. Various methods of estimating x_c (others include density measurements, thermal analysis,

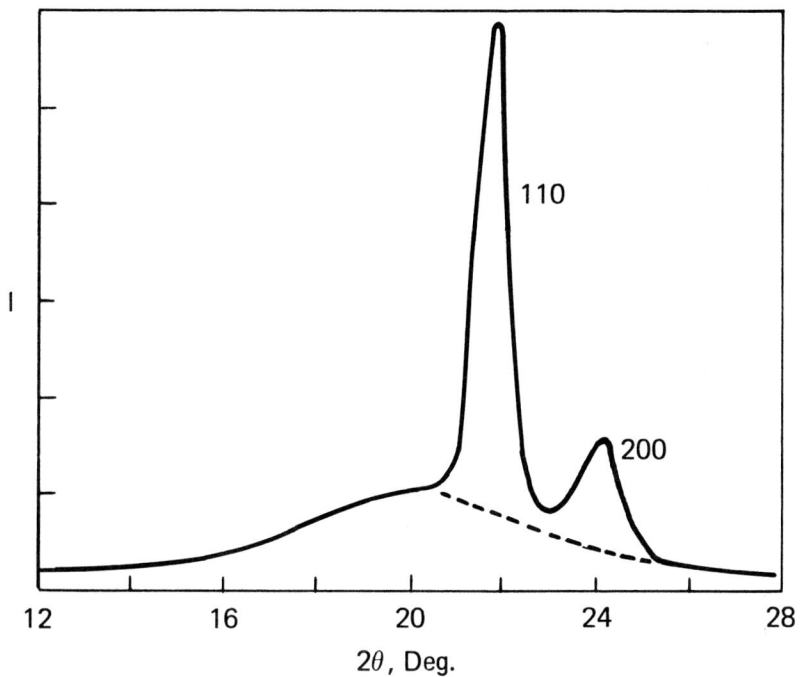

I

12   16   20   24   28

2θ, Deg.

Figure 4.   X-ray diffraction from a polyethylene specimen.

infrared spectroscopy) may yield somewhat different values since different properties of the semicrystalline specimen are examined.

To determine x_c from an unoriented sample, one of several methods of treating I versus 2θ data are employed. (For oriented samples, a technique for averaging intensity at a fixed Bragg angle while varying azimuthal angles may be used[9]). If an entirely amorphous specimen is available, its halo is compared to that of the sample of interest, where at each angle,

$$x_a = 1 - x_c = \left(\frac{I_{sample}}{I_{amorphous\ sample}} \right). \tag{1}$$

Corrections for unequal scattering mass must be included. More frequently, the peak height and peak area A_c and A_a (areas of crystalline and amorphous peaks, respectively) are found to be proportional to x_c and x_a. From samples with different levels of crystallinity, A_c is plotted versus A_a as in Figure 5. x_c is then given by

$$x_c = \frac{A_c}{A_{c,\ 100\%}} = 1 - (slope) \frac{A_a}{A_{c,\ 100\%}}, \tag{2}$$

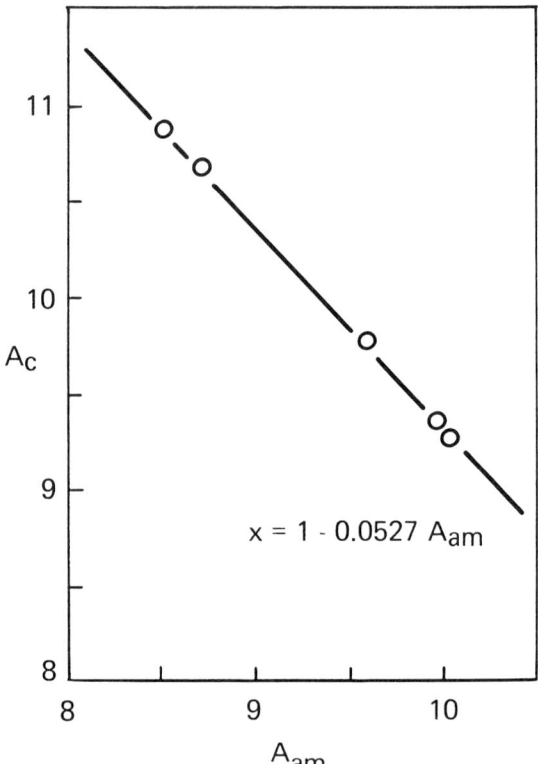

Figure 5. Crystalline peak area (110 and 200) vs. amorphous peak area for 5 polyethylene samples.

For these methods, contributions of defects to the amorphous halo are neglected.

One difficulty in the above methods is resolving the scattering curve into crystalline and amorphous peaks. One recent publication discusses a method for peak resolution for polyethylene terephthalate fibers[10].

No quantitative measurements of x_c were made on the polyisocyanate samples. However, after slight swelling and subsequent drawing, the degree of crystallinity increased markedly, as is seen by comparing the halos in Figure 6.

Molecular and Crystal Structure

One difficulty encountered in the determination of polymer crystal structure is the lack of large single crystals, prohibiting the use of modern

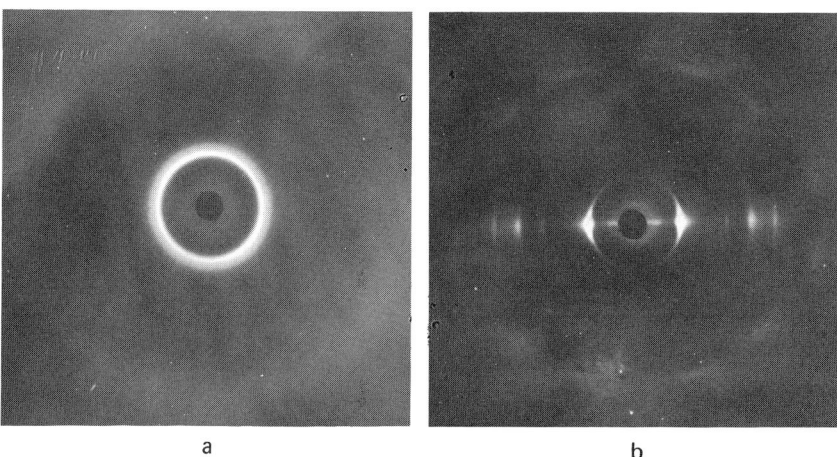

Figure 6. X-ray diffraction from (a) unoriented and (b) drawn samples of poly(n-hexyl isocyanate).

single-crystal techniques for the determination of three-dimensional structure. On the other hand, sometimes information on chain configuration can be obtained rather easily by orienting the molecules. In a recent determination of the crystal structure of α-Gutta Percha, rolling (double orienting) the specimen helped in the data analysis[11]. A more important example of using orientation is the use of helical diffraction theory [12–13]. The helical configuration can be determined from the intensity of layer lines in an X-ray fiber diagram, and the projection of the repeat unit length onto the chain direction is obtained from the layer line spacing. For a proposed u_t helix, the equation

$$tn + um = \ell \qquad (3)$$

is solved for each ℓ (layer line number): u is the number of repeating motifs in t turns of the helix, and n and m are integers; n is the order of the Bessel function controlling the intensity of layer ℓ. The lower the value of n, the higher the intensity.

Using the above procedures, Shmueli et al.[14] determined that the chains in poly(n-butyl isocyanate) form an 8_3 helix. For poly(n-hexyl isocyanate), Table I is constructed from Equation (3) for a 12_5 helix. The high intensity of layers 0, 2, 5, 7, and 12 in Figure 6(b) indicate that this polymer forms a 12_5 helix (when drawn while slightly swollen in toluene and dried). The spacing of the repeat motif (projection along the chain direction) is 3.1 Å, and there are 2 monomer units per repeat motif. The equatorial spacings show that the lateral chain packing is not hexagonal as

TABLE I
n and m Values for 12_5 Helix
$(tn + um = \ell)$

ℓ	m	n
0	0	0
1	−2	5
2	1	−2
3	−1	3
4	2	−4
5	0	1
6	−2	6
7	1	−1
8	−1	4
9	2	−3
10	0	2
11	−2	−5
12	1	0

found for the n-butyl polymer. The spacings observed are in column 1 of Table II.

Crystallization under some conditions led to polymorphism. A low-molecular-weight sample crystallized from tetrahydrofuran gave only the second structure. These spacings are listed in the second column of Table II.

Orientation

Recent papers have discussed the determination of extreme orientation in drawn and extruded polymers[15,16]. Polyethylene crystallized under orientation and pressure in a capillary rheometer had the polymer chains virtually aligned parallel to the flow direction, and had an accompanying high modulus.

TABLE II
Spacings for Two Crystal Structures
of Poly(n-Hexyl Isocyanate)

1	2
(hkO) reflections	
d, A	d, A
16.1	13.6
8.08	7.99
6.42	6.88
5.38	6.31
	5.07

Dynamic X-Ray Experiments

Measurements of the response of polymers to various types of deformation is important in understanding and predicting their behavior in various applications. X-ray diffraction during sinusiodal tensile strain[17] and during relaxation after stretching[18] leads to information on the type and rate of crystal movement. The latter reference indicates that after stretching, first the noncrystalline regions orient, followed by relaxation involving crystals.

Small-Angle X-Ray Scattering (SAXS)

Both photographic and counting methods are widely used in polymer structural studies to determine repeat periods in the range approximately 50–500 Å. The periods of the lamellar structure of single crystals[19], drawn films and fibers[20], and bulk polymer[21] have been estimated from the angle of a scattering peak by using Bragg's Law. See Figure 7. Further, the dimensions of microphases in noncrystalline ABA-type block copolymers have been determined by this technique[22]. Intensity measurements allow the estimation of the specific surface area of a dispersed phase, and electron-density differences of two phases and the correlation function[24]. The presence of SAXS was taken as evidence of some microphase segregation of hard and soft segments in segmented polyurethane elastomers[23]. A few examples of current topics related to bulk polymers follow.

One problem in the use of SAXS to determine the lamellae thickness of bulk samples is the presence of a supposed second-order peak at $\theta_2 \neq 2\theta_1$, and a discrepancy between values obtained by X-ray scattering and electron microscopy[25]. Crist[26] has emphasized the need for careful experimentation and data analysis. His work with annealed samples of polyethylene and polyoxymethylene (POM) yielded values close or equal to 2.0 for the ratio $\theta_2{:}\theta_1$. Quenched samples of POM had a ratio of 2.5, which could indicate coexisting structures or a large skew in the distribution of lamellae thickness.

Quantitative measurement of the X-radiation scattered from block

Figure 7.   Small angle X-ray diffraction from a nylon 66 monofilament.

copolymers (*e.g.*, styrene–butadiene–styrene) yields information regarding the microphase separation. From the shape of the tail of the scattering curve, Kim found that there were sharp interfaces between phases in his samples[27]. Further, the volume fraction of each phase and the shape of the dispersed phase were determined. Skoulios[28] realized similar results from very precise photographic data on ABA block copolymers, and further determined the partitioning of solvents between the phases.

In another type of problem, the absolute intensity of SAXS from randomly chlorinated polyethylene was measured[29]. From the scattering power and the mean-square fluctuation of electron density, the partitioning of Cl atoms between the crystalline and amorphous phases was determined. About 15 percent of the Cl atoms were found in the crystalline phase, indicating incomplete rejection of these atoms foreign to the lattice.

Electron Microscopy and Diffraction

One advantage the polymer morphologist has when working with thin single crystals or ultrathin films is the ease of specimen preparation. The only necessary preparation is a method of gaining contrast between different phases or structures. This may be achieved by shadowing techniques (borrowed from ceramists and metallurgists) or staining (borrowed from biologists). For thick bulk specimens, it becomes necessary to either prepare sections by ultramicrotomy, or to replicate the surface.

In the preparation for Figure 1, a dilute dispersion of crystals was placed on a carbon substrate. After evaporation of the solvent, the crystals were shadowed with Cr metal, providing the contrast at the borders. The electron beam destroys the crystal lattice, but the overall structure remains for photography.

This latter point makes electron diffraction by polymer crystals difficult; the diffraction pattern (Figure 8a) disappears in a few seconds or less at normal beam intensities. It is necessary to work with extremely weak beams in order to preserve the pattern. Advances in image intensification instrumentation have aided in this problem. Figure 8b shows a polyethylene diffraction pattern on a monitor after a 60-minute exposure using image intensification. Despite this problem, in one case electron diffraction has been used exclusive of X-ray diffraction to determine polymer single-crystal structure[30]. Small-angle electron diffraction has been used to study macrolattice structure in extruded triblock polymer[31].

Another technique useful for crystal studies is that of gold decoration[32]. Upon evaporation onto the crystal surface, gold nucleates on edges and discontinuities. Metal falling on the smooth surfaces is mobile, and migrates to the nucleation sites. Blundell and Keller have also concluded from the technique that the mobility on the surfaces of single

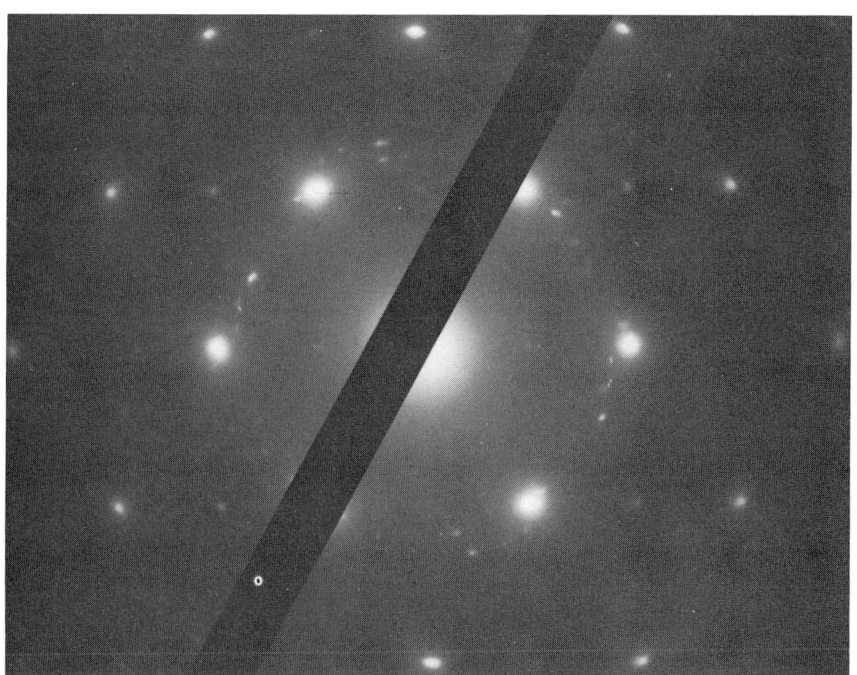

a

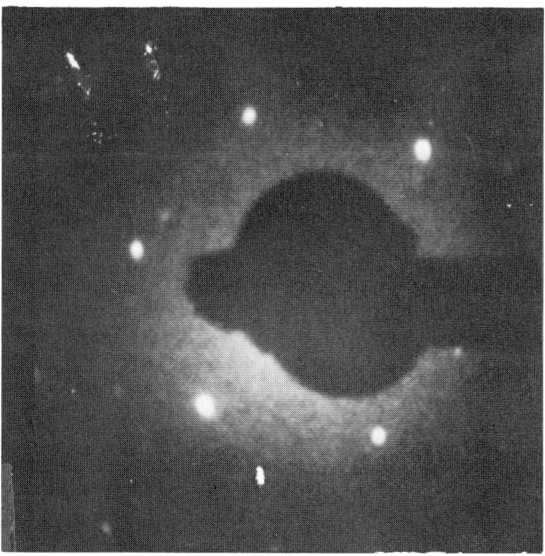

b

Figure 8. (a) Electron diffraction from polyethylene single crystal. (b) Polyethylene diffraction pattern after 60 min exposure using image intensification.

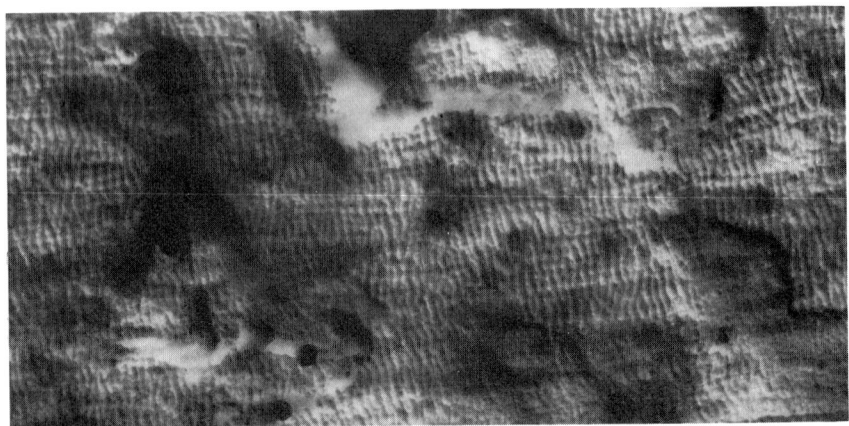

Figure 9. Surface replica of etched polyethylene film.

crystals is due primarily to cilia, and only to a lesser extent to loose loops at the chain folds.

Surface replication has had application in the study of extended chain crystals[3], drawn film and fiber morphology[33,34], and rubber-reinforced plastics[35]. Figure 9 shows a replica of the surface of a highly oriented polyethylene film after etching the noncrystalline portions with fuming nitric acid to enhance demarcation of the crystalline areas. A lamellar structure normal to the horizontal stretching direction is observed. In addition, a superimposed structure aligned with the orientation direction indicates the presence of a microfibrillar structure.

Contrast improvement by staining with OsO_4 has had wide application in the elastomeric copolymer field[36]. Reaction of the stain with the double bonds of one component causes greatly diminished transmission in regions primarily composed of this block, if the other polymeric chain is a saturated one. Figure 10 shows the dark rubbery regions (polybutadiene) embedded in a glassy matrix. The sample was sectioned with a diamond knife prior to staining. This technique has been especially helpful in studying microphase separation in ABA block copolymers, as discussed at the 19th Sagamore Conference.

Figure 11 shows the crystals resulting when poly(n-hexyl isocyanate) is crystallized from a dilute dichloroethane solution. Ribbons about 300 Å in width serve as a backbone for the lateral growth of other ribbon-like crystals. In some areas of observation, the backbone crystals had only little overgrowth. Selected-area electron diffraction should yield information on the orientation of the chains within the crystals, but these experiments have yet to be completed. It is likely that the chains fold as in the case of other stiff-chain polymers. Bittiger assumed that the chains are normal to the fibril axis in the case of crystals of mannan, but did not have the diffraction evidence to verify this[37].

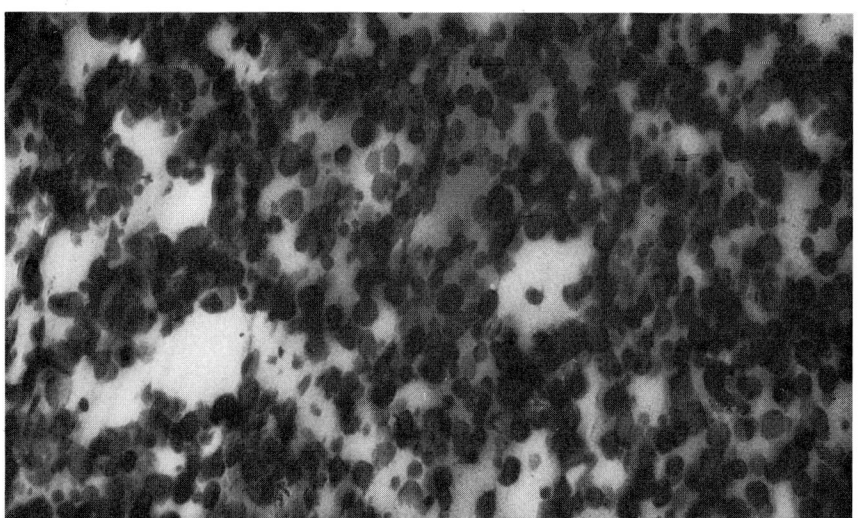

Figure 10.   Stained rubbery regions in a rubber-reinforced glassy polymer.

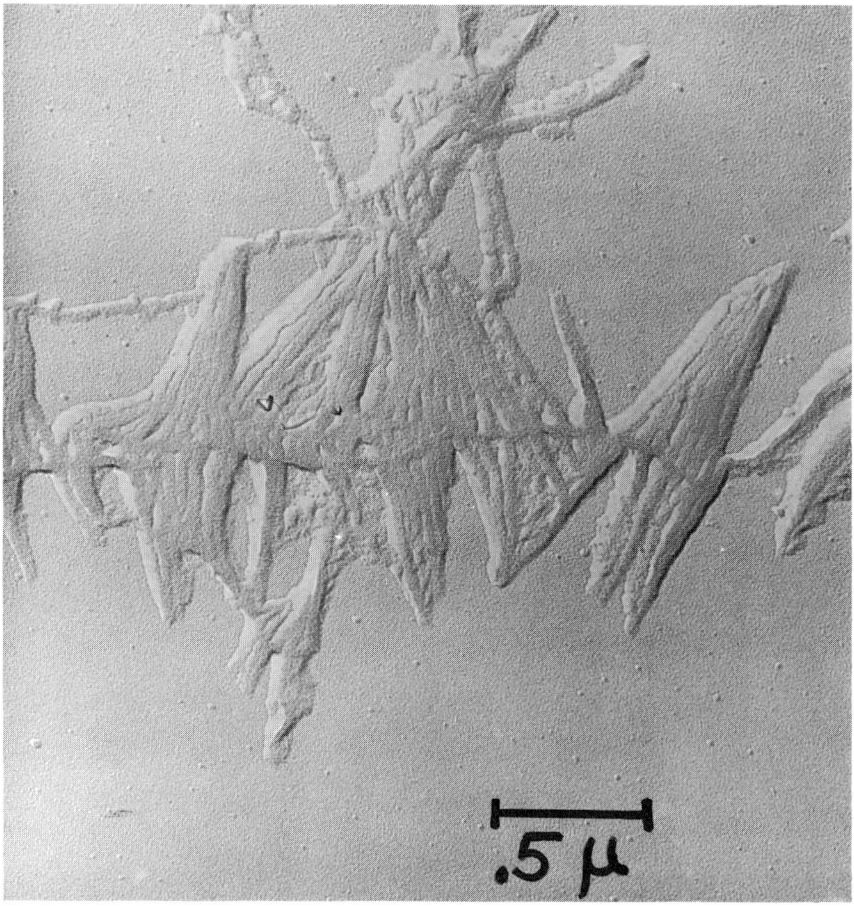

Figure 11.   Electron micrograph of poly(n-hexyl isocyanate) crystals.

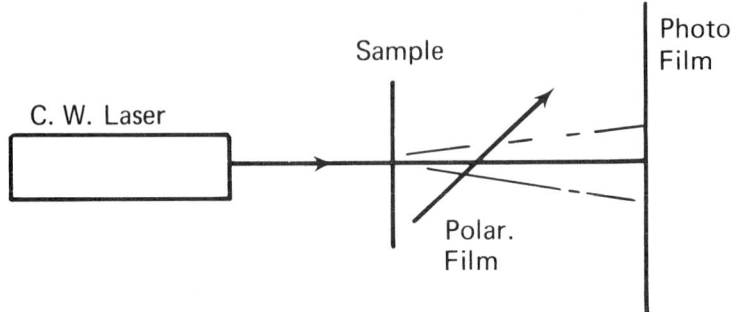

Figure 12. Small-angle light-scattering apparatus for photographic recording.

Light Scattering

The scattering of light by bulk polymer arises primarily from fluctuations in density and in orientation of anisotropic elements. There are two general approaches to the experiments for the study of morphology of polymer films. In the first case, the scattering is recorded photographically at low angles (up to 10°), and the resulting pattern is compared to those calculated from various models of crystal aggregation. Alternatively, the intensity of scattered light is recorded photometrically, and correlation functions for density and orientation are determined. Scattering at larger angles usually requires this quantitative method.

Small-Angle Light Scattering

Stein and his coworkers have pioneered and widely employed the photographic technique. The required apparatus can be easily assembled (Figure 12), or an optical microscope may be used[38]. A film containing spherulites gives a pattern as shown in Figure 13a when the polarizers are crossed (H_v). This is qualitatively similar to patterns calculated from single two- or three-dimensional optically anisotropic spherulites (Figure 13b)[39–41]. The intensity dependence on the radial angle θ and azimuthal angle ξ is given by equations such as

$$(I_{H_v})^{1/2} = KCE_0(\alpha_1 - \alpha_2) A \sin 2\xi \left(\frac{1}{w^2}\right)(2[1 - J_0(w)] - wJ_1(w)) \quad (4)$$

where

$w = kR\sin\theta$

K, C, and E_0 are constant
A, α_1, α_2, and R are parameters describing the spherulite
k is the wave number

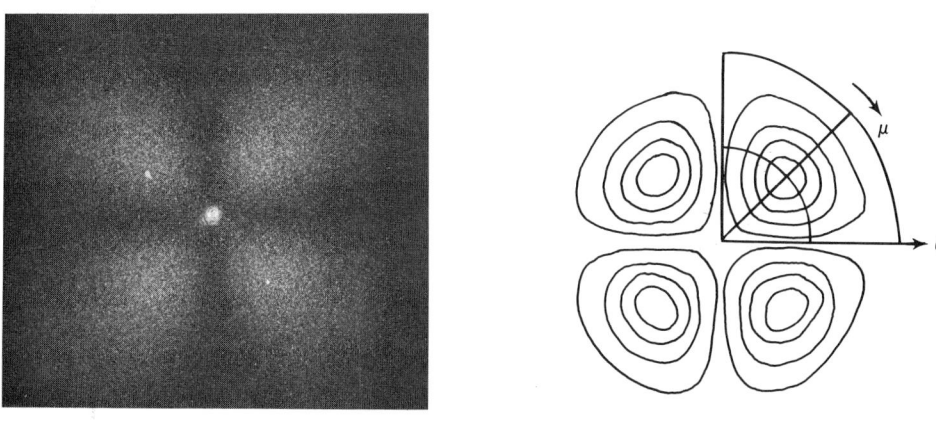

a.   b.

Figure 13. (a) Small-angle light-scattering pattern from spherulitic sample. (b) Calculated pattern.

The average radius of spherulites in a polymer film can be determined from the size of the pattern. The changes in the patterns are useful in determining how spherulites deform when the specimen is stretched[39]. Other models examined have been collections of anisotropic rods[42], single spherulites comprised of biaxial crystals[43], collections of spherulites [44], and spherulites with some disordering of crystals[45]. Also, the theory and experiments using circularly polarized light have been worked out[46], but these appear to be of less utility than when linearly polarized light is used.

Recent studies have extended the technique to the study of cystallization kinetics. During early stages of development, spherulites have a sheaflike structure. Calculated patterns from models of sheaves are similar to those observed at the start of the crystallization process[47]. Further, the Avrami exponent is $n = 5$ for the growth of these structures, and reduces to $n = 3$ for the subsequent spherulitic growth. In another application to kinetics, an apparatus for scanning and continuous recording of the H_V pattern has been developed[48].

Further theoretical studies have considered two-dimensional spherulite models with the incident beam at an angle (other than 90°) to the disk [49]. This allows one to distinguish between two- and three-dimensional spherulites in a sample. Also, for the two-dimensional case, one can get the degree of planar orientation of the optic axes of the scattering elements.

Photometers allowing measurement to low angles have been used to obtain quantitative data. These data have been related using the correlation function method described below to examine the spherulitic organization in shock-cooled polypropylene samples[50].

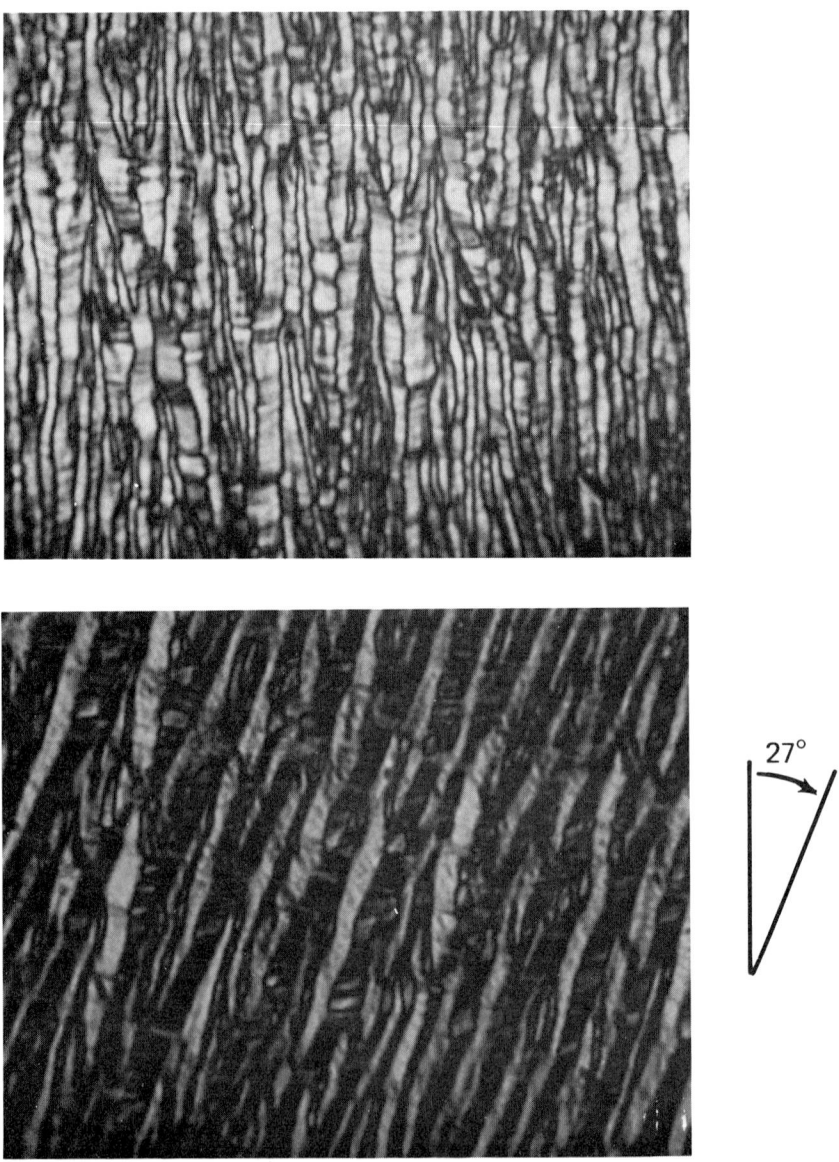

Figure 14. Optical micrographs of rod structure in poly(n-hexyl isocyanate) films. One field is rotated to cause extinction of alternate rods.

Stein has recently extended the technique to dynamic experiments, *i.e.*, the scattered intensity is measured during periodic vibration of the samples[51]. The change in intensity with strain change is related to parameters describing crystal reorientation within the spherulite.

Small-angle scattering patterns from films of poly(n-hexyl isocyanate) prepared by solvent evaporation have monotonically decreasing intensity with increasing scattering angle, typical of scattering from rodlike structures. Further, from the azimuthal scattering angle, it is determined that the optic axes of the crystals lie at an angle to the rods. Birefringence measurements on drawn samples show the optic axis to be aligned with (or perpendicular to) the chain direction. Other birefringence measurements using an optical microscope showed the optic axes to be perpendicular to a fibril structure, with the fibrils at an angle of about 25° to the long direction of the rods. Thus the chains lie at an angle to the rod structure. Figure 14 shows two optical micrographs of a section of film with parallel rods. The extinction of alternate structures in the polarizing microscope when the film is rotated about 25° is consistent with optic axes perpendicular to fibrils oriented at this angle to the long direction of the rods. If the optic axis is along the helix direction, the chains are perpendicular to the fibril axis.

These structures are similar to those observed in films of poly(γ-benzyl-L-glutamate)[52]. The helix of these polymer chains was found to be perpendicular to fibrils which were oriented at an angle (on the order of 20°) to "super rods."

Wide-Angle Light Scattering

The scattering at larger angles originates from fluctuations of refractive index and orientation of anisotropic elements with dimensions smaller than the spherulitic or other superstructure. The photometric method is used to obtain quantitative data, which is transformed into correlation functions (*e.g.*, correlation of refractive index as a function of distance r between two points in the sample). The correlation functions are frequently approximated be combinations of exponential functions of r. Correlation functions for orientation can be calculated from models of crystal aggregation, and the predicted scattered intensity compared with experiment and with patterns calculated by the methods used for small-angle scattering[53].

Acknowledgment

The polyisocyanate samples were received from Dr. G. Hagnauer and synthesized by D. Macione, both of the Army Materials and Mechanics Research Center, Watertown, Mass.

References

1. Wilkes, G.L., and Samuels, S.L., "Rheo-Optical Studies of Block and Segmented Copolymers," in *Block and Graft Copolymers,* Proceedings of the 19th Sagamore Army Materials Research Conference, J.J. Burke and V. Weiss, eds., Syracuse: Syracuse University Press (1973), 225-77.

2. Geil, P.H., *Polymer Single Crystals,* New York: Interscience Publishers (1963), 79-183.

3. Wunderlich, B. and Davidson, T., "Extended Chain Crystals," *J. Polym. Sci., Part A-2,* 7 (1969), 2043-50.

4. Schneider, N., Furusaki, S. and Lenz, R., "Chain Stiffness in Polyisocyanates," *J. Polym. Sci., Part A,* 3 (1965), 933-48.

5. Fetters, L. and Yu, H., "Equilibrium Conformation and 'Worm-Like Coil' Configuration of Poly(n-alkyl isocyanates)," *Macromolecules,* 4 (1971), 385-89.

6. Bur, A. and Roberts, D., "Rodlike and Random Coil Behavior of Poly(n-butyl isocyanates) in Dilute Solutions," *J. Chem. Phys.,* 51 (1969), 406-20.

7. Han, C. and Yu, H., "Chain Configuration of Poly(n-alkyl isocyanates)," *Polym. Prepr., Amer. Chem. Soc., Div. Polym. Chem.,* 14 (1973), 121-26.

8. Alexander, L.E., *X-Ray Diffraction Methods in Polymer Science,* New York: John Wiley & Sons (1969), 137-97.

9. Desper, C.R. and Stein, R., "Randomization of Orientation of Films and Fibers," *J. Polym. Sci., Part B,* 5 (1967), 893-900.

10. Wlochowicz, A. and Jeziorny, A., "Determination of Crystallinity in Polyester Fibers by X-Ray Methods," *J. Polym. Sci., Polym. Phys. Ed.,* 10 (1972), 1407-14).

11. Takahashi, Y., Sato, T., Tadokoro, H. and Tanaka, Y., "Crystal Structure of α-Gutta Percha," *J. Polym. Sci., Polym. Phys. Ed.,* 11 (1973), 233-48.

12. Alexander, L.E., *X-Ray Diffraction Methods in Polymer Science,* New York: John Wiley & Sons (1969), 357-420.

13. Cochran, W., Crick, F. and Vand, V., "The Structure of Synthetic Polypeptides," *Acta Crystallogr.,* 5 (1952), 581-86.

14. Shmueli, U., Traub, W. and Rosenheck, K., "Structure of Poly(n-butyl isocyanate)," *J. Polym. Sci., Part A-2,* 7 (1969), 515-24.

15. Desper, C.R., "Structural Characterization of Isotactic Polypropylene Films of Ultrahigh Orientation," *J. Macromol. Sci., Phys.,* B7 (1973), 105-20.

16. Desper, C.R., Southern, J., Ulrich, R. and Porter, R., "Orientation and Structure of Polyethylene Crystallized Under the Orientation and Pressure Effects of a Pressure Capillary Viscometer," *J. Appl. Phys.,* 41 (1970), 4284-89.

17. Stein, R., "Rheo-Optics on Two Continents," *Polym. Eng. Sci.,* 9 (1969), 320-30.

18. Oda, T. and Stein, R., "X-Ray Diffraction Relaxation of Polyethylene," *J. Polym. Sci., Part A-2,* 10 (1972), 685-92.

19. Burmester, A., Dreyfuss, P., Geil, P. and Keller, A., "On the Annealing of Polyamide Crystal Mats," *J. Polym. Sci., Part B,* 10 (1972), 769-75.

20. Balta-calleja, F. and Peterlin, A., "Plastic Deformation of Polypropylene," *J. Macromol. Sci., Phys.,* B4 (1970), 519-40.

21. Hoffman, J., "Theoretical Aspects of Polymer Crystallization with Chain Folds: Bulk Polymers," *SPE (Soc. Plast. Eng.) Trans.,* 4 (1964), 315-62.

22. Keller, A., Pedemonte, E. and Willmouth, F., "Macro Lattice from Segregated Amorphous Phases of a Three Block Copolymer," *Kolloid 2.2 Polym.,* 238 (1970), 385-89.

23. Clough, S., Schneider, N. and King, A., "Small Angle X-Ray Scattering from Poly-
    urethane Elastomers," *J. Macromol. Sci., Phys.*, B2 (1968), 641–48.

24. Alexander, L.E., *X-Ray Diffraction Methods in Polymer Science,* New York: John
    Wiley & Sons (1969), 280–353.

25. Geil, P., "Small Angle Scattering from Bulk Crystalline Polymers," *J. Polym. Sci.,
    Part C,* No. 13 (1969), 149–63.

26. Crist, B. and Morosoff, N., "Small Angle X-Ray Scattering of Semicrystalline
    Polymers," *J. Polym. Sci., Polym. Phys. Ed.,* 11 (1973), 1023–45.

27. Kim, H., "Morphology of Styrene-Butadiene-Styrene Triblock Copolymers by a Small
    Angle X-Ray Scattering Technique," *Macromolecules,* 5 (1972), 594–97.

28. Skoulios, A.E., "Organizational and Structural Problems in Block and Graft Copoly-
    mers," in *Block and Graft Copolymers,* Proceedings of the 19th Sagamore Army Ma-
    terials Research Conference, J.J. Burke and V. Weiss, ed., Syracuse: Syracuse Uni-
    versity Press (1973), 121–39.

29. Roe, R. and Gieniewski, G., "Small Angle X-Ray Diffraction Study of Chlorinated
    Polyethylene Crystallized from the Melt," *Macromolecules,* 6 (1973), 212–17.

30. Frank, F., Keller, A. and O'Connor, A., "Observations on Single Crystals of an
    Isotactic Polyolefin," *Phil. Mag.,* 4 (1959), 200–14.

31. Dlugosz, J., Folkes, M. and Keller, A., "Macrolattice Based on a Lamellar Mor-
    phology in an SBS Copolymer," *J. Polym. Sci., Polym. Phys. Ed.,* 11 (1973), 929–38.

32. Blundell, D. and Keller, A., "Surface Studies of Polyethylene Crystals as Revealed by
    Surface Decoration," *J. Macromol. Sci., Phys.,* B7 (1973), 253–78.

33. Peterlin, A., "Molecular Mechanism of Plastic Deformation of Polyethylene," *J.
    Polym. Sci., Part C,* No. 18 (1967), 123–32.

34. Crystal, R. and Hansen, D., "Morphology of Cold Drawn Nylon 66," *J. Polym. Sci.,
    Part A-2,* 6 (1968), 981–93.

35. Bucknall, C., Drinkwater, I. and Keast, W., "An Etch Method for Microscopy of
    Rubber-Toughened Plastics," *Polymer,* 13 (1972), 115–18.

36. Matsuo, M., Sagae, S. and Asai, H., "Fine Structures of Styrene-Butadiene Block
    Copolymer Films Cast from Toluene Solution," *Polymer,* 10 (1969), 79–88.

37. Bittiger, H. and Husemann, E., "Crystal Structure of Precipitated Mannan," *J. Polym.
    Sci., Part B,* 10 (1972), 367–72.

38. Stein, R., Erhardt, P., Clough, S., van Aartsen, J. and Rhodes, M., "The Scattering of
    Light by Solid Polymer Films," in *Electromagnetic Scattering,* R. Rowell and R. Stein,
    eds., New York: Gordon and Breach, Science Publishers, Inc. (1967), 339–412.

39. Clough, S., van Aartsen, J. and Stein, R., "Scattering of Light by Two-Dimensional
    Spherulites," *J. Appl. Phys.,* 36 (1965), 3072–85.

40. Stein, R. and Rhodes, M., "Photographic Light Scattering by Polyethylene Films,"
    J. Appl. Phys., 31 (1960), 1873–84.

41. Clough, S., Stein, R. and Picot, C., "Low Angle Light Scattering Equations for
    Polymer Spherulites" *J. Polym. Sci., Part A-2,* 9 (1971), 1147–48.

42. Rhodes, M. and Stein, R., "Scattering of Light from Assemblies of Oriented Rods,"
    J. Polym. Sci., Part A-2, 7 (1969), 1539–58.

43. Clough, S. and Stein, R., "Scattering of Light by Ringed Spherulites Containing
    Biaxial Crystals," *J. Appl. Phys.,* 38 (1967), 4446–50.

44. Stein, R. and Picot, C., "The Effect of Interference Between Anisotropic Scattering
    Entities on the Light Scattered from Polymer Films," *J. Polym. Sci., Part A-2,* 8 (1970),
    1955–69.

45. Hashimoto, T. and Stein, R., "Scattering of Light by Disordered Spherulites," *J. Polym. Sci., Part A-2*, 9 (1971), 1747-67.

46. Holoubek, J., "Small Angle Scattering of Circularly Polarized Light from an Anisotropic Sphere," *J. Polym. Sci., Polym. Phys. Ed.*, 11 (1973), 683-91.

47. Stein, R. and Misra, A., "Kinetics of Growth of Developing Spherulites," *J. Polym. Sci., Polym. Phys. Ed.*, 11 (1973), 109-16.

48. van Antwerpen, F. and van Krevelen, D., "Light Scattering Method for Investigation of the Kinetics of Crystallization of Spherulites," *J. Polym. Sci., Polym. Phys. Ed.*, 10 (1972), 2409-21.

49. Hashimoto, T., Todo, A. and Kawai, H., "Light Scattering from Tilted Two-Dimensional Spherulites," *J. Polym. Sci., Polym. Phys. Ed.*, 11 (1973), 149-73.

50. Keijzers, A. and van Aartsen, J., "Light Scattering by Shock-Cooled Isotactic Polypropylene Film," *J. Appl. Phys.*, 36 (1965), 2874-79.

51. Hashimoto, T., Prud'homme, R. and Stein, R., "Dynamic Light Scattering of Polyethylene," *J. Polym. Sci., Polym. Phys. Ed.*, 11 (1973), 693-736.

52. Wilkes, G., "Unique Highly Developed Superstructure in Films of Poly (γ-Benzyl-4-Glutamate) Cast in a Magnetic Field," *J. Polym. Sci., Part B*, 10 (1972), 935-40.

53. Stein, R., Erhardt, P., Clough, S. and Adams, G., "Scattering of Light by Films Having Nonrandom Orientation Fluctuations," *J. Appl. Phys.*, 37 (1966), 3980-90.

WILLIAM W. GRAESSLEY
Northwestern University
Evanston, Illinois

Chapter 15

Rheological Characterization
of Polymers

ABSTRACT

The rheological properties of polymer melts affect both processing behavior and final properties, the latter through the effects of flow history on solid-state microstructure. The current methods of characterization are linear viscoelastic response and response in steady shearing flow. The current objectives are to correlate these forms of rheological behavior with molecular structure, and to develop simple methods for relating fundamental rheological parameters to response in the complex flows encountered in processing equipment. Transient flow response and the fundamental connections between flow history and microstructure are still relatively unexplored areas.

Introduction

Most thermoplastics pass through the melt state in the course of being processed into useful forms by extrusion or molding. Behavior during melt processing can be very complicated due to the viscoelastic nature of polymer systems. Rapid and efficient processing requires careful control of flow properties, and these in turn depend on molecular characteristics, especially molecular weight, molecular-weight distribution, and long-chain branching. The mechanical properties of the final product are also related to the flow properties and to the flow and temperature history of the process. These variables, together with the local chain structure and postsolidification deformation, control the solid-state microstructure, that is, the proportion of the various phases and their orientation and morphology. Processing effects on final properties are, of course, not unknown in other types of materials. In polymers, however, they tend to be accentuated because chain molecules are readily oriented by flow and configurational relaxation in the melt state is slow.

437

The purpose of rheological characterization is to obtain information from a few experiments which can be used to correlate the response in as many flow situations as possible. Correlations with molecular structure play a useful role in guiding the polymerization step. Two principal methods are used for rheological characterization: linear viscoelastic measurements and measurements of the components of stress developed in shearing flows. The latter, of course, approximate more closely the situation during processing, but a surprising amount of information relevant to flow can also be extracted from small deformation experiments.

This chapter emphasizes characterization by shearing flows, with brief remarks on linear viscoelasticity. Related subjects not covered here are the application of viscoelastic data to mechanical properties of cross-linked elastomers[1]; combined X-ray, optical, and rheological techniques for studying deformation mechanisms in crystalline polymers[2]; and correlation of ultimate properties with low-temperature dielectric and mechanical response[3].

Nature and Characterization of Polymer Melt Behavior

General Features

Pressure-driven flow through a small-diameter capillary is one of the classical methods for measuring viscosity. If the fluid is Newtonian, the volumetric flow rate Q is proportional to the pressure difference $\Delta P = P - P_0$, and the proportionality constant depends on capillary geometry and the viscosity η:

$$Q = \frac{\pi \eta D_0^4}{128 L} \Delta P \qquad \frac{L}{D_0} \gg 1 \qquad (1)$$

The diameter of the exit stream D is very nearly the same as that of the capillary, apart from gravitation and surface effects, and is practically independent of flow rate. Careful measurements show in fact a slight dif-

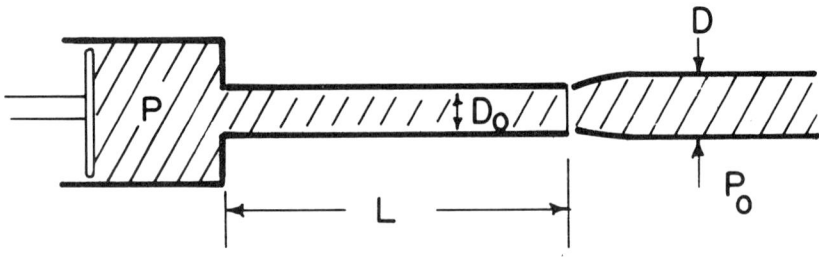

ference between D and D_0 depending on the Reynolds number of the flow $R_e = D_0 \rho Q / L^2 \eta$ (ρ = fluid density, and $R_e < 2,000$ to assure laminar flow). A slight contraction takes place at high flow rates ($D/D_0 \approx 0.86$ for $R_e > 50$) and a slight expansion at low ($D/D_0 \approx 1.15$ for $R_e < 1$)[4].

If capillary flow is used with polymer melts, a number of features appear which have no counterpart in Newtonian fluids. First, $Q/\Delta P$ is no longer a constant for a given capillary but increases with ΔP. Second, if a constant flow rate is maintained but different capillary lengths are used, the ratio $\Delta P/L$ is not constant. Experimentally, $\Delta P \propto L + L_e$, and the end correction L_e increases with Q[5]. Third, the diameter of the exit stream depends strongly on flow rate, sometimes reaching values as large as $3D_0$. This behavior, called "die swell" or the Barus effect, appears to be a general characteristic of viscoelastic fluids.

These observations are explained as follows. The viscosity of polymer melts is a decreasing function of the shear rate. In a given capillary, this means that the apparent viscosity decreases with increasing flow rate. Continuous deformation of polymer melts results in elastic energy storage as well as a viscosity-related dissipation of energy. The length correction L_e in capillary flow corresponds to storage of some of the pressure energy, making it unavailable for driving the fluid through the capillary[5]. The increase in stream diameter at the exit corresponds to a kind of elastic recoil that takes place when the wall constraints are left behind and the stored energy is released.

Accompanying these non-Newtonian flow properties is a characteristic time-dependent behavior. Rapid deformations produce an elasticlike response. This is followed by eventual relaxation back to a stress-free state if the imposed deformation is held constant. Depending on the specimen and the temperature, the process of stress relaxation may require many decades of time to complete. Such behavior may be either desirable or undesirable from the practical point of view, depending on the process and desired properties of the final product.

Measurement of Fundamental Rheological Properties

Laminar flow in a long capillary is an example of steady shearing flow. The physical situation is most easily seen, however, by considering the flow that takes place between parallel plates, one fixed in space and the other moving with constant velocity V:

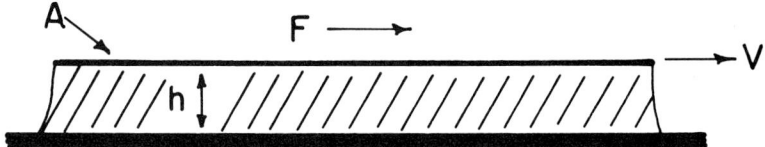

The fluid velocity is proportional to distance from the lower plate x_2 and can be expressed as

$$v_1 = \dot{\gamma} x_2$$
$$v_2 = 0 \tag{2}$$
$$v_3 = 0$$

in which 1 is the direction of flow and 2 is the direction normal to the shearing planes. The shear rate $\dot{\gamma}$ is the velocity gradient V/h, the shear stress transmitted through the fluid is the shearing force per unit area F/A, and the ratio of shear stress to shear rate is the viscosity of the fluid. The shear stress defines only part of the forces exerted on any element of fluid through its contacts with neighboring elements. Consider the components of force per unit area acting instantaneously on the faces of a very small cubic element of flowing fluid, selected out for examination at some instant of time.

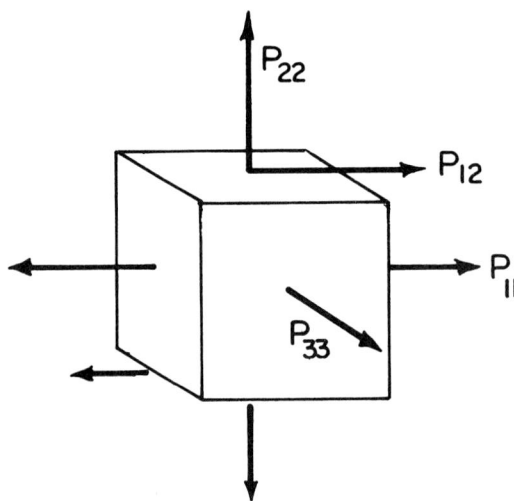

As in the analysis of solids, the balance of forces and torques reduces 18 potential components to 6—3 shear components and 3 normal components—which together comprise the stress tensor. In simple shearing flow, two of the shear components are automatically zero, leaving the shear stress p_{12} which is defined above, and the components, p_{11}, p_{22} and p_{33}, which act normal to the faces of the fluid element.

In Newtonian fluids, the normal components are the same and, with the sign convention that tensions are positive, are equal to $-P_0$, where P_0 is the pressure at that point in the fluid. In polymer melts, the normal components are not equal. It is this property which is connected with elasticity and memory effects. In steady shearing flows, γ is independent of time for each fluid element, and[6,7]

$$p_{12} = \eta(\dot{\gamma})\dot{\gamma}$$
$$N_1 = p_{11} - p_{22} = \theta_1(\dot{\gamma})\dot{\gamma}^2 \tag{3}$$
$$N_2 = p_{22} - p_{33} = \theta_2(\dot{\gamma})\dot{\gamma}^2$$

N_1 and N_2 are the first and second normal stress differences, and $\theta_1(\dot{\gamma})$ and $\theta_2(\dot{\gamma})$ are the normal stress functions. The forms in Equation (3) are chosen because at sufficiently low shear rates $\eta(\dot{\gamma})$, $\theta_1(\dot{\gamma})$, and $\theta_2(\dot{\gamma})$ reduce to constants

$$p_{12} = \eta_0\dot{\gamma}$$
$$N_1 = (\theta_1)_0\dot{\gamma}^2 \tag{4}$$
$$N_2 = (\theta_2)_0\dot{\gamma}^2$$

in which η_0 is the zero-shear viscosity of the melt. Experimentally, N_1 is positive, corresponding to a tension along the lines of flow which is superimposed on the pressure. The coefficient $(\theta_1)_0$ at low shear rates is related to J_e^0, the steady-state shear compliance[8], an important parameter in linear viscoelasticity which characterizes the elastic component of melt response[1]:

$$(\theta_1)_0 = 2J_e^0\eta_0^2 \tag{5}$$

Much less is known about N_2. It appears to be negative and smaller than N_1[9]. Typically, $N_2/N_1 \approx -0.1$ to -0.3. It is difficult to measure, and it is not even certain that it reduces to the form in Equation (4) at low shear rates. Its connections with molecular structure are unknown.

Plate-Cone Rheometers

Both $\eta(\dot{\gamma})$ and $\theta_1(\dot{\gamma})$ can be determined from steady shearing measurements in a plate-cone viscometer. Several commercial instruments are now available.

The polymer melt is placed in the gap between a flat circular plate and a cone. One member is rotated at constant angular velocity ϕ, and, if the gap angle α is small ($\sim 1° - 4°$), the melt experiences a shear rate which is everywhere the same:

$$\dot{\gamma} = \dot{\phi}/\alpha \tag{6}$$

The torque M transmitted to the other member can be measured and used to calculate the viscosity at that shear rate:

$$\eta(\dot{\gamma}) = \frac{3M}{2\pi R^3\dot{\gamma}} \tag{7}$$

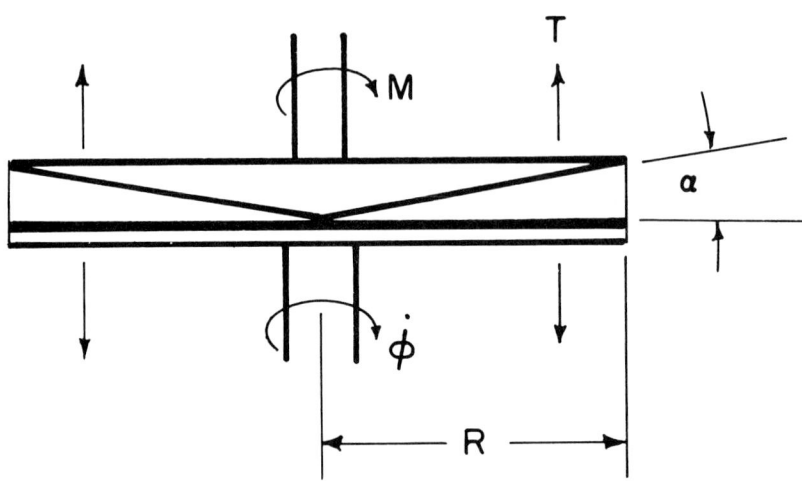

A thrust, tending to force the plate and cone apart, is also developed in viscoelastic fluids. The thrust is a direct result of the tension along the flow lines. Each ring of fluid squeezes upon the inner rings, producing a boundary pressure which increases towards the rotation axis. The resultant total vertical force T can be measured and used to calculate the normal stress function.

$$\theta_1(\dot{\gamma}) = \frac{2T}{\pi R^2 \dot{\gamma}^2} \tag{8}$$

Capillary Rheometers

Capillary instruments are inherently more complicated from the standpoint of data analysis than plate-cone devices. The shear rate, although constant with time for any element of fluid, varies with radial position. This happens because shear stress depends on radial position,

$$p_{12} = \frac{r}{2L} \Delta P \tag{9}$$

so the shear rate, and hence the viscosity in polymer systems, must also vary with radial position. Thus, the velocity profile and resulting volumetric flow rate depend on the values of viscosity at the various radial positions. It is a remarkable fact, however, that $\dot{\gamma}_w$, the shear rate at the capillary wall ($r = D_0/2$), can be evaluated from data on Q versus ΔP alone, regardless of the form of $\eta(\dot{\gamma})$:

$$\dot{\gamma}_w = \frac{8Q}{\pi D_0^3}\left[3 + \frac{d \log Q}{d \log \Delta P}\right] \tag{10}$$

With the shear stress at the wall p_w, obtained with Equation (9) from ΔP, the viscosity is obtained as a function of shear rate

$$\eta(\dot{\gamma}_w) = \frac{p_w}{\dot{\gamma}_w} \qquad (11)$$

The length correction L_e may affect the results, of course. The length correction is established by runs with different capillary lengths, L_e is determined as a function of Q and applied in Equation (9) to obtained corrected values of the shear stress at the wall.

Unfortunately, there appears to be no direct method for determination of the normal stress functions from capillary flow data. Indirect information on elastic response has been deduced from the length correction[5], extrudate swelling ratio D/D_0[10,11,12,13], recoil forces developed on the capillary[14], and pressure on the capillary wall[15] during extrusion. The first three methods depend on the development or release of extra stresses which are then related by various theories to the values of $p_{11} - p_{22}$ inside the capillary. The theories depend on assumptions which are probably incorrect, at least from a rigorous standpoint. However, the values obtained are in fair agreement with normal stress data obtained directly in plate-cone instruments. The fourth method depends on information gathered from pressure taps mounted in the capillary wall, and some question remains about the proper interpretation of such data[16].

Linear Viscoelastic Techniques

The time-dependent response of melts can be measured by stress-relaxation experiments. The tensile stress is monitored as a function of time after the sudden imposition of a small constant tensile strain. The tensile stress relaxation modulus $E(t)$ is the time-dependent ratio of stress to imposed strain. In order to conform to the dictates of linear viscoelasticity, $E(t)$ must be independent of the initial strain, which places an upper limit on the allowable strains.

All relaxation processes in polymer melts typically have the same temperature-dependence. Measurements at several temperatures are usually combined through time–temperature superposition[1]. The resulting stress-relaxation master curve spans many more decades of time than could ever be covered in a single isothermal experiment. Methods are available for converting such information to a distribution of relaxation times[1].

A characteristic mean time τ_0 for relaxation can be defined:

$$\tau_0 = \frac{\displaystyle\int_0^\infty tE(t)\, dt}{\displaystyle\int_0^\infty E(t)\, dt} \qquad (12)$$

The theory of linear viscoelasticity leads directly to the result

$$\tau_0 = \eta_0 J_e^0 \tag{13}$$

A number of commercial instruments are available for measuring the dynamic shear moduli $G'(\omega)$ and $G''(\omega)$ of melts. These provide the same information as stress relaxation with somewhat greater experimental convenience and precision for the ranges of viscosity in polymer melts. From linear viscoelasticity theory:

$$\eta_0 = \lim_{\omega \to 0} \frac{G''}{\omega} \tag{14}$$

$$\tau_o = \lim_{\omega \to 0} \frac{G'(\omega)}{\omega G''(\omega)} \tag{15}$$

Thus, linear viscoelastic data provides information relevant to shearing flow, and are especially valuable in evaluating elastic parameters such as the recoverable compliance J_e^0.

Limitations on Rheological Instruments

The range of shear rates presently available to plate-cone instruments is restricted on the low side to about $\dot\gamma = 10^{-3}$ sec^{-1}. The instruments themselves place lower limits on the angular velocity, and the waiting time for achievement of steady-state response becomes inconveniently long at lower shear rates. The upper limit is set by a flow instability whose origin is still uncertain, but which appears to develop when $p_{11} - p_{22}$ becomes comparable in magnitude to p_{12}. The shear rate at instability can be increased somewhat by the use of smaller gap angles, but eventually a practical limit is reached, usually preventing a very deep penetration into the non-Newtonian region. Capillary instruments can go as low as $\dot\gamma = 1$ sec^{-1}, and with care can be made to reach 10^{-1} sec^{-1}. The problem is instrumental, caused by lower limits on pressure measurement in the case of devices driven hydrostatically and a combination of detection problems, reservoir pressure contributions, and plunger-speed limitation in piston-driven instruments. The upper limit is much greater than in plate-cone systems, with shear rates as high as 10^4 sec^{-1} being not uncommon.

A variety of flow instabilities also develop in capillary instruments, usually displaying themselves by irregularities in the extrudate[17,18]. Sometimes the viscosity–shear rate data obtained with irregular extrudates appears to be a very reasonable continuation of the curve obtained at lower shear rates and smooth extrudates; in other cases discontinuities appear. Prudence dictates the acceptance of viscosity data only when smooth extrudates are formed.

Some types of instability appear to originate at the inlet region to the capillary (inlet fracture). Proper design of the inlet region can prolong stable flows to higher shear rates. Other instabilities appear to be initiated at the capillary wall (land fracture) and are relatively insensitive to capillary geometry. Capillary-flow instabilities turn out to be of considerable interest from the practical point of view. The upper limit on processing rate is governed in many cases by the onset of instability[19]. Some types of processes, such as wire coating by high-density polyethylene, are actually operated at shear rates well beyond the first appearance of instability. For unknown reasons, a second stable region appears for this polymer at very high shear rates.

Commercial instruments for determining dynamic moduli, typically plate-cone viscometers with oscillatory drive attachments, are presently limited to the frequency range $10^{-1} \, \text{sec}^{-1} < \omega < 10^2 \, \text{sec}^{-1}$. Plentiful use of frequency–temperature superposition is therefore required to develop information on the relaxation-time distribution. Special techniques are usually required to measure phase angles with an accuracy sufficient to apply Equation (15)[20].

Examples of Experimental Results

Figure 1 shows viscosity–shear rate data on a sample of commercial polystyrene, deduced from measurements with a plate-cone instrument

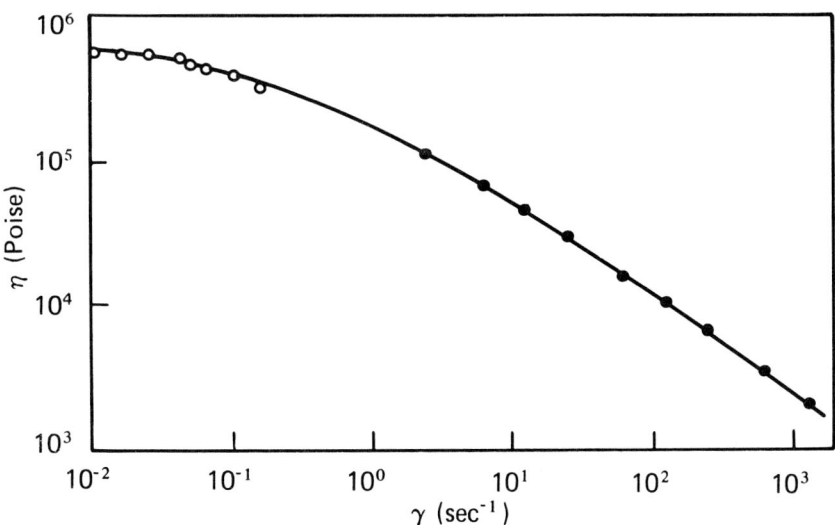

Figure 1. Viscosity–shear rate behavior for a commercial polystyrene ($\overline{M}_w = 220,000$). The temperature is 190°C; the filled circles are data from a capillary viscometer, and the open circles are data from a plane-cone viscometer.

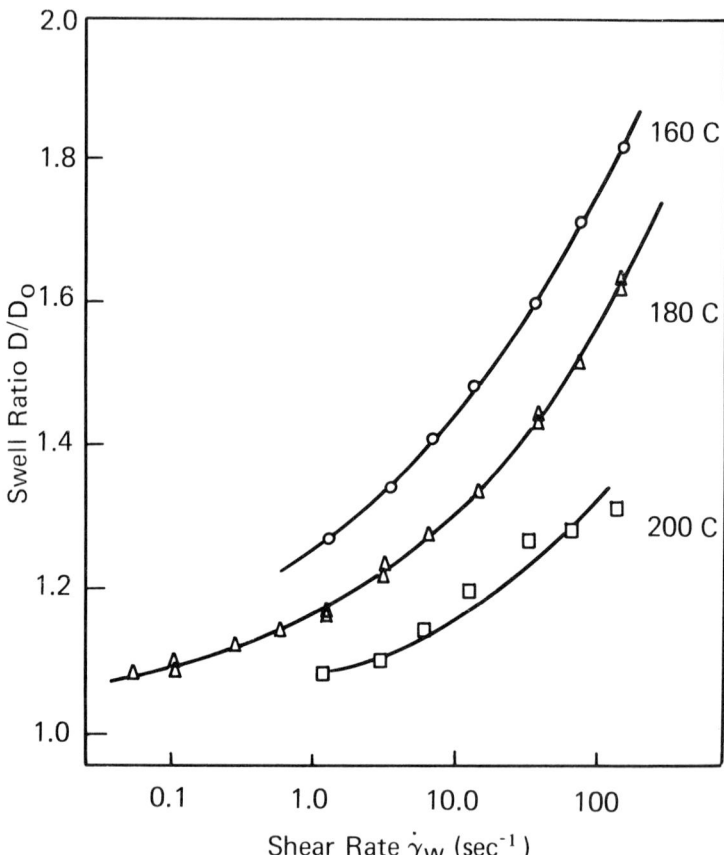

Figure 2.   Die swell as a function of shear rate for a commercial polystyrene ($\overline{M}_w$ = 220,000). Data was gathered on nonannealed extrudates; D_0 = 0.070'', L/D_0 ranged from 27 to 56.

and a capillary instrument. It is evident that the two instruments deal with somewhat different portions of the η versus $\dot{\gamma}$ curve. Figure 2 shows the extrudate swelling ratio D/D_0 as a function of shear rate for different temperatures, while Figure 3 shows the same data as a function of shear stress at the capillary. As is evident from this example, D/D_0 is primarily a function of shear stress. Raising the temperature lowers viscosity and D/D_0 at a given flow rate, but when the data are compared at the same pressure drop (or shear stress), D/D_0 is practically independent of temperature.

Figure 4 shows the normal stress function $p_{11}-p_{22}$ as a function of

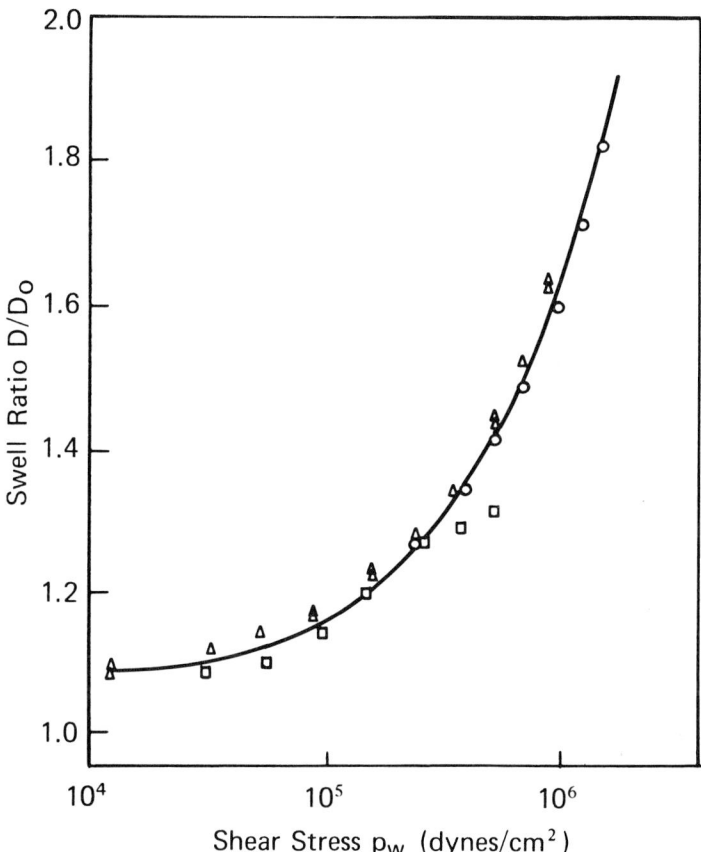

Figure 3. Die swell as a function of shear stress at the capillary wall. The data are the same as shown in Figure 2.

shear stress. The data were obtained from total thrust measurements in plate-cone flow. The lines represent $p_{11}-p_{22}$ calculated from D/D_0 data according to a number of theories[11,12,13,21,22]. The order of magnitude is generally correct, and agreement with calculations based on the Tanner theory is quite good. Figure 5 shows both viscosity and swelling behavior as functions of shear rate. At low shear rates where the viscosity is nearly constant, D/D_0 is small. As the shear rate increases, D/D_0 begins to rise in the same region where η begins to fall. As we shall see, this is a rather general characteristic of polymer systems, dictated by an underlying connection between the normal stress coefficient $(\theta_1)_0$ and the shear-rate dependence of viscosity.

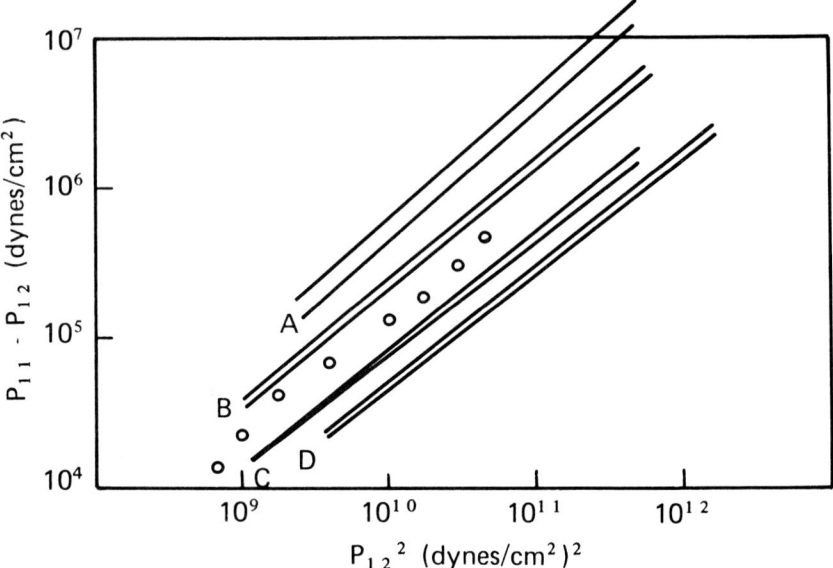

Figure 4. Normal stress difference as a function of shear stress for a commercial polystyrene ($\overline{M}_w$ = 220,000). The points are data measured in a plate-cone instrument. The lines are values calculated from die-swell behavior according to several theories: Curve A - Graessley, Glasscock, and Crawley; Curve B - Tanner; Curve C - Mori; Curve D - Bagley and Duffey.

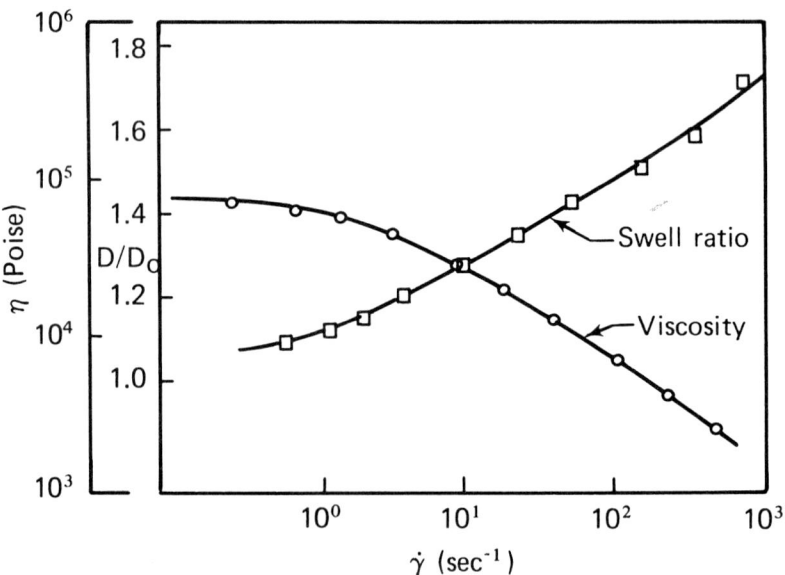

Figure 5. Comparison of viscosity and die-swell dependences on shear rate for a commercial polystyrene ($\overline{M}_w$ = 220,000). The temperature is 180°C.

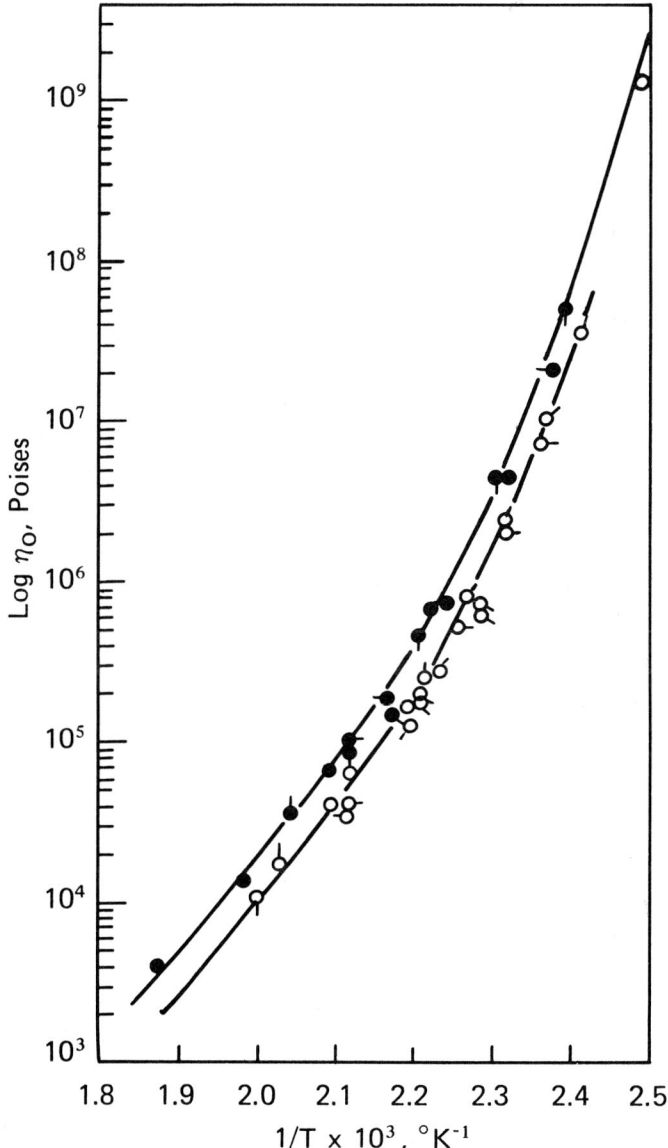

Figure 6.   Zero-shear viscosity as a function of temperature for two narrow-distribution samples of polystyrene. The various symbols indicate different investigators.

Structure-Property Relations in Melts

Zero-Shear Viscosity

The viscosity at zero shear rate is controlled by the product of two parameters, a friction factor ζ_0 which depends on the local molecular properties that govern the viscosity of monomeric liquids, and a structure factor F which depends on the large-scale polymeric features[23].

$$\eta_0 = \zeta_0 F \qquad (16)$$

The Vogel equation,

$$\zeta_0(T) = A e^{+B/(T-T_0)} \qquad (17)$$

governs the temperature-dependence of viscosity in the vicinity of the glass temperature T_g. The parameters A, B, and T_0 are characteristic constants of the polymer, and $T_0 \approx T_g - 60°C$. The Vogel form is replaced by the usual Arrhenius expression

$$\zeta_0 = A' e^{E/RT} \qquad (18)$$

at temperatures in excess of about $T_g + 100°C$. Figure 6 is an Arrhenius plot for samples of polystyrene, showing the typical increase in apparent activation energy for flow as the temperature approaches T_g, 100°C in this case. The temperature-dependence is insensitive of molecular weight, except for very short chains ($M < 10^4$).

The structure factor is the same function of molecular weight for all linear polymers. At constant temperature

$$
\begin{aligned}
\eta_0 &= KM & M &< M_c \\
\eta_0 &= K'M^{3.4} & M &> M_c
\end{aligned}
\qquad (19)
$$

in which M_c is a characteristic molecular weight for the polymer. Typically, M_c corresponds to a chain length of 300–700 main chain atoms; values for several common polymers are given in Table I. The same form

TABLE I
Characteristic Molecular Weights for the Viscosity
Behavior of Linear Undiluted Polymers[1]

Polymer	M_c
Polyethylene	3800
Polyethylene Oxide	4400
Polybutadiene	5900
Polyisoprene	10,000
Polyisobutylene	15,200
Polyvinyl Acetate	24,500
Polydimethyl Siloxane	24,500
Polystyrene	33,000

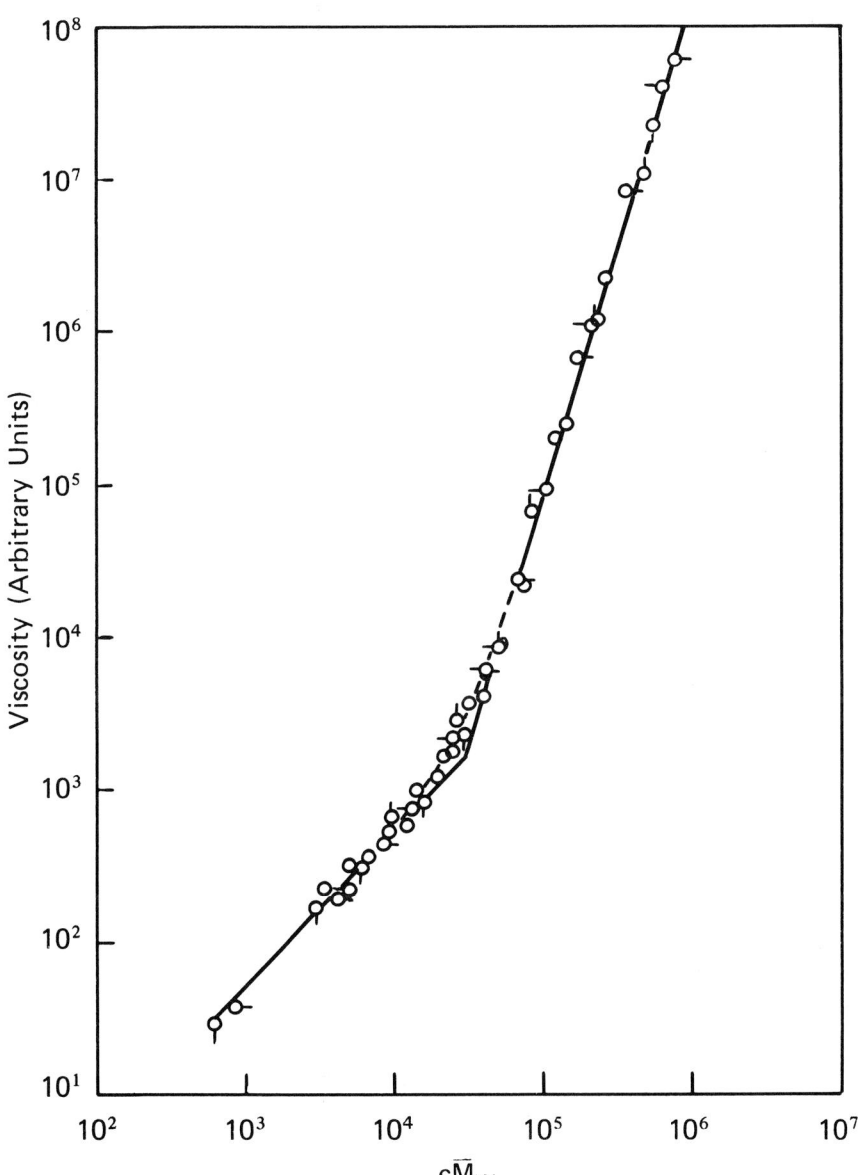

Figure 7. Zero-shear viscosity as a function of the product of polymer concentration and molecular weight for polystyrene. Concentrations range from 25% to 100% polymer; the viscosities have been adjusted to constant friction factor ζ.

applies for concentrated solutions, except that M is replaced by cM/ρ, ρ being the density of the polymer and c its concentration (wt/volume). The friction factor also changes with concentration, principally due to changes in T_0. Figure 7 shows η_0 versus cM behavior for polystyrene. Concentrated solution viscosities have been adjusted to accomodate for changes in friction factor.

In polydisperse systems M is replaced by $\overline{M}_w$. The molecular weights of most commercial polymers are well above M_c, so it is clear that the limiting viscosity in most practical cases is extremely sensitive to molecular weight.

Steady-State Shear Compliance

Molecular theories provide an equation for J_e^0:

$$J_e^0 = \frac{2}{5} \frac{M}{\rho R T} \tag{20}$$

For polydisperse systems:

$$J_e^0 = \frac{2}{5} \frac{\overline{M}_w}{\rho R T} \frac{\overline{M}_z \overline{M}_{z+1}}{\overline{M}_w^2} \tag{21}$$

The same equations apply to concentrated solutions if ρ is replaced by c. These equations are surprisingly good for order of magnitude estimates of J_e^0 based on molecular information alone. However, departures from Equation (20) begin to appear at molecular weights greater than about $3.5\,M_c$. Figure 8 is a correlation of J_e^0 for narrow-distribution polystyrene systems. Other linear polymers show the same behavior; like the viscosity, the compliance appears to be expressable in an approximate universal form:

$$\frac{J_e^0 c R T}{M} = 0.5 \qquad cM < 3.5\,\rho M_c$$

$$\frac{J_e^0 c R T}{M} = 0.5 \left(\frac{3.5\,\rho M_c}{cM} \right) \quad cM > 3.5\,M_c$$

The precise effect of polydispersity is a matter of current debate. The form suggested by theory

$$J_e^0 = (J_e^0)^* \frac{\overline{M}_z \overline{M}_{z+1}}{\overline{M}_w^2} \tag{22}$$

where $(J_e^0)^*$ is the compliance for a narrow-distribution polymer of molecular weight $\overline{M}_w$, predicts a very strong dependence on the properties of the high-molecular-weight tail. This is qualitatively correct, but somewhat

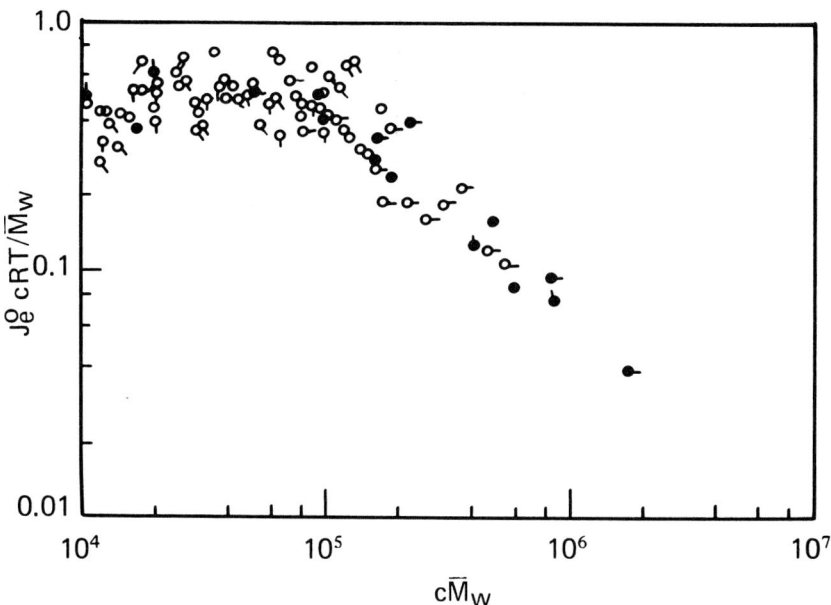

Figure 8. Reduced steady-state recoverable compliance as a function of the concentration-molecular weight product for narrow distribution polystyrene. The filled circles are undiluted polymer, $T \approx 150-200°C$; the open circles are concentrated solutions, $T \approx 25°C$.

better numerical agreement with experimental data has been obtained with polydispersity factors of the form $(\overline{M}_z / \overline{M}_w)^a$, where a is approximately 3.0.

In contrast to η_0, the compliance is insensitive to temperature, being merely inversely proportional to the product of density and absolute temperature. This is useful in practice because J_e^0 can be measured for a polymer sample at experimentally convenient conditions by any one of several methods, and, neglecting the minor temperature correction, treated as essentially a characteristic constant for that sample.

Viscosity–Shear Rate Behavior

In many polymers the onset of shear-rate dependence in the viscosity is governed by τ_0, the characteristic time that governs stress relaxation. Furthermore, the *form* of the η-$\dot{\gamma}$ curve depends in quite a direct way on the large-scale structure of the polymer, principally the molecular weight distribution in the case of linear polymers. These characteristics are expressed by the master relationship

$$\eta = \eta_0 f(\tau_0 \dot{\gamma}) \tag{23}$$

where the viscosity master function $f(\ )$ depends only on the breadth of the molecular weight distribution, and η begins to depart from η_0 at shear rates in the range $\dot{\gamma}_c = 1/\tau_0$.

According to Equations (13), (21), and (23), viscosity data for the same polymer sample at different temperatures should form a single curve when plotted as η/η_0 versus $\eta_0 \dot{\gamma}/\rho T$. Lacking values of η_0 and ignoring the weak temperature-dependence of ρT, log-log plots of η versus p_{12} at different temperatures should be superposable by shifts along the viscosity axis alone.

Figure 9 shows η versus p_{12} obtained at two temperatures in a capillary rheometer. Figure 10 is the same data, shifted to a single temperature and then replotted as η versus $\dot{\gamma}$. The combining procedure thus extends the portion of the master curve which can be obtained by a single instrument. Superposition is by no means a general principle, however, so it is generally desirable to achieve more overlap than that shown in Figure 10 (by making measurements at more intermediate temperatures, for instance). Perhaps the best way to state the situation is the following. The temperature coefficient of viscosity, measured at constant shear stress, should be the same for all values of shear stress according to the superposition principle given above. By contrast, the temperature coefficient at

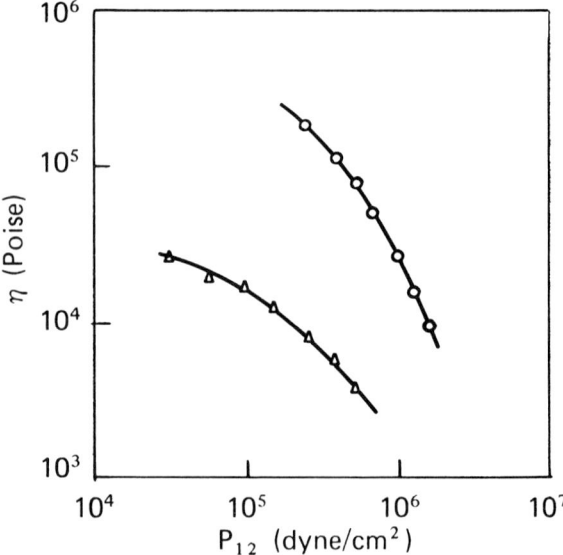

Figure 9. Viscosity as a function of shear stress for a commercial polystyrene ($\overline{M}_w$ = 220,000). The triangles are measurements at 200°C; the hexagons are measurements at 160°C.

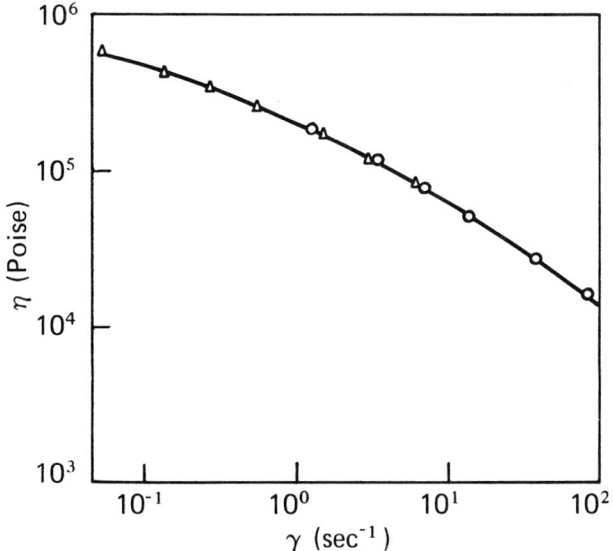

Figure 10. Viscosity-shear rate behavior for a commercial polystyrene ($\overline{M}_w$ = 220,000). The data in Figure 9 were combined by shear stress-temperature superposition and replotted at 160°C.

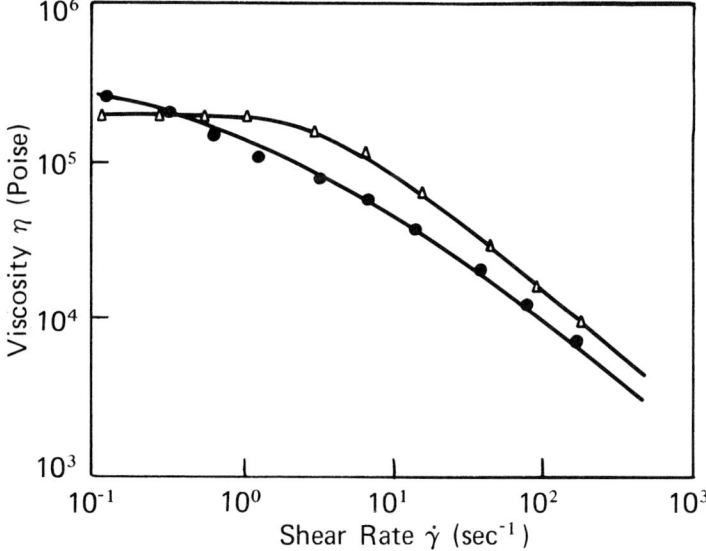

Figure 11. Comparison of viscosity-share rate behavior for a commercial poly-styrene ($\overline{M}_w$ = 261,000) and a narrow-distribution polystyrene ($\overline{M}_w$ = 160,000). The temperature is 180°C; the data were measured in a capillary viscometer.

constant shear rate should vary with shear rate. Experimentally, the temperature coefficient at constant stress is indeed much less variable than that at constant shear rate, but not necessarily independent of stress. The residual variation could be caused either by a change in the form of $f(\ )$ with temperature, in which case the flow curves would be nonsuperposable, or by an anomalous temperature-dependence in the scaling factor τ_0.

The effect of molecular weight distribution on the flow curve is illustrated by data on polystyrene melts in Figure 11. Broadening the distribution widens the transition between Newtonian behavior at low shear rates and approximate power law behavior ($\eta \mid \dot{\gamma} \mid^{-b}$) at high shear rates. The onset of non-Newtonian behavior takes place at lower shear rates as the distribution broadens. The latter behavior follows, of course, from the increased compliance of broad distribution systems, causing $\dot{\gamma}_c$ to decrease. The viscosity becomes rather insensitive to molecular structure at very high shear rates.

High-density polyethylenes have very broad distributions and hence require many decades of shear rate to span the region between Newtonian

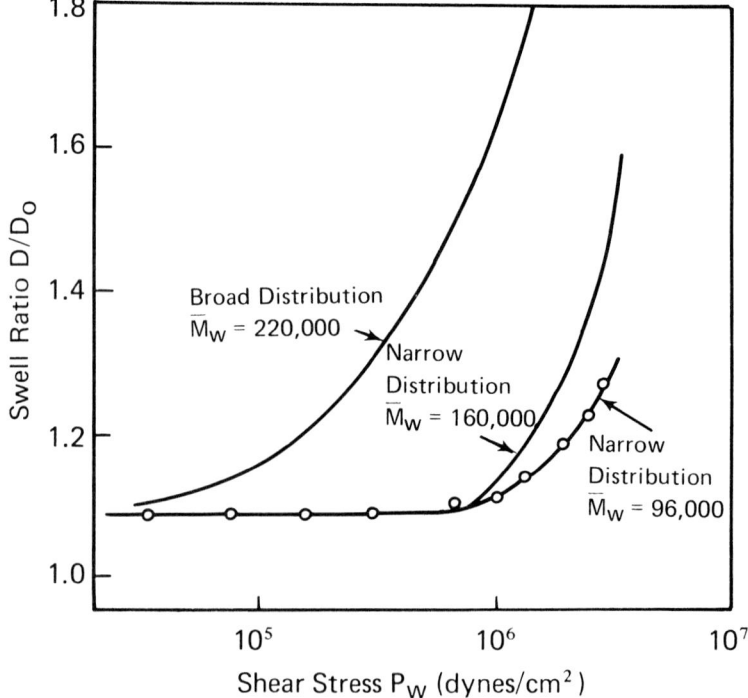

Figure 12. Comparison of die-swell behavior for narrow-distribution polystyrene and a commercial polystyrene. Curves represent a composite of results from 160°C to 200°C as well as a variety of capillary diameters and L/D_0 values.

and power law behavior. Commercial polystyrenes have narrower distributions, but are still broad compared to the very narrow distributions which can be formed by special techniques such as anionic polymerization.

Die-Swell Behavior

Distribution effects are also evident in die-swell behavior. Figure 12 shows D/D_0 versus p_w for narrow-distribution polystyrenes and for a commercial polystyrene. As with the viscosity, D/D_0 remains constant out to much higher shear stresses (or shear rates) for narrow-distribution polymers, but then it rises very steeply, appearing to approach values comparable to the broad-distribution sample at high shear rates.

REFERENCES

1. Ferry, J.D., *Viscoelastic Properties of Polymers,* 2d ed., New York: John Wiley & Sons, 1970.

2. Oda, T. and Stein, R.S., "X-ray Diffraction Relaxation of Polyethylene," *J. Polym. Sci., Part A-2,* 10 (1972), 685–91; and Uemura, Y. and Stein, R.S., "Change in Infrared Dichroism during and following High-Speed Elongation of Polyethylene," *J. Polym. Sci., Part A-2,* 10 (1972), 1691–98.

3. Ward, I.M., *Mechanical Properties of Solid Polymers,* London and New York: Wiley-Interscience, 1972; and McCrum, N.G., Read, B.E. and Williams, G., *Anelastic and Dielectric Effects in Polymeric Solids,* London and New York: John Wiley & Sons, 1967.

4. Gavis, J. and Modan, M., "Expansion and Contraction of Jets of Newtonian Liquids in Air: Effect of Tube Length," *Phys. Fluids,* 10 (1967), 487–97.

5. Bagley, E.B., "The Separation of Elastic and Viscous Effects in Polymer Flow," *Trans. Soc. Rheol.,* 5 (1961), 355–68; and Bagley, E.B. and Schreiber, H.P., "Elasticity Effects in Polymer Extrusion," in *Rheology,* Vol. 5, *Theory and Applications,* F.R. Eirich, ed., New York: Academic Press (1969), 93–125.

6. Lodge, A.S., *Elastic Liquids,* London: Academic Press, 1964.

7. Coleman, B.D., Markovitz, H. and Noll, W., *Viscometric Flow of Non-Newtonian Fluids,* New York: Springer-Verlag, 1966.

8. Coleman, B.D. and Markovitz, H., "Normal Stress Effects in Second-Order Fluids," *J. Appl. Phys.,* 35 (1964), 1–9.

9. Tanner, R.I., "A Correlation of Normal Stress Data for Polyisobutylene Solutions," *Trans. Soc. Rheol.,* 17 (1973), 365–73.

10. Metzner, A.B., Houghton, W.T., Sailor, R.A. and White, J.L., "A Method for the Measurement of Normal Stresses in Simple Shearing Flow," *Trans. Soc. Rheol.,* 5 (1961), 133–47.

11. Graessley, W.W., Glasscock, S.D. and Crawley, R.L., "Die Swell in Molten Polymers," *Trans. Soc. Rheol.,* 14 (1970), 519–44.

12. Bagley, E.B. and Duffey, H.J., "Recoverable Shear Strain and the Barus Effect in Polymer Extrusion," *Trans. Soc. Rheol.,* 14 (1970), 545–53.

13. Tanner, R.I., "A Theory of Die-Swell," *J. Polym. Sci., Part A-2,* 8 (1970), 2067–78.

14. Shertzer, C.R. and Metzner, A.B., "Measurement of Normal Stresses in Viscoelastic Materials at High Shear Rates," in *Proceedings of the Fourth International Congress on Rheology, Part 2,* E.H. Lee, ed., New York: Wiley-Interscience (1965), 603–18.

15. Han, C.D., Charles, M. and Philippoff, W., "Rheological Implications of the Exit Pressure and Die Swell in Steady Capillary Flow of Polymer Melts. I. The Primary Normal Stress Difference and the Effect of L/D Ratio on Elastic Properties," *Trans. Soc. Rheol.,* 14 (1970), 393–408.

16. Kaye, A., Lodge, A.S. and Vale, D.G., "Determination of Normal Stress Differences in Steady Shear Flow. II. Flow Birefringence, Viscosity, and Normal Stress Data for a Polyisobutene Liquid," *Rheol. Acta,* 7 (1968), 368–79.

17. Tordella, J.P., "Unstable Flow of Molten Polymers," in *Rheology, Vol. 5, Theory and Applications,* F.R. Eirich, ed., New York: Academic Press (1969), 57–92.

18. Ballenger, T.F., Chen, I-J., Crowder, J.W., Hagler, G.E., Bogue, D.C. and White, J.L., "Polymer Melt Flow Instabilities in Extrusion: Investigation of the Mechanism and Material and Geometric Variables," *Trans. Soc. Rheol.,* 15 (1971), 195–215.

19. Brydson, J.A., *Flow Properties of Polymer Melts,* New York: Van Nostrand-Reinhold, 1971.

20. Prest, W.M. Jr., Porter, R.S. and O'Reilly, J.M., "Non-Newtonian Flow and the Steady-State Shear Compliance," *J. Appl Polym. Sci.,* 14 (1970), 2697–2706.

21. Vlachopoulos, J., Horie, M., and Lidorikis, S., "An Evaluation of Expressions Predicting Die Swell," *Trans. Soc. Rheol.,* 16 (1972), 669–85.

22. Mori, Y., "On Die Swell in Molten Polymers," *Appl. Polym. Sym.* No. 20 (1973), 209–20.

23. Berry, G.C. and Fox, T.G., "The Viscosity of Polymers and Their Concentrated Solutions," *Adv. Polym. Sci.,* 5 (1968), 261–357.

C. RICHARD DESPER
Army Materials and Mechanics Research Center
Watertown, Massachusetts

Chapter 16

Characterization of Molecular and Crystalline Orientation of Anisotropic Solid Polymers

ABSTRACT

Polymers in the bulk state are often subjected to treatments, such as drawing or rolling, to produce a preferred orientation of the polymer molecules on crystals. Such preferred orientation results in materials which may be highly anisotropic with respect to a number of properties, including physical, mechanical, optical, and spectroscopic properties. To obtain useful models for predicting the mechanical behavior of such materials, it is necessary to examine the degree of orientation of the structural elements comprising the polymer system at the submicroscopic level.

Orientation functions, which are second- or fourth-power moments of the orientation distribution, provide a sound mathematical basis for characterization of the structural elements of a solid polymer. Since a solid polymer may be crystalline, paracrystalline, amorphous, or a mixture of phases, detailed information may be required for full characterization of a given system. X-ray diffraction is highly satisfactory as a measure of orientation in crystalline and paracrystalline phases, but not for amorphous phases. Small-angle X-ray diffraction is used to determine the preferred orientation of the lamellar crystals commonly encountered in semicrystalline polymers. Two other properties, birefringence and sonic velocity, give a measure of overall molecular orientation which may be somewhat biased when more than one phase is present. In such cases, methods have been developed to use birefringence or sonic velocity to determine amorphous orientation by difference, using X-ray diffraction to measure crystalline orientation.

Spectroscopic methods, notably infrared absorption and Raman scattering, are sensitive to particular chemical groups within a polymer. Thus, these methods offer additional information which would otherwise be unavailable. A particular band may, however, be relatively insenitive to orientation because of mathematical problems arising from symmetry considerations.

Orientation parameters may be correlated with mechanical properties if a suitable morphological and mechanical model is used. Examples are given for the correlation of orientation with modulus, dynamic mechanical properties, and ultimate tensile properties.

Introduction

It is well known that preferred molecular orientation can be induced in solid polymers through proper mechanical and thermal treatment. This molecular orientation can have a profound effect on mechanical properties. A basic understanding of structure–property relationships in polymers demands, in many instances, an intimate knowledge of the state of orientation. Over the years, a number of techniques have been found useful for characterizing orientation in polymers, but as research in this area has progressed, our original naive conception of molecular orientation has given way to more complex theoretical formulations. We are no longer content with a simple measure of molecular orientation, such as a birefringence value, but must ask more searching questions into the state or orientation of various structural elements, such as chemical groups, crystal lattice planes, crystal lamellae, etc.

Research in orientation has its own nomenclature and concepts which must be established as a basis for further discussion. Preferred orientation is fundamentally a nonrandom distribution in the orientation of a certain basic type of structural unit with respect to a fixed coordinate system. A given polymer sample may possess a number of preferred orientations, such as the preferred orientation of crystal axes, of amorphous chain segments, and of side groups. However, one unifying concept—the orientation function—provides a common basis for treatment of a wide variety of orientation problems and for formulating the interrelationships between orientation of different structural elements. The basic principle of the orientation function, introduced by Hermans et al. [1] and used extensively by Stein[2], is that of characterizing an orientation distribution by a second-order moment. For example, let us consider an orientation distribution $N(\alpha)$, where α is the angle between a fiber axis and some individual structural element. The orientation function f is defined by Hermans as:

$$f = \frac{(3\overline{\cos^2\alpha} - 1)}{2} \tag{1}$$

where the bar over $\cos^2\alpha$ indicates that this quantity is to be averaged over the entire orientation distribution. Depending upon the nature of the distribution $N(\alpha)$, the function f can range between $-1/2$ and $+1$, the limiting cases being:

   (a) perpendicular orientation, α assuming only the value $\pi/2$, which gives $f = -1/2$;

(b) random orientation, with all α values equally probable, which gives $f = 0$; and

(c) parallel orientation, where α is restricted to a value of zero, which gives $f = 1$.

Partially oriented systems will have intermediate values of the orientation function, the exact value characterizing the degree of orientation.

The above discussion has presumed uniaxial orientation, implying the existence of an axis of cylindrical symmetry so that the orientation distribution may be expressed as a function of a single angle. The present discussion will be limited to the uniaxial orientation case. Techniques exist for treating orientation problems in cases of lower symmetry, and the reader is referred to a more comprehensive review[3] for a more complete discussion. Since the original introduction of the orientation function, the concept has been extended to include fourth-order moments[4–8]. Fourth-order orientation functions have been found useful in dealing with experimental techniques including fluorescence, Raman scattering, and nuclear magnetic resonance, but they will not be included in the present discussion.

Various polymers exist in the solid state in a variety of forms. Certain polymers crystallize either slowly or not at all, and are found in a single amorphous phase. For the semicrystalline polymers, two phases are present: a crystalline phase, exhibiting the usual crystal diffraction effects, and a noncrystalline, or amorphous phase, which gives rise to a more diffuse X-ray diffraction pattern. A third type of order may also be found in polymers, intermediate in perfection between the highly ordered crystal phase and the highly disordered amorphous phase. This level of order includes the paracrystalline structure of polyacrylonitrile[9] and the "smectic" structure of quenched polypropylene[10].

It is evident that the two-phase semicrystalline class of polymers offers the greatest complexity in studying orientation, since one must determine separately the degree of orientation of the crystalline and amorphous phases, an overall average degree of molecular orientation being inadequate for an understanding of structure–property relationships. On the other hand, semi-crystalline polymers offer greater promise in terms of mechanical properties. The high-tenacity synthetic fibers—the polyamides, polypropylene, and poly(vinyl alcohol)—are all highly oriented semicrystalline polymers. Thus, the present discussion will be framed in terms of semicrystalline polymers, with the understanding that the methods of study may be reduced to the less complex single-phase case if required. Experimental data illustrating the principles involved will be drawn, for the most part, from the author's work with oriented polyproplylene.

The techniques for determining preferred orientation include X-ray diffraction, optical birefringence, sonic velocity, polarized infrared absorption, polarized fluorescence, and polarized Raman scattering. Each method is sensitive to a particular aspect of orientation, and thus a combination of methods is usually required for a complete characterization of orientation in a semicrystalline polymer.

Methods of Characterization

Characterization of Crystalline Orientation by X-Ray Diffraction

The X-ray diffraction experiment offers essentially complete information on the orientation of the crystalline phase in semicrystalline polymers. Several recent reviews[3,11,12] offer comprehensive surveys of the experimental techniques, so the present work will be addressed mainly to methods of interpretation, notably the orientation function approach.

Crystal diffraction of X-rays arises from interference between parallel planes of atoms in the crystal lattice, and may be detected using photographic film or through use of a diffractometer equipped with a photon counter detector. While the film method is used extensively for qualitative survey work, the counting method is required for accurate quantitative characterization with orientation functions. The instrument used in the present work is shown in Figure 1. The diffractometer has been interfaced to a small computer and has been programmed[13] for preferred orientation studies, although it was originally designed for single-crystal structure studies.

The technique for examining uniaxially oriented samples is illustrated in Figure 2. Two diffractometer angles are used:

(a) the Bragg angle 2θ, which selects the diffraction plane to be examined in accordance with Bragg's Law,

$$\lambda = 2d_{hk\ell} \sin\theta_{hk\ell} \tag{2}$$

and (b) the latitude angle χ, which defines the orientation of the crystallites contributing to a given diffraction intensity datum. [Note that χ is the complement of the angle α previously defined.] From a series of intensity measurements at a fixed 2θ value, the orientation function $f_{hk\ell}$ for a particular diffraction plane is determined[14] by integration:

$$f_{hk\ell} = \frac{\int_0^{\pi/2} P_2(\sin\chi) \, I(2\theta_{hk\ell},\chi) \, \cos\chi \, d\chi}{\int_0^{\pi/2} I(2\theta_{hk\ell},\chi) \, \cos\chi \, d\chi} \tag{3}$$

where the weighting function $P_2(\sin\chi)$ is the second Legendre poly-

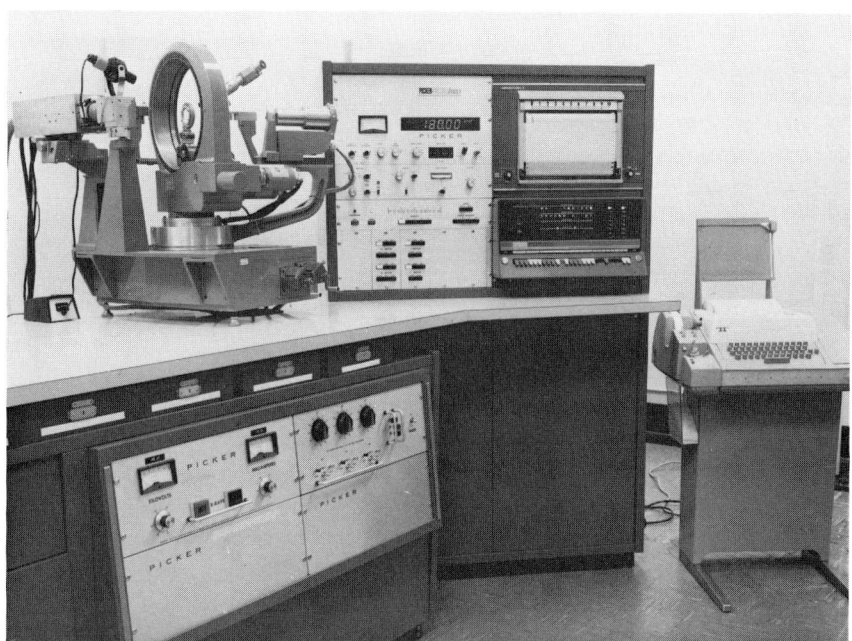

Figure 1. The Picker FACS-1 Computer-Controlled X-ray Diffraction System (Desper[13]).

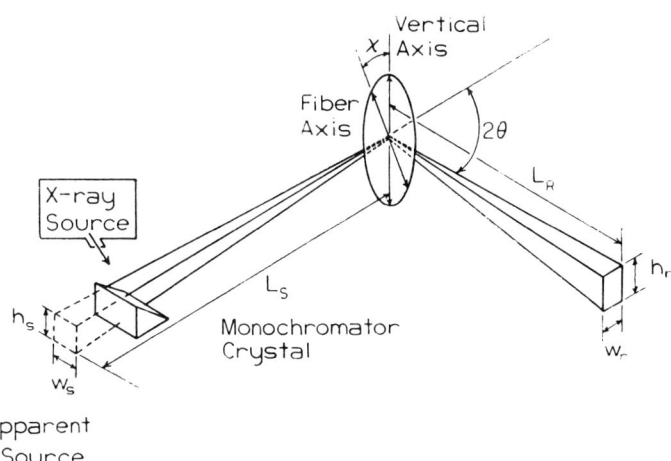

Figure 2. X-ray diffractometer arrangement for uniaxial orientation studies (Desper, Southern, Ulrich, and Porter[58]).

nomical function, defined by:

$$P_2(\sin\chi) = \frac{(3\sin^2\chi - 1)}{2} \tag{4}$$

Thus the experiment will yield an orientation function for any diffraction plane possessing sufficient experimental intensity. However, no more than five diffraction planes may have independent orientation functions[15], because of the geometric relationships between the orientation of various diffraction planes within the unit cell. In practice, the orientation function of greatest interest that of the crystallographic C axis, which for most polymer crystals is parallel to the polymer chain. This is denoted f_c, which may be thought of as the "C-axis orientation function" or as the "crystalline orientation function." The function fc is usually not determined directly, since the diffraction plane perpendicular to the C-axis is often weak or nonexistent. Instead, fc may be calculated[1-3,11,15,16] from orientation functions f_{hk0} for one, two, or three planes parallel to the C-axis, the number of independent orientation functions required depending on crystal symmetry[15]. In the case of the monoclinic polypropylene unit cell, for instance, f_c is given by:

$$f_c = \frac{(1 - 2\sin^2\rho_{040})f_{110} - (1 - 2\sin^2\rho_{110})f_{040}}{(\sin^2\rho_{040} - \sin^2\rho_{110})} \tag{5}$$

where the angles ρ_{040} and ρ_{110} are fixed by unit-cell geometry[17]. Inserting the values for ρ_{040} and ρ_{110} into Equation (5), it reduces to:

$$f_c = -0.9006 f_{040} - 1.0994 f_{110} \tag{6}$$

Characterization of Amorphous Orientation

Birefringence Method

Birefringence, sometimes called double refraction, is a phenomenon in which a body displays different refractive indices for light with plane polarization in two perpendicular directions. Birefringence is a manifestation of preferred orientation of polymer molecules, arising from the anisotropy of polarizability of the various chemical bonds constituting a polymer chain. In an oriented polymer solid, the nonrandom orientation of chemical bonds causes the macroscopic polarizability to become a tensor instead of a simple scalar; *i.e.*, the polarizability becomes dependent upon the direction of polarization of light. The specific polarizability P is related to index of refraction n through the well-known Lorenz–Lorentz equation

$$\frac{n^2 - 1}{n^2 + 2} = \frac{4}{3}\pi P \tag{21}$$

Thus the index of refraction is, in turn, dependent upon direction of polarization.

For a uniaxially oriented sample, there will be two principle indices of refraction $n_\parallel$ and $n_\perp$, corresponding to polarization of light parallel to and perpendicular to the sample symmetry axis. The birefringence Δ is defined as the difference between the two indices:

$$\Delta = n_\parallel = n_\perp \tag{7}$$

and is used as a measure of the degree of polymer orientation.

Recent reviews of Samuels[18], Wilkes[19], and Stein[20] have given details of the interpretation of birefringence data in terms of molecular parameters. We shall therefore discuss the interpretation of birefringence at a fundamental level and show how it may be used to measure amorphous orientation. The birefringence of a semicrystalline polymer is given by:

$$\Delta_t + x_c \Delta_c + (1 - x_c) \Delta_a + \Delta_f \tag{8}$$

where Δ_t is the total (experimental) birefringence, Δ_c and Δ_a are crystalline and amorphous phase birefringences, x_c is the degree of crystallinity, and Δ_f is form birefringence. The form birefringence is usually small and is often neglected. The crystal birefringence Δ_c depends upon the crystal orientation functions f_a, f_b, and f_c for the three crystal axes; thus this term may be calculated from X-ray diffraction data, using an x_c value determined from density or X-ray diffraction. When the individual crystals are uniaxial, i.e., when $n_a = n_b$, the crystal birefringence Δ_c is given by:

$$\Delta_c = f_c \Delta_c^0 \tag{9}$$

where Δ_c^0 is the intrinsic birefringence of the crystal phase. Similarly, the amorphous phase birefringence may be expressed as the product of an orientation function and the intrinsic amorphous phase birefringence:

$$\Delta_a = f_{am} \Delta_a^0 \tag{10}$$

Solving the Equations (8)–(10) for f_{am} and neglecting form birefringence, we obtain:

$$f_{am} = \frac{\Delta_t - x_c f_c \Delta_c^0}{(1 - x_c) \Delta_a^0} \tag{11}$$

Equation (11) gives the amorphous orientation function f_{am} in terms of density, birefringence, and X-ray diffraction data (for f_c). The constants Δ_c^0 and Δ_a^0 remain to be determined. Samuels[21,22] provides an ingenious method for determining these quantities involving experimental sonic velocity values in addition to the above data. His values for intrinsic crystalline and amorphous birefringence in polypropylene are 29.1×10^{-3} and 60.0×10^{-3}, respectively[21].

Sonic Velocity Method of Orientation Characterization

Preferred orientation in polymers is of interest mainly as a technique for improving mechanical properties, notably modulus and tensile strength. One of the best ways to observe this effect is by observing the velocity of sound in solid oriented polymers. There are two prime advantages of sonic measurements over other determinations of mechanical properties: (a) the frequency is high (10 kHz), so it is more likely to observe the elastic region of response in these viscoelastic materials; and (b) the strain is quite small and reversible. Because of these factors, it is possible to correlate sonic velocity directly with molecular orientation.

Moseley proposed[23,24] a quite simple but effective theory for sonic velocity which treats the oriented polymer as a single phase. Moseley proposes an experimental parameter α defined by:

$$\alpha = 1 - \left(\frac{C_u}{C}\right)^2 \tag{12}$$

where C_u and C are the sonic velocities for unoriented polymer and for the oriented sample respectively. In polymer systems where C_u is not dependent upon the degree of crystallinity (which means that the amorphous regions are below their glass-transition temperature), α may be identified with the average degree of molecular orientation f_{ave}:

$$\alpha = f_{ave} = x_c f_c + (1 - x_c) f_{am} \tag{13}$$

Samuels[18,21,22] found it necessary to modify the theory for polypropylene at room temperature to take into account the two-phase nature of the system. The problem is that the amorphous phase, when it is not in the glassy state, will inherently have a much lower modulus and sonic velocity than the crystal phase. The Samuels theory is expressed in terms of modulus rather than velocity, but the two are readily interconverted by the relationship:

$$E_{or} = \rho C^2 \tag{14}$$

where E is the sonic modulus, ρ is density, and C is sonic velocity. Samuels defines two key constants $E_{t,c}^0$ and $E_{t,am}^0$, which are the intrinsic transverse moduli of the crystalline and amorphous phases. When the two constants are unequal, Moseley's α is no longer a valid measure of f_{ave}, since the sonic velocity will be biased to be more sensitive to the lower modulus (amorphous) phase, assuming a series mechanical model for amorphous and crystalline phases (Figure 3). The Samuels theory has been solved[17] to give amorphous orientation explicitly:

$$f_{am} = 1 + \frac{x_c(1 - f_c)E_{t,am}^0}{(1 - x_c)E_{t,c}^0} - \frac{3 E_{t,am}^0}{2(1 - x_c)E_{or}} \tag{15}$$

where E_{or} is the experimental modulus of the oriented sample. The need

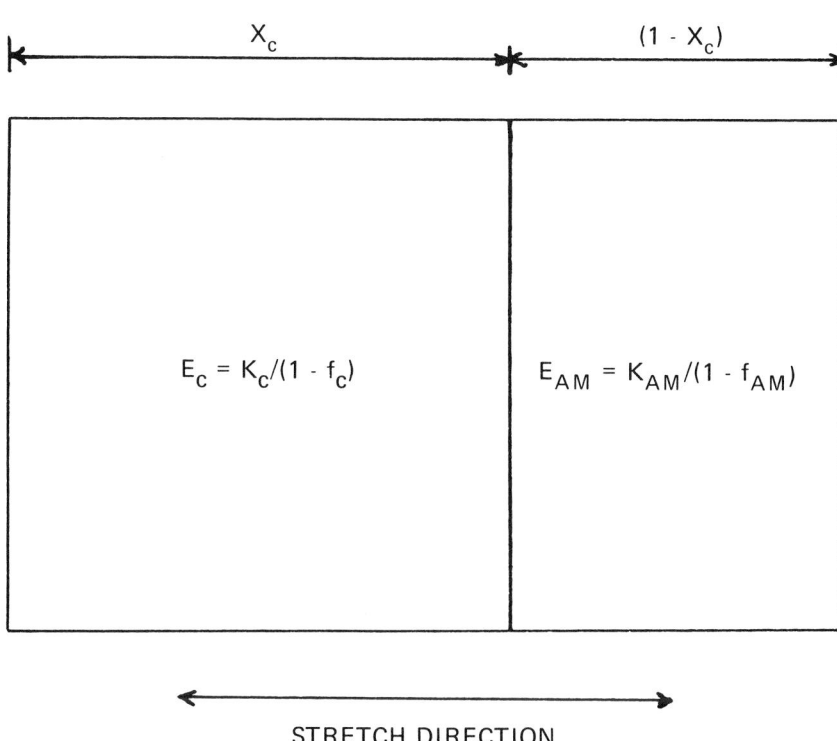

Figure 3. Two-phase mechanical model of oriented semicrystalline polymers (Desper, Lewis, Lopata, and Roylance[36]).

for the more complex Samuels theory becomes evident for polypropylene when the intrinsic modulus values

$$E^0_{t,am} = 1.06 \times 10^{10} \text{ dynes/cm}^2 \qquad (16)$$

and

$$E^0_{t,c} = 3.96 \times 10^{10} \text{ dynes/cm}^2 \qquad (17)$$

are found to differ[21] by a factor of 3.7.

Vibrational Spectroscopy of Oriented Systems

Polarized Infrared Absorption

Polarization was first used in infrared spectroscopy to resolve questions arising in the assignment of vibrational frequencies. Having established the assignment for a particular band, one may reverse the procedure and measure the degree of orientation of the vibrating species using absorption intensities at different polarizations. The experimental methods have been well established and may be found in a number of references[18–20, 25–27]. Infrared absorption arises from interaction between the electric field vector $\mathbf{E}$ of the light beam and the transition moment vector M for the vibrating polymer segment. The absorbance A for a particular band and a particular light polarization is given by:

$$a = k(\overline{\mathbf{M} \cdot \mathbf{E}})^2 \qquad (18)$$

where k is a proportionally constant and $\mathbf{M} \cdot \mathbf{E}$ is the dot product of the two vectors. For a uniaxially oriented polymer, there will be two independent absorbances a_y and a_z, corresponding to polarization perpendicular to and parallel to the axis of symmetry z. The anisotropy of absorption may be analyzed in terms of the dichroic ratio r_{zy}.

$$r_{zy} = \frac{a_z}{a_y} \qquad (19)$$

or in terms of the orientation function

$$f_{mz} = \frac{a_z - a_y}{a_z + 2a_y} \qquad (20)$$

This orientation function characterizes the degree of orientation of the transition moment vectors $\mathbf{M}$ of the individual polymer segments with respect to the fiber axis z.

The difficulty is that one must establish the orientation of $\mathbf{M}$ with respect to the local polymer chain (see Figure 4). The vector $\mathbf{M}$ may be parallel, perpendicular, or tilted at an angle to the polymer chain, so measuring $f_{m,z}$ does not necessarily establish the molecular orientation factor $f_{c,z}$. When there is threefold or higher symmetry about the chain direction c, the two orientation functions are related by:

$$f_{c,z} = \frac{f_{m,z}}{f_{c,m}} \qquad (21)$$

where $f_{c,m}$ is the orientation function for the angle $\phi_{m,c}$ in Figure 4. This third orientation factor is fixed for a particular polymer band and need only be measured once.

Raman Spectroscopy

Until the advent of the continuous wave laser, Raman spectroscopy was not a practical quantitative tool. Within the past five years there has been pioneering work in the Raman spectroscopy of oriented systems, but this field is still in its infancy. The theoretical problem is a level greater in intricacy than the infrared case since (a) fourth-order orientation functions,

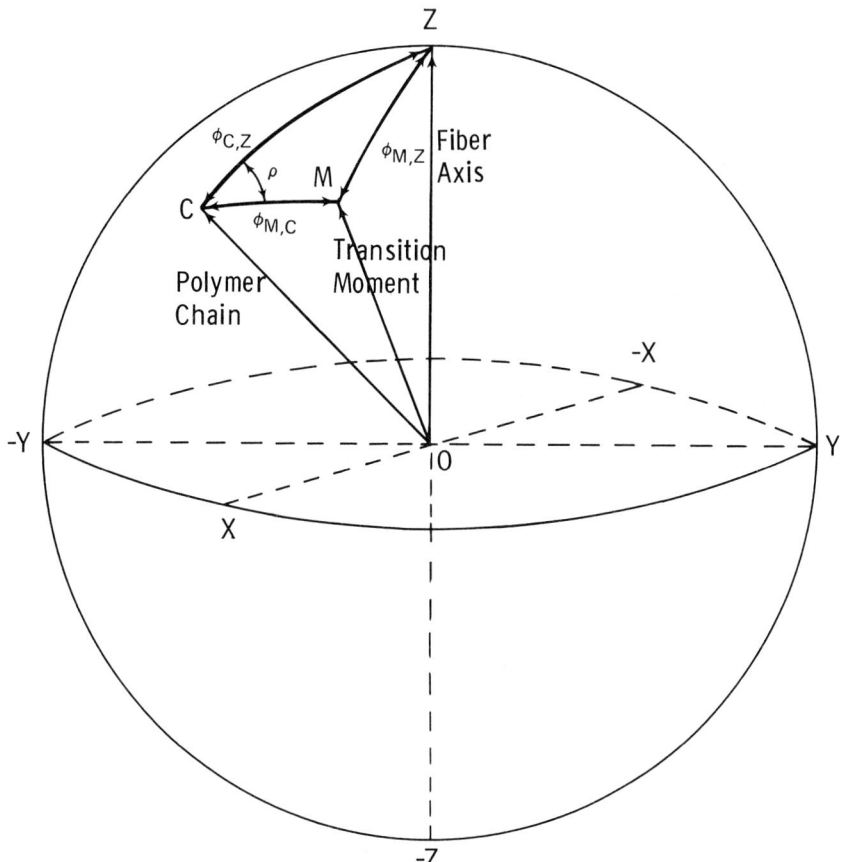

Figure 4. Relationship between the infrared transition movement vector, the polymer chain, and the fiber axis (Desper[3]).

rather than second-order, are required to describe the process; and (b) there are two experimental polarization directions (incident-beam and scattered-beam polarization) instead of the one direction in infrared absorption.

Limiting our discussion to the uniaxial orientation case, it may be shown that there are four independent intensities for any Raman vibrational frequency[28]. These arise from four possible polarization combinations, and may be denoted I_{zz}, I_{yy}, I_{xy}, and I_{yz}, respectively. The two subscripts denote the polarization directions of the incident and scattered beams. The x and y directions are perpendicular to the axis of symmetry, z, as in the previous infrared discussion.

As in the infrared case, however, there are complicating factors. Raman scattering arises from a change in polarizability α with the vibrational frequency, and this polarizability change may be parallel to z, perpendicular to z, or have both parallel and perpendicular components. In the latter case (which is the most common), even a perfectly oriented sample would exhibit non-zero values for all four intensities, thus mitigating the orientation effect to a large extent[3].

In the particular case of polypropylene, which is a three-fold helix in the solid state, the fundamental vibrations are divided into two symmetry classes[29–31]. For the A modes, the "cross-term" intensity I_{yz} will be zero for perfect molecular orientation, while for the E modes, the "parallel-parallel" intensity I_{zz} will vanish for perfect orientation. Any vibrational frequency may be purely A or purely E in character, or may be of mixed character. Occasionally both will occur, but with a splitting in frequency between the two modes.

The A and E species analysis also applies to infrared absorption frequencies. In this instance the A mode has a $\phi_{m,c}$ value of $0°$, and is termed the "parallel band," while the E mode has a $\phi_{m,c}$ value of $90°$, and is termed the "perpendicular band."

Correlation of Orientation Parameters with Processing Variables and Mechanical Properties

Process Parameters Versus Structure Parameters

The orientation parameters of a given polymer subjected to processing treatment (such as stretching, rolling, or annealing) will depend upon, and thus may be correlated with, the values of the process variables. These orientation parameters (along with other structural variables) will, in turn, exert a strong influence on the useful mechanical properties of the final polymer. A phenomenological evaluation of the process may be made in terms of correlating the mechanical properties with the process parameters without taking the structural response of the polymer into account. This empirical approach has been used in large part in the development of polymer processes. However, advances in our knowledge of the fundamentals of polymer solid-state structure, along with development of newer techniques for characterizing structure, have, in the last 15 years, made possible a more rational approach to orientation processes.

The empirical approach to polymer processing is illustrated in Figure 5. The tenacity is given as a function of draw ratio for polypropylene films and fibers at several drawing temperatures. (Tenacity is a specific tensile strength given in textile units of grams per denier; it may be converted to a conventional tensile strength in psi by multiplying by $1.28 \times 10^4 \times$ specific gravity.) The figure illustrates that there is no single relationship between tenacity and draw ratio; tenacity also depends upon

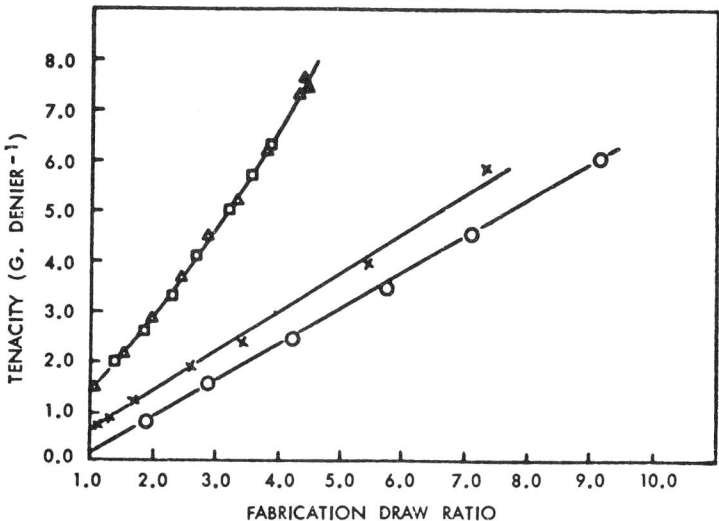

Figure 5. Relation between tenacity and fabrication draw ratio for isotactic polypropylene fibers and films drawn at a strain rate of 50%/min. (0) Film, draw temperature 135°C; (X) film, draw temperature 110°C; (□) fiber, draw temperature 90°C; (△) fiber, heat set (Samuels[32]).

drawing temperature, as shown, and may well depend upon other factors not illustrated in the figure, such as strain rate and initial state of orientation of the polymer. Draw ratio itself is a processing parameter, not a structural parameter, and tells very little about the degree of orientation of the polymer.

The more fundamental approach, using polymer structure parameters instead of process parameters, is illustrated in Figure 6. Samuels[32] has characterized each of the 31 samples of Figure 5 in terms of the average molecular orientation function, as defined by Equation (13). When the tenacity data is plotted against this parameter f_{ave} all of the data points fall on a single curve despite the various draw ratios and draw temperatures involved. This indicates that f_{ave} is one of the fundamental structural parameters of polypropylene, and that the tenacity depends upon the f_{ave} value and not on the process by which it was obtained.

This essential point cannot be emphasized too strongly. In looking for materials with particular properties for a particular application, one should specify the material structure, rather than the process by which the material is made. A given process may not be the optimum process, or it may possess unknown and uncontrolled process variables. By focusing our attention on structural parameters, competition is left open to any new process which may produce a better material, and deficiencies in the existing process in terms of uncontrolled variables are more readily detected.

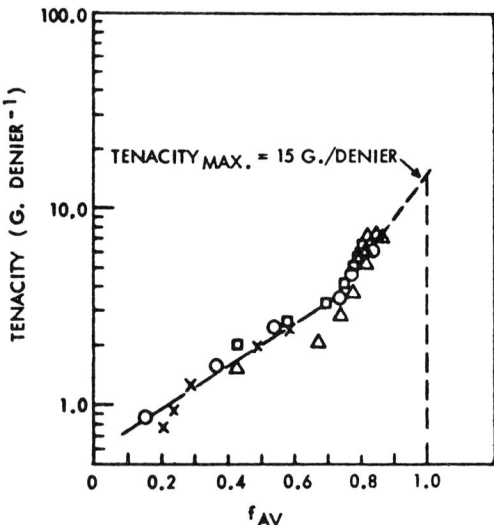

Figure 6. Comparative tenacities of fibers and films from Figure 5 as a function of average fabricated molecular orientation (Samuels[32]).

High-Tenacity Polypropylene Fibers

The Departments of the Army and of the Navy have had an interest in high-tenacity polypropylene fibers as a potential material for personel armor applications, and have supported research in this area at the Southern Research Institute[33–35]. Sheehan et al.[34] investigated the correlation of fiber tenacity with overall molecular orientation, crystalline orientation, crystallinity, molecular weight, and molecular weight distribution. The empirical relationship obtained is:

$$\overline{TS} = 10^4[-0.16 + 0.76(1 - \alpha)^{-1.5} - 0.18Q + 5.14 \times 10^{-6}\overline{M_w}] \quad (22)$$

where $\overline{TS}$ is tensile strength in lbs/in^2; α is Moseley's molecular orientation parameter, equivalent to f_{ave}, determined by sonic velocity and defined by Equations (12) and (13); Q is the ratio of weight-average to number-average molecular weight, and is therefore an indicator of the breadth of the molecular weight distribution; and $\overline{M_w}$ is the weight-average molecular weight. Correlation with degree of crystallinity was attempted, but was dropped when it was found to give only slight improvement in the degree of correlation. The orientation factor α has the greatest effect on tensile strength, while Q and $\overline{M_w}$ have relatively mild influence on tensile strength. The correlation of Equation (22) is shown in Figure 7 as a plot of $(1 - \alpha)^{-1.5}$ against an ordinate which consists largely of tensile strength, with minor corrections for Q and $\overline{M_w}$.

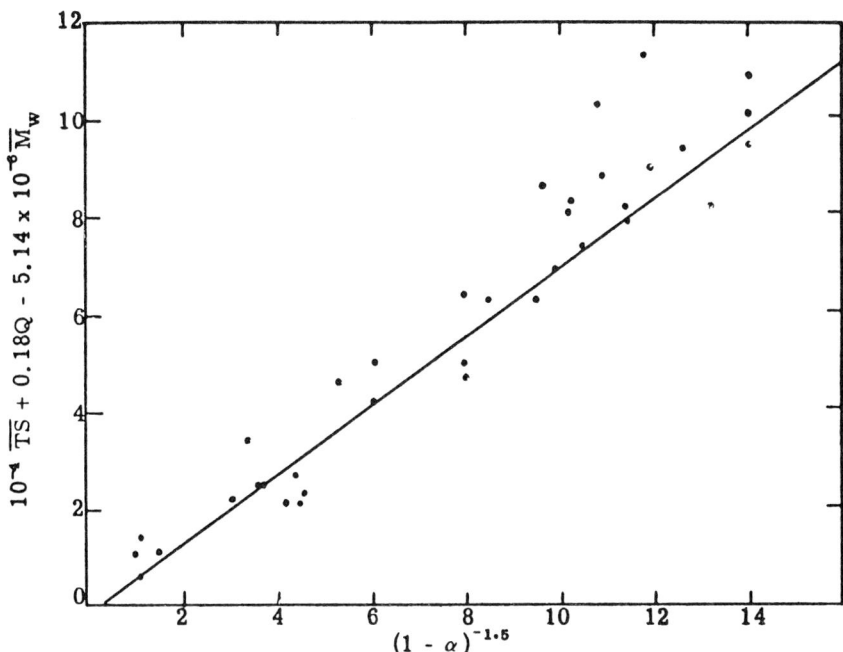

Figure 7. Relationship between combined structural parameters and tensile strength of polypropylene monofilaments (Sheehan, Wellman, and Cole[34]).

The direct correlations between α and fiber tenacity and modulus are given in Figures 8 and 9. The modulus values in Figure 9 were measured on a tensile tester, and will be considerably lower than the sonic modulus values because of irreversibility and rate effects. Both tenacity and modulus increase with molecular orientation, the greatest increase occurring above an α value of 0.70.

Sheehan also reported a correlation between f_c and tenacity (Figure 10), but discounts this effect as only an apparent correlation which appears because of the correlation between both tenacity and f_c and average orientation α. Sheehan does not attempt any correlation with the amorphous orientation function f_{am}, and did not even calculate f_{am} values. This fact arises largely from his method of taking the sonic velocity data. Sheehan's data was measured at $-18°C$, where the amorphous regions are in a glassy state and Equation (13) is valid. Thus, sonic velocity was used as a direct measure of f_{ave}, and the additional calculation of f_{am} was not performed. In contrast, the parallel studies of Samuels[18,21,22,32] relied on sonic-velocity measurements at room temperature, where the amorphous material is inherently much more compliant than the crystalline material and Equation (15) must be used in place of Equation (13).

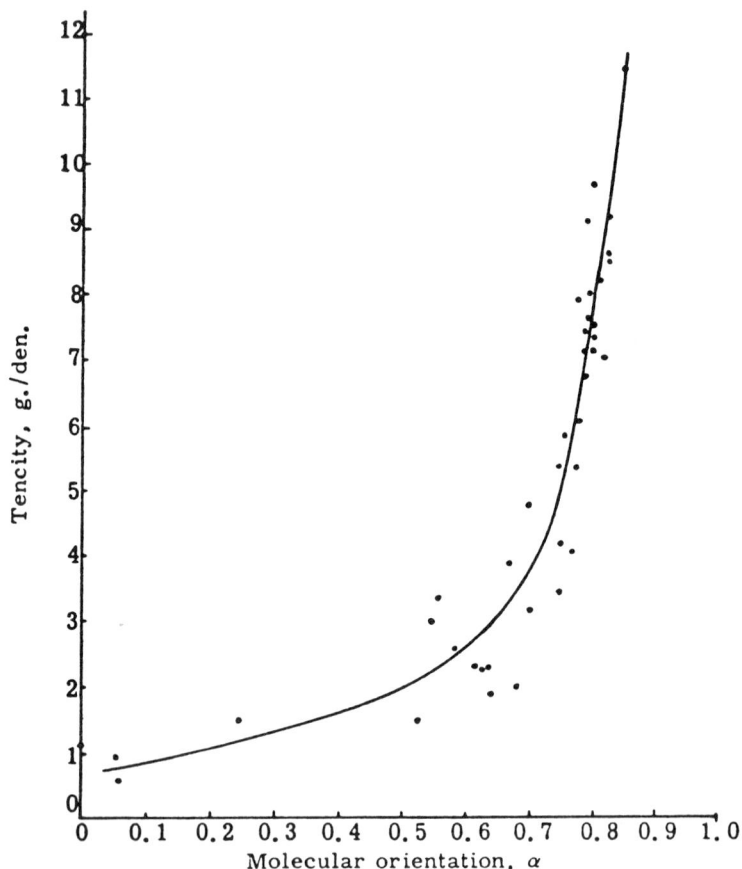

Figure 8. Relationship between molecular orientation and tenacity of polypropylene monofilaments of different molecular weights (Sheehan, Wellman, and Cole[34]).

This equation gives the amorphous orientation factor f_{am}. The function
f_{ave} no longer appears as a simple result of the sonic velocity measurement but can be calculated only after X_c, f_c, and f_{am} are all determined.
As a consequence, Samuels uses the amorphous orientation function f_{am}
as an important orientation parameter while Sheehan does not consider
this variable.

The Samuels approach results in a dramatic improvement in the correlation of tenacity with orientation. Comparing Samuels' data in Figure 6 with Sheehan's data in Figure 8, both authors show a definite
correlation of tenacity with average molecular orientation. However,
Samuels uses log tenacity as the ordinate and fits the data to two straight
lines intersecting at a break point of $f_{ave} = 0.76$, which he interprets as

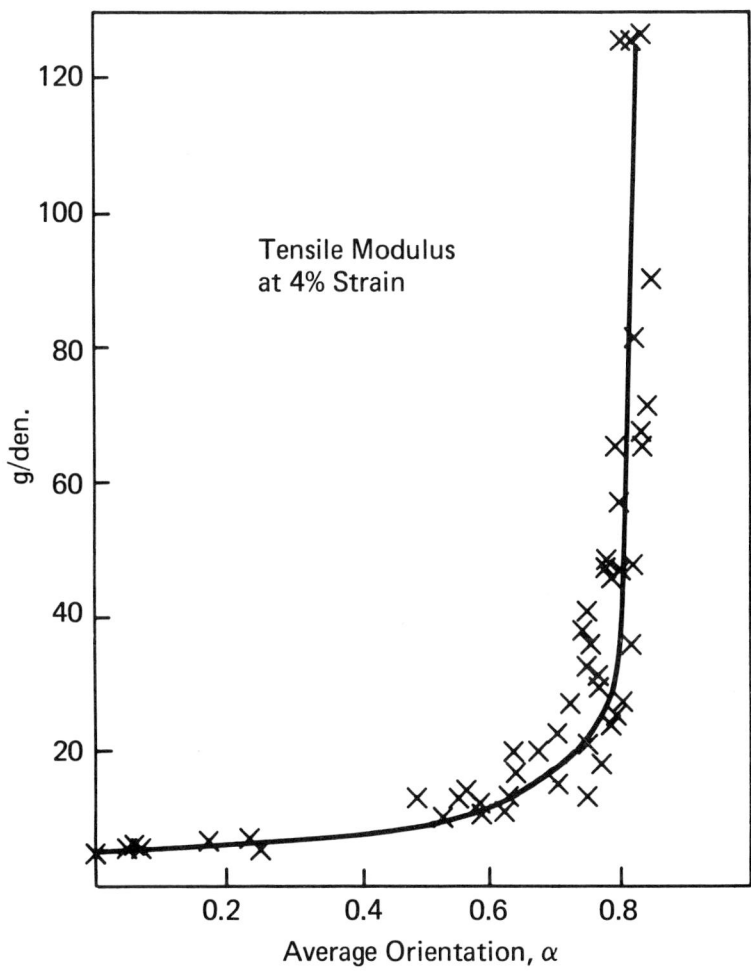

Modulus of Polypropylene Fibers Versus Orientation

Figure 9. Relationship between molecular orientation and modulus at 4% strain of polypropylene monofilaments. Plotted from the tabulated data of Sheehan *et al.*[34].

the transition from spherulitic to microfibrillar structure. Even more striking results are shown in Figure 10, where Samuels has plotted log tenacity against amorphous orientation function. This time a single straight line is obtained with no break point. This fact reveals that the amorphous orientation function is the fundamental factor controlling the tenacity of polypropylene fibers. The apparent correlation in Figure 6 arises from the correlation of f_{ave} with f_{am}; the real correlation is that with

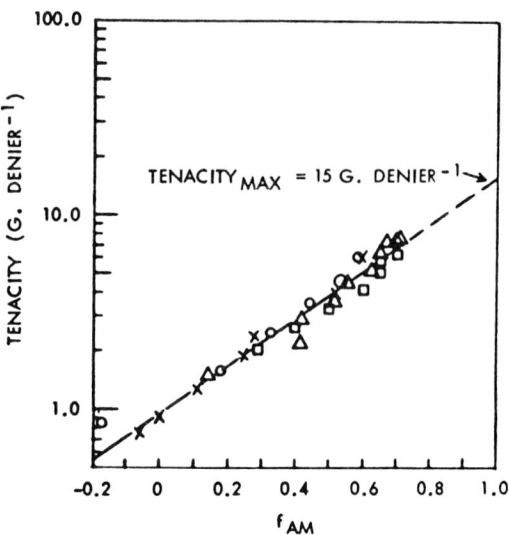

Figure 10. Comparative tenacities of fibers and films from Figure 5 as a function of amorphous orientation (Samuels[32]).

f_{am} in Figure 10. The break point in Figure 6 results from the differing mix of crystalline and amorphous orientation in spherulitic and microfibrillar structures. In the spherulitic region (below f_{ave} = 0.76), most of the orientation is crystalline orientation; the amorphous regions remain at a much lower state of orientation. In the microfibrillar region (above f_{ave} = 0.76), the crystalline regions are close to the limiting value f_c = 1, and *changes* in orientation are occurring more rapidly in the amorphous regions. Thus the tenacity, which is really responding to amorphous orientation, changes more rapidly with f_{ave} in the microfibrillar region, because more of the overall increase in orientation is going into amorphous regions. Samuels' data in Figures 6 and 10 also predict a limiting tenacity value of 15 grams per denier (approximately 175,000 psi) by extrapolation to f_{ave} = 1 or f_{am} = 1. For comparison, the highest tenacity reported for polypropylene fibers[33] was 13.1 grams per denier. This indicates that very little further improvement in fiber tenacity can be expected without drastic changes in microstructure, such as by somehow attaining the structure of a perfect extended chain crystal.

Orientation and Properties of Highly Stretched Thin Film Polypropylene

It was previously noted that the Department of the Army's interest in stretched polypropylene is as a potential personnel-armor material. Recent work in this area has employed the concept of thin, highly oriented polypropylene film, assembled into thick pads by stacking up multiple layers of film. Alternate layers of film are usually placed with orientation axes at right angles to each other. It has been found, of course, that the degree of orientation of the original film controls the ballistic performance of the final pad[36]. Extensive pilot-plant studies of these materials have been conducted and a wide variation in ballistic performance with process variables have been found. Consequently, these films have undergone considerable scrutiny to characterize the physical state and microstructure.

One process for preparing films of this type is illustrated in Figure 11. The film is stretched between two motor-driven rolls, passing through a radiant-heat oven between the rolls. The surface temperatures of the radiant heaters are controlled in the range 600–900°F (316–482°C), well above the 165–170°C melting range of polypropylene. However, the actual temperature of the film is undoubtedly well below that of the radiant heaters, but cannot be measured directly because of the nature of the process. Measurement of the film temperature as it exits the oven gives a value close to 170°C, using an infrared pyrometer.

As shown in Figure 11, the film deforms by extension in a localized neck region. The film typically shows a pronounced tendency to "fibrillate," *i.e.,* to break up by splitting parallel to the machine direction into narrow fibrils. (Indeed, this type of process is used commercially to pro-

Figure 11. Radiant-oven film-stretching unit at AMMRC, showing oven in retracted position.

duce a film which is deliberately fibrillated, and the resulting fibrils are twisted to form a binding twine.) Electron microscopy (Figure 12) reveals a microfibrillar structure, the smallest microfibrils being approximately 100 Å in diameter. The microfibrillar structure explains the tensile properties of the samples: quite strong in the machine direction, very weak and brittle in the transverse direction. Normal tensile testing gives very erratic results because of the tendency of specimens to fibrillate, followed by piecemeal breaking of fibrils at different strains. The tensile strengths

Figure 12.   Electron micrograph of a surface replica of highly stretched polypropylene film, showing microfibrillar structure. Line = 1 micron.

TABLE I

Morphological Parameters for Polypropylene Films of Ultrahigh Orientation (Desper[17]).

Parameter	Symbol	Units	Sample A	Sample B	Sample C	Sample D	Precision
Density	ρ	g/cm^3	0.9102	0.9104	0.9092	0.9080	±0.0004
Crystallinity	x_c		0.688	0.691	0.676	0.661	±0.005
Birefringence	ΔT		0.032	0.034	0.026	0.026	±0.002
Thickness	t	mm	0.020	0.035	0.041	0.052	±0.002
Sonic modulus	E_{or}	dyn/cm^2	14.4×10^{10}	16.2×10^{10}	7.89×10^{10}	7.48×10^{10}	$\pm 0.4 \times 10^{10}$
Crystalline orientation:							
f_{110}			−0.4984	−0.4981	−0.4959	−0.4961	±0.0001
f_{040}			−0.4984	−0.4982	−0.4958	−0.4962	±0.0001
f_{130}			−0.4986	−0.4984	−0.4961	−0.4965	±0.0001
f_c			+0.9968	+0.9963	+0.9917	+0.9922	±0.0002
χ 1/2 (110)		deg	1.9	1.8	3.3	3.0	±0.1
χ 1/2 (040)		deg	1.9	1.9	3.0	2.8	±0.1
χ 1/2 (130)		deg	1.7	1.8	3.0	2.8	±0.1
Amorphous orientation:							
Sonic, f_{am}			0.65	0.69	0.38	0.38	±0.01
Birefringence, f_{am}			0.64	0.75	0.33	0.34	±0.1
Average orientation:							
Sonic, f			0.89	0.90	0.80	0.78	±0.01

observed (40,000–70,000 psi) are below the true tensile strength of the individual fibrils since they do not break simultaneously.

Extremely high degrees of crystalline orientation are achieved by this process. The author[17] has reported f_c values up to 0.9968 (see Table I), as high as any previously reported values. Stated another way, X-ray diffraction indicates that the majority of the crystallites are oriented closer than 2° to the machine direction. Yet marked differences are possible even at this high level of crystalline orientation. Referring to Table I, the f_c values of samples A through D vary from 0.9917 to 0.9968, but samples A and B are found to have superior ballistic resistance, while C and D are much poorer. It is not reasonable to attribute the different ballistic performance to the small difference in f_c. A better explanation lies in the large difference in f_{am}, which ranges from 0.38 to 0.69, determined from the sonic velocity using Equation (15). Sonic velocity is particularly valuable as a method of characterizing these highly oriented films for two reasons:

(a) for f_c values close to 1, reference to Figure 3 and Equation (15) indicates that the sonic modulus is sensitive to f_{am} but not to f_c; and

(b) the sonic velocity is intimately related to a basic parameter of the ballistic impact event; i.e., the velocity of the shock wave generated in the material upon impact.

The correlation of sonic modulus with process variables is indicated in Figure 13. The highest sonic modulus value, 20×10^{10} dynes/cm^2, corresponds to 59 percent of the limiting modulus of the polypropylene crystal, 34×10^{10} dynes/cm^2[36]. The starting material is a tubular extruded polypropylene film with a total thickness (two layers) of approximately 0.11 mm. Three zones of radiant heaters heat the film from above; the middle-zone heaters are held at a higher surface temperature than the other two, to localize and thus stabilize the drawing of the film.

Film input is fixed at 15 ft/min, while the draw ratio, taken as the ratio of draw-roll to feed-roll speed, was changed by varying the draw-roll velocity. It is evident that this process would be difficult to specify in terms of process variables, but the more important goal is to specify the product in terms of structural variables.

Examination of the data in Figure 13 indicates a complex behavior pattern. Films were stretched at four heater temperatures and five draw ratios. For a given heater temperature, there is a minimum and maximum draw ratio for stable drawing. The minimum draw at three temperatures is 8:1, but at the highest heater temperature (900°F) a 9:1 draw is the minimum usable value. At the other end of the graph, draw ratios up to 14:1 are attained except at the lowest heater temperature (750°F), at which the maximum draw is 12:1. At any given draw ratio, the highest amorphous orientation (indicated by sonic velocity) occurs at 800°F, one of the intermediate values of heater temperature.

Analysis in terms of orientation functions indicates that all of these samples are above $f_{ave} = 0.76$, and thus fall into the fibrillar region by

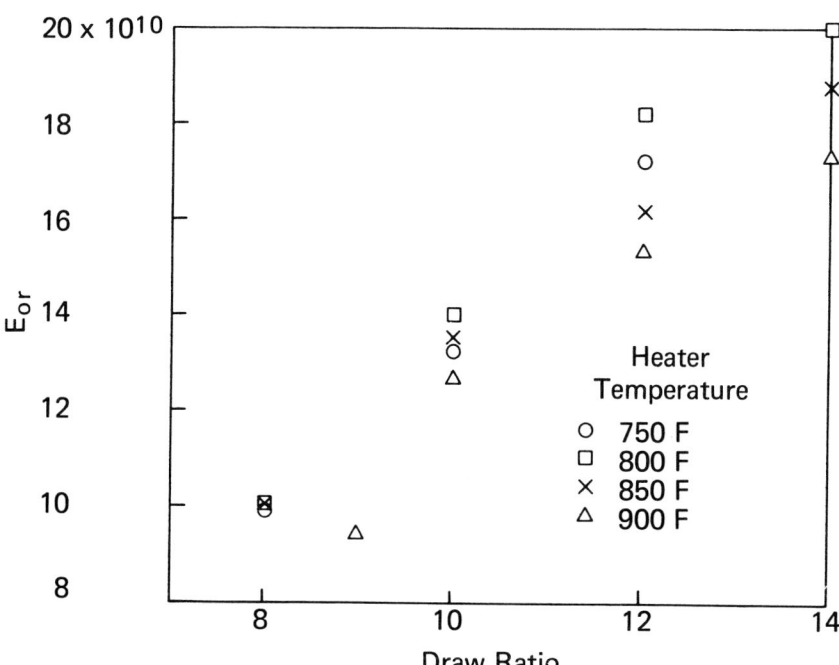

Figure 13. Correlation of sonic modulus with process variables in the radiant-oven stretching of tubular polypropylene film.

Samuels' criterion[32]. The crystalline orientation is above $f_c = 0.99$, while the amorphous orientation function f_{am} ranges from 0.54 to 0.78.

The orientation functions cannot be correlated with tensile strength because of the previously noted difficulties in tensile testing these materials. The amorphous orientation can be correlated with another mechanical property, the orientation release stress. This property is measured in a tensile tester by clamping a specimen at fixed extension under a small preload, then heating the specimen gradually to a point above its melting point[37]. The amorphous orientation frozen in at room temperature results in a build-up in stress at elevated temperatures until the onset of melting permits flow to relieve the stress. Figure 14 shows that orientation release stress is correlated with f_{am}. Because of this correlation, orientation release stress may be used as a material specification, although it is an empirical property rather than a fundamental material parameter. Sonic velocity is a better parameter for specifying material properties, since it is more accurate, can be related to structural parameters, and is more readily measured—provided the required apparatus is available. Standard procedures have been established for sonic-velocity testing[39].

It is well known that polymeric solids exhibit viscoelasticity, which manifests itself in terms of creep or stress relaxation in quasi-static test-

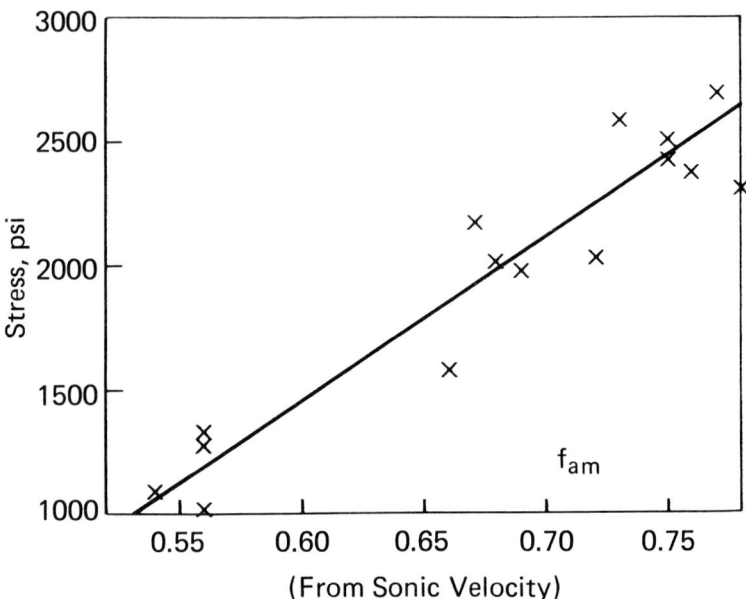

Figure 14. Correlation of orientation release stress with amorphous orientation for films of Figure 13.

ing, or as a phase lag causing mechanical losses in oscillatory testing. This aspect of the mechanical behavior of highly oriented polypropylene has been experimentally examined using the Rheovibron Model DDV-II direct reading dynamic viscoelastometer. This instrument subjects a polymer to a sinusoidal tensile deformation (the present work uses a frequency of 110 Hz) and measures the dynamic modulus E' and the loss factor tan δ, where δ is the phase lag between stress and strain. Typical E' and tan δ curves for oriented polypropylene films, taken from an earlier report[36], are shown in Figures 15 and 16. A good check is established in Figure 15 between the Rheovibron curve and the sonic modulus. The E' curve crosses the sonic modulus value at 15°C. The 5° difference between this temperature and that of the sonic experiment is explained by the different testing frequencies involved. The shapes of the tan δ curves are also strongly influenced by orientation. The undrawn material (curve A' in figure 16) shows a strong peak in the tan δ around 15°C, which is the so-called β-relaxation attributed to the onset of motions in the amorphous regions[39]. With increasing degrees of amorphous orientation this peak tends to flatten out and nearly vanishes for samples D and E having the highest f_{am} values. An attempt has been made to correlate the area of this β-relaxation peak with amorphous orientation and is shown in Figure 17. A correlation is indicated, although the scatter leaves much to be desired.

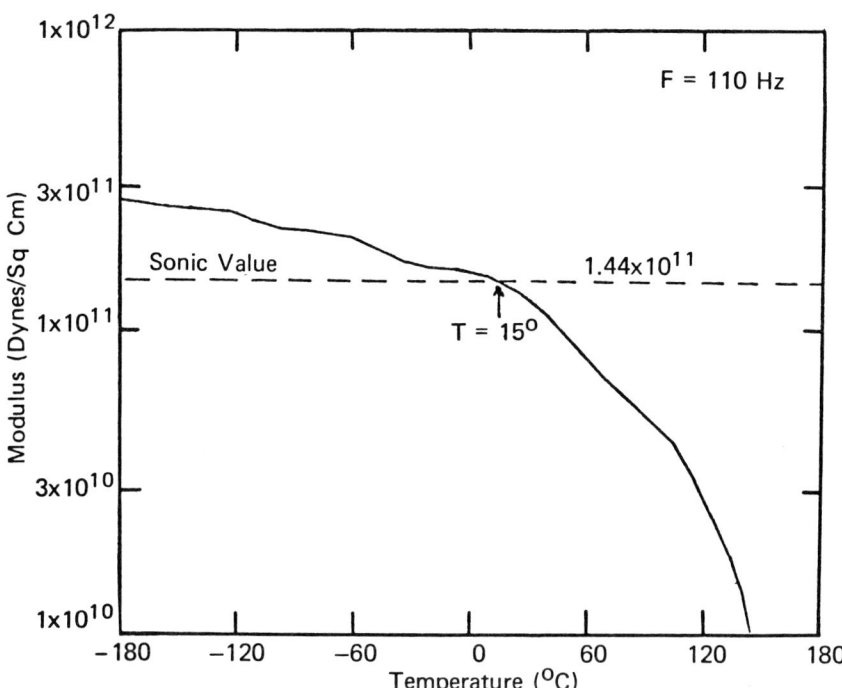

Figure 15. Modulus (at 110 Hz) versus temperature for highly oriented polypropylene sample 775-11-3 (see column A, Table I, for properties) (Desper, Lewis, Lopata, and Roylance[36]).

One must be very careful in analyzing mechanical-loss data in oriented polymers. The data in Figures 16 and 17 support a general statement that increasing amorphous orientation tends to suppress the intensity of the β-relaxation in polypropylene when tested in the machine direction. One may question[39] the validity of using tan δ as a quantitative measure of relaxation magnitude. Also, the weakening of the β-peak in the present test may indicate that the fundamental process of this relaxation, although still present in the sample, is not effectively excited when tested in the machine direction. To be more explicit, Ward and co-workers[40–42] attribute a similar β relaxation in polyethylene to interlamellar shear, i.e., to simple shear of viscoelastic amorphous material between the flat lamellar polyethylene crystals. The drawing process turns the lamellar crystals normal to the machine direction, as shown in Figure 3. With perfect lamellar orientation, tension along or perpendicular to the machine direction would not result in a resolved shear stress for interlamellar shear; tension at a 45° angle would be far more favorable for inducing interlamellar shear. Ward's experimental data with oriented polyethylenes supports this viewpoint; the β-peak is often more pronounced for tension at 45° than at the 0° or 90° positions.

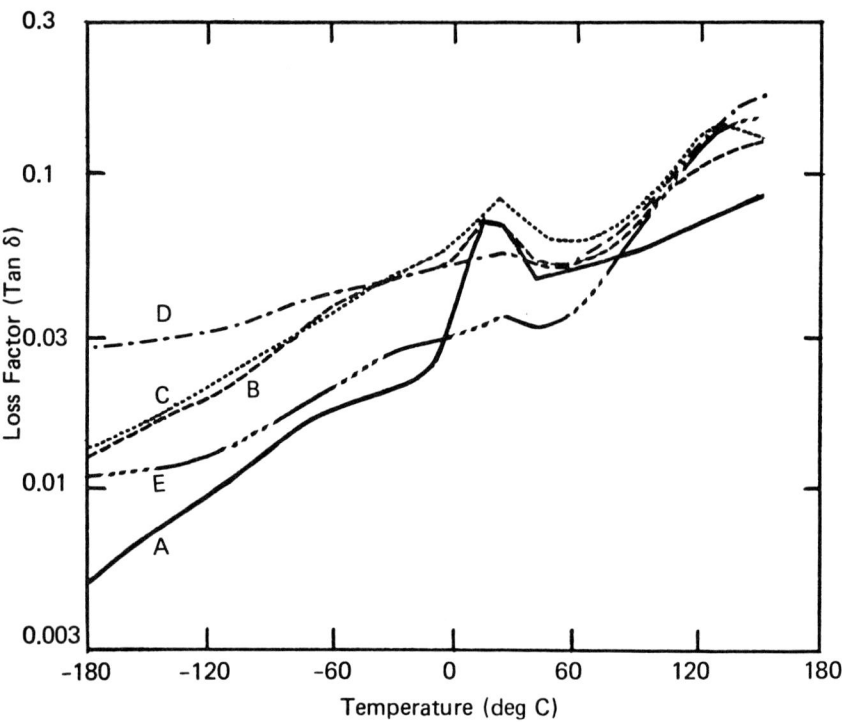

Figure 16. Loss factor tan δ vs. temperature at 110 Hz for five polypropylene films of different degrees of orientation (Desper, Lewis, Lopata, and Roylance[36]).

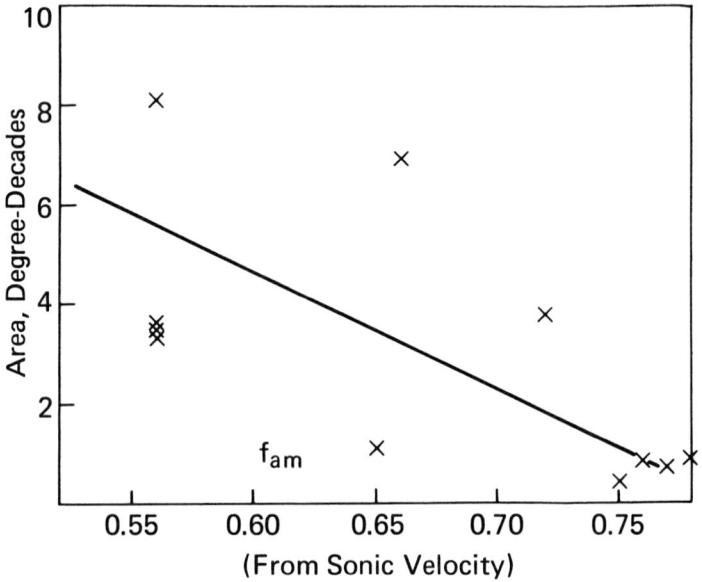

Figure 17. Area of β-relaxation peak versus amorphous orientation for samples from Figures 13 and 16.

These concepts could also apply to polypropylene, depending upon the accuracy of the analogy between the β-peaks in the two materials. The required experimental work to more closely establish the nature of the β-peak in polypropylene has not been done. There are obvious difficulties associated with testing a material with such a tendency to fibrillate in any direction other than the machine direction. The point to be made at this juncture is that the disappearance of the β-peak need not, as we had earlier thought, indicate a basic change in the amorphous material. The potential for exciting the β loss mechanism could still be present if one could test the sample in a more favorable direction. A second point is that lamellar orientation should be taken into account. The lamellae can have their own degree of orientation, independent of crystalline and amorphous orientation, which may be examined by small angle X-ray diffraction[43,44]. Until these effects are assessed, our knowledge of mechanical loss mechanisms in oriented polypropylene will be incomplete.

Application of Vibrational Spectroscopy to Oriented Polymers

Characterization of Polyurethane Oriention

The discussion of the application of vibrational spectroscopy in the characterization of oriented polymers will comprise two parts. First, the method will be illustrated with recent results obtained with segmented polyurethane elastomers to illustrate its versatility and sensitivity to structural details. Second, the application to oriented polypropylene will be discussed to illustrate how unsolved fundamental problems can lead to difficulties in this particular system.

Seymour, Cooper, and Allegrezzi[45] studied segmented polyurethane elastomers made from p,p'-diphenylmethane diisocyanate (MDI), butanediol, and prepolymer segments consisting of a low-molecular-weight polyether. The molecular structure is:

$$H - [- ET - (UG)_x - U -]_m - ET - H$$

where U represents the MDI residue, G is the butanediol residue, and ET is the "soft segment," a poly(tetramethylene oxide) of molecular weight 1,000 or 2,000. Details of the chemical structure are given in an earlier publication[46]. The formula shows the segmented nature of the molecule: segments of polyether-type polymer (ET) alternate with segments of polyurethane-type structures $[-(UG)_x - U -]$. The former are quite mobile at room temperature and are termed "soft segments," while the latter chain segments are stiff and are termed "hard segments." The two segments are not compatible and undergo a microphase separation in the solid state, yielding soft- and hard-segment domains of a 50 Å typical size. The resulting material is an elastomer, the soft segments conferring the rubbery property and the hard-segment domains acting as both physical cross links and filler.

Polarized infrared absorption allows one to follow the degree of orientation of the soft and hard segments. The vibrational modes used by Seymour, Cooper, and Allegrezzi are: (a) the CH_2 stretching vibration, used to follow soft-segment orientation since most of the CH_2 groups are in the soft segments; (b) the NH stretching vibration, which reflects hard-segment orientation since all NH groups are in the hard segments; and (c) the C=O stretching modes, which again is confined totally to hard-segment material. Two C=O frequencies are observed, one for carbonyls associated by hydrogen bonding to NH groups, the other for nonbonded carbonyls. None of these vibrations has a transition moment **M** vector parallel to the polymer chain. The values of $\phi_{m,c}$ (see Figure 4) are 90° for the CH_2 and NH vibrations and 79° for the C=O vibration. Equations (20) and (21) are used to calculate molecular orientation functions for all of the vibrations.

Figure 18 shows the experimental orientation functions for a polyurethane as a function of tensile strain. The four different curves indicate molecular orientation of four different populations of chain segments as measured from the four vibrational modes. The lowest curve represents the orientation of soft segments as measured by the CH_2 absorption. The remaining three curves indicate that hard segments are at a higher state of orientation than the soft segments. Of these, the hydrogen-bonded hard segments are at a higher state of orientation than the nonbonded hard segments at high strain, indicated by the two C=O curves. The

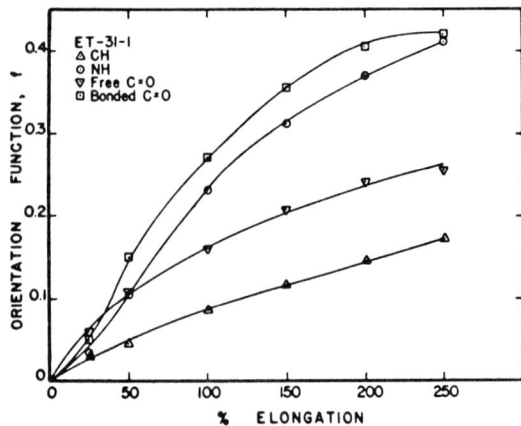

Figure 18. Molecular orientation functions in a polyether-type segmented polyurethane elastomer for various chemical species (Seymour, Cooper, and Allegrezzi[45]).

NH absorption gives a measure of both free and bonded hard segments, but will be weighted towards the latter since the free NH group has a significantly lower absorption coefficient[46]. Moreover, infrared data also shows that when stress is released, the soft segments return to a state of zero orientation, while the hard segments retain part of their orientation. This type of information can be obtained only by spectroscopy, since such methods as birefringence or sonic velocity yield only an average over the various species present.

Characterization of Oriented Polypropylene by Vibrational Spectroscopy: Theoretical Aspects

The vibrational spectra of polypropylene are far more complicated to interpret. The basic molecular structure of crystalline or smectic (quenched) polypropylene is the threefold helix shown in Figure 19, and the fundamental frequencies have been analyzed in terms of this structure. Early workers[47,48] noted that a number of infrared bands become quite weak above the crystalline melting point. These bands were dubbed "crystalline bands," and their intensities were correlated with other measures of crystallinity. This viewpoint encountered a major stumbling block when these crystalline bands were found to be quite strong in quenched polypropylene, a smectic but noncrystalline state[49]. In the smectic state, the polymer is predominantly in the threefold helix but are not packed in a crystal lattice. This suggested that these bands arise from helical material, and Brader[50] proposed use of the ratio of two infrared intensities to measure helical content. This helical content was proposed as a quantity which could vary with quenching or annealing conditions, would always equal or exceed the crystalline content, and would, after sufficient annealing to develop maximum helical content, give a measure of the isotactic (as opposed to atactic) content of a particular resin.

Thus was born the notion that, for purposes of vibrational spectroscopy, polypropylene divides into two categories: helical material and non-helical material, and that only the former gives rise to the helical bands. A more sophisticated viewpoint is taken[51] by Zerbi, Gussoni, and Ciampelli. The helical bands are strongest for a regular helix of infinite length, and fall off in intensity for shorter lengths of helical sequences. However, the cut-off is not sharp, and the helical band at 998 cm^{-1} may still be detected, although weakened, in the melt or in ethylene-propylene copolymers where the helix length may be no more than 4–7 monomer units. Thus the helical bands, called "regularity bands" by Zerbi, are sensitive in intensity to an average helix length, the weighting function becoming quite weak for short helices, and increasing in intensity with increasing helix length up to a saturation value. Thus, Equation (18) should be modified to read: $a_\eta = k_\eta (\overline{\mathbf{M} \cdot \mathbf{E}})^2$, where k_η is a weighting func-

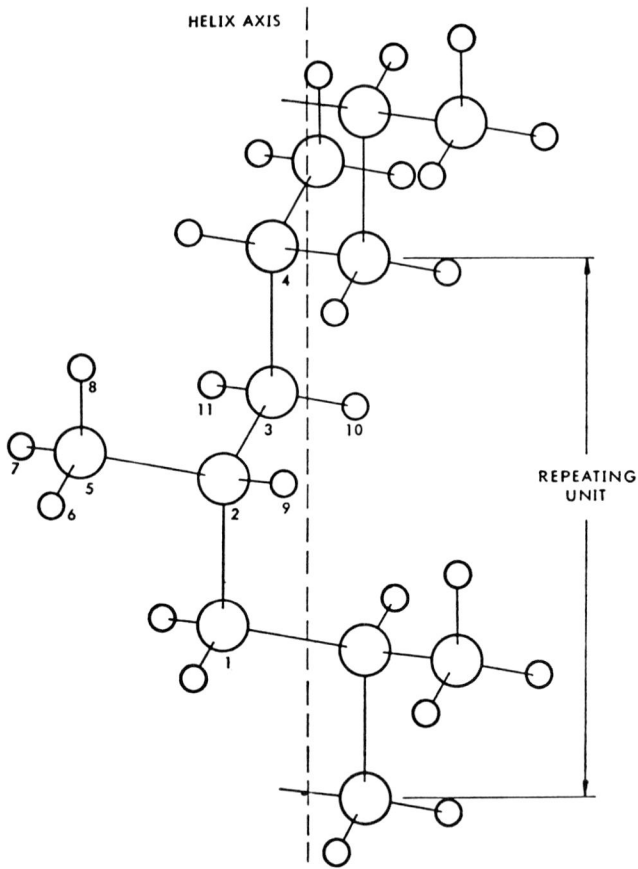

Figure 19. The threefold polypropylene helix. Small spheres are hydrogens, large spheres are carbons (Snyder and Schachtschneider[59]).

tion for the band in question depending upon the number of monomers η in a particular helical segment.

The concept of dividing polypropylene into "helical" and "non-helical" material is challenged by recent statistical mechanical models which consider the molecule in solution and possibly in the amorphous solid to be a series of left- and right-handed helical segments[52]. Separating the adjacent helical segments is only a hand-reversal, which could consist of half of a monomer unit. Such a hand-reversal does not have a vibrational spectrum of its own. Thus in melting polypropylene, no new bands appear, and no existing bands get stronger, in contrast to the case in polyethylene, where *gauche* bands are characteristic of the amorphous material. Thus, in polypropylene it is not useful to speak of the amount of nonhelical material. A more useful model is to consider the hand-reversals as defects in a one-dimensional lattice which, under suitable conditions, can travel along the chain, coverting material from one hand to the other, and which will annihilate each other if an adjacent pair of defects move to the same point in the chain. The concentration of defects determines a number-average helix length. When crystalline material is present, the helix length in the crystal phase is identical with the c axis dimension up to the onset of melting, which occurs when these defects can propagate inside the crystal.

Infrared Data on Oriented Polypropylenes

The infrared absorption data in Tables II and III may be explained in terms of this concept. Table II shows dichroic ratios, average absorbances, and $f_{m,z}$ values for a number of bands in a highly oriented polypropylene film made by the radiant-oven method. The average orientation function was previously reported[17] to be 0.89 for this sample, which fixes the limits for those bands insensitive to helix length. Referring to Equation (21), if the $f_{c,z}$ value is 0.89, the possible $f_{m,z}$ values range from -0.445 to $+0.89$ since the possible range of $f_{c,m}$ values is $-1/2$ to $+1$. The $f_{m,z}$ values falling outside this range are at frequencies of 808,

TABLE II
Infrared Absorption Data for Polypropylene
Sample 775-11-3 (Sample A in Table I)

Frequency, cm^{-1}	R_{zy}	$\overline{A}$	$f_{m,z}$	Species
808	0.00	0.17	-0.50	E
839	∞	0.47	1.00	A
898	0.00	0.28	-0.50	E
939	0.00	0.04	-0.50	E
973	13.51	0.51	0.81	A + E
998	115.44	0.50	0.97	A (E?)
1043	∞	0.05	1.00	A
1101	0.00	0.07	-0.50	E
1151*	≈ 0	0.10	$\approx -1/2$	E (A?)
1168	∞	0.48	1.00	A
1218	0.00	0.05	-0.50	E
1255	30.77	0.13	0.91	A (E?)
1296	0.00	0.07	-0.50	E
1304	∞	0.13	1.00	A
1326	0.29	0.07	-0.31	E + A
1359	0.08	0.29	-0.44	E (A?)
1379	0.33	2.27	-0.29	E + A

*The 1151 band is not sufficient resolved from the stronger 1168 band to allow determination of R_{zy} and $f_{m,z}$.

Note: Where two species are indicated, the stronger component is listed first; (A?) indicates a species whose presence is in doubt.

TABLE III

Comparison of $f_{m,z}$ Values for Selected Regularity Bands
of Three Oriented Polypropylenes

Sample Designation	775-11-3	813-29-1	813-10-2
f_c	0.9968	0.9972	0.9854
f_{am}	0.65	0.75	0.40
f_{ave}	0.89	0.91	0.78
$f_{m,z}$ values at:			
808 cm^{-1}	−0.50	−0.50	−0.41
839 cm^{-1}	1.00	1.00	0.86
898 cm^{-1}	−0.50	−0.47	−0.37
973 cm^{-1} *	0.81	0.81	0.61
998 cm^{-1}	0.97	0.94	0.79
1168 cm^{-1}	1.00	1.00	0.70
1218 cm^{-1}	−0.50	−0.50	−0.46
1255 cm^{-1} *	0.91	0.89	0.79

*Bands insensitive to helix length.

839, 898, 939, 998, 1043, 1101, 1168, 1218, 1296, and 1304 cm^{-1}. These bands are known[47,48,50] to be sensitive to helix length since they become quite weak in the melt. Of the remaining bands at 973, 1151, 1255, 1326, 1359, and 1379 cm^{-1}, only the 1326 cm^{-1} band is sensitive to helix length[47]. This one band could have a low negative $f_{c,m}$ value which brings its experimental $f_{m,z}$ value within the abovementioned range. The remaining helical bands all show orientation functions weighted more towards the orientation of the crystalline phase (which contains the longest helices) than to that of the amorphous phase (which is a mixture of shorter helices). The actual $f_{m,z}$ value observed for these bands would depend upon two factors: the degree of sensitivity to helix length (*i.e.*, the form of the k_n function in Equation 23) and the transition moment orientation (the $\phi_{m,c}$ value in Figure 4). These two factors cannot be independent assessed from the present data in a quantitative manner.

Table III presents further evidence, however, of this concept of the polypropylene vibrational bands. Experimental orientation functions are shown for three samples, comparinng f_c, f_{am}, f_{ave}, and $f_{m,z}$ values for six "regularity bands" and two regularity-insensitive bands. These samples are particularly significant because although all three have nearly perfect crystal orientation, two have high f_{am} values and one does not. Thus, if a band were truly a crystalline band, it would show the same $f_{m,z}$ value for all three samples. A lower (in absolute value) $f_{m,z}$ value for the third sample is an indication that the band is sensitive to material outside of the crystalline phase. The data show that none of the bands is a purely "crystalline band" by this criterion. In the present context, this means that all of the bands have some intensity for helices shorter than 100 Å (46 monomer units), the c-axis crystal size indicated by three different methods.

The band at 1218 cm^{-1} appears to be the least sensitive to helices shorter than 100 Å. This is one of the weaker bands, however, and the data in Table III should not be taken as definitive. This particular band was reported by Samuels[21] to be a crystalline band, which may not be a bad approximation. However, Samuels' value of $\phi_{m,c}$ = 72° is definitely in error, since such a band could never exhibit a greater degree of orientation than $f_{m,z}$ = −0.36 because of Equation (21). Table III would indicate a $\phi_{m,c}$ value close to 90° for the 1218 cm^{-1} band. Samuels also reported that the 1255 cm^{-1} band is a good measure of f_{ave}. The data in Table III give a practically quantitative check of this contention when the f_{ave} and $f_{m,z}$ values are compared for the three samples, provided one uses a $\phi_{m,c}$ value of 0° for this vibration. Again, this is inconsistent with Samuels' value of 38.5° for $\phi_{m,c}$, but there is no way that Samuels' $\phi_{m,c}$ value could result in the high $f_{m,z}$ values indicated in Table III at 1255 cm^{-1}. The discrepancy between the present data and Samuels' data for both bands may be instrumental in origin. His work was performed before the highly efficient wire-grid infrared polarizers were generally available. The earlier polarizers were not totally efficient in rejecting the unwanted component and would thus tend to underestimate the degree of orientation.

The polarized infrared absorption data on oriented polypropylene can be understood in a qualitative manner, but a more sophisticated approach than that used in the past has been found necessary. The most useful bands are the 1218 and 1255 cm^{-1} bands, sensitive to f_c and f_{ave}, respectively; none of the bands are useful for direct characterization of f_{am}. On the whole, the results have been more informative in understanding the nature of the vibrational frequencies than in determining degree of orientation, which may be more readily assessed by other means.

Raman Data on Oriented Polypropylene

Raman intensity data obtained on a highly oriented polypropylene made by the radiant-oven method is shown in Table IV. Three spectra were obtained: the I_{yy}, I_{yz}, and i_{zz} spectra, where the subscripts indicate the polarization of the Raman and exciting lines. The Y and Z axes are perpendicular to and parallel to the machine direction as previously noted. (The Raman frequencies in Table IV are generally higher than the infrared frequencies in Table II. This is undoubtedly an instrumental effect which need not concern us in the present discussion.)

Intensities are given in Table IV for 26 vibrational frequencies, and species are assigned using the criterion that I_{yz} vanishes for A species, while I_{zz} vanishes for E species. This criterion strictly holds only for the perfect orientation case, so in some cases a particular intensity may fail to vanish. This leads to a certain degree of ambiguity: The required intensity value may not vanish even for a band of pure A or pure E character be-

TABLE IV
Polarized Raman Intensities for Polypropylene Sample 775-11-3
(Sample A in Table I)

Frequency, cm^{-1}	I_{yy}	I_{yz}	I_{zz}	Species
814	2.6	1.3	24.2	A (E?)
846	4.4	0.1	3.3	A (E?)
904	0.6	0.2	0.0	E
942	0.3	1.1	0.2	E (A?)
976	0.8	0.3	12.6	A (E?)
1003	1.6	0.1	0.0	E
1047	0.7	1.3	0.9	A + E
1105	1.3	0.3	0.0	E
1158	2.5	1.0	0.0	E
1172	0.6*	0.5*	15.4	A (E?)
1221	0.6	2.2	0.2	E (A?)
1261	0.3	0.2	2.7	A (E?)
1297	0.2	0.5	0.0	E
1307	0.1	0.0	2.3	A
1330	2.0	4.1	1.2	A + E
1363	0.5	2.3	0.5	A + E
1379	0.5	0.6*	0.5	A + E
1444	0.9	1.6	1.2*	E (A?)
1464	2.3	4.6	1.9	A + E
2722	0.7	0.1	2.1	A + E
2842	4.5	0.5	8.5	A (E?)
2872	6.2*	0.0	12.5*	A
2886	9.3*	2.3	16.0*	A + E
2907	7.8*	0.0	13.7*	A
2927	4.6*	0.1	10.6*	A (E?)
2954	4.5	0.5	0.0	E
2962	0.0	2.2	22.1	A + E

*Indicates intensities which are not accurate because of overlap with a nearby band. Number includes the contribution above baseline of the interfering band.

(?) Indicates a doubtful species. Failure of a particular intensity to vanish may be due to lack of perfect orientation rather than presence of this species.

cause of imperfect orientation. In such a case the strong species can be established as being present but the weak species cannot be excluded. Such instances are indicated by a "?" in the table. In other instances, strong values of both I_{yz} and I_{zz} may be taken as definite indication of the presence of both species (e.g., 1464 cm^{-1}).

If this method of assigning species sounds ambiguous, it is a great improvement over the previous method. One would measure the depolarization ration ρ for a randomly oriented sample, then compare the value with theory, which states that $0 \leq \rho \leq 0.75$ for an A-species mode, while $\rho = 0.75$ for an E-species mode. By this criterion, it is evident that there is no way of establishing the presence of an E species; a ρ value of 0.75 is ambiguous. A ρ value of 0 would establish a band as being purely

A, but such a ρ value is quite rare. An intermediate value indicates the presence of A species but does not exclude the possibility of some E-species intensity.

Table V reconciles the species assignments obtained here with depolarization ratios measured by Vasko and Koenig[31]. The only discrepancies are the 904 and 1003 cm^{-1} bands which are assigned as E species but have quite low ρ values. Both bands are quite weak in Table IV, and either the present assignment or the ρ value could be in error.

The species assignments from Raman (Table IV) and infrared data (Table II) are readily reconciled. The only serious discrepancies are the bands at 814 (808) cm^{-1} and 1003 (998) cm^{-1}. The former has strong A character in Raman but strong E character in infrared, while the situation is reversed in the latter case. The 814 cm^{-1} band is strong in both Raman and infrared so this discrepancy must be taken to be real. The only remaining explanation is that this is a degenerate frequency, but for some reason the A species is silent in infrared and the E species is silent in Raman. Certainly the two methods of detecting molecular vibrations are quite different, and the intensity mix could shift between the two processes.

This particular behavior—a band showing A-species character in Raman and E species in infrared—also appears in the CH stretch region. Table VI compares vibrational frequencies for A and E species observed in Raman and infrared[52–56]. The Raman line at 2872 cm^{-1} shows pure A character, while its counterpart in infrared (2868 cm^{-1}) is an E band. The Raman line at 2924 cm^{-1} is

TABLE V
Comparison of Raman Species Assignments from Table IV
with Depolarization Ratio ρ

Frequency, cm^{-1}	Species	ρ	Frequency	Species	ρ
814	A + E	0.29	1307	A	0.44
846	A (E?)	0.13	1330	A + E	0.73
904	E	0.51	1363	A + E	0.82
942	E (A?)	0.88	1379	A + E	—
976	A (E?)	0.68	1444	E (A?)	0.76
1003	E	0.26	1464	A + E	0.80
1047	A + E	0.68	2842	A + E	0.08
1105	E	0.72	2872	A	0.17
1158	E	0.74	2886	A + E	0.22
1172	A (E?)	0.73	2907	A	0.18
1221	E (A?)	0.85	2927	A (E?)	0.13
1261	A (E?)	0.34	2954	E	0.44
1297	E	—	2962	A (E?)	

Theoretical ρ values:
A modes $0 \leq \rho \leq 0.75$
E modes $\rho = 0.75$
Experimental ρ values are from Vasko and Koenig[31].

TABLE VI
Observed Vibrational Frequencies of Polypropylene in the CH Stretch Region

Frequency (cm^{-1}):	ν_1	ν_2	ν_3	ν_4	ν_5	ν_6	ν_7
A Species							
Raman	2842	2872	2886	2907	2924	—	2962
IR,LLB	2836	—	—	2900	—	—	2960
IR,PF	2838	—	—	2906	—	—	2959
IR,MW	2843	—	—	—	—	—	2956
IR,T	2840	—	—	—	—	—	2965
E Species							
Raman	2842?	—	2886	—	2927?	2954	2962
IR,LLB	2836	2868	2875	—	2920	2947	—
IR,PF	2838	2868	2879	—	2921	2950	—
IR,MW	—	2868	2880	2907*	2925	2951	2956
IR,T	2840	2869	2877	—	2921	2953	—

Sources of data:
Raman—present work
Infrared—
LLB: Liang, Lytton, and Boone[53]
PF: Peraldo and Farina   [54]
MW: McDonald and Ward   [55]
T: Tadokoro et al.   [56]
*Probably a misprint. Appears to be an A band in the figure.

strongly A, with a questionable E component; in contrast, its infrared counterpart is an E band. Again, the frequencies are degenerate, but the A species appears in Raman while the E species appears in infrared.

One of the interesting features of Table VI is the fact that seven frequencies appear instead of six. For polypropylene, one would predict six fundamental modes of vibration: three CH_3 modes, two CH_2 modes, and one CH mode. Each of these modes can be degenerate or can split into A and E species occurring at distinct frequencies. The occurrence of seven frequencies indicates that five modes are degenerate while the sixth splits into distinct A and E species frequencies.

The value of obtaining the Raman data on oriented polypropylene becomes evident when one tries to sort out the CH stretch frequencies. First of all, the various workers do not all report seven frequencies. The infrared workers often miss the 2907 cm^{-1} band, which definitely appears in Raman[31]. In contrast, the Raman work of Vasko and Koenig[31] reports a single band at 2957 cm^{-1}, while the present Raman data and several infrared papers report two bands, 2954 and 2962 cm^{-1}. The explanation is that Vasko and Koenig's Raman data was for an unoriented sample, for which the weak 2954 line is overshadowed by the stronger 2962 line. The weak line is resolved for the oriented sample at yy and yz polarization, while most of the 2762 intensity appear in zz polarization where it does not interfere. Moreover, Vasko and Koenig's data can only report that the six frequencies observed (excluding ν_6) are all polarized, but this is ambiguous in terms of defining the species present. A purely A mode will be polarized, but a polarized mode need not be purely A. By obtaining Raman data on an oriented sample, it is established that six frequencies have definite A character present, while the seventh (ν_6, 2954 cm^{-1}) is a purely E band. On the other hand, infrared data shows that only two bands, ν = 2842 and ν_7 = 2862, have definite A character present, while five (ν_1, ν_2, ν_3, ν_5, and ν_6 in Table VI) have definite E character present. By examining both infrared and Raman data in Table VI, the five degenerate (A + E) frequencies are readily identified as ν_1, ν_2, ν_3, ν_5, and ν_7. By elimination, the nondegenerate mode splits into ν_4 = 2907 cm^{-1}, an A frequency; and ν_6 = 2954 cm^{-1}, an E frequency. (Note: McDonald and Ward[55] tabulate 2907 cm^{-1} as an E band, but their absorption curves indicate that it is an A band.)

The Raman data also clears up the question of overtones. Two frequencies in Table VI, ν_3 and ν_4, have been assigned as overtones by infrared workers[53,55]. These assignments are repudiated by the Raman data, in which these two frequencies are quite strong (see Table IV). Overtones are notoriously weak in the Raman experiment. One overtone/combination band does appear close to this region at 2722 cm^{-1}, but this is weaker than any of the CH stretch bands. This rather broad band probably results from the accidental degeneracy of several combinations and overtones. The author counts two overtones and six combinations which fall within 15 cm^{-1} of 2722 cm^{-1}.

The author has not interpreted the data in terms of degree of orientation because one of the four independent spectra, the I_{xy} spectrum, has not been measured because of experimental limitations. To date, only one experimental paper has appeared[57] relating polarized Raman data to orientation functions. Purvis, Bower, and Ward have related Raman intensities in oriented poly(ethylene terephthalate) for a p-disubstituted benzene band to second- and fourth-order molecular orientation functions. The interpretation is far simpler than in the polypropylene case, since the benzene rings act as independent oscillators and the direction of the atomic vibrational motions is easily established.

Summary

The various methods of characterizing preferred orientation in polymers are shown to be sensitive to different aspects of the structure. The complexity of structure in semicrystalline polymers requires careful interpretation of orientation data to yield significant results. The tensile strength of polypropylene is seen to be mainly correlated with amorphous orientation, despite apparent correlations with crystalline and overall orientation. The effect is dramatically observed in polypropylene films of ultrahigh orientation, where large differences in mechanical properties and ballistic performance, attributable to varying degrees of amorphous orientation, are seem among films possessing nearly perfect crystal orientation.

Vibrational spectroscopy reveals a great deal of intramolecular ordering even in the noncrystalline regions of polypropylene. The noncrystalline regions are seen, not as a random mixture of rotational conformers, but as a collection of short helical runs, separated from each other by no more than a hand reversal. The concept of "nonhelical" and "helical" fractions thus has no usefulness. Infrared dichroism of many so-called "helical" bands is found to be sensitive to amorphous as well as crystalline orientation, depending upon the sensitivity of a particular band to the shorter helical segments.

Acknowledgment

The author wishes to acknowledge the assistance of S. L. Lopata with the X-ray diffraction experiments, R. Tumminelli with the film stretching and characterization experiments, M. E. Roylance with the dynamic mechanical experiments, and D. Tabb and J. L. Koeing with the infrared and Raman spectroscopy.

References

1. Hermans, J.J., Hermans, P.H., Vermaas, D. and Weidinger, A., "Quantitative Evaluation of Orientation in Cellulose Fibers from the X-ray Fiber Diagram," *Rec. Trav. Chim. Pays-Bas,* 65 (1946), 427–43.

2. Stein, R.S., "The X-ray Diffraction, Birefringence, and Infrared Dichroism of Stretched Polyethylene.   II. Generalized Uniaxial Crystal Orientation," *J. Polym. Sci.,* 31 (1958), 327–30.

3. Desper, C.R., "Technique for Measuring Orientation in Polymers," *Crit. Rev. Macromol. Sci.,* 1 (1973), 501–43.

4. Desper, C.R. and Kimura, I., "Mathematics of the Polarized Fluorescence Experiment," *J. Appl. Phys.,* 38 (1967), 4225–33.

5. Nomura, S., Kawai, H., Kimura, I. and Kagiyama, M., "General Description of Optical Dichroic Orientation Factors for Relating Optical Anisotropy of Bulk Polymer to Orientation of Structural Units," *J. Polym. Sci., Part A-2,* 5 (1967), 479–95.

6. Stein, R.S., "The Polarization of Fluorescent Light from a Uniaxially Oriented Rubber," *J. Polym. Sci., Part A-2,* 6 (1968), 1975–85.

7. Kimura, I., Kagiyama, M., Nomura, S. and Kawai, H., "General Description of Optical (Dichroic) Orientation Factors for Relating Optical Anisotropy of Bulk Polymer to Orientation of Structural Units. Part II. Fourth Moments of Orientation Distribution and Polarized Fluorescence," *J. Polym. Sci., Part A-2*, 7 (1969), 709–24.

8. Nomura, S., Kawai, H., Kimura, I. and Kagiyama, M., "General Description of Orientation Factors in Terms of Expansion of Orientation Distribution Function in a Series of Spherical Harmonics," *J. Polym. Sci., Part A-2*, 8 (1970), 383–400.

9. Bohn, C.R., Schaefgen, J.R. and Statton, W.O., "Laterally Ordered Polymers: Polyacrylonitrile and Poly(vinyl Trifluoroacetate)," *J. Polym. Sci.*, 55 (1961), 531–48.

10. Gezovich, D.M. and Geil, P.H., "Morphology of Quenched Polypropylene," *Polym. Eng. and Sci.*, 8 (1968), 202–209.

11. Wilchinsky, Z.W., "Recent Developments in the Measurement of Orientation in Polymers by X-ray Diffraction," in *Advances in X-ray Analysis*, Vol. 6, W.M. Mueller and M. Fay, eds., New York: Plenum Press (1963), 231–41.

12. Alexander, L.E., *X-ray Diffraction Methods in Polymer Science*, New York: Wiley-Interscience (1969), 200–79.

13. Desper, C.R., "A Computer-Controlled X-ray Diffractometer for Texture Studies of Polycrystalline Materials," in *Advances in X-ray Analysis*, Vol. 12, C.S. Barrett, G.R. Mallett, and J.B. Newkirk, eds., New York: Plenum Press (1969), 404–24.

14. Desper, C.R., "Computer Programs for Reduction of X-ray Diffraction Data for Oriented Polycrystalline Specimens," Army Materials and Mechanics Research Center, Watertown, Mass. Technical Report No. AMMRC-TR-72-34, November 1972 (AD 753-920).

15. Wilchinsky, Z.W., "On Crystalline Orientation in Polycrystalline Materials," *J. Appl. Phys.*, 30 (1959), 792.

16. Sack, R.A., "Indirect Evaluation of Orientation in Polycrystalline Materials," *J. Polym. Sci.*, 54 (1961), 543–50.

17. Desper, C.R., "Structural Characterization of Isotactic Polypropylene Films of Ultrahigh Orientation. I. Determination of Crystalline and Amorphous Orientation," *J. Macromol. Sci. Phys.*, B7 (1973), 105–19.

18. Samuels, R.J., "Characterization of Deformation in Polycrystalline Polymer Films," in *The Science and Technology of Polymer Films*, Vol. 1. O.J. Sweeting, ed., New York: Interscience (1968), 255–364.

19. Wilkes, G.L., "The Measurement of Molecular Orientation in Polymeric Solids," *Advan. Polym. Sci.*, 8 (1971), 91–136.

20. Stein, R.S., "Optical Methods of Characterizing High Polymers," in *Newer Methods of Polymer Characterization*, B. Ke, ed., New York: Interscience (1964), 155–206.

21. Samuels, R.J., "Morphology of Deformed Polypropylene. Quantitative Relations by Combined X-ray, Optical, and Sonic Methods," *J. Polym. Sci., Part A*, 3 (1965), 1741–57.

22. Samuels, R.J., "Quantitative Characterization of Deformation in Drawn Polypropylene Films," *J. Polym. Sci., Part A-2*, 6 (1968), 1101–15.

23. Moseley, W.W. Jr., "Measurement of Molecular Orientation in Fibers by Acoustic Methods," *J. Appl. Polym. Sci.*, 3 (1960), 266–73.

24. Charch, W.H. and Moseley, W.W. Jr., "Structure-Property Relationships in Fibers. I. Structure as Revealed by Sonic Observations," *Text. Res. J.*, 29 (1959), 525–38.

25. Krimm, S., "Infrared Spectra of High Polymers," *Fortschr. Hochpolymer Forsch.*, 2 (1960), 51–103.

26. Zbinden, R., "Orientation Measurements," in *Infrared Spectroscopy of High Polymers*, R. Zbinden, ed., New York: Academic Press (1964), 166–233.

27. Koenig, J.L., Cornell, S.W. and Witenhafer, D.E., "Infrared Technique for Measurement of Structural Changes During the Orientation Process in Polymers," *J. Polym. Sci., Part A-2*, 5 (1967), 301–23.

28. Bower, D.I., "The Investigation of Molecular Orientation Distributions by Polarized Raman Scattering and Polarized Fluorescence," *J. Polym. Sci., Polym. Phys. Ed.*, 10 (1972), 2135–49.

29. Snyder, R.G., "Raman Scattering Activities for Partially Ordered Molecules," *J. Mol. Spectrosc.*, 37 (1971), 353–65.

30. Koenig, J.L., "Raman Scattering of Oriented Systems," Case-Western Reserve University, Cleveland, Ohio, Division of Macromolecular Science, Report No. 209, April 1971.

31. Vasko, P.D. and Koenig, J.L., "Raman Scattering and Polarization Measurements of Isotactic Polypropylene," *Macromolecules*, 3 (1970), 597–604.

32. Samuels, R.J., "Quantitative Structural Characterization of the Mechanical Properties of Isotactic Polypropylene," *J. Macromol. Sci., Phys.*, B4 (1970), 701–59.

33. Sheehan, W.C. and Cole, T.B., "Production of Super-Tenacity Polypropylene Filaments," *J. Appl. Polym. Sci.*, 8 (1964), 2359–88.

34. Sheehan, W.C., Wellman, R.E. and Cole, T.B., "Relationship Between Structural Parameters and Tenacity of Polypropylene Monofilaments," *Text. Res. J.*, 35 (1965), 626–37.

35. Sheehan, W.C., Wellman, R.E. and Cole, T.B., "Structural Parameters and Tenacity of Polypropylene Fibers," Proceedings of Symposium on Polypropylene Fibers, 17–19 September 1964, Birmingham, Ala.: Southern Research Institute (1964), 73–99.

36. Sakurada, I. and Kaji, K., "Relation Between the Polymer Conformation and the Elastic Modulus of the Crystalline Region of Polymer," *J. Polym. Sci., Part C*, 31 (1970), 57–76.

37. ASTM Test Standard D-1504-70, "Determining Orientation Release Stress of Plastic Sheeting," *1973 Annual Book of ASTM Standards, Part 26*, Philadelphia: American Society for Testing and Materials (1973), 112–13.

38. ASTM Test Standard F-89-68, "Modulus of a Flexible Barrier Material by Sonic Method," *1968 Annual Book of ASTM Standards, Part 15*, Philadelphia: American Society for Testing and Materials (1968), 898–904.

39. McCrum, N.G., Read, B.E. and Williams, G., *Anelastic and Dielectric Effects in Polymer Solids*, New York: John Wiley & Sons (1967), 377–87.

40. Davies, G.R., Owen, A.J., Ward, I.M. and Gupta, V.B., "Interlamellar Shear in Anisotropic Polyethylene Sheets," *J. Macromol. Sci., Phys.*, B6 (1962), 215–28.

41. Stachurski, Z.H. and Ward, I.M., "Mechanical Relaxations in Polyethylene," *J. Macromol. Sci., Phys.*, B3 (1969), 445–94.

42. Stachurski, Z.H., and Ward, I.M., "β-Relaxations in Polyethylene and their Anisotropy," *J. Polym. Sci., Part A-2*, 6 (1968), 1817–33.

43. Hay, I.L. and Keller, A., "A Study on Orientation Effects in Polyethylene in the Light of Crystalline Texture. Part 2. Correlation of Molecular Orientation with that of the Textural Elements," *J. Mater. Sci.*, 2 (1967), 538–58.

44. Point, J.J., Homes, G.A., Gezovich, D. and Keller, A., "A Study on the Orientation Effects in Polyethylene in the Light of Crystalline Texture. Part 4. Truly Single Texture on Unidirectional Rolling," *J. Mater. Sci.*, 4 (1969), 908–18.

45. Seymour, R.W., Cooper, S.L. and Allegrezzi, A., "Orientation Studies of Polyure-

thane Block Copolymers," *Polym. Prep., Amer. Chem. Soc., Div. Polym. Chem.,* 14 (1973), 1046–50.

46. Estes, G.M., Seymour, R.W. and Cooper, S.L., "Infrared Studies of Segmented Polyurethane Elastomers. II. Infrared Dichroism," *Macromolecules,* 4 (1971), 452–47.

47. Heinen, W., "Infrared Determination of the Crystallinity of Polypropylene," *J. Polym. Sci.,* 38 (1959), 545–47.

48. Quynn, R.G., Riley, J.L., Young, D.A. and Noether, H.D., "Density, Crystallinity, and Heptane Insolubility in Isotactic Polypropylene," *J. Appl. Polym. Sci.,* 2 (1959), 166–73.

49. Miller, R.L., "On the Existence of Near-Range Order in Isotactic Polypropylenes," *Polymer,* 1 (1960), 135–43.

50. Brader, J.J., "Determination of the Degree of Helical Structure and Related Isotacticity of Polypropylene by an Infrared Method," *J. Appl. Polym. Sci.,* 3 (1960), 370–71.

51. Zerbi, G., Gussoni, M. and Ciampelli, F., "Structure of Liquid Isotactic Polypropylene from its Vibrational Spectrum," *Spectrochim. Acta,* 23A (1967), 301–11.

52. Luisi, P.L., "Calculation of the Conformational Properties of Isotactic Poly-(α-olefins) in Solution," *Polymer,* 13 (1972), 232–39.

53. Liang, C.Y., Lytton, M.R. and Boone, C.J., "Infrared Spectra of Crystalline and Stereoregular Polymers. II. Carbon-Hydrogen and Carbon-Deuterium Stretching Frequencies of Polypropylene and Deuterated Polypropylenes," *J. Polym. Sci.,* 47 (1960), 139–48.

54. Peraldo, M. and Farina, M., "Spettri Infrarossi di Polipropileni Deuterati," *Chim. Ind. (Milan),* 42 (1960), 1349–53.

55. McDonald, M.P. and Ward, I.M., "The Assignment of the Infra-red Absorption Bands and the Measurement of Tacticity in Polypropylene," *Polymer,* 2 (1961), 341–55.

56. Tadokoro, H., Kobayashi, M., Ukita, M., Yasufuku, K. and Murahashi, S., "Normal Vibrations of the Polymer Molecules of Helical Conformation. V. Isotactic Polypropylene and its Deuteroderivatives," *J. Chem. Phys.,* 42 (1965), 1432–49.

57. Purvis, J., Bower, D.I. and Ward, I.M., "Molecular Orientation in Poly(ethylene Terephthalate) Studied by Polarized Raman Scattering," *Polymer,* 14 (1973), 398–400.

58. Desper, C.R., Southern, J.H., Ulrich, R.D. and Porter, R.S., "Orientation and Structure of Polyethylene Crystallized under the Orientation and Pressure Effects of a Pressure Capillary Viscometer," *J. Appl. Phys.,* 41 (1970), 4284–89.

59. Snyder, R.G. and Schachtschneider, J.H., "Valence Force Calculation of the Vibrational Spectra of Crystalline Isotactic Polypropylene and Some Deuterated Polypropylenes," *Spectrochim. Acta,* 20 (1964), 853–69.

CASE HISTORIES

Moderator: Kenneth J. Smith

New York State College of Environmental Science and Forestry
Syracuse, New York

K. VEDAM
The Pennsylvania State University
University Park, Pennsylvania

Chapter 17

Characterization of Surfaces

ABSTRACT

The potentialities and limitations of the various techniques such as FIM, SEM, AES, ESCA, SIMS, ISS, LEED, HEED, etc., that are currently being used to characterize surfaces will be briefly reviewed. Emphasis will be placed on the morphological, chemical, structural, and defects aspects of characterization by these techniques. It is shown that ellipsometry is yet another tool for nondestructive characterization of surfaces, yielding information on the optical properties of (1) the substrate material, (2) the contaminant film on the surface, and (3) the damaged surface layers of the substrate which is dependent on the preparative history of the specimen.

Introduction

Surfaces play an extremely important role in many fields of materials science and engineering, industrial and natural catalysis, chemistry of solids and sediments, etc. This importance of the surfaces arises from the fact that solids communicate with the rest of the universe only through their surfaces. Consequently, any imperfection or departure from the ideal surface can and does influence the various properties of the material. Characterization of surfaces thus becomes a necessary and important step before a thorough understanding of the behavior, as well as the desired fabrication, of the material is attempted.

Before we begin to discuss the various methods of characterization, it is worthwhile to consider the essential features of the "real" surface that we have to contend with. An ideally perfect surface would correspond to the plane surface terminating an ideally perfect semi-infinite solid as in Figure 1(a). But in practice the surface is never truly planar but has cleavage steps and/or fracture tear lines and other blemishes, depending on its preparative history. Further, it is well known that monolayers of con-

503

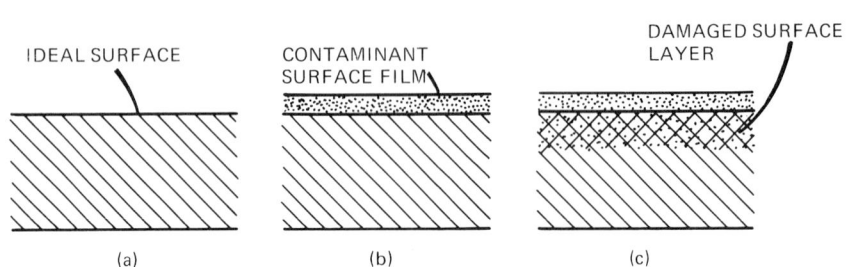

Figure 1. Schematic drawing of (a) ideal surface, (b) surface with a contaminant layer, and (c) surface with both a damaged layer and a contaminant layer.

tamination can form on atomically clean surfaces in about two hours even in ultrahigh vacuum of 10^{-10} torr pressure. Thus, unless special precautions are observed, the surfaces under study generally are covered with contaminant layers, whose chemical composition may also possess some degree of variation with respect to the distance from the boundary [Figure 1(b)]. Furthermore, the outermost layers of the substrate material are usually damaged or deformed or even disordered [Figure 1(c)], depending on its preparative history.

In general, a complete characterization of surface involves studies that can be grouped under four headings—morphological, chemical, structural, and defect characterization. Numerous techniques and instruments have been developed during the last few years that are particularly suited for this purpose. This chapter is an attempt to bring out the highlights of some of these techniques and to point out their advantages and disadvantages as we understand them at this time, and also to provide recent appropriate references to the literature. Finally, a brief description of ellipsometry is given and it is shown that this is yet another nondestructive tool for the characterization of surfaces, yielding information on the optical properties of (1) the substrate material, (2) the contaminant film on the surface, and (3) the damaged surface layers of the substrate.

Morphological Characterization

Contact Stylus Instrument

The traditional contacting-stylus instrument[1] similar to a high-fidelity phonograph needle has been the standard method for surface-profile measurements. A fine diamond stylus coupled to a transducer scans the surface, providing closely spaced profiles that can be displayed on an X-Y recorder. The diameter of the stylus point determines the width and depth of the crevices it will measure. Stylus instruments range in elevation sensitivity from 25 Å to 250 Å with a horizontal resolution between 1 and 10 microns. The elevation sensitivity is excellent but, unfortunately, the stylus sometimes obliterates the original surface features.

Topografiner

This instrument[2] recently developed at the National Bureau of Standards uses a noncontacting field-emission probe to measure surface elevation with resolution in the 30 Å range and horizontal resolution in the 200 Å range. In this instrument, two X and Y piezo drivers scan the emitter parallel to the specimen surface and slightly above it. Emitter-to-specimen probe voltage is determined by the constant current passing between them, and by their spacing. The Z piezo driver is controlled by a servo system that holds this voltage, and hence the emitter-specimen spacing, constant. Thus the Z-piezo voltage gives the surface profile directly. Figure 2 is a map of a 180-line-per-mm diffraction grating replica obtained with this device. The use of the field-emission probe requires a vacuum of about 10^{-9} torr, and the ultimate resolution is limited by the mechanical vibration and the electrical noise in the system.

Optical Microscopy

Among the various types of optical microscopes, the interference microscope is the one most frequently used to obtain quantitative information about surface topography[3,4]. In particular, the technique of Tolansky's multiple-beam interferometry including Fringes of Equal Chromatic Order is very well known and has been exploited for over 25 years and hence will not be discussed further here. Suffice it to say that steps in the surface topography as small as 5 Å can be resolved by this technique. However, this method suffers from a drawback that the surface must be flat to within one or two fringes and, furthermore, highly reflecting film(s) must be overlaid on the surface under examination, unless the surface itself has high reflectivity.

The recently developed Nomarski technique[5,6], based on the interference between polarized beams produced by the method of wave shearing, enables one to see and photograph irregularities a few tens of angstroms in height with an optical microscope. It is completely nondestructive, and does not require any elaborate preparation. With the recent availability of the Nomarski interference attachment to polarizing microscopes, the full potentialities of this simple and powerful technique is just beginning to be realized. Figure 3 shows photographs[7] of a surface of sodium chloride taken with a Nomarski optical interference microscope, a dark-field optical microscope, and a scanning electron microscope using two different display modes. The photographs taken in the optical micro-

Figure 2. Topographic map of a 180-line-per-mm diffraction-grating replica, obtained with the "topografiner," a noncontacting field-emission probe.

(a)
(b)
(c)
(d)

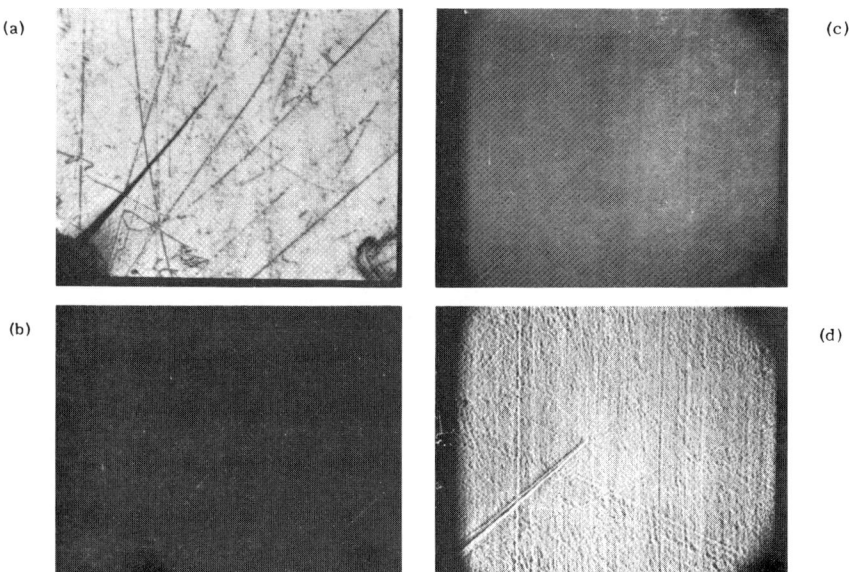

Figure 3. Micrographs showing surface scratches in sodium chloride window. (a) Nomarski optical interference microscope (320X); (b) dark-field optical microscope (320X); (c) scanning electron microscope, Z-modulation (800X); (d) scanning electron microscope, Y-modulation (800X).

scopes are of the same area of the specimen but at a lower magnification than the SEM photographs. All photographs show the presence of scratches approximately 1 μm wide on the surface, with varying degrees of clarity and resolution. The outstanding capabilities of the Nomarski technique are quite evident.

Transmission Electron Microscopy

In this technique, a monolayer or so of gold is evaporated on the surface (say of cleaved sodium chloride crystal) such that the gold atoms form nuclei which collect along the edges of steps on the surface. When a carbon replica is subsequently deposited, the gold nuclei are incorporated in the carbon replica and appear as step decorations in the microscope image. Figure 4 shows such an image of sodium chloride surface[8] showing single and double atom steps 2.8 Å and 5.6 Å high, respectively.

Scanning Electron Microscopy

The scanning electron microscope[9] combines some of the best features of the optical microscope, electron microscope, and electron microprobe into an instrument of outstanding performance, high reliability, and ease

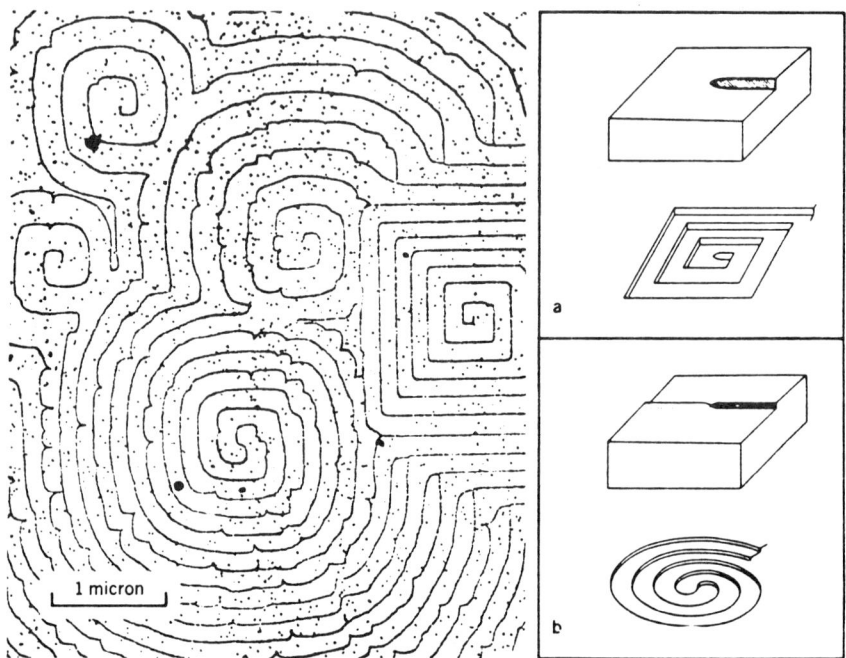

Figure 4. Atomic steps on a cleaved and evaporated sodium-chloride crystal, revealed by a transmission electron microscope by the gold-decoration-carbon-replication technique. Screw dislocations can lead to round or square spirals; single steps forming the edges of round spirals (b) combine in pairs to form the double steps at the edges of square spirals (a).

of operation. It is uniquely suited to a large variety of research problems and, in fact, during the last five years it has actually caused a revolution in the study of surfaces of ceramics, semi-conductors, metals, and biological systems.

In the typical scanning electron microscope, a beam from an electron gun is focused to a spot about 100 Å in diameter on the specimen. Secondary electrons, backscattered electrons, photons, or X-rays from the specimen can be detected and amplified, and the resulting signal used to modulate the brightness of a cathode-ray tube display system. The electron beam is scanned across the specimen as a raster synchronous with the cathode-ray tube scan. As the beam only strikes one spot of the surface at a time, the X-rays and photons emitted can be used to obtain chemical information at each point on the surface.

Magnifications range between 15X and 10^5X, bridging the gap between light and electron microscopes. The ease of changing magnification makes it easy to zoom from gross image to fine detail. Scanning electron

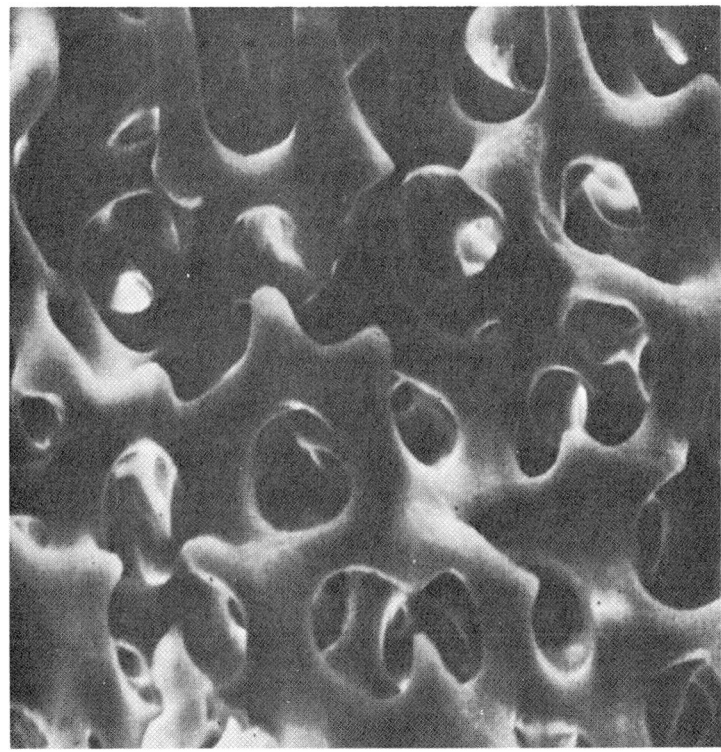

10µ

Figure 5. High-strength biogenic material (skeletal ossicle of the asteroid, Acanthaster Planci): single-crystal high-purity magnesium calcite.

microscope manufacturers guarantee resolutions of 100–200 Å. Compared to conventional electron or light microscopes, scanning electron microscope pictures exhibit remarkable three-dimensional effects due to the extra-ordinary depth of field made possible by the wavelength and apertures involved (Figure 5). Depth of field is 1 cm at 100X and 1 µm at 10,000X.

Connecting a computer directly to the SEM makes it possible to do quantitative petrography on a scale and with a precision inconceivable a decade ago. White *et al.*[10] have recently demonstrated the wealth of information that can be obtained by such a technique, as shown in Figure 6 on a sample of Lucalox gas-polished to AA finish of 3.2 µm.

A comparison of the typical resolution capabilities of the various surface microtopographical instruments discussed above is given in Table I.

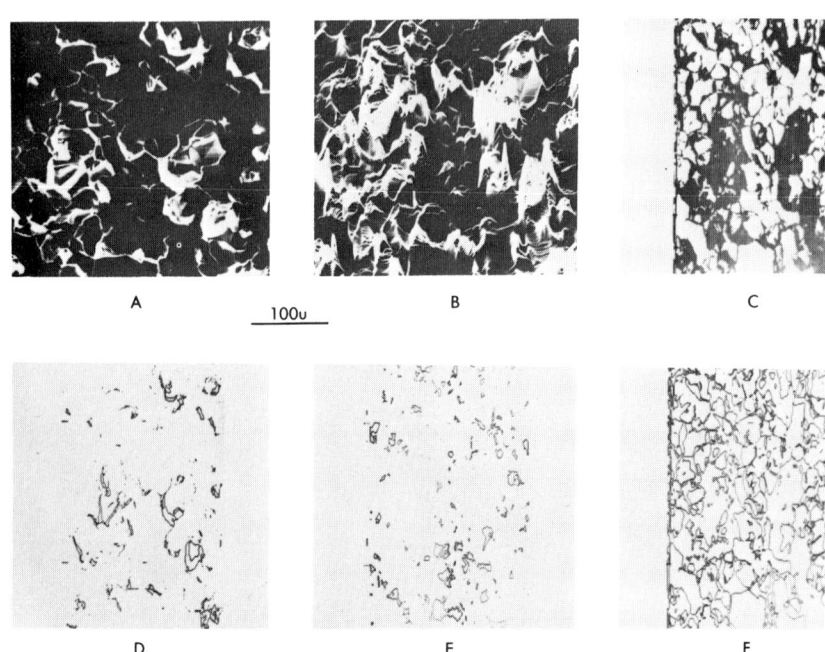

Figure 6.   Lucalox gas-polished to AA finish of 3.2 μm. (A) Secondary-electron images
with detector conditions set to minimize shadowing effects. Regions of steepest slope have
greatest brightness. (B) Y-axis modulation of same area as A. An instrumental distortion
has elongated this image toward the top so that the registration is not accurate with respect to A. (C) Computer-generated binary map at the 50% level. All data points whose
intensity is greater than the average value are indicated by a black dot. (D) Computer-
generated perimeter at the 95% level. The computer printer introduces an elongation in
the top-bottom direction. The bright areas in A can be matched to their corresponding
configurations. (E) Computer-generated perimeter map at the 50% level. (F) Computer-
generated perimeter map at the 10% level (inverted).

TABLE I
Typical Resolutions of Surface-Microtopographical Instruments [11]

Instrument	Approx. vertical resolution, Å	Approx. horizontal resolution, Å
Transmission electron microscope	1500	50
Scanning electron microscope	1500	100
Optical interference	5	25000
Stylus instrument	25	10000
Topografiner	30	4000

Field Ion Microscopy

The ultimate sensitivity in the study of surface geometry is the imaging of individual atoms on the surface, and this has been achieved by Professor E. W. Mueller[12] using the field ion microscope. This method is based on the strong electric fields that can be created at the tip of a very sharp

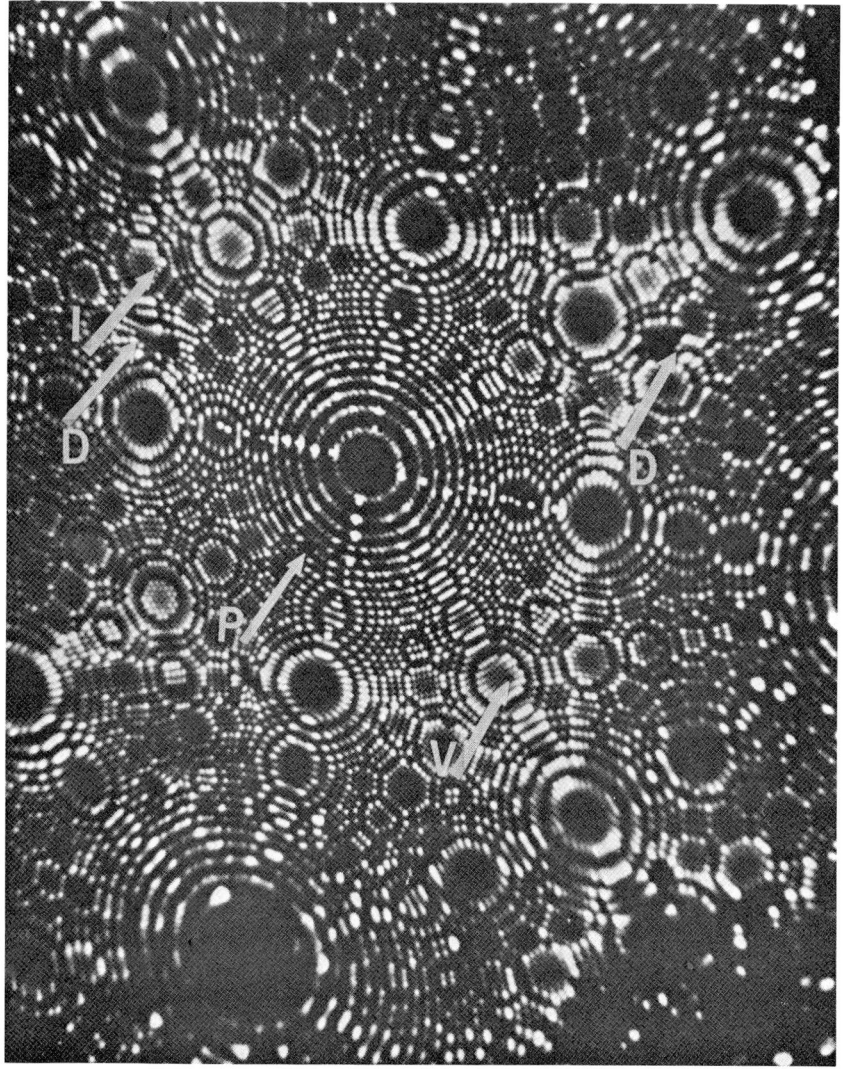

Figure 7.   Field-ion microscope image of a $\langle 001 \rangle$ oriented iridium crystal, showing a vacancy (V), dislocation (D), and the atom-probe hole (P) (Courtesy E. W. Mueller).

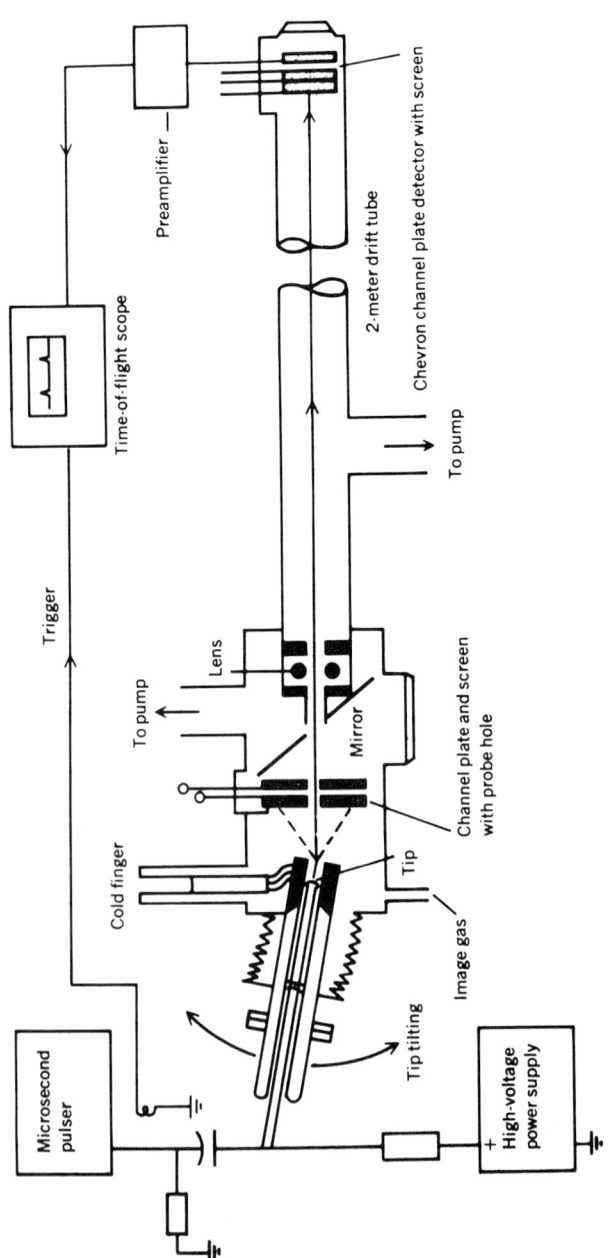

Figure 8.   Schematic drawing of the atom-probe field ion microscope.

metal point. Helium atoms ionized by these fields are accelerated to a fluorescent screen and produce an image of the metal tip, in which spots identified with individual surface atoms are clearly resolved. Figure 7 shows such an image of a ⟨001⟩ oriented iridium crystal, where dislocation cores (D), a vacancy (V), and an impurity atom (I) can easily be recognized. Very recently Mueller[13] has adapted a time-of-flight–mass spectrometer to the field ion microscope, so that any atom corresponding to a given image spot can be selected, field-evaporated through a small probe hole (P in Figure 7) on the screen, and analyzed. The mass-to-charge ratio of the ion passing through this hole is determined from its time of flight. Figure 8 depicts a schematic diagram of this atom-probe field ion microscope, and Figure 9 shows typical time-of-flight traces of desorbed species of 20 consecutive pulses, indicating the emergence of ions such as Rh^{++}, RhH^{++} and $RhHe^{++}$.

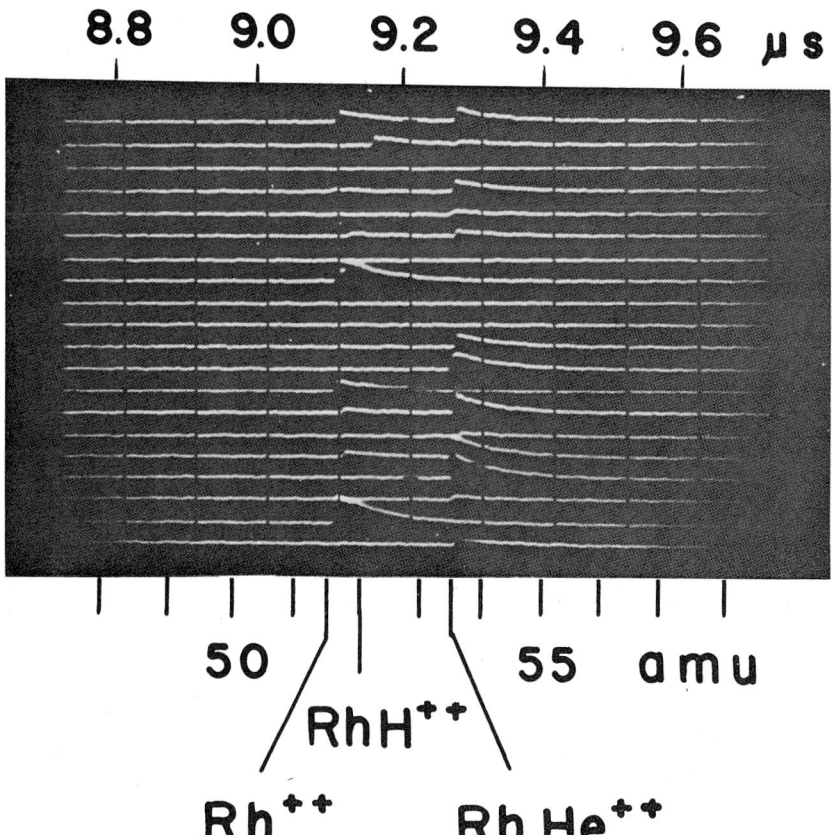

Figure 9. Time of flight traces of desorbed species for 20 consecutive pulses indicate Rh^{++}, RhH^{++} and $RhHe^{++}$ ions (Courtesy E. W. Mueller).

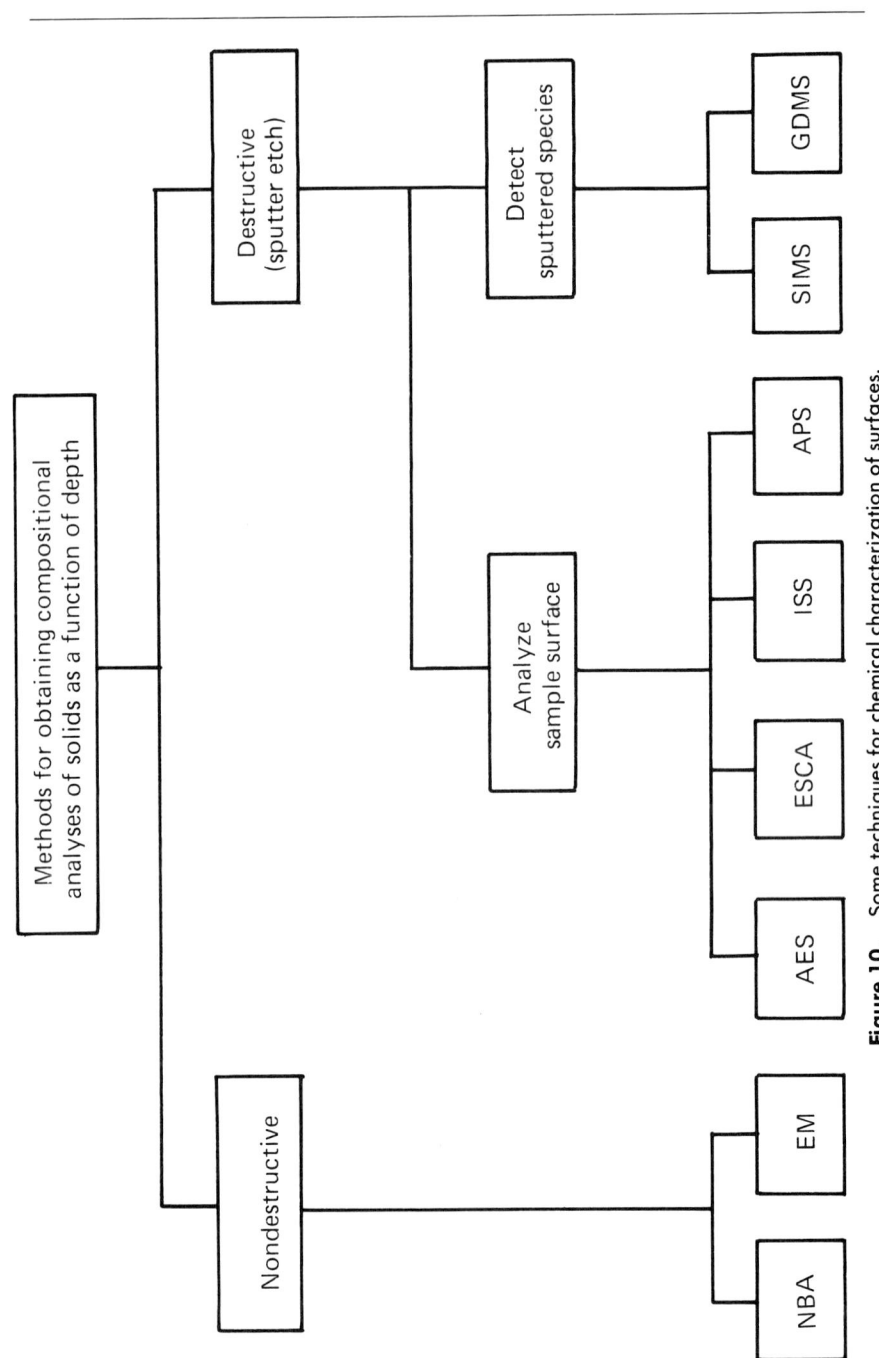

Figure 10.   Some techniques for chemical characterization of surfaces.

Thus, it is no exaggeration to state that the atom-probe field ion microscope is the ultimate in sensitivity, both in surface geometry and composition. But, unfortunately, this method is rather limited with respect to the materials that can be studied. Only good-conducting metals can be used and then they must be capable of being thinned down to a sharp point. However, recently some work has been done on molecules adsorbed on the metal tip.

Before we begin discussing the chemical aspects of characterization, it should be mentioned that no attempt has been made here to review the techniques of surface-area measurement[14–16] such as the use of the classical BET adsorption isotherm for the interpretation of the measurements on the physical adsorption of gases on the surface.

Chemical Characterization

Over eight different techniques are currently being used for the compositional analysis of the surface layers of a solid. Most of the analytical schemes capable of good depth resolution are destructive and rely on microsectioning by sputter-etching. Chemical etching is a common microsectioning technique, but sputter-etching has the advantages of being able to microsection essentially all materials in a clean controllable manner, and of being compatible with modern surface-analysis techniques and various methods of observing the removed species. This discussion will consider only sputter-etching as a microsectioning technique.

Some of these techniques that can yield information about the composition of a solid as a function of depth can be grouped (Figure 10) as either nondestructive, or destructive—as sputter-etching with analysis of the sample surface, and sputter-etching with analysis of the sputtered particles, as has been done by Coburn and Kay[17]. The term "nondestructive" technique has been used here in a rather loose manner: the disorder introduced by the exciting irradiation is being ignored at this stage, but will be discussed later in this chapter. Again, the relationship between the various forms of excitation energy used in these techniques as probes, and the emitted energy detected, can be shown (Table II).

Nuclear Backscattering Technique (NBS)

In this nondestructive method[18,19] the sample is bombarded with H^+ and He^+ ions of approximately 2 MeV energy and the energy of the backscattered ions is recorded. The energy lost by these ions can be related to the type and depth of various species in the sample up to a depth of a few microns. The depth of resolution is of the order of 100 Å, but it is rather difficult to detect low-mass impurities in a high-mass matrix.

TABLE II
Analytical Techniques based on Photons, Electrons, and
Ions as the Probing and Detected Particles

Energy Out \ Energy In	Electromagnetic Radiation (hν)	Electrons (e)	Ions	Electrostatic
hν	X-ray fluorescence	Electron microprobe APS	IES	
e	ESCA Auger electrons Photoelectrons	AES	Ion neutralization	FIM Atom probe
Ions			Ion microprobe SIMS ISS GDMS	

AES = Auger electron spectroscopy; APS = Appearance potential spectroscopy; ESCA = Electron spectroscopy for chemical analysis; FIM = Field-ion microscopy; IEX = Ion X-ray spectroscopy; ISS = Ion-scattering spectrometry; SIMS = Secondary-ion mass spectrometry; GDMS = Glow discharge mass spectrometry.

Electron Microprobe (EM)

This most useful and nondestructive technique for the elemental chemical analysis of solids[20] on a microscale can also be used with advantage for the analysis of surface layers. This technique based on the electron-beam stimulation of the characteristic X-rays in an extremely small area (~ 1 μm^2) also yields information on the chemical bonding[21]. For example,

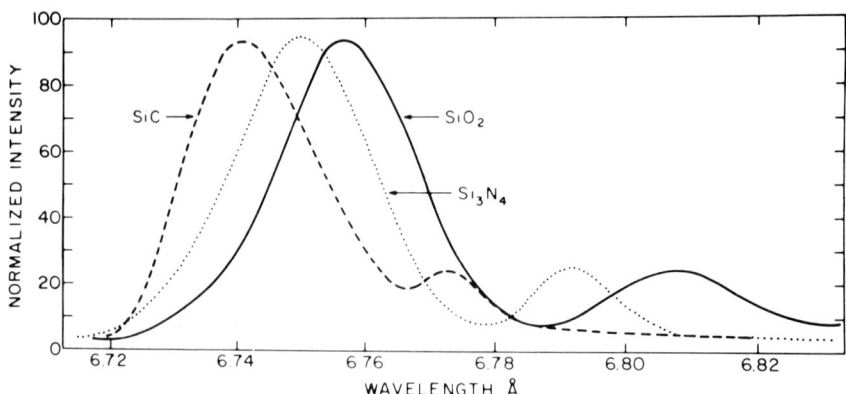

Si K_β, K_β' for SiC, Si_3N_4, and α-SiO_2.

Figure 11.   Si K_β, K_β' for SiC, Si_3N_4 and α-SiO_2.

Figure 11 shows that SiK_β line together with its satellite K'_β on the low energy side, for the SiC, Si_3N_4 and α-SiO_2. It is seen that each compound yields a K'_β peak position characteristic of each ligand. Since each K'_β peak can be completely resolved, it is easy to use K'_β intensity ratios to establish the relative population of oxygen, carbon, and nitrogen combined with silicon in each situation.

The electron-microprobe technique is particularly useful for analyzing surface layers in situations when (1) the elements in the surface layers are different from those in the bulk; (2) lateral variations in composition

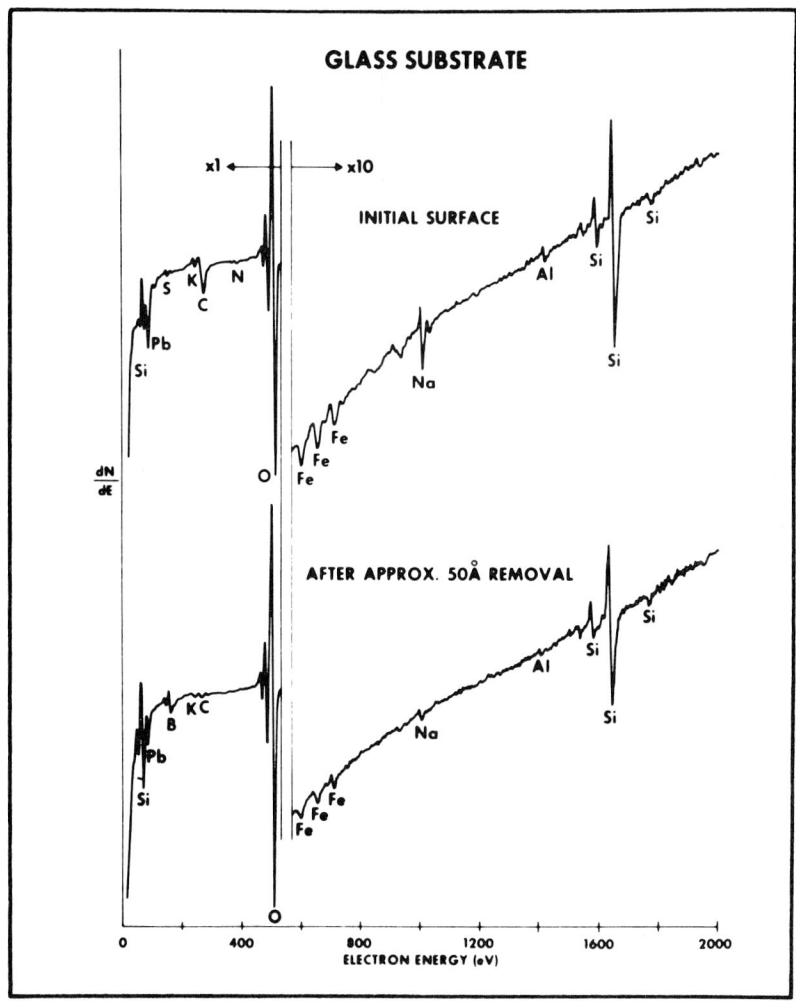

Figure 12. Auger spectrum of glass obtained before and after rare-gas sputter etching (Courtesy P. W. Palmberg).

or thickness are of importance, taking advantage of relatively high spatial resolution ($\sim 1\ \mu$m) of the electron microprobe; (3) one wishes to determine the valency or coordination of elements in the layer; and (4) the analysis must be nondestructive.

Auger Electron Spectroscopy (AES)

In AES[22,23] the surface of the solid is irradiated typically with electrons having a few keV energy, and the energy distribution of the secondary electrons is measured. The location of the Auger peaks in this energy distribution identifies the nature of the surface species, while the intensity of these peaks is a measure of the abundance of the species on or near the surface. It is extremely sensitive, and can detect 10^{-2} to 10^{-3} monolayers of material on a surface. It can detect all the elements in the periodic table except H and He, with almost equal sensitivity within a factor of 10. The technique is capable of excellent spatial resolution in the plane of the sample surface and can be used with a scanning electron microscope. Since AES involves bombarding the material with a beam of electrons and performing the energy analysis of electrons ejected from it, the complex charge build-up in insulating specimens can pose a problem, but this can usually be overcome by using the incident beam at a grazing angle of incidence. Figure 12 shows the Auger spectra from a glass substrate, obtained by the grazing-angle incidence technique.

Electron Spectroscopy for Chemical Analysis (ESCA)

In this technique[24–26] the sample is bombarded with monochromatic X-rays of approximately 1 keV energy, and the energy distribution of the ejected photoelectrons is measured. As in AES, the energy of a peak in the photoelectron energy distribution serves to identify the elemental species whereas the intensity of the peak is a measure of the abundance. The sensitivity is somewhat less than that of AES, possibly by a factor of 10. An important feature of ESCA is its ability to obtain detailed chemical bonding in the solid. For example, Figure 13 depicts the binding-energy shifts of the lead 4f electrons as oxide layers form on a freshly evaporated lead surface[27].

In such photoelectron spectroscopic studies, one can also use photons of lower energy or electrons as the exciting radiation instead of X-rays, and, in fact, investigators have coined many new names to describe their experimental arrangements. However, no accepted terminology has as yet been established. Following Siegbahn[24] and Carlson[25] all these variations can be grouped under the category of ESCA, and Figure 14 shows the overall connection between these experiments. In fact, even AES also can be brought under the broad umbrella of ESCA; however, we have

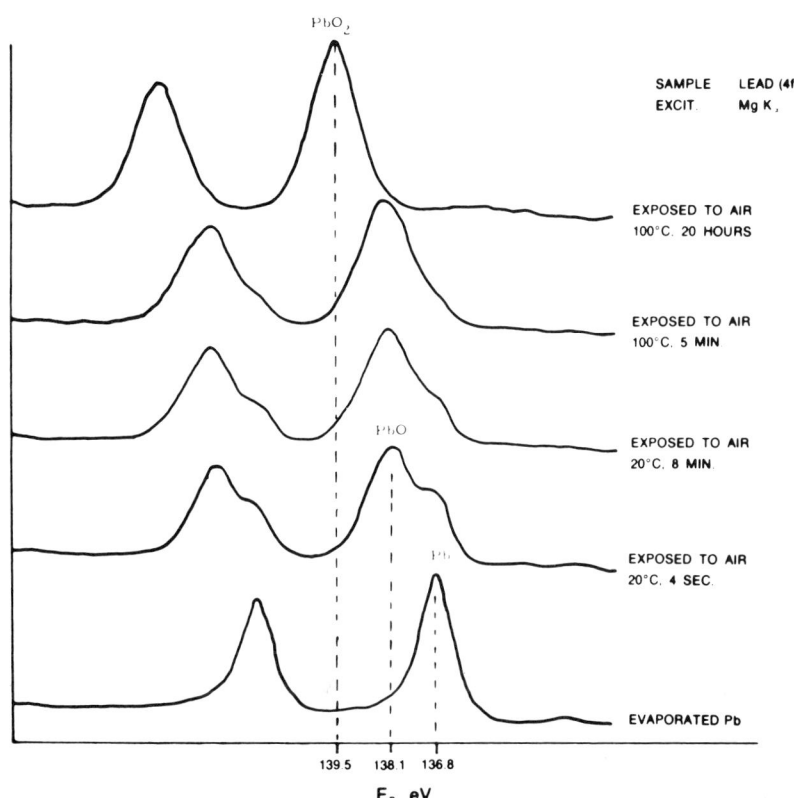

Figure 13. Binding-energy shifts of the lead 4f electrons as oxide layers form on a pure lead sample as revealed by ESCA.

chosen to list it separately in this chapter in view of its extensive use in the literature as AES.

Ion Scattering Spectrometry (ISS)

In this technique[28,29], the sample is bombarded with an ion beam (typically 1 keV He^+ ions) and the energy distribution of the reflected ions is measured. The identification of the species in the outer-most monolayer is provided by the energy of the reflected ions, using the simple binary collision model. The intensity of the peaks provides a measure of the abundance of a particular species. It is capable of observing 10^{-2}–10^{-3} monolayers of material on a surface. Isotopes can be observed, but the mass resolution is poor at high masses even when higher mass probing ions such as argon are used.

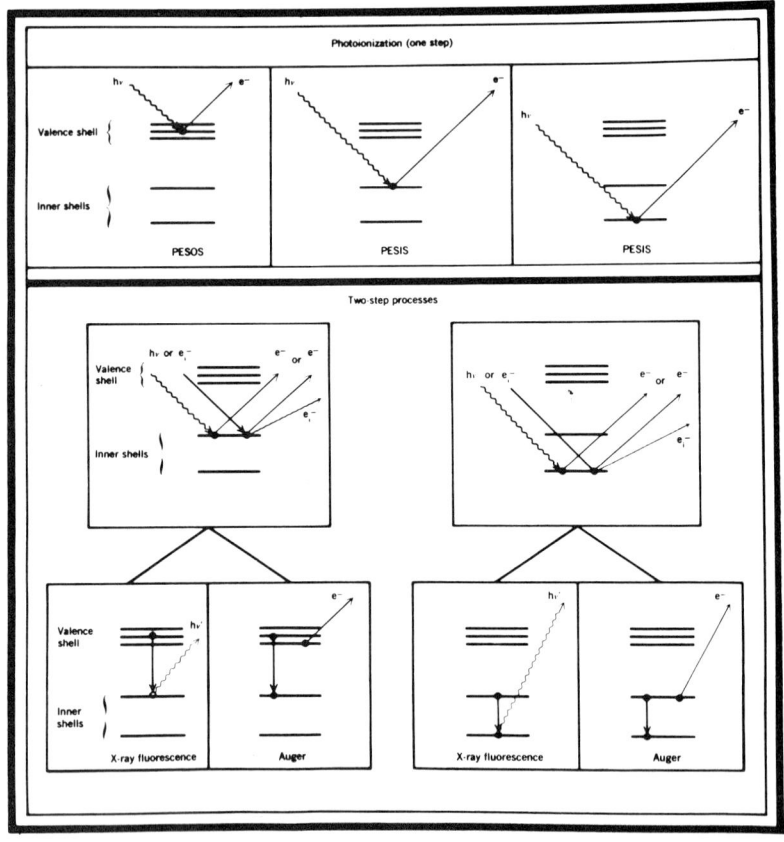

Figure 14. Mechanism involved in ESCA, AES, and X-ray fluorescence. PESOS = photoelectron spectroscopy of the outer shell electrons; PESIS = photoelectron spectroscopy of the inner shell electrons.

Secondary-Ion Mass Spectrometry (SIMS)

In this technique[30–32], the sample is bombarded with a beam of 1–20 keV ions. These primary ions cause the upper atomic layers to be sputtered off. Most of the material leaves as neutral atoms or molecules, but a small fraction is ejected as positive or negative ions. These secondary ions are collected and mass-analyzed.

Control and localization of the sputtering process permit chemical analysis with x,y resolutions of approximately 1 μm, the examination of fractional monolayers, an in-depth analysis with 50–100 Å resolution, and the acquisition of secondary-ion images. The mass spectrometer provides elemental converage from hydrogen to uranium, isotopic characterization, and sensitivities of 10^{-15}–10^{-19} gram, depending on the element under con-

HELVITE $(Mn,Fe)_4 Be_3 Si_3 O_{12} S$

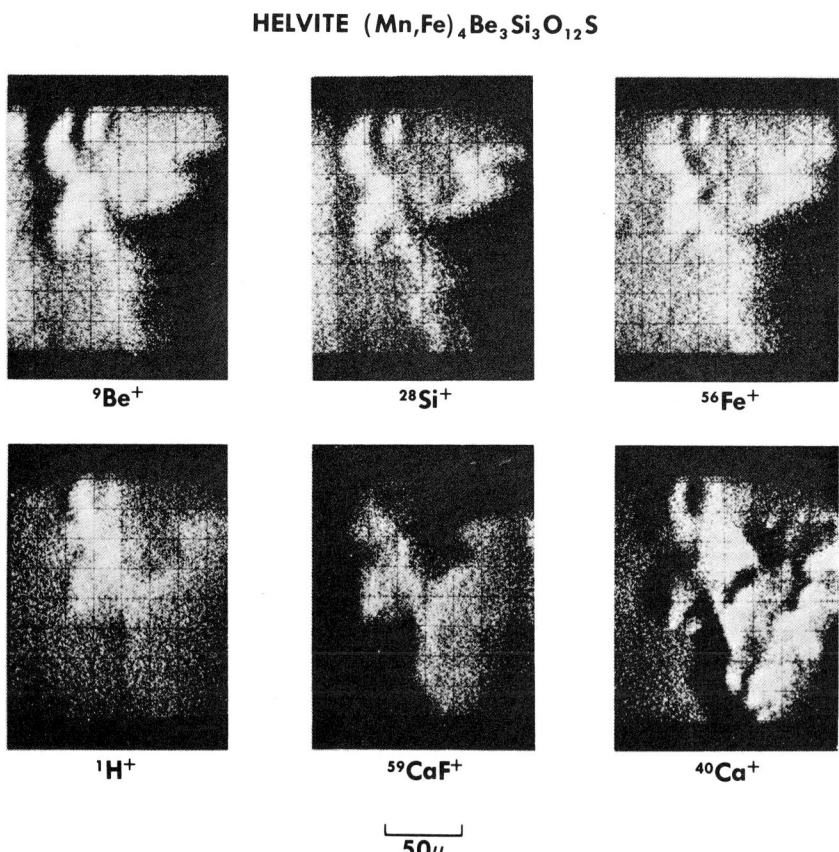

Figure 15.   Ion-microprobe beam scanning images of a sample of mineral Helvite.

sideration. As Evans[30] states, there is probably no other instrumental analytical technique that can make these claims. As against this, some of the disadvantages of this technique are the severe influence of the matrix on the sensitivity and the nonuniform sputter-etching of the ion beam which adversely affects the depth resolution. The versatality of this technique can be seen from Figure 15, the beam-scanning images[33] of a mineral sample (Helvite).

Glow-Discharge Mass Spectrometry (GDMS)

In this technique[34] the neutral ions which are sputtered from the sample kept in the rf glow discharge sputtering system are Penning-ionized and detected mass-spectrometrically. The sensitivity permits analysis rates as low as one monolayer per minute with 100 ppm trace detection capability, and is also independent of the matrix. But the spatial resolution is poor in this technique; further depth resolution is adversely influenced by the nonuniform sputtering over the large sample area.

Appearance Potential Spectroscopy (APS)

Some workers prefer to use the term "soft X-ray appearance potential spectroscopy" (SXAPS) for this technique[35,36], wherein the sample is bombarded with electrons of variable energy and the total X-ray emission intensity is measured as a function of incident-electron energy. The elements can be identified easily by the energy at which abrupt changes in intensity occur. The energy values correspond to the core-level binding energies for the atom involved. The sensitivity of this technique to chemical effects is clear from Figure 16, where the L_3 and L_2 spectrum of a clean titanium surface is compared with that of the same surface following oxidation[35].

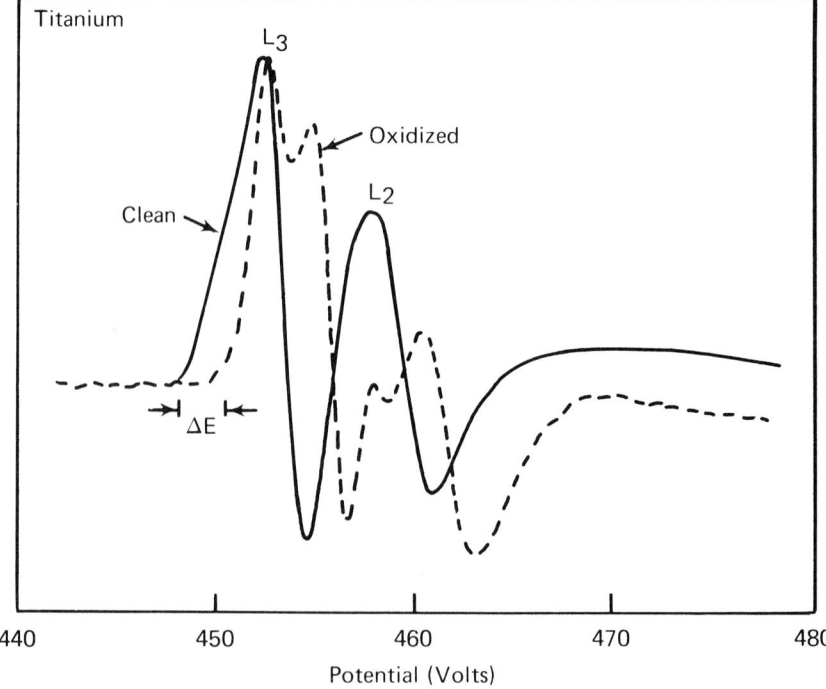

Figure 16. Soft X-ray appearance potential spectrum of a clean polycrystalline titanium surface before (solid curve) and after (dashed curve) oxidation. Redistribution of valence electrons produces a "chemical" shift ΔE in the threshold.

TABLE III
Comparative Table for the Various Techniques
Used for the Chemical Characterization of Surfaces

	NBS	EM	AES	ESCA	ISS	SIMS	GDMS	APS
Destructive to sample (in general)	No	No	No	No	No	Yes	Yes	No
Elements that can be detected	heavy	$Z \geq 4$	$Z \geq 3$	$Z \geq 3$	$Z \geq 3$	All	All except He, Ne	$Z \geq 3$
Elemental identifica-tion [a]	F	G	E	E	E	G	G	E
Sensitivity (typical, in monolayers)	50	5	~0.01	<0.01	~0.01	<1	~1	≤0.1
Detectability (i.e., ppm) [b]	NA	100	<1	NA	NA	1	100	NA
Results are (in principle) [c]	Abs	Abs	Abs	Abs	Abs	Abs	Abs	Abs
Depth probed (in Å)	10^4	10^4–10^5	15–20	15–75	3	$\sim 5 \times 10^4$	10–10^4	~10
Depth distribution of elements [d]	Yes	Yes	Y/d	No	Yes	Yes	Yes	Y/d
Chemical (i.e., binding) information	No	Yes	Yes	Yes	No	No	No	Yes

a—E, Excellent; G, Good; F, Fair   c—Rel, Relative; Abs, Absolute
b—NA, Not applicable   d—Y/d, Yes, if destructive

Infrared Spectroscopy (IR)

Infrared spectroscopy is the classical but powerful tool that is becoming increasingly utilized in surface studies[37,38]. The method is sufficiently sensitive to permit identification and measurement of a fraction of a monolayer of chemisorbed molecules. The sensitivity of the infrared measurement can be increased by the Multiple Internal Reflection (MIR) technique; and by such technique Gilby et al.[39] have shown that even monolayers can be measured. The utilization of polarized light for infrared MIR studies permits an indication of the molecular orientation of the deposited material on the crystal surface.

A rather crude summary of the potentialities and limitations of the various techniques discussed above for the chemical characterization of the surfaces is given in Table III.

Structural Characterization

When once the chemical composition of the surface layers is known, then the next requirement is the "structure" or the relative positions of the atoms in a surface, before any attempt can be made to a detailed interpretation of the various physical and chemical properties. Electron diffraction can provide this information on the microscopic surface structure, and the following is a brief summary of the present status of this important field.

For historical as well as technical reasons, a distinction is usually made between low-energy electron diffraction, LEED, which typically employs electrons of energy 10–500 eV, and high-energy electron diffraction, HEED, which uses electrons in the 10–100 keV range. Both methods rely on observations of the elastic electrons backscattered ("reflected") from the surface. For high-energy electrons, this is sometimes emphasized by writing RED or RHEED, to distinguish the method from the more common transmission electron diffraction (TED or THEED) which is used extensively for the examination of the bulk structure of thin films [40]. Both LEED[41,42] and RHEED[43] have been the subject of recent surveys. Of the two methods, LEED is used much more widely at present and the following discussion will therefore be concerned chiefly with this method.

Surface analysis by electron scattering involves measurements done in ultrahigh vacuum at pressures of 10^{-10} torr. This condition gives the observer working times of at least 10–30 minutes in which the surface does not change appreciably due to the adsorption of residual gases in the vacuum chamber. LEED experiments require a primary-electron source capable of producing a nearly monoenergetic and well-collimated electron beam (beam size ~ 1 mm, beam current ~ 1 μa), a crystal holder, and an electron detector with provision for rejecting inelastically scattered electrons. The intensity of the elastically diffracted beam is measured by a Faraday collector over an angular range for a fixed energy of the incident electrons, or the entire diffracted pattern can be photographed with the help of a fluorescent screen. RHEED experiments also require almost similar conditions. The necessity for ultrahigh-vacuum environment makes conventional high-energy diffraction instruments and electron microscopes of limited value in surface studies and hence systems of special design are normally used.

Even though the symmetry and the two-dimensional periodicity of the surface can be derived readily from the LEED pattern, unambiguous assignment of atomic positions within the unit mesh has not been possible until now. Only recently[44–46] with the proper understanding of the low-energy electron scattering process in surface layers, i.e., by considering both the effects of multiple scattering and the severe attenuation of the elastically scattered electrons, has it been possible to obtain reasonably good agreement between experimentally measured and theoretically calculated intensities from the low index surfaces of a few metals such as aluminum, copper, silver, nickel, and beryllium. An example of such good agreement in the peak width, shapes, and relative peak intensities experimentally measured and calculated for the case {001} surface of

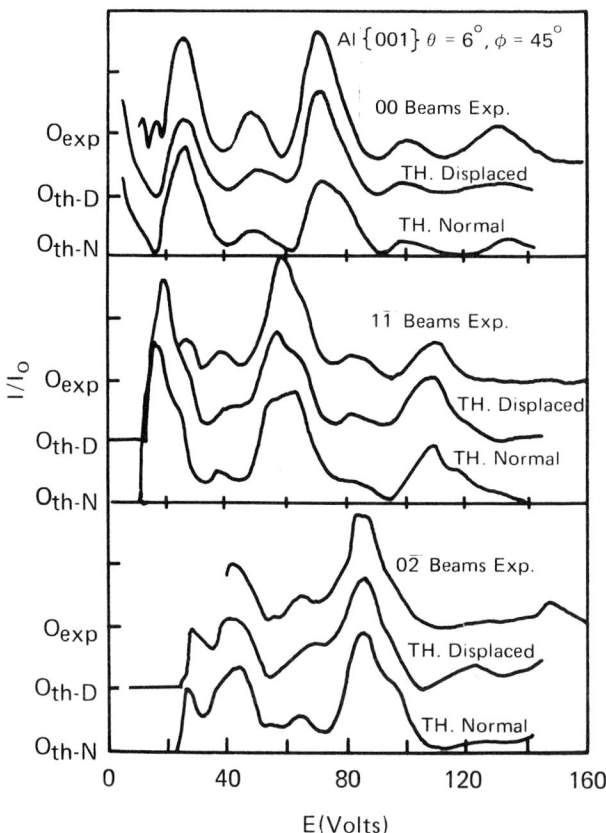

Figure 17. Experimental and theoretical reflected flux LEED spectra for three beams of Al {001} over E = 0 to 150 V showing effect of perpendicular displacement of top layer. TH = theoretically computed curve.

aluminum is shown in Figure 17, where the effect[47] of displacing the topmost layer by 10 percent is also included. The first structure of over-layers of the chemisorbed O, S, Se, and Te on Ni (001) has also been just reported[48]. However, many problems such as the structure of recon-structed surface layers of materials (such as the semiconductors Si, Ge, etc.; the metals Pt, Au, etc.) are yet to be delineated. Yet it appears that a breakthrough in our understanding of the structures of the surface layers of solids is imminent. This is a field of enormous potential. The structure determination of surfaces is in its infancy, compared to the X-ray crys-tallography of the bulk.

A quantitative structure analysis of the Si (111) surface structures by HEED has been attempted recently by Menadue[49].

Defect Characterization

No unique method or technique has been reported for the specific characterization of defects in the surface layers. However, defects such as
vacancies, interstitials, impurity atoms, dislocations, grain boundaries,
twin boundaries, and other disorders can easily be identified in the field-
ion microscope images and studied. In fact, Mueller[12] and his co-workers have exploited this unique power of this instrument for numerous
studies on the intrinsic as well as induced defects in various refractory
metals. However, as mentioned earlier, this technique suffers from some
severe limitations as to the nature of the material (i.e., only good conduc-
tors) and the type of the surface (only the surface of a fine-pointed tip)
that can be studied.

It has often been implied in the pertinent literature that the observation of well-developed low-energy electron diffraction (LEED) pattern
with low background and high contrast is equivalent to stating that the
surface under examination is atomically clean and perfectly crystallized.
This statement is often not even approximately correct. Diffraction measurements tend to emphasize the periodic aspects of structures, and second-order effects produced by surface imperfections may often go un-
recognized in a subjective evaluation of the spot pattern. To be more
specific, MacRae[50] showed that well-developed LEED patterns characteristic of a clean surface could be obtained even when the surface was
contaminated with a *visible* film of gallium. Similarly, Jona[51], by a careful study of the LEED intensities from a Si {111} 7 structure on which was
deposited progressively increasing amounts of disordered Si atoms, concluded that a surface producing a well-developed LEED pattern is well
crystallized and ordered over *at least* 80 percent of its area, and that the
remaining 20 percent can be disordered.

An often recommended method of producing a truly clean and well-
ordered surface of any specimen is to employ repeated cycle of argon-ion
bombardment followed by high-temperature annealing procedures. It
may be recalled that this recipe was suggested by Farnsworth *et al.*[52] on
the basis of LEED patterns obtained from such surfaces. Here it is relevant to point out the recent FIM observations of Walls *et al.*[53] on the
damage produced by the bombardment of low-energy inert gas ions.
Using the techniques and procedures developed by Mueller[12] for the
FIM study of ion bombarded surfaces, Walls *et al.* have shown that the
surface layers of a tungsten crystalline tip are completely disordered by
bombardment of low-energy rare gas ions. Figure 18(a–f) shows a sequence of FIM pictures of tungsten tip before [Figure 18(a)] and after
[Figure 18(b)] bombardment with 700 ev xenon ions and after field-evaporating 1 surface layer [Figure 18(c)], 2 layers [Figure 18(d)], 5 layers
[Figure 18(e)] and 12 layers [Figure 18(f)]. Similar experiments with 100
ev argon ions revealed that the topmost 4 layers of a (110) face of a tungsten tip get totally disordered by bombardment. Annealing such a totally

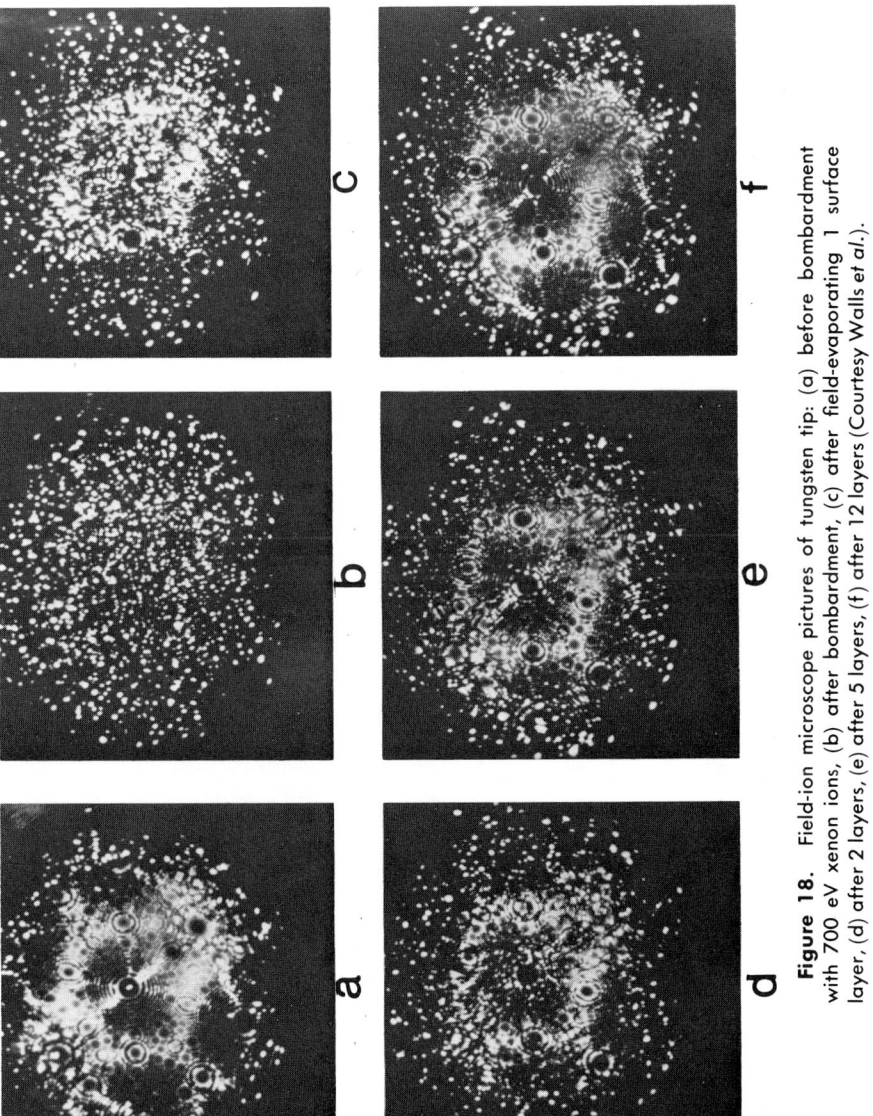

Figure 18. Field-ion microscope pictures of tungsten tip: (a) before bombardment with 700 eV xenon ions, (b) after bombardment, (c) after field-evaporating 1 surface layer, (d) after 2 layers, (e) after 5 layers, (f) after 12 layers (Courtesy Walls *et al.*).

disordered surface may not recrystallize the surface layers in its entirety to perfection and, as mentioned in the previous paragraph, the LEED pattern often fails to reveal the presence of any disorders of even as much as 20 percent in the surface layers.

Besides these direct techniques, one can infer information about the defects in the surface layers by a study of the physical properties of the surfaces. One such method is ellipsometry—a nondestructive optical method of characterization of surfaces.

Ellipsometry

Ellipsometry, or the study of the state of polarization of light reflected at nonnormal incidence from the surface under investigation, is an extremely sensitive technique that can detect and measure the contaminate or corroded layer on the surface even if it is only a few atomic layers thick. Detailed accounts of the theory of ellipsometry are given in the literature [54–55] and hence only the basic essentials are presented here.

If the material is a dielectric and no surface film is present on it, then if a plane-polarized light polarized at 45° to the plane of incidence is incident on it, the reflected light also will be plane-polarized. Then the ratio of r_p, the reflected amplitude polarized parallel to the plane of incidence to r_s, that polarized perpendicular to it gives the tangent of the ellipsometric parameter ψ:

$$\tan \psi = \frac{r_p}{r_s}$$

If there is a surface film or if the substrate itself is absorbing, then r_p and r_s will no longer be in phase and the reflected light will be elliptically polarized. The phase difference $\Delta = \delta_p - \delta_s$ is the second ellipsometric parameter, where δ_p and δ_s are the phases of r_p and r_s respectively.

Exact equations of ellipsometry relating the measured quantities Δ and ψ to the optical parameters (the optical constants and the thickness) of the substrate and the film were derived by Drude[56] at the turn of the century. Since these exact equations of ellipsometry are rather involved, they cannot be inverted analytically and thus it is only with the recent advent of computer technology that the equations could be solved[57] by numerical methods.

The experimental apparatus is extremely simple, as shown in Figure 19. A beam of monochromatic collimated light is sent through a polarizer, a quarter-wave plate, and then reflected from the sample and analyzed with a nicol. In our apparatus, the fast axis of the quarter-wave plate is set at either 45° or −45° with respect to the plane of incidence. For this arrangement, the positions of the analyzer and polarizer which yield minimum intensity of the transmitted light can be used to directly determine Δ and ψ.

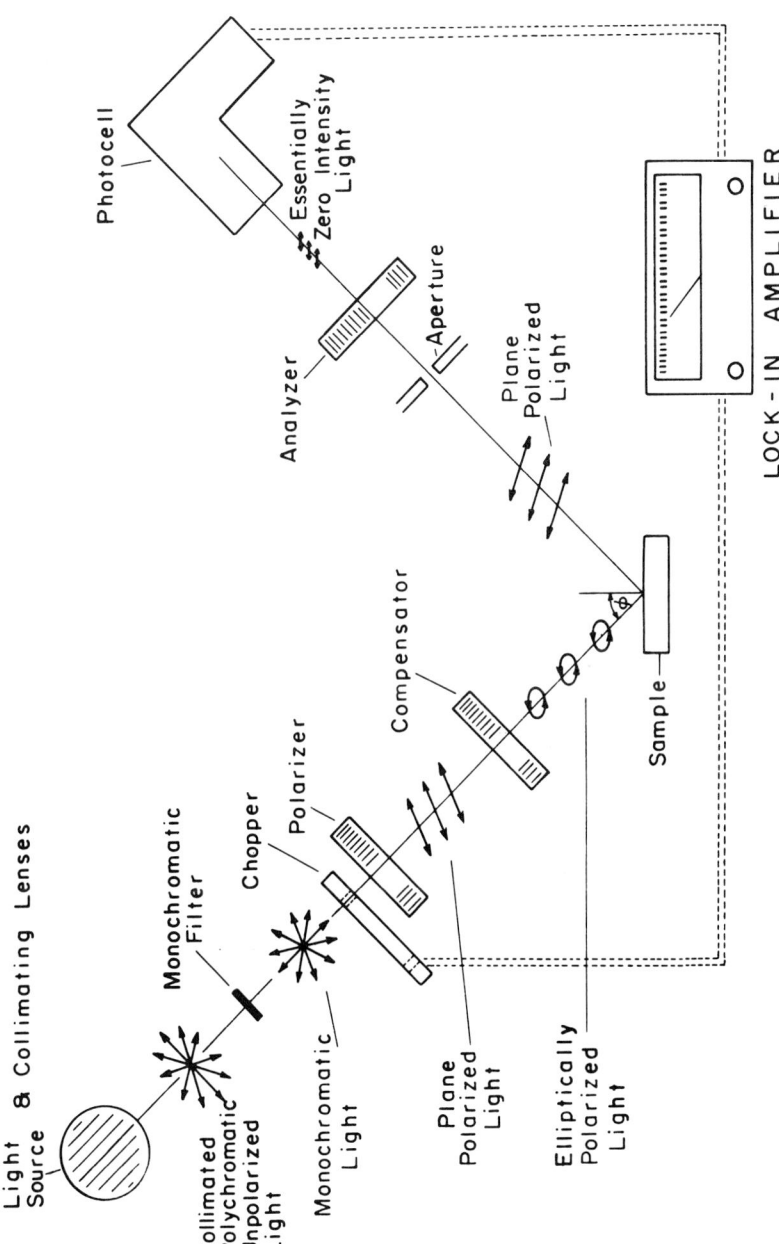

Figure 19. Schematic diagram of the experimental arrangement for ellipsometry.

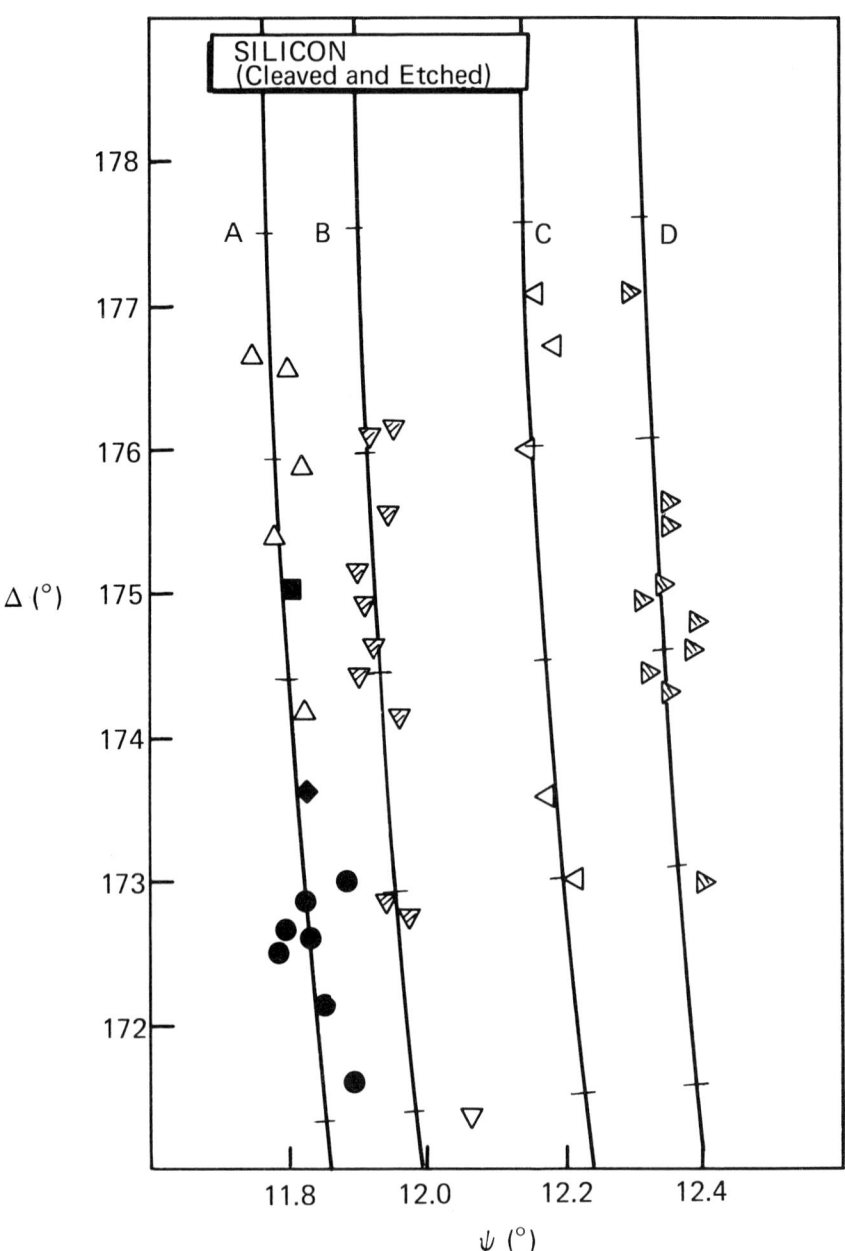

Figure 20. Δ, ψ curves for four freshly cleaved silicon samples during oxidation. The observed data of (Δ, ψ) denoted by triangles are explained on the basis of different effective refractive index for each sample. The data obtained by various workers on chemically polished samples are also included as full or dark points.

It can be shown with the help of the exact ellipsometric equations that, in general, at very low film thickness that the ellipsometric parameter Δ is a function of the film thickness and that ψ is essentially constant. Many workers have used this property to study oxidation or corrosion or the growth of the film at the early stages, etc., by following the variation of Δ. Recently Vedam and So[58] have shown that the variation in ψ can be used to characterize quantitatively the surface layer of the substrate itself, whether it is damaged (or strained) or not. For example, Figure 20 shows the data observed on four freshly cleaved samples of high-purity silicon single crystals. The data obtained by various workers on chemically polished samples are also included in the figure. It is seen that while the values of ψ obtained by different workers in different laboratories on chemically polished specimens agree with each other very well, the data on freshly cleaved high-purity specimens cleaved from the same boule in our laboratory do not agree amongst themselves. This is because all the damaged surface layers in the chemically polished specimens have been removed during polishing, and thus the observed ψ represents the true intrinsic value of silicon. On the other hand, in the case of cleaved samples, the discrepancy can be attributed to the influence of dislocations present on or near the cleaved surfaces which are introduced during the process of cleavage. The effect of these fresh dislocations can be taken into account by an effective value of the refractive index of silicon substrate or, more meaningfully, by an effective thin damaged surface layer with different optical parameters, on the bulk silicon substrate. Computation using the equations of ellipsometry reveals that the effective thickness of the damaged layer of the four specimens A, B, C, D are 0, 3.2, 9.5, and 12 Å, respectively, with the optical constants corresponding to those of amorphous silicon.

Table IV lists the values of the ellipsometric parameters Δ and ψ mea-

TABLE IV
Summary on Effects of Surface Treatments of Single-Crystal Silicon
in Ellipsometric Measurements

Sample	Δ (deg)	ψ (deg)	$\bar{n}_2$	$\bar{k}_2$	$\bar{R}$ (%)
Mechanically polished	160.54	16.35	4.520	0.924	42.30
Sputtered by Ar$^+$	174.04	15.92	4.710	0.292	42.36
Cleaved	174.60	12.34	4.123	0.169	37.23
Cleaved and annealed	173.17	11.90	4.055	0.141	36.57
Chemically polished	172.48	11.83	4.038	0.219	36.49
			$\bar{n}_2$	$\bar{k}_2$	$\bar{R}$ (%)
Amorphous*			4.360	1.300	42.67

*Results of Beaglehole and Zavetova[59]

TABLE V
Vitreous Silica

Surface history	Polished with Diamond Paste ψ (in degrees) Δ		Polished with Cerium Oxide ψ (in degrees) Δ	
Ideal case	3.79	360.00	3.79	360.00
Mech. polish	0.52	290.00	3.79	359.64
2 min etch in 10% HF	2.55	334.00	3.77	359.93
4 min etch in 10% HF	3.93	4.00	3.79	0.11
6 min etch in 10% HF	4.16	3.04	3.80	0.29

sured on silicon samples with different preparative histories, as well as the pseudoreflectivities calculated from these Δ and ψ. As explained before, the variation in Δ is attributable to different thicknesses of SiO_2 film on Si, whereas the variation in ψ (which can be determined to $\pm$ 0.02°) is attributable to the damaged surface layer.

The most recent work in our laboratory has been focused on the corrosion of glass surfaces by different ambients as studied by ellipsometry. Obviously, measurements on such transparent samples are inherently more difficult than on highly reflecting surfaces of metals or semiconductors. However, with the use of phase-sensitive lock-in-amplifier devices and thus with the improved signal-to-noise ratio, meaningful and reproducible results on vitreous silica and other glasses have been obtained.

Table V lists the values of Δ and ψ observed on two samples of vitreous silica, which were respectively polished with diamond paste and cerium oxide. It is seen that the value of ψ obtained on the specimen polished with diamond paste is 0.52° instead of 3.79°, corresponding to the ideal strain-free surface. Computations with the ellipsometric equations reveal that the surface layers of the diamond-polished specimen have been permanently densified during mechanical polishing, with the optical parameters of the densified layer as a refractive index of 1.530 and a thickness of 950 Å.

On the other hand, the specimen polished with cerium oxide appears to be free of the densified layer, as can be seen from the value of ψ measured on this specimen. It may be recalled that the hardness of vitreous silica and cerium oxide are 4.9 Mohs and 5.0 Mohs, respectively, and this may be the reason why polishing with cerium oxide does not permanently damage the surface layers of vitreous silica.

The results of the removal of the densified surface layers by etching the vitreous silica specimen polished with diamond paste are also shown in Table V. It is seen that both ψ and Δ increase rapidly toward the values corresponding to the ideal case, but on continued etching they overshoot the ideal values. This may be due to the effect of surface roughness on a microlevel which develops[60] on vitreous silica specimens polished with diamond paste.

TABLE VI
Glasses Polished with CeO$_2$
λ 5461 Å ϕ = 59°

Glass	ψ_0	ψ_{oil}	ψ_{water}	n	Damaged Layer $n'_{2\,oil}$	$n'_{2\,water}$
	in degrees					
Soda Lime (73:17)						
0080	3.87	3.84	3.50	1.5121	1.513	1.520
Borosilicate (66:24)						
9741	5.06	4.80	4.95	1.4703	1.476	1.475
Borosilicate (65:18)						
7052	4.68	4.53	4.60	1.4833	1.490	1.493
Aluminosilicate (62:17)						
1720	3.21	3.13	3.16	1.530	1.539	1.539

Finally, Table VI lists some preliminary results obtained on four Corning glasses which were polished with cerium oxide with oil/water as lubricant. When oil is used as a lubricant, in every case the surface layers are densified, as can be seen from the decrease in the value of ψ from ψ_0 corresponding to the ideal case. On the other hand, when water is used as a lubricant, some of the cations in the surface layers of the glass are leached out, with consequent decrease in the refractive index. In other words, this leaching out of cations has the opposite effect of densification and what one observes is the overall effect. One additional factor that must be considered in this experiment is the influence of the load during polishing—the higher the load, the greater the densification, and vice versa[61].

Summary

This brief survey has been an attempt to discuss in general terms some of the techniques that can be used to obtain morphological, chemical, structural, and defect characteristics of the surface layers of a solid. Detailed comparisons of the various techniques will have to wait experimental studies of identical samples by each method. No one method is available for universal applicability with high sensitivity, selectivity, and range. Hence, it is advisable to design experiments so that two or more techniques, which are complimentary in their performance capabilities, can be utilized on the same sample at the same time. It is shown that ellipsometry is yet another tool for the nondestructive characterization of both opaque and transparent materials, yielding information on the optical properties of (1) the substrate, (2) the contaminant film, and (3) the damaged surface layers of the substrate, which is dependent on the preparative history of the specimen.

Acknowledgment

This work was supported by the Office of Naval Research, Metallurgy Program. The author's sincere thanks are due to Professor E. W. Mueller of The Pennsylvania State University, Dr. P. W. Palmberg of Physical Electronics Industries Inc., and to Dr. J. M. Walls of the University of Aston, Birmingham, for permission to use their unpublished data and to Professors Rustum Roy and Bruce Knox of The Pennsylvania State University and Professor F. Jona of the State University of New York at Stony Brook for many useful discussions. The author's thanks are also due to M. Malin, and M. Lovette for their painstaking ellipsometric measurements on vitreous silica and glasses reported here.

References

1. Farago, F.T., *Handbook of Dimensional Measurement*, New York: Industrial Press (1968), 367–96; and Good, C.H., "Surface Texture Measurement," in *Handbook of Industrial Metrology*, ASTME Manufacturing Engineering Series, Englewood Cliffs, N.J.: Prentice-Hall, Inc. (1967), 359–77.

2. Young, R.D., "Field Emission Ultramicrometer," *Rev. Sci. Instrum.*, 37 (1966), 275–78.

3. Francon, M., *Progress in Microscopy*, Oxford: Pergamon Press, Ltd., 1961.

4. Tolansky, S., *Multiple-Beam Interferometry of Surfaces and Films*, Oxford: Clarendon Press, 1948.

5. Nomarski, G. and Weill, A.R., "Sur l'observation des figures de Croissance des cristaux par les methodes interferentielles a deux ondes," *Bull. Soc. Franc. Mineral. Crystallogr.*, 77 (1954), 840–68.

6. Allen, R.D., David, G.B. and Nomarski, G., "The Zeiss-Nomarski Differential Interference Equipment for Transmitted-Light Microscopy," *Z. Wiss. Mikrosk. Mik.*, 69 (1969), 193–221.

7. Bernal, E.G., Anderson, R.H., Chaffin, J.H., Koepke, B.G. and Stokes, R.J., "Preparation and Characterization of Polycrystalline Halides for Use in High Power Laser Windows," Quarterly Report No. 5, 1 April–30 June 1973. Honeywell, Inc., Corporate Research Center, Bloomington, Minn., Advanced Research Projects Agency Contract Report No. HR-73-266:5-26, June 1973 (AD 912 555L).

8. Bethge, H., "Oberflachenstruckturen und Kristalloaufehler im Elektronenmikroskopischen Bild, Untersucht am NaCl," *Phys. Status Solidi*, 2 (1962), 3–27.

9. Oatley, C.W., Nivon, W.C. and Pease, R.F.W., "Scanning Electron Microscopy," in *Advances in Electronics and Electron Physics*, Vol. XXI, L. Marton, ed., New York: Academic Press, 1956.

10. White, E. W., McKinstry, H.A. and Diness, A., "Quantitative Surface Finish Characterization by CESEMI," in *The Science of Ceramic Machining and Surface Finishing*, NBS Special Publication 348, J.S. Schneider, Jr. and R.W. Rice, eds., Washington, D.C.: Government Printing Office (1972), 309–16.

11. Young, R.D., "Surface Microtopography," *Phys. Today*, 24, No. 11 (1971), 42–49.

12. Mueller, E.W. and Tsong, T.T., *Field Ion Microscopy: An Introduction to Principles, Experiments, & Applications*, New York: American Elsevier Publishing Company, 1969.

13. Mueller, E.W., "Atom-Probe Field Ion Microscopy," *J. Vac. Sci. Technol.*, 8 (1971), 1-89.

14. Emmett, P.H., "Measurement of the Surface Area of Solid Catalysts," in *Catalysis*, Vol. 1. P.H. Emmett, ed., New York: Reinhold Publishing Corporation (1954), 31–74.

15. Dubinin, M.M., "The Potential Theory of Adsorption of Gases and Vapors for Adsorbents with Energetically Non-Uniform Surfaces," *Chem. Rev.*, 60 (1960), 235–41.

16. Kindl, B., Negri, E. and Cerefolini, G.F., "Adsorption Isotherms of Noble Gases on Glasses," *Surface Sci.*, 23 (1970), 299–310.

17. Coburn, J.W. and Kay, E., "Surface Analysis Today," *Res. and Develop.*, 23, No. 12 (1972), 35–44.

18. Meyer, O., Gyulai, J. and Mayer, J.W., "Analysis of Amorphous Layers on Silicon by Backscattering and Channeling Effect Measurements," *Surface Sci.*, 22 (1970), 263–76.

19. Zeigler, J.F. and Baglin, J.E.E., "Determination of Surface Impurity Concentration Profiles by Nuclear Backscattering," *J. Appl. Phys.*, 42 (1971), 2031–40.

20. Anderson, C.A., ed., *Microprobe Analysis*, New York: John Wiley & Sons, 1973.

21. White, E.W., "Application of Soft X-Ray Spectroscopy to Chemical Bonding Studies with Electron Microprobe," in *Microprobe Analysis*, C.A. Anderson, ed., New York: John Wiley & Sons (1973), 349–69.

22. Chang, C.C., "Auger Electron Spectroscopy," *Surface Sci.*, 25 (1971), 53–79.

23. Palmberg, P.W., "Use of Auger Electron Microscopy and Inert Gas Sputtering for Obtaining Chemical Profiles," *J. Vac. Sci. Technol.*, 9 (1972), 160–63.

24. Siegbahn, K., *et al.*, *ESCA: Atomic, Molecular and Solid State Structure Studied by Means of Electron Spectroscopy*, Uppsala: Almquist & Wiksells Boktryckeri Ab, 1967.

25. Carlson, T.A., "Electron Spectroscopy for Chemical Analysis," *Phys. Today*, 25 No. 1 (1972), 30–39.

26. Shirley, D.A., *Electron Spectroscopy*, Amsterdam: North Holland Publishing Company, 1972.

27. Rendina, J.F., "Electron Spectroscopy for Chemical Analysis," *Amer. Lab.*, 3 No. 2 (1972), 17–25.

28. Smith, D.P., "Analysis of Surface Composition with Low Energy Back-Scattered Ions," *Surface Sci.*, 25 (1971), 171–91.

29. Ball, D.J., "Investigation of Low-Energy Ion Scattering as a Surface Analytical Technique," *Surface Sci.*, 30 (1972), 69–90.

30. Evans, C.A. Jr., "Secondary Ion Mass Analysis: A Technique for Three-Dimensional Characterization," *Anal. Chem.*, 44, No. 13 (1972), 67A–80A.

31. Morabito, J.M. and Lewis, R.K., "Secondary Ion Emission for Surface and In-Depth Analysis of Tantalum Thin Films," *Anal. Chem.*, 45 (1973), 869–80.

32. Benninghoven, A. and Loebach, E., "Tandem Mass Spectrometer for Secondary Ion Studies," *Rev. Sci. Instrum.*, 42 (1971), 49–52.

33. Bayard, M., "The Ion Microprobe," *Amer. Lab.*, 3 No. 4 (1971), 15–20.

34. Coburn, J.W. and Kay, E., "A New Technique for the Elemental Analysis of Thin Surface Layers of Solids," *Appl. Phys. Lett.*, 19 (1971), 350–52.

35. Park, R.L. and Houston, J.E., "Analysis of Solid Surfaces by Soft X-Ray Appearance Potential Spectroscopy," in *Advances in X-Ray Analysis*, Vol. 15. K.F.J. Heinrich, C.S. Barrett, J.B. Newkirk, and C.O. Ruud, eds., New York: Plenum Press (1972), 462–69.

36. Musket, R.G., "Direct Comparison of Auger Electron Spectroscopy with Appearance Potential Spectroscopy," *J. Vac. Sci. Technol.*, 9 (1972), 603–07.

37. Varadi, P.F., "Infrared Spectroscopy in Vacuum Surface Studies," *Res. and Develop.*, 24, No. 4 (1973), 46–52.

38. Bradshaw, A.M. and Pritchard, J., "Infrared Spectra of Carbon Monoxide Chemisorbed on Metal Films: A Comparative Study of Copper, Silver, Gold, Iron, Cobalt and Nickel," *Proc. Roy. Soc. London, Ser. A*, 316A (1970), 169–83.

39. Gilby, A.C., Casseb, J. and Wills, P.A. Jr., "Internal Reflectance Spectroscopy," *Appl. Spectrosc.*, 24 (1970), 539–43.

40. Hirsch, P.B., Howie, A., Nicholson, R.B., Pashley, D.W. and Whelan, M.J., *Electron Microscopy of Thin Crystals*, New York: Plenum Press, 1965.

41. Estrup, P.J. and McRae, E.G., "Surface Studies by Electron Diffraction," *Surface Sci.*, 25 (1971), 1–52.

42. Ignatiev, A., "Surface Analysis by Low Energy Electron Diffraction," *Res. and Develop.*, 24, No. 3 (1973), 12–22.

43. Bauer, E., "Reflection Electron Diffraction," in *Techniques of Metals Research*, Vol. II, Pt. 2, R.F. Bunshah, ed., New York: John Wiley & Sons (1969), 501–639.

44. Jepsen, D.W., Marcus, P.M. and Jona, F., "LEED Spectra from [001] Surfaces to FCC Metals: Theory and Experiment," *Phys. Rev., B*, 5 (1972), 3933–52.

45. Tait, R.H., Tong, S.Y. and Rhodin, T.N., "New Perturbative Approach to the Application of LEED—the T-Matrix Formalism," *Phys, Rev. Lett.*, 28 (1972), 553–56.

46. Strozier, J.A. and Jones, R.O., "LEED Intensity Calculations for Beryllium with a Realistic Crystal Potential," *Phys. Rev., B*, 3 (1971), 3228–43.

47. Marcus, P.M., Jepsen, D.W. and Jona, F., "Application of the Solution of the Dynamical Scattering Problem for Electrons to Surface Structure Analysis," *Surface Sci.*, 31 (1972), 180–97.

48. Demuth, J.E., Jepsen, D.W. and Marcus, P.M., "Chemisorption Bonding of c(2x2) Chalcogen Overlayers on Nickel (001)," *Phys. Rev. Lett.*, 31 (1973), 540–42.

49. Menadue, J.F., "Si(111) Surface Structures by Glancing Incidence High-Energy Electron Diffraction," *Acta Crystallogr., Sect. A*, A28 (1972), 1–11.

50. MacRae, A.V., "Low Energy Electron Diffraction Study of Polar[111] Surfaces of GaAs and GaSb," *Surface Sci.*, 4 (1966), 247–64.

51. Jona, F., "A Note on the Sensitivity of LEED to Surface Perfection," *Surface Sci.*, 8 (1967), 478–84.

52. Farnsworth, H.E., Schlier, R.E., George, T.H. and Burger, R.M., "Ion Bombardment-Cleaning of Germanium and Titanium as Determined by Low Energy Electron Diffraction," *J. Appl. Phys.*, 26 (1955), 252–53.

53. Walls, J.M., Braun, E. and Southworth, H.N., "Field Ion Microscope Observations of Low Energy Ion Bombardment," paper presented at 20th Field Emission Symposium, Pennsylvania State University, University Park, 20–23 August 1973; and Walls, J.M., unpublished Ph.D. dissertation, University of Aston, Birmingham, England, 1973.

54. Heavens, O.S., *Optical Properties of Thin Solid Films*, New York: Dover Publications, Inc., 1966.

55. Vasicek, A., *Optics of Thin Films*, Amsterdam: North Holland Publishing Company, 1960.

56. Drude, P., *The Theory of Optics*, New York: Dover Publications, Inc., 1959.

57. So, S.S. and Vedam, K., "Generalized Ellipsometric Method for the Absorbing Substrate Covered with a Transparent-Film System. Optical Constants of Silicon at 3655 Å," *J. Opt. Soc. Amer.*, 62 (1972), 16–23.

58. Vedam, K. and So, S.S., "Characterization of Real Surfaces by Ellipsometry," *Surface Sci.*, 29 (1972), 379–95.

59. Beaglehole, D. and Zavetova, M., "The Fundamental Absorption of Amorphous Ge, Si, and GeSi Alloys," *J. Non-Cryst. Solids,* 4 (1970), 272–78.

60. Homer, P.N. and Crawford, B.J., "The Microstructure of Etched Glass Surfaces," *Glass Technol.,* 11 (1970), 10–14.

61. Yokota, H., Sakata, H., Nishibori, M. and Kinosita, K., "Ellipsometric Study of Polished Glass Surfaces," *Surface Sci.,* 16 (1969), 265–74.

F. ROBERT BARNET and
MARRINER K. NORR
U.S. Naval Ordnance Laboratory
Silver Spring, Maryland

Chapter 18

Characterization of Carbon Fibers by Oxygen Plasma Etch

Introduction

Carbon fibers are one of the more important reinforcement materials for advanced composites because of their excellent strength-to-density and stiffness-to-density characteristics. However, tensile strengths of carbon fibers are much lower than theoretically expected and often drop off with increasing modulus. Also, the fibers are very brittle. For these reasons there has been considerable effort made to characterize the crystallinity and general structure of these fibers in order to explain their mechanical properties and give direction for possible improvements[1].

A number of fiber structural models have been proposed. One of these, the "onion skin" model, for both rayon-based and PAN-based fibers[2,3], pictures the carbon-layer planes as lying parallel to the fiber surface, like the layers in an onion. Larsen[4] confirmed this model for intermediate- and high-modulus rayon-based fibers. According to his model, the crystallites are circumferentially oriented in the outer layers of the fiber and are randomly oriented in the interior. He also suggested that the thickness of the circumferential layer increases with the modulus of the fiber.

However, for high-modulus PAN-based fibers, Knibbs[5], using polarized light microscopy, deduced that the outer layers of these fibers are circumferentially oriented, and that the interior layers are radially oriented. This "circumferential-radial" model was also confirmed by Larsen[4], as well as by Barnet and Norr[6,7].

A variety of techniques have been used to study carbon-fiber structure. These include X-ray, optical microscopy, and transmission and scanning electron microscopy. So far as the literature of carbon-fiber morphology is concerned, the technique used in our studies is new. (In contrast, ion etching has proved to be ineffective.) The specimens are etched in an oxygen plasma to remove the "softer" parts of the fiber and leave behind the more resistant parts. This gives etch patterns with new information on the internal macrostructure of the fiber. It is these patterns that may be used to characterize the various types of carbon fibers.

Experimental

Carbon fibers are formed into unidirectional fiber composites for ease of handling. The fiber ends and sides, as well as the plastic binder, are etched in the plasma and then examined in a scanning electron microscope. A brief description of the experimental procedure is given below. Details may be found in references [6] and [7].

Materials

The carbon fibers selected for our study were chosen to represent the three main precursor types: amorphous, single-chain crystalline, and folded-chain crystalline, and to give a wide range of moduli, as shown in Table I.

Specimen Preparation

In order to hold the fibers in place for etching and subsequent viewing in the microscope, they are embedded in an epoxy resin to form bars of unidirectional composites. Centimeter-long specimens are cut from the bars, ground flat, and mechanically polished on one end and on one face.

Fiber Etching

The specimens are etched in a radio-frequency plasma under oxidizing conditions in the apparatus sketched in Figure 1. The system is evacuated by a mechanical vacuum pump, and oxygen bled into the reaction chamber through a leak valve set to maintain a pressure of about 150–200 μm. Radio-frequency energy of 100 mHz is coupled into the system through the two bands around the tube. This energy causes a continuous glow discharge or plasma within the chamber. The plasma contains oxygen atoms, oxygen ions, and possibly various other activated species which react with the fibers and resin of the specimens, gently oxidizing them at a low temperature, probably about 200°C. In a typical run, 2–4 specimens are etched for a period of either 4 or 8 hours.

TABLE 1
Carbon Fiber Identification

	Carbon Fiber Characteristics			
	Ultimate Tensile Strength (psi $\times 10^{-3}$)	Modulus (nominal) (psi $\times 10^{-6}$)	Density (g/cc)	Interlayer Spacing (Å)
A. Carbon Fibers from Amorphous Precursor (Petroleum Pitch)				
KCF-100	150*	10*	1.58	~ 3.8
KGF-200	150*	16*	1.80	~ 3.6
B. Carbon Fibers from Single Chain Crystalline Precursor (Cellulose)				
VYB	210	6*	1.53*	~ 3.8
Thornel 25	194	25*	1.43	—
Thornel 40	219*	40*	1.52	—
Thornel 50	280	50*	1.62	3.41
Thornel 75	406	75*	1.84	3.37
C. Carbon Fibers from Folded Chain Crystalline Precursor (Polyacrylonitrile)				
Thornel 400	423*	31*	1.80	3.48
Modmor II	310	35*	1.77	3.50
Modmor I†	250	55*	1.97	3.38

*Manufacturer's data
†Meter-length fiber

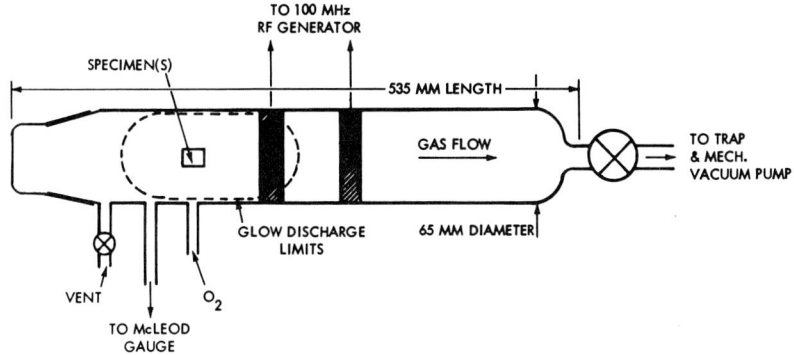

Figure 1. Schematic of plasma etching apparatus.

Scanning Electron Microscopic Examination

The etched specimens are mounted on aluminum specimen holders and coated with a thin layer of gold in a vacuum evaporator to make their surfaces conducting. The end and face that have been polished prior to etching are examined in the microscope and photographed using magnifications up to 10,000X.

Results and Discussion

The most characteristic, or unique, etch patterns are found on the ends of the fibers. In contrast, the side patterns are essentially alike for all fibers examined except Thornel 25 (see Figure 2). For this reason, our studies have been concentrated on the etch patterns formed on the ends of the fibers.

Petroleum Pitch-Based Fibers

KCF-100 (low modulus)

The etch pattern is uncomplicated (Figure 3). The ends are circular and tapered, with rounded flat tops. Tapering develops because the resin etches preferentially, leaving the ends of the fibers exposed to the plasma for a longer period of time than the parts of the fiber closer to the resin surface. Shallow holes can be seen on the ends of a few of the fibers.

The etched surfaces of the fibers give no indication of long-range crystal structure. This is what we would expect from a low-modulus fiber, particularly one derived from an amorphous precursor.

KGF-200 (low modulus)

A relatively thick layer of resin has been eaten away, leaving long fibers standing above the resin surface (Figure 4). The fibers taper gently to sharp points.

Roundish pits occur in the core regions of the fibers. These may indicate the existence of cavities present prior to etching or the presence of pockets of easily etchable material. In contrast, the surface layers show almost no anomalies.

In some areas, a considerable amount of fibrous material can be seen attached to the fiber ends.

The etch patterns of this fiber give no evidence of long-range crystallinity even though the fibers were heat treated to 2000°C during manufacture.

Cellulose-Based Fibers

VYB (low modulus)

The fibers have flat tops and very little taper; the sides are almost parallel to each other (Figures 5, 6). We believe this indicates that there is a uniformity of microstructure in all directions, which is what we would expect in a fiber having random crystallinity and only a small amount of long-range order.

The pattern also implies a lack of internal stress fields or other defects within the fiber. If such stress fields or other defects had been present, we would have expected that they would have been etched preferentially. The lack of such stress fields is to be expected, however, since the fibers were only moderately heat treated and their crystalline structure has not yet developed to any degree.

Bumpy ends and occasional fibrous material may be seen in Figure 6. These bumpy ends may indicate the presence of axial structure and suggest that the fiber has a minor degree of long-range order.

Thornel 25 (intermediate modulus)

The fiber ends are roughly cone-shaped (Figures 7–9). There is considerable riblike structure of a rather uniform diameter running lengthwise on the fiber surface. It looks as if the interior of the fiber is being eaten away, and as this occurs, the riblike units collect together at the end of the fiber. We believe that these riblike units are the long-range crystalline portion of the fiber and not the remains of the flutes originally present on the unetched fiber. No obvious correlation between the size and distribution of this structure and that of the original surface flutes can be established. A similar but finer structure is also present on another etched fiber, Modmor II, which was no surface fluting. It is suggested that this riblike structure on Thornel 25 represents clusters of fibrils from the surface sheath of the basic fiber.

Viewed from the side (Figure 9) the riblike structure shows a waviness one would expect to see when stress within the fiber is gradually released as the softer parts of the fiber are eaten away. A threadlike transverse structure can be seen connecting some of the ribs (Figures 8, 9).

Thornel 40 and 50 (intermediate modulus)

There are no real differences between the etch patterns of Thornel 40 and Thornel 50. (Figures 10–12; and compare Figures 10 and 11). The fibers etch to sharp, rugged points.

Some definite ridgelike formations can be seen running lengthwise along the sides of the fibers (see, especially, Figure 12). Again, there is no evidence that these ridges are derived from the flutes present on the unetched fiber. First, there are instances where more structure is apparent than can be attributed solely to the shape of the unetched fiber. Second, there is no other visible evidence for long-range crystalline order found in these fibers, yet long-range order is present, especially in Thornel 50, because of the relatively high modulus of the fiber. The fact that there are no riblike structures (like those on Thornel 25) separating from the main part of Thornel 40 and Thornel 50 is taken as evidence that these two fibers have sheaths of a more perfect crystalline structure than does Thornel 25.

Thornel 75 (high modulus)

After 4 hours of etching, evidence of long-range crystalline order was found (Figures 13–14). First, the fibers appear to have a surface shell which other studies have indicated is a layer of circumferentially oriented crystallites. Second, in the centers of the fibers, numerous nodes can be seen. It is suggested that these nodes may be the ends of fibril clusters.

After 8 hours of etching, the general fluted shape of the fiber is still clearly visible (Figures 15, 16). Although no fibrous or tissuelike material was seen after the 4-hour etch, a considerable amount is now visible. The tissuelike caps above the fiber ends appear to be connected to the fibers by threads. Some of these threads come off the ends of the fiber flutes and seem to represent the remains of these flutes.

The centers of the fibers appear to be softer or less etch resistant than the sheath. From this we conclude that the crystalline structure in the interior is less well developed than in the sheath. The fibers are believed to have high built-in stress fields, which is not surprising, because cellulose loses as much as 85 percent of its weight when fully carbonized, which causes a considerable shrinkage of the fiber. During fiber heat treatment, as long-range circumferential crystallinity develops, normal fiber shrinkage becomes inhibited by these crystallites, and severe stresses develop within the fiber. Such stresses make it more vulnerable to etching.

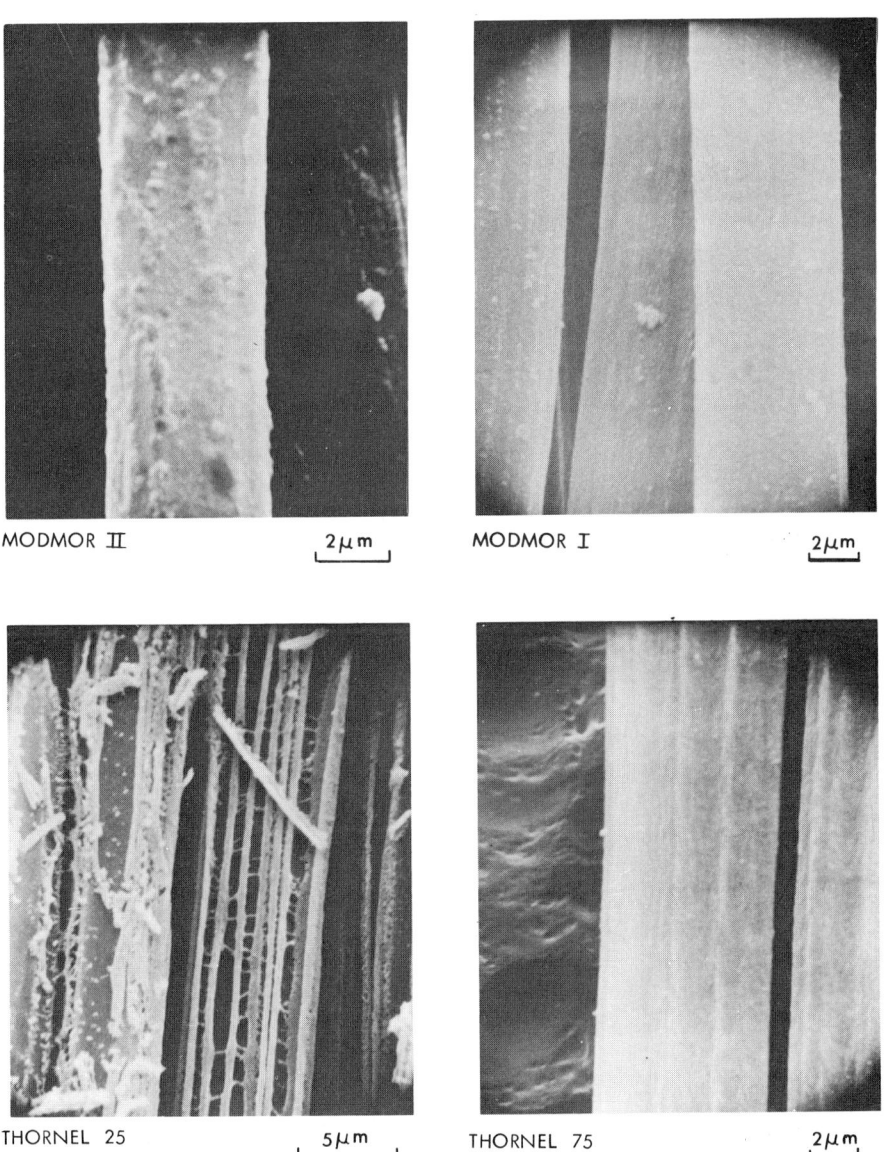

MODMOR Ⅱ   2 μm

MODMOR Ⅰ   2 μm

THORNEL 25   5 μm

THORNEL 75   2 μm

Figure 2. Selected fibers etched 8–9 hours.

Figure 4.   KGF-200 fibers etched 8 hours.

Figure 3.   KCF-100 fibers etched 8 hours.

Figure 6. VYB fibers etched 4 hours.

Figure 5. VYB fibers etched 4 hours.

Figure 8. Thornel 25 fibers etched 9 hours.

Figure 7. Thornel 25 fibers etched 9 hours.

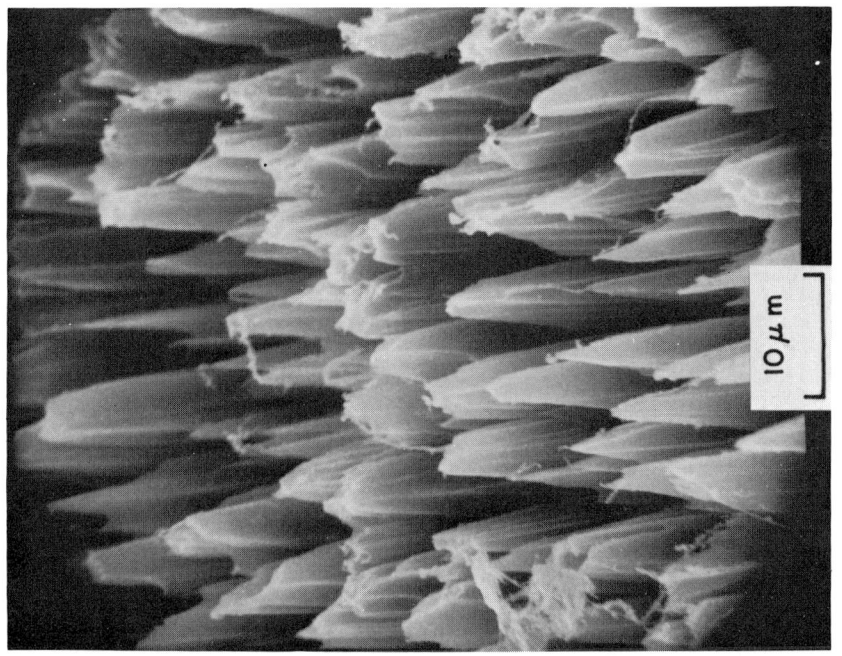

Figure 10. Thornel 40 fibers etched 9 hours.

Figure 9. Thornel 25 fibers etched 9 hours (side-etch pattern).

Figure 12. Thornel 50 fibers etched 8 hours.

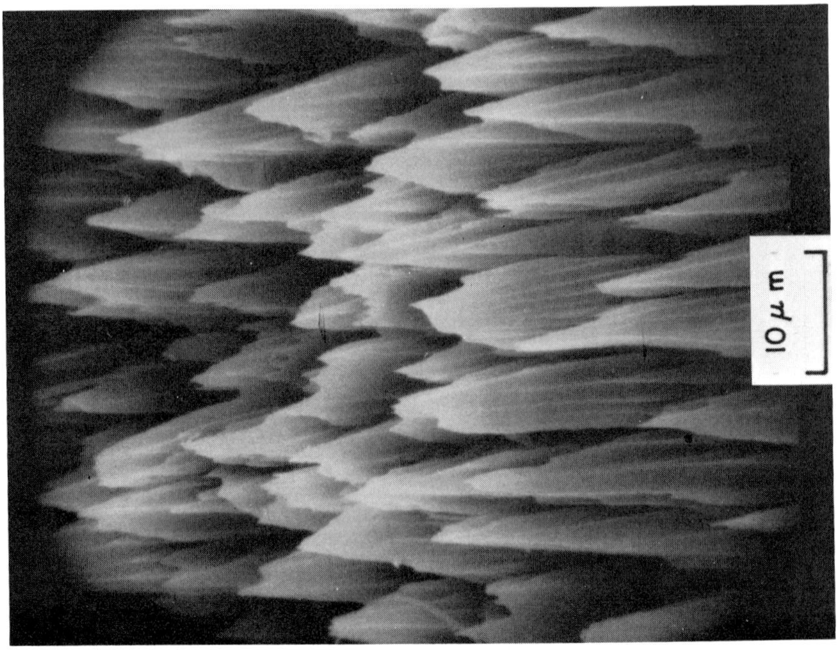

Figure 11. Thornel 50 fibers etched 8 hours.

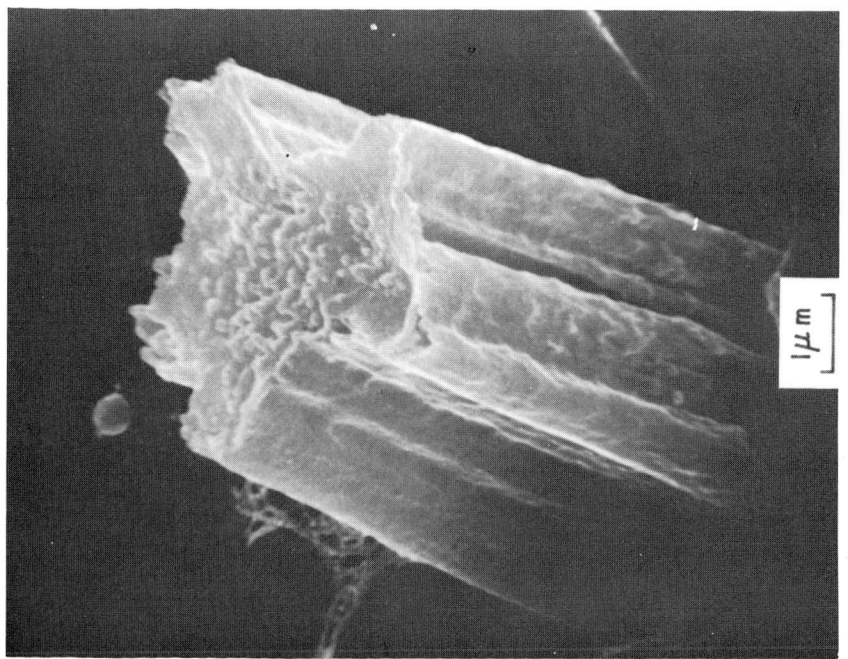

Figure 14. Thornel 75 fiber etched 4 hours.

Figure 13. Thornel 75 fibers etched 4 hours.

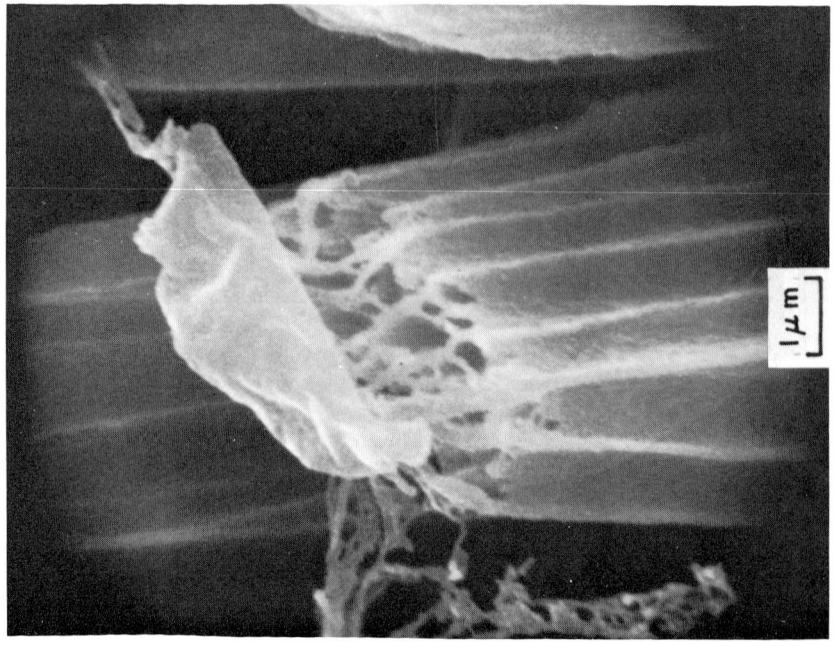

Figure 16.   Thornel 75 fiber etched 8 hours.

Figure 15.   Thornel 75 fiber etched 8 hours.

Figure 18. Thornel 400 fibers etched 8 hours.

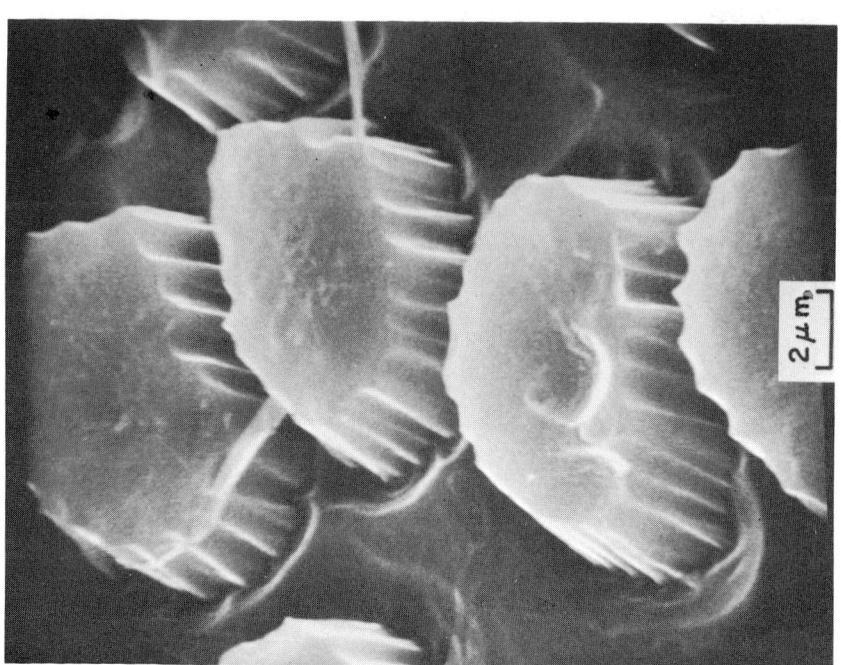

Figure 17. Thornel 400 fibers etched 8 hours.

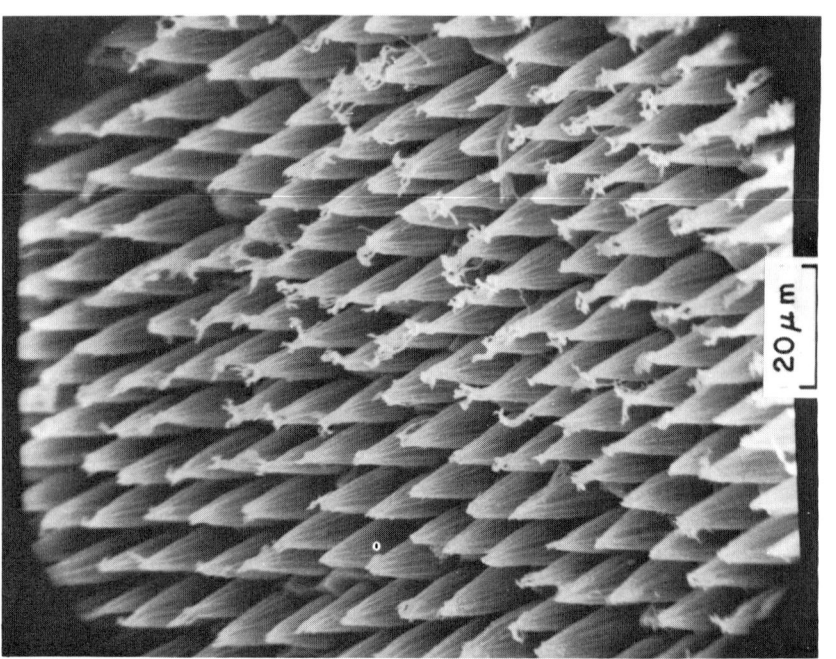

Figure 20.   Modmor II fibers etched 4 hours.

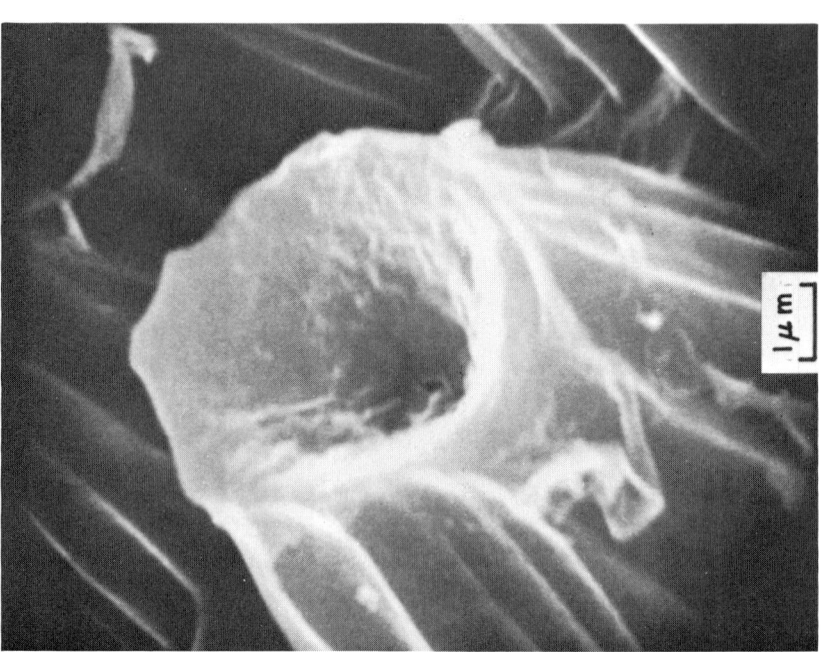

Figure 19.   Thornel 400 fiber etched 8 hours.

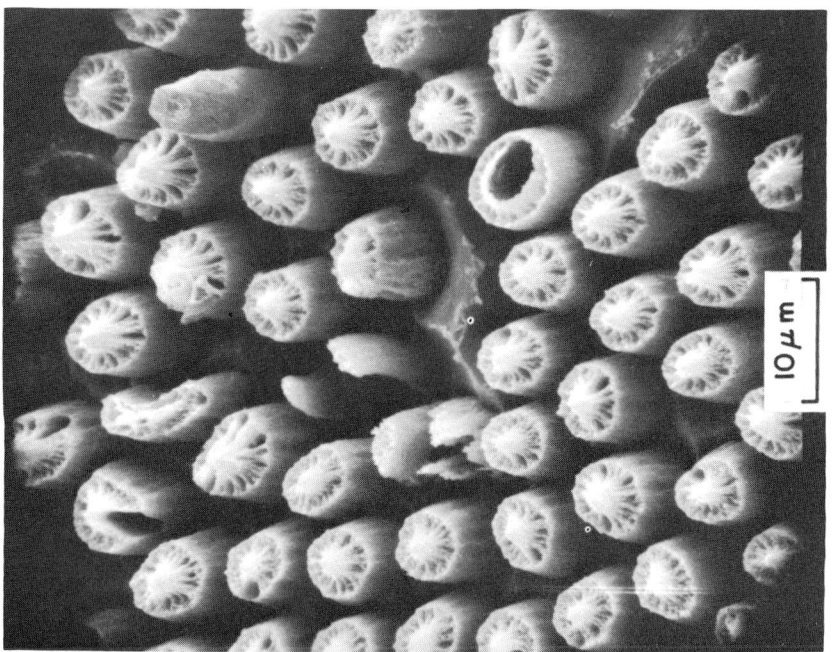

Figure 22. Modmor I fibers etched 8 hours.

Figure 21. Modmor II fibers etched 4 hours.

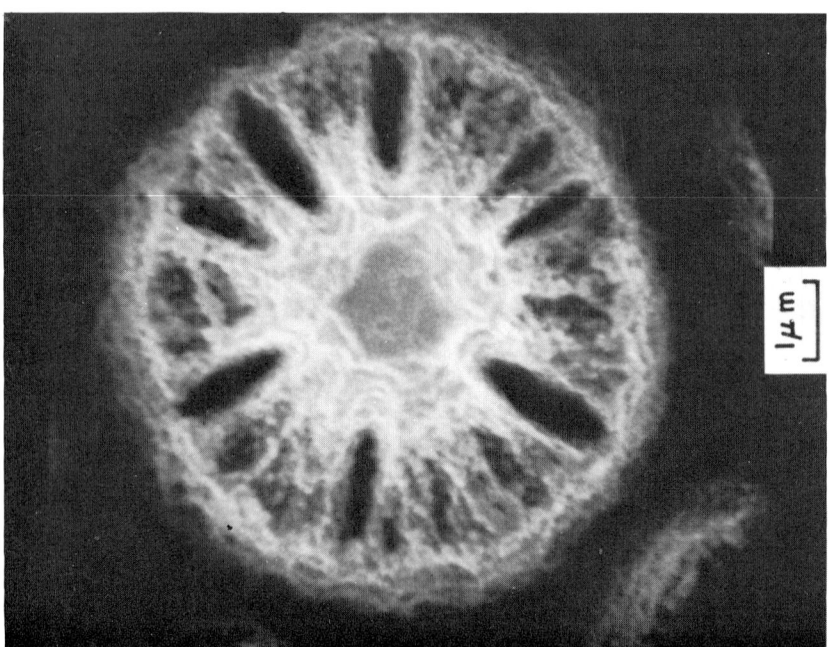

1 μm

Figure 24.   Modmor I fiber etched 8 hours.

2 μm

Figure 23.   Modmor I fiber etched 8 hours.

PAN-Based Fibers

Thornel 400 (intermediate modulus)

The etched specimen of this fiber shows two regions, each with a different etch pattern (Figures 17–19). In the first pattern, the fibers have flat tops and slightly tapering sides, like a previous fiber, VYB. Faint evidence of structure can be seen on the flat tops and occasional holes may also be seen.

In the region of the other pattern, the specimen appears to have been etched more deeply. The fibers have a greater taper, with volcanolike depressions in their cores.

The bumpy end of the fiber pictured in Figure 19 may indicate some long-range crystalline structure. We have concluded that in this fiber, the carbon layer planes are oriented circumferentially in the sheath and these layers have a well-developed crystalline structure. As one moves toward the center of the fiber, crystalline perfection decreases and the inner region is strained, which makes this part of the fiber more vulnerable to etching.

Modmor II (intermediate modulus)

The fibers have been etched to uniform, ogival-shaped points, which suggests that the overall macrostructure is rather uniform (Figures 20, 21). No evidence of flaws was seen on these ogival-shaped fibers.

Note the ridges, or threadlike structure, running lengthwise along the fibers and extending a little beyond the tips. This material seems to be the most resistant part of the fiber. As the main part of the fiber is etched away, these more resistant threads remain and collect together in the form of a tuft. This structure is analogous to that seen on Thornel 25. However, the Modmor II structure is finer and seems to be more firmly attached to the fiber, suggesting that the long-range crystalline order of Modmor II may be more advanced than in Thornel 25.

Modmor I (high modulus)

A group of Modmor I fibers is seen in Figure 22. There is a large variation in the fiber cross sections with the shapes ranging from circles to ovals to crescents to split fibers: the latter two are relatively rare. Based on the very uniform shape of the intermediate-modulus PAN-based fiber, Modmor II, it is believed that the unusual shapes seen here develop, in part, when the fiber is heat treated above 2000°C and are due to fiber inhomogeneities and flaws.

The etched fiber ends have distinctive shapes. The center is the most etch-resistant; the sheath is somewhat less resistant; and between these two regions, there is a pronounced radial pattern of ridges and valleys.

It is generally agreed that the crystallites in the sheath are circumferentially oriented and are larger than those in the interior and that there is an interweaving of the crystal layers here to form a three-dimensional structure. Also, there is a measurable flaw population on the fiber surface.

The crystal orientation in the core has not yet been determined. It may be that its orientation is radial, or perhaps it has no clear-cut preferred orientation.

Ridges can be seen running radially between the core and the sheath. Several workers, using optical microscopy have concluded that in this region the crystallites are radially oriented. The ridges on the etched fiber ends are consistent with the arguments for such a radially oriented structure. Valleys or pits occur between the ridges of the radially oriented interior crystallites and are bounded on the ends by the circumferentially oriented crystallites of the sheath and by the core. The tendency for the plasma to etch preferentially in this area would be due to internal stresses caused by fiber shrinkage during manufacture, and to internal stresses caused by differential thermal contraction occurring as the fiber cools down from its graphitization temperature.

The pattern of radial ribs and pits may have its origin, in part at least, from imperfections in the precursor fiber itself. Figure 24 is an end-on view of an etched Modmor I fiber. The dark oval spots are actually pits similar to those seen in Figure 23. This pattern of pits is strikingly similar to the pattern seen on cross sections of as-spun, uncollapsed, PAN fiber[8,9]. Upon heating and stretching the fiber, these void-like regions disappear—that is, they can no longer be seen under a microscope. However, they are probably still present in the collapsed fiber and are very likely still present even after the precursor fiber has been carbonized and graphitized. In the graphitized fiber, these areas are believed to have a lower degree of long-range crystalline order than that found in the surrounding regions. Upon plasma etching, these regions of poorer crystallinity are preferentially etched and pits form, while the more crystalline regions between the flaws are less attacked, leaving the ridgelike structure standing between the pits.

A three-dimensional model has been proposed[10] for the macrostructure of meter-length Modmor I fiber (Figure 25). In this model, the crystallites are circumferentially oriented in the sheath and radially oriented in the interior annulus between the core and sheath. Dispersed within the radial structure are pockets of carbon having a relatively poor crystallinity. These pockets are random in size, but tend to be teardrop in shape and are oriented radially within the fiber. Little can yet be said about the crystal structure and orientation of the core. The patterns of the various regions change randomly from cross section to cross section as one proceeds along the fiber length, but the crystalline part of the fiber forms a continuum extending the length of the fiber. In addition to

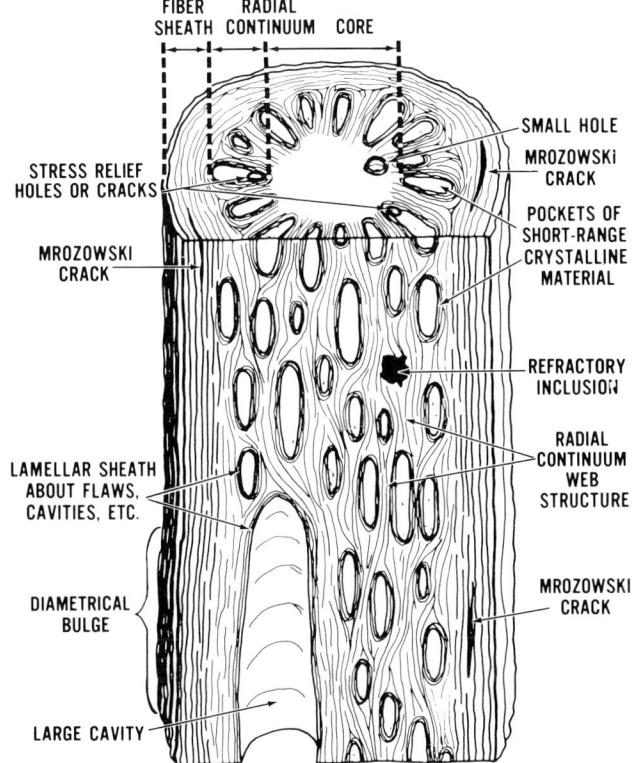

Figure 25. Three-dimensional model for meter-length Modmor I fibers.

regions of poor crystallinity, the fibers also contain cavities, cracks, inclusions, and surface flaws.

Conclusions

When a carbon-fiber composite is oxidized in a radio-frequency plasma, distinctive etch patterns are produced on the ends of the fibers. These patterns are reproducible and are related to the precursor fiber—its chemical nature, shape, and imperfections, the conditions under which the carbon fiber was manufactured, and the shape of the unetched carbon fiber. The plasma-etch procedure thus distinguishes the macrocrystalline structure of the fiber.

This technique is a valuable tool for characterizing carbon fibers, both for monitoring fibers under production, and for giving guidance in the development of improved fibers.

References

1. Bacon, R., "Carbon Fibers from Rayon Precursors," in *Physics and Chemistry of Carbon*, Vol. 9, P.L. Walker, Jr. and P.A. Thrower, eds., New York: Marcel Dekker, Inc. (1973), 1–102.

2. Butler, B.L. and Diefendorf, R.J., "Microstructure of Carbon Fibers," in *Summary of Papers of the Ninth Biennial Conference on Carbon*, Boston College, 16–20 June 1969, University Park, Pa.: American Carbon Committee, Pennsylvania State University (1969), 161–63.

3. Butler B.L. and Diefendorf, R.J., "Graphite Filament Structure," paper presented at the Tenth Annual ASME Carbon Composite Technology Symposium, University of New Mexico, 29–30 January 1970.

4. Larsen, J.V., "Carbon Fiber Structure, Surface and Surface Treatments," unpublished Ph.D. dissertation, University of Maryland, 1971.

5. Knibbs, R.H., "The Use of Polarized Light Microscopy in Examining the Structure of Carbon Fibers," *J. Micros. (Oxford)*, 94 (1971), 273–81.

6. Barnet, F.R. and Norr, M.K., "Carbon Fiber Microstructure," Naval Ordnance Laboratory, White Oak, Md., Technical Report No. NOLTR-72-32, March 1972 (AD 740 315).

7. Barnet, F.R. and Norr, M.K., "Carbon Fiber Etching in an Oxygen Plasma," *Carbon (Oxford)*, 11 (1973), 281–88.

8. Takashi, M. and Watanbe, M., "Relation Between Coagulating Forces of Spinning Baths and Properties of Fibers," *Sen'i Kikai Gakkaishi*, 15 (1959), 951–59 (in Japanese).

9. Craig, J.P., Knudsen, J.P. and Holland, V.F., "Characterization of Acrylic Fiber Structure," *Text. Res. J.*, 32 (1962), 435–48.

10. Barnet, F.R. and Norr, M.K., "A Three-Dimensional Structural Model for a High Modulus PAN-Based Carbon Fiber," Naval Ordnance Laboratory, White Oak, Md., Technical Report No. NOLTR-73-154, to be published.

Index

561